ENDOCRINE DISRUPTION

ENDOCRINE DISRUPTION
Biological Bases for Health Effects in Wildlife and Humans

Edited by

David O. Norris

James A. Carr

OXFORD

UNIVERSITY PRESS

2006

OXFORD
UNIVERSITY PRESS

Oxford University Press, Inc., publishes works that further
Oxford University's objective of excellence
in research, scholarship, and education.

Oxford   New York
Auckland   Cape Town   Dar es Salaam   Hong Kong   Karachi
Kuala Lumpur   Madrid   Melbourne   Mexico City   Nairobi
New Delhi   Shanghai   Taipei   Toronto

With offices in
Argentina   Austria   Brazil   Chile   Czech Republic   France   Greece
Guatemala   Hungary   Italy   Japan   Poland   Portugal   Singapore
South Korea   Switzerland   Thailand   Turkey   Ukraine   Vietnam

Published by Oxford University Press, Inc.
198 Madison Avenue, New York, New York 10016

www.oup.com

Oxford is a registered trademark of Oxford University Press

Library of Congress Cataloging-in-Publication Data
Norris, David O.
Endocrine disruption : biological bases for health effects in wildlife and humans / edited by
David O. Norris and James A. Carr.
p. cm.
Includes bibliographical references and index.
ISBN-13 978-0-19-513749-1
ISBN 0-19-513749-3
1. Endocrine toxicology.   2. Reproductive toxicology.   3. Environmental toxicology.
I. Carr, James A.   II. Title.
RA1224.2.N675 2005
571.9'51—dc22   2004053111

9 8 7 6 5 4 3 2 1

Printed in the United States of America
on acid-free paper

Preface

This book actually is about an old problem: pollution. However, it is about a subtle kind of pollution that can enter organisms and alter biological processes without being acutely toxic. Endocrine-disrupting chemicals (EDCs), sometimes in minute quantities, can mimic or inhibit the actions of internal messengers called hormones. As such, many consider EDCs to represent a potential threat to the health of wildlife and humans. The purpose of this book is to bring together in one place the pertinent information on endocrine physiology, toxicological methods, and descriptions of potential and known EDCs to help identify reasonable biomarkers for determining the risks of EDC contamination to wildlife and humans. Although the focus is biased toward vertebrates, the potential impact of EDCs on invertebrate species that form the nutritional base for vertebrate populations cannot be ignored. The principles discussed apply equally well to both vertebrate and invertebrate systems.

We hope this volume will allow scientists from different disciplines (including endocrinologists, toxicologists, and ecologists), environmentalists, politicians, and government officials to grasp the fundamentals of unfamiliar areas and to better understand their interrelationships—and thus broaden their understanding of the overall, complex problem of regulation and the health of wildlife and humans. Each of these groups views the subject of endocrine disruption from a different perspective, with a different focus, and often with different goals, in part due to a limited understanding of the diverse factors that are involved. Perhaps the endocrinologist will better understand the science of toxicology, and the toxicologist will have a better understanding of endocrine systems and their implications for animal and human health. This book should also be a useful primer for students interested in gaining a comprehensive introduction to the problem of endocrine disruptions. The long-term consequences as well as the solutions to pollution seem obvious, but the remedies are indeed difficult.

Rachel Carson (1962) first awakened the general public to the ecological ramifications of endocrine disruption through her book *Silent Spring*. This awareness led eventually to the ban on use of the insecticide DDT in the United States and the use of diethylstilbestrol for treating animals and humans. Scientific awareness of the toxic effects of chemicals and their ability to alter animal populations caused the field of toxicology to develop as the major scientific discipline for assessing toxicity of chemical pollutants during the latter half of the 20th century. Pioneering work of toxicologists developed methods for assessing chemical impacts and standards for the maximal allowable presence in the environment (see chapter 7). Application of these standards led to the subsequent recovery of bird populations as well as to the general rejuvenation of the North American Great Lakes.

With the crisis past, biologists generally returned to the business of being ordinary biologists; that is, until another enlightening book appeared: *Our Stolen Future* by Theo Colborn and co-workers (1996). This book again focused our attention on the widespread dangers of chemical pollution, but this time the emphasis was on how anthropogenic chemicals have been subtly altering the development, behavior, physiology, and ultimately the well-being and survival of natural populations, including our own. Fishes in our rivers that contain mixtures of estrogenic and other endocrine-active chemicals are exhibiting feminization, masculinization, or other abnormal endocrine responses (see Kime, 2001). Gonadal differentiation in some amphibian species is extremely sensitive to alterations in hormone status and can be influenced by a wide number of potentials EDCs under laboratory conditions including pesticides, polychlorinated biphenyls (PCBs), alkylphenols, and the thyroid-disrupting contaminant perchlorate (Goleman et al., 2002; Hayes et al., 2002; Mosconi et al., 2002; Carr et al., 2003; Qin et al., 2003, Levy et al., 2004). Alligator reproduction has been severely affected in exposed populations (Guillette et al., 2000). Recently reported increases in steroid-related cancers in human populations could be exacerbated by estrogenic EDCs (Toppari and Skakkebaek, 2000; Damstra et al., 2002), although more data are needed to establish this link (see Damstra et al., 2002; Soto and Sonnenschein, 2000). Increases in birth defects (e.g., Sherman, 1995) and decreased learning performance in children also have been linked to EDCs as a possible cause (see Guillette, 2000). Although the incidence of various clinical disorders such as autism also has risen dramatically, whether this increase is related to improved diagnosis and reporting or to environmental factors (e.g., high perchlorate levels in the water or exposure to chemicals such as pesticides or dioxins) is still being debated (see Croen et al., 2002a,b). For example, the dramatic rise in Alzheimer's disease during recent years in the United States may be simply a consequence of population demographics (Hebert et al., 2003) and may not be related to environmental factors. Although there remain many uncertainties regarding the adverse effects of potential EDCs at the population and ecosystem levels, there is general agreement that EDCs have the potential to impact reproductive and developmental fitness in humans and wildlife.

As simply stated on the e.hormone website (http://e.hormone.tulane.edu), endocrine disruption is a "process by which an exogenous substance causes adverse health effects consequent to changes in endocrine function." This broad definition covers many types of organic chemicals. Many of these compounds are industrial and domestic pollutants such as dioxins (chapter 10), PCBs, polycyclic aromatic hydrocarbons, or PAHs (chapter 11), and alkylphenols (chapter 14). Many phytoestrogens (chapter 15) as well as androgenic compounds are natural substances released from processing of wood that are concentrated and released into aquatic systems by pulp mills. Insecticides such as estrogenic DDT and its metabolites still persist in our environment (chapter 12), and herbicides, such as triazines (chapter 16), also are potential EDCs. Furthermore, chemicals with EDC activity are not limited to organic compounds. Ions such as perchlorate ion released from rocket fuels can interfere with thyroid function (chapter 4), and heavy metals such as cadmium (chapter 13) have been shown to disrupt both reproductive and adrenal function. Humans have introduced thousands of additional chemical compounds into the environment (Krimsky, 2000), and any number of them could have EDC activity. Assessing the potential risk of thousands of chemicals to humans and wildlife represents a daunting task.

The scientific assessment of environmental contamination has traditionally been the focus of toxicology. Using dose–response analyses of laboratory animals and clearly defined endpoints such as death of half of the animals within a certain time period (e.g., the lethal concentration, LC, or the lethal dose, LD, that will kill 50% of the population within 96 h of initial exposure; LC_{50} or LD_{50}, respectively), toxicologists have helped federal agencies establish levels for these contaminants that are unlikely to cause death or toxicity (see chapter 7). From toxicological studies, scientists and regulatory agencies could define no-observable-effect levels or concentrations (NOELs or NOECs) or no-observable-adverse-effect levels or concentrations (NOAELs or NOAECs) for any chemical with respect to toxicity or its ability to induce cancer (as a carcinogen) as well as additional safeguards for humans by limiting exposures to levels that are much lower than the observed NOEL/NOAEL. The possibility of chemicals that mimic hormone actions or prevent secretion or action of endogenous hormones poses a new problem for the field of toxicology, requiring a new understanding of potential impacts. Because they can add to or subtract from endogenous levels of compounds that normally function at extremely low concentrations, virtually any amount of an EDC added to the environment and absorbed into an animal via the digestive tract, respiratory system, or skin could affect normal functions. Furthermore, the possibility of several compounds producing additive effects due to similar endocrine activity (e.g., a large number of chemicals are estrogenic and produce effects by interacting with estrogen receptors) suggests that establishment of minimal safe levels may be difficult (Ramamoorthy et al., 1997; Rapjapakse et al., 2002; Tollefsen, 2002).

One of the things we have learned is that endocrine systems often exhibit paradoxical effects at low doses and often do not show linear dose–response relationships typical of those found in traditional toxicological studies, making it difficult if not impossible to identify a NOEL/NOAEL. A recent example is the effect of the androgenic growth promoter 17-β-trenbolone on fecundity and reproductive endocrinology of the fathead minnow (Ankley et al., 2003). The existence of these U-shaped or J-shaped dose–response relationships is termed "hormesis" and may be the norm in physiological responses to EDCs as well as to cancer induction (Kayajanian, 2002; Calabrese and Baldwin, 2003). In fact, endocrinologists have long known that different doses of hormones can affect different pathways and hence produce paradoxical results.

Because the effects of EDCs are subtle, differences in mean values (a measure of central tendency) of measured parameters may not be observed, yet the population may be affected negatively. In support of this notion, Orlando and Guillette (2001) argue that animals exposed to these chemicals frequently show much larger variances around the mean, suggesting a possibly meaningful disturbance of population health. In other words, this argument questions the accepted practice of scientists of regarding mean differences of greater than 0.05 probability as not being statistically significant (implying that there is less than 95% confidence in these data) for ecologically based studies of wildlife and humans and suggests that increased variability in a trait is sufficient to establish an effect of a chemical.

The majority of EDCs identified to date exhibit estrogenic or antiestrogenic activity and have been named "xenoestrogens." Others have proven to exhibit androgenic or antiandrogenic action. However, labeling specific compounds as estrogenic after examining its effects in one species must be done with caution because in another species it may have an antiestrogenic or androgenic effect. To date, most research has been directed toward the effects of xenoestrogens on reproduction, sex determination, and sex differentiation (feminization, masculinization, sex reversals). However, it is becoming more obvious that many compounds may be interfering with thyroid or adrenal functions, resulting in important developmental effects (both thyroid and adrenal function) and normal responses to stressors (adrenal function). Furthermore, both normal thyroid and adrenal function are essential for successful reproduction, and hence their dysfunction may alter reproduction indirectly. Endocrine systems, then, do not function in isolation, and effects on one hormone axis can alter the activity of others.

An additional complication in endocrine systems stems from the reserve capacity of the endocrine system to absorb environmental insults. For example, simply measuring circulating cortisol levels may not tell us much about the dynamics of the adrenal system. Two brown trout populations may have the same level of plasma cortisol, but the population exposed to high levels of cadmium requires more pituitary corticotropin to maintain that level, indicating endocrine disruption that is not reflected when measuring a single endocrine parameter (Norris, 2000).

The task for endocrinologists, toxicologists, environmentalists, and regulatory agencies is to examine collectively what Krimsky (2000) has called the "environmental endocrine hypothesis." This hypothesis "asserts that a diverse group of industrial and agricultural chemicals in contact with humans and wildlife have the capacity to mimic or obstruct hormone function" (p. 2).

References

Ankley, G.T., Jensen, K.M., Makynen, E.A., Kahl, M.D., Korte, J.J., Hornung, M.W., Henry, T.R., Denny, J.S., Leino, R.L., Wilson, V.S., Cardon, M.C., Hartig, P.C., Gray, L.E., 2003.

Effects of the androgenic growth promoter 17-beta-trenbolone on fecundity and reproductive endocrinology of the fathead minnow. Environ. Toxicol. Chem. 22, 1350–1360.

Calabrese, E.L., Baldwin, L.A., 2003. Toxicology rethinks its central belief. Nature 421, 691–692.

Carr, J.A., Gentles, A., Smith, E.E., Goleman, W.L., Urquidi, L.J., Thuett, K., Kendall, R.J., Giesy, J.P., Gross, T.S., Solomon, K.R., van der Kraak, G., 2003. Response of larval Xenopus laevis to atrazine: assessment of growth, metamorphosis, and gonadal and laryngeal morphology. Environ. Toxicol. Chem. 22, 396–405.

Carson, R., 1962. Silent Spring. Houghton Mifflin, Boston.

Colborn, T., Dumanoski, D., Myers, J.P., 1996. Our Stolen Future. Penguin Books, New York.

Croen, L.A., Grether, J.K., Hoogstrate, J., Selvin, S., 2002a. The changing prevalence of autism in California. J. Autism Dev. Discord. 32, 207–215.

Croen, L.A., Grether, J.K., Selvin, S., 2002b. Descriptive epidemiology of autism in a California population: Who is at risk? J. Autism Dev. Discord. 32, 217–224.

Damstra, T., Barlow, S., Bergman, A., Kavlock, R., Van Der Kraak, G., 2002. Global Assessment of the State-of-the-Science of Endocrine Diosruptors. World Health Organization, Geneva.

Goleman, W.L., Carr, J.A., Anderson, T.A., 2002. Environmentally relevant concentrations of ammonium perchlorate inhibit thyroid function and alter sex ratios in developing Xenopus laevis. Environ. Toxicol. Chem. 21, 590–597.

Guillette, E.A., 2000. An anthropological interpretation of endocrine disruption in children. In: Guillette, L., Jr., Crain, D.A. (Eds), Environmental Endocrine Disrupters: An Evolutionary Perspective. Taylor and Francis, New York, pp. 322–338.

Guillette, L.J., Jr., Crain, D.A., Gunderson, M.P., Kools, S.A.E., Milnes, M.R., Orlando, E.F., Rooney, A.A., Woodward, A.R., 2000. Alligators and endocrine disrupting contaminants: A current perspective. Am. Zool. 40, 438–452.

Hayes, T.B., Collins, A., Lee, M., Mendoza, M., Noriega, N., Ali Stuart, A., Vonk, A., 2002. Hermaphroditic, demasculinized frogs after exposure to the herbicide atrazine at low ecologically relevant doses. Proc. Natl. Acad. Sci. USA 99, 5476–5480.

Hebert, L.E., Scherr, P.A., Beinias, J.L., Bennett, D.A., Evans, D.A., 2003. Alzheimer disease in the US population. Arch. Neurol. 60, 1119–1122.

Kayajanian, G.M., 2002. The J-shaped dioxin dose response curve. Ecotoxicol. Environ. Safety 51, 1–4.

Kime, D.E., 2001. Endocrine Disruption in Fish. Kluwer Academic Publishers, Dordrecht, the Netherlands.

Krimsky, S., 2000. Hormonal Chaos: The Scientific and Social Origins of the Environmental Endocrine Hypothesis. The Johns Hopkins University Press, Baltimore, MD.

Levy, G., Lutz, I., Kruger, A., Kloas, W., 2004. Bisphenol A induces feminization in Xenopus laevis tadpoles. Environ Res. 94, 102–111.

Mosconi, G., Carnevali, O., Franzoni, M.F., Cottone, E., Lutz, I., Kloas, W., Yamamoto, K., Kikuyama, S., Polzonetti-Magni, A.M., 2002. Environmental estrogens and reproductive biology in amphibians. Gen. Comp. Endocrinol. 126, 125–129.

Norris, D.O., 2000. Endocrine disruptors of the stress axis in natural populations: How can we tell? Am. Zool. 40, 393–401.

Qin, Z.F., Zhou, J.M., Chu, S.G., Xu, X.B., 2003. Effects of Chinese domestic polychlorinated biphenyls (PCBs) on gonadal differentiation in Xenopus laevis. Environ. Health Perspect. 111, 553–556.

Orlando, E.F. and Guillette, L.J. Jr., 2001. A re-examination of variation associated with environmentally stressed organisms. Hum. Reprod. Update 7, 265–272.

Rajapakse, N., Silva, E., Kortenkamp, A., 2002. Combining xenoestrogens at levels below individual no-observed-effect concentrations dramatically enhances steroid hormone action. Environ. Health Perspect. 110, 917–921.

Ramamoorthy, K., Vyhlidal, C., Wang, F., Chen, I.-C., Safe, S., McDonnell, D.P., Leonard, L.S., Gaido, K.W., 1997. Additive estrogenic activities of a binary mixture of 2',4',6'-trichloro- and 2',3',4',5'-tetrachloro-4-biphenylol. Toxicol. Appl. Pharmacol. 147, 93–100.

Sherman, J.D., 1995. Chlorpyrifos (Dursban)-associated birth defects: A proposed syndrome, report of four cases, and discussion of the toxicology. Intl. J. Occup. Med. Toxicol. 4, 417–431.

Soto, A.M., Sonnenschein, C., 2000. Xenoestrogens in the context of carcinogenesis. In: Guillette, L., Jr., Crain, D.A. (Eds.), Environmental Endocrine Disrupters: An Evolutionary Perspective. Taylor and Francis, New York, pp. 291–321.

Tollefsen, K.-E., 2002. Interaction of estrogen mimics, singly and in combination, with plasma sex steroid-binding proteins in rainbow trout (*Oncorhynchus mykiss*). Aquat. Toxicol. 56, 215–225.

Toppari, J., Skakkebaek, N.E., 2000. Endocrine disruption in male human reproduction. In: Guillette, L., Jr., Crain, D.A. (Eds.), "Environmental Endocrine Disrupters: An Evolutionary Perspective. Taylor and Francis, New York, pp. 269–290.

Contents

Contributors

Dieldrich S. Bermudez
Department of Zoology
University of Florida
Gainesville FL USA

Alan L. Blankenship
Department of Zoology
Michigan State University
East Lansing MI USA

James A. Carr
Department of Biological Sciences
Texas Tech University
Lubbock TX USA

Richard L. Dickerson
Department of Environmental
    Toxciology
Texas Tech University
Lubbock TX USA

Kenneth R. Dixon
Institute of Environmental and
    Human Health
Texas Tech University
Lubbock TX USA

Lynn Frame
Department of Pharmacology
Texas Tech University HSC
Lubbock TX USA

John P. Giesy
Department of Zoology
Michigan State University
East Lansing MI USA

Timothy S. Gross
US Geological Survey
Gainesville FL USA

Louis J. Guillette, Jr.
Department of Zoology
University of Florida
Gainesville FL USA

Mark P. Gunderson
Department of Zoology
University of Florida
Gainesville FL USA

Alice Hontela
Department of Biological Sciences
   and Water Institute for Semi-arid
   Ecosystems
University of Lethbridge
Lethbridge Alberta Canada

Paul D. Jones
Department of Zoology
Michigan State University
East Lansing MI USA

K. Kannan
Department of Zoology
Michigan State University
East Lansing MI USA

Werner Kloas
Humboldt University
Berlin Germany

Stefan A.E. Kools
Department of Zoology
University of Florida
Gainesville FL USA

Alexandra Lacroix
Departement des Sciences
   Biologiques
Universite du Quebec a Montreal
Montreal Quebec Canada

Clyde F. Martin
Texas Tech University
Lubbock TX USA

J.L. Newsted
Department of Zoology
Michigan State University
East Lansing MI USA

David O. Norris
Department of Integrative
   Physiology
University of Colorado
Boulder CO USA

Miles Orchinik
Department of Biology
Arizona State University
Tempe AZ USA

Catherine Propper
Department of Biological
   Sciences
Northern Arizona University
Flagstaff AZ USA

R. Heath Rauschenberger
US Geological Survey
Gainesville FL USA

Ernest E. Smith
Institiute of Environmental and
   Human Health
Texas Tech University
Lubbock TX USA

Alan M. Vajda
Department of Integrative
   Physiology
University of Colorado
Boulder CO USA

Mary K. Walker
College of Pharmacy
University of New Mexico HSC
Albuquerque NM USA

Part I

Vertebrate Chemical Regulation

1

Introduction to Endocrinology

David O. Norris
James A. Carr

Chemical regulators govern every aspect of the life of vertebrate animals, beginning with early development and differentiation of tissues and organs and culminating in reproduction. Hormones are responsible for controlling homeostasis, the ability to maintain a balanced equilibrium in physiological processes in the face of environmental pressures. Reproduction would not be possible without the close interactions of several chemical regulators. Many behaviors are initiated or are influenced by the internal chemical environment of the body. Discovery of disruptive effects by anthropogenic chemicals accumulating in aquatic and terrestrial environments on endocrine mechanisms associated with development and reproduction (for reviews, see Kendall et al., 1998; Di Giulio and Tillitt, 1999) led to the designation of these substances as endocrine-disrupting chemicals or contaminants (EDCs). However, this phenomenon of chemical interference is much broader in scope, affecting endocrine systems not directly associated with reproduction (e.g., thyroid, adrenal), as well as other types of chemical regulation in animals.

The field of endocrinology classically dealt with a series of ductless glands that secreted their products, hormones, directly into the blood vascular system through which they traveled to distant target tissues, where they produced specific effects. Later it was discovered that the nervous system also released chemical regulators into the circulation, and these chemicals were dubbed "neurohormones." In more recent years, scientists have identified cell-to-cell chemical regulation among classical endocrine cells, neural cells, and in the immune system, as well as in other tissues. Furthermore, many tissues and organs not originally classified as "endocrine" also produce and release chemical regulators into the blood (e.g., adipose tissue, the heart, liver, and kidneys). A listing of the major endocrine regulators,

3

their sources, and their functions is provided in table 1-1 (see also general endocrinology references at the end of this chapter).

Currently, we consider that chemical regulation involves the nervous system, the endocrine system, and the immune system, as well as some general tissues, all of which secrete specific chemical regulators (figure 1-1). Although our focus for this book is on endocrine disruption, we provide a general overview of chemical regulation that covers some other types of chemical regulation. The generalized account in this chapter can be supplemented easily from the readings provided at the end of this chapter.

Vertebrates (fishes, amphibians, reptiles, mammals, and birds) as well as many diverse groups of invertebrates (e.g., insects, crustaceans, mollusks, annelid worms, sea stars) employ specific hormones in development, growth and maturation, and reproduction. Invertebrate endocrine systems are built on a similar plan to that of vertebrates, so that many of the principles discussed here for vertebrates are applicable to invertebrates. Recent research has shown that many of the same or closely related chemicals occur as regulators among vertebrate and invertebrate groups. Consequently, invertebrates also are potential targets for EDCs. The reader is referred to references at the end of this chapter that address regulation of endocrine systems in invertebrates and known actions of EDCs.

Types of Chemical Regulators

Endogenous chemical regulators that communicate between tissues can be placed in three major categories: hormones, neurocrines, and cytocrines (figure 1-2; table 1-2). As already mentioned, hormones are secreted by endocrine cells into the blood. Neurocrines are chemical regulators secreted by the nervous system and can be divided into two broad subdivisions. The first subdivision includes neurohormones that are secreted into the general circulation. The second subdivision consists of the neurotransmitters and neuromodulators that are released into the extracellular fluid of the synaptic connections between neurons and their postsynaptic target cells, which include other neurons, muscle cells, or glandular cells (even endocrine cells may be innervated). After binding to the postsynaptic receptor, a neurotransmitter evokes or inhibits an action potential in the postsynaptic cell. A neuromodulator alters the responsiveness of a postsynaptic cell to the normal neurotransmitter by rendering it more or less responsive to the neurotransmitter.

The final category of chemical regulators consists of local chemical regulators, called cytocrines, that are released into the extracellular fluid of tissues and diffuse short distances to their target cells. These cytocrines can be further separated into paracrine regulators that affect other cell types and autocrine regulators that affect the secreting cell type. The cytokines of the immune system can be classified as cytocrine regulators.

Chemical messengers in many groups of animals carry information between organisms and control the physiology and/or behavior of other individuals. These signal chemicals have been named semiochemicals and may be species-specific or trans-specific. Species-specific semiochemicals are called pheromones and vary

Table 1-1. Major vertebrate chemical regulators, sources, targets, and effects.

Source and Name	Target Examples	Effects
Hypothalamic releasing hormones		
Gonadotropin-releasing hormone (GnRH)	Gonadotrope in pituitary	Secretion of FSH and LH
Galanin	Gonadotrope in pituitary	Enhances LH release
Thyrotropin-releasing hormone (TRH)	Thyrotrope in pituitary	Secretion of TSH
Corticotropin-releasing hormone (CRH)	Corticotrope in pituitary[a]	Secretion of ACTH
Somatostatin (SS = GH-RIH)	Somatotrope in pituitary	Inhibits secretion of GH
Somatocrinin (GH-RH)	Somatotrope in pituitary	Secretion of GH in absence of SS
Prolactin release-inhibiting hormone (PRIH = dopamine, DA)	Lactotrope in pituitary	Inhibits PRL secretion
Vasoactive intestinal peptide (VIP; PRH)	Lactotrope in pituitary	Secretion of PRL in absence of DA
Hypothalamic nonapeptides		
Arginine vasopressin (AVP)	Kidney tubules	Increases water reabsorption
Oxytocin (OXY)	Oviduct/uterus/vas deferens	Gamete transport
Arginine vasotocin (AVT)[b]	Kidney/reproductive ducts	Water balance/gamete transport
Pineal organ		
Melatonin	Hypothalamus/pigment cells	Inhibitory actions
Pituitary (adenohypophysis)		
Follicle-stimulating hormone (FSH)	Gonad	Gamete formation, E_2 synthesis
Luteinizing hormone (LH)	Gonad	Gamete release, androgen synthesis
Thyrotropin (TSH)	Thyroid	Thyroid hormone synthesis
Growth hormone (GH)	Liver, muscle, bone	Synthesis of IGFs, proteins
Prolactin (PRL)	Mammary glands, epididymis	Stimulates protein secretion
Corticotropin (ACTH)	Adrenocortical cells	Synthesis of corticosteroids
Melanotropin (MSH)	Melanin-producing cells	Synthesis/translocation of melanin
Ovary		
Estrogens (e.g., estradiol-17β = E_2)	Uterus, mammary gland	Proliferation
Progesterone	Uterus	Protein secretion
Inhibin	Gonadotrope in pituitary	Inhibits FSH release to GnRH

(continued)

5

Table 1-1. Continued

Source and Name	Target Examples	Effects
Testis		
Testosterone, dihydrotestosterone (DHT)	Sex accessory structures	Stimulates development
Inhibin	Gonadotrope in pituitary	Inhibits FSH release to GnRH
Thyroid gland		
Tetraiodothyronine (T_4 = thyroxine) and Triiodothyronine (T_3)	Most tissues in body	Influences development, metabolism
Adrenal gland: Adrenocortical tissue (cortex)		
Cortisol (F) and/or corticosterone (B)	Liver/muscle	Gluconeogenesis/protein hydrolysis
Aldosterone[c]	Kidney	Sodium reabsorption
Adrenal gland: Chromaffin tissue (medulla)		
Epinephrine (E) and norepinephrine (NE)	Liver/muscle/heart	Raises blood glucose/increases contractions
Liver		
Insulinlike growth factors (IGFs)	Cartilage	Stimulates synthesis of matrix
Kidney		
Erythropoietin in response to hypoxia	Bone marrow	Increase red blood cell production
1,25-dihydroxycholecalciferol (DHC)	Small intestine	Production of carrier protein for calcium
Renin (an enzyme)	Renin substrate in blood	Release of peptide, angiotensin 1
Endocrine pancreas (islets)		
Insulin	Liver/adipose tissue	Lowers blood glucose and fatty acids
Glucagon	Liver	Raise blood glucose levels
Somatostatin	Other cell types in islets	Inhibits secretion

6

Parathyroid glands
  Parathyroid hormone (PTH) — Bone/kidney — Bone resorption/calcium reabsorption
Ultimobranchial bodies
  Calcitonin — Bone — Blocks action of PTH
Gastrointestinal tract
  Gastrin (stomach) — Parietal cell of stomach — Secretion of HCl into stomach lumen
  Secretin (small intestine) — Exocrine pancreas — Release of basic juice
  Cholecystokinin (small intestine) — Pancreas/gall bladder — Release of enzymes/contraction
  Glucose-dependent insulinotropic peptide (GIP) (small intestine) — Endocrine pancreas — Stimulates insulin release
  Vasoactive intestinal peptide (VIP) — Dilates visceral blood vessels — Increases flow of blood to intestine
Heart
  Atrial natriuretic peptide (ANP) — Kidney/brain — Inhibits aldosterone action/prevents AVP release
Thymus
  Thymosins — Spleen, lymph nodes — Increase lymphocyte production
Immune system macrophages
  Interleukin 1 — Helper T-cells — Activation
  Interleukin 2 — Cytotoxic T-cell — Activation

[a]Also stimulates TSH release in fishes and amphibians.
[b]Replaces AVP in nonmammals; performs roles of AVP and OXY.
[c]Function assumed by thyroid hormones and cortisol in fishes.

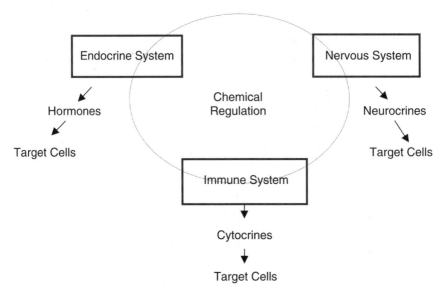

Figure 1-1. What is chemical regulation? Chemical regulation encompasses the separate secretions of the classical endocrine system (hormones), the nervous system (neurocrines = neurotransmitters, neurohormones, neuromodulators), and the immune system (cytocrines) all working through similar mechanisms. These systems are not isolated from one another but influence each other in a variety of ways.

from simple uses such as the attraction of a sperm to an oocyte to the attraction of a mate from miles away. Many of the pheromones involved in reproduction are modified steroid hormones or are produced under the regulation of steroid hormones, making them susceptible to alteration by reproductively active EDCs. In special cases, semiochemicals are used to communicate between species and are called allelomones. Examples of allelomones include the chemical odors secreted by skunks and the sexual attractant of a wasp that also attracts a predator.

How Chemical Regulators Work

In all cases, the secretion of a chemical regulator, its action, and its ultimate degradation follows a similar pattern (figure 1-3). Secretion includes both synthesis and release, which may be under separate control in some systems. Once it is released, a chemical regulator travels to its target cell. In some cases, the regulator may travel to its target while loosely bound to a carrier or binding protein, whereas in other cases it may be free. Regulators are bound in an equilibrial (reversible) fashion to such binding proteins so that there is also some level of free regulator present in the medium. The amount of free regulator is determined by the strength of the bond between the regulator and the binding proteins. After traveling through blood, lymph, or extracellular fluids to reach its target cell, the regulator binds to a specific recep-

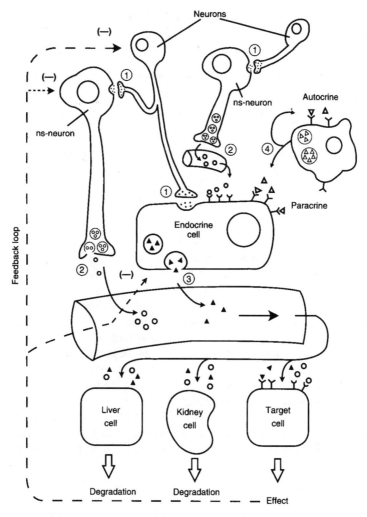

Figure 1-2. Patterns of chemical regulation via hormones, neurocrines, and cytocrines. Reprinted with permission from Norris (1997).

tor molecule within or on the surface of the target cell, much in the same way it binds to a binding protein. This will initiate, enhance, or inhibit particular biochemical events in the target cell, resulting in the typical response called the effect of the regulator. The nature of the equilibrium interaction of a regulator with its specific receptor and the mechanisms whereby it alters target cell function is the subject of chapter 2.

The presence of a specific receptor for a given regulator defines a cell as a target cell. The regulator molecule will bind to a receptor with high affinity and specificity. Because of the reversible nature of the regulator–receptor interaction, it is possible to develop synthetic molecules that can mimic the action of the regulator (i.e.,

Table 1-2. Cytocrines, neurocrines, and endocrines.

Type and Source	Subtype	Where Released	Target
Cytocrine			
Any cell type	Autocrine	Into extracellular fluid (ECF)	The same secreting cell
Any cell type	Paracrine	Into ECF	Any different local cell type
Neurocrine			
Neuron	Neurotransmitter	Into synaptic space (ECF)	Neuron, muscle, or gland cell
Neuron	Neuromodulator	Into synaptic space (ECF)	Neuron
Neuron	Neurohormone	Into the blood	Endocrine glands, smooth muscle, special epithelia
Endocrine			
Endocrine cell	Hormone	Into the blood	Any cell with receptor

an agonist). In contrast, an antagonist can reduce the regulator's effectiveness through competition for the regulator's binding site, but it fails to activate the characteristic biochemical response in the target cell.

An important aspect of the action of a chemical regulator is the production of a signal that informs the secretory cells that they have achieved their objective (i.e., produced an effect). This process is referred to as feedback, and it is an essential component of chemical regulation (figure 1-4). In the endocrine system, feedback usually occurs in the form of a bloodborne substance that tells the secreting cell whether it has produced sufficient regulator to make the necessary changes in the system to maintain homeostasis or change the target system to operate at a different level. To do this, the secretory center must be able to compare information on circulating levels of the feedback substance in the blood with a preprogrammed set point that is somehow encoded in the cell. As shown in chapter 3, feedback may take a variety of forms in different systems and may occur at more than one level in complicated regulatory systems such as neuroendocrine pathways that involve both neural and endocrine components. These neuroendocrine pathways involve the integrated actions of neurotransmitters, neurohormones, and hormones.

For target cells not to be stimulated constantly by regulators, elaborate biochemical pathways have evolved for metabolism and inactivation of regulators. Whereas feedback turns off regulator production once the desired effect has been achieved, the remaining hormone in the system must be inactivated to prevent overshoot of the response. This inactivation may occur in the target cell or in liver and/or kidney and may consist of actual degradation (e.g., conversion of peptides to amino acids) or the conjugation of the regulator with another substance to make it more water soluble and increase its excretion by the kidneys or by other excretory routes.

Patterns of Secretion

Release of regulators from cells is phasic or pulsatile rather than steady or tonic. Typically, regulators are secreted in discrete patterns or rhythms over time. Some

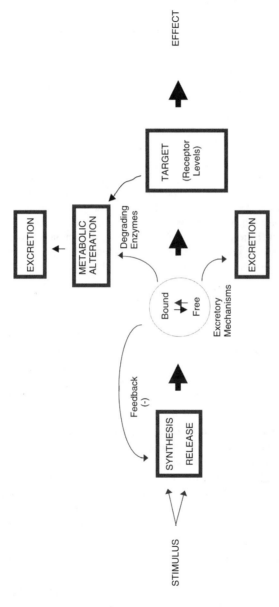

Figure 1-3. Life history of a hormone. By classical definition, a hormone is synthesized in an endocrine cell and released into the blood (synthesis + release = secretion), where it may circulate through the organism in free and protein-bound forms. Free hormone may be excreted and/or metabolized or may bind to a highly specific receptor on the surface of or within a target cell. After the hormone binds to its receptor, biochemical changes occur in the target cell, resulting in some effect specific to that hormone. After binding, the hormone typically is metabolized and excreted.

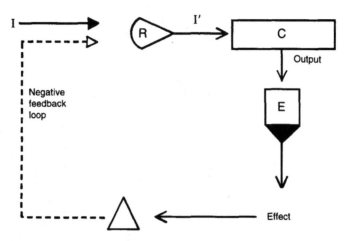

Figure 1-4. Feedback mechanisms for hormone, neurocrine, and cytocrine regulators. Information (I) could be the level of a hormone in the blood, a neurotransmitter release from a neuron, a metabolite in the blood, or even an external agent such as light. Information is detected by some sort of receptor (R) which translates or transduces the signal into a biochemical event (I') that causes the cell (C) to release a regulator such as a hormone (output) that acts upon a target cell (E) to produce an effect such as reabsorption of more sodium, excretion of potassium, release of glucose into the blood, and so on. This effect cause a change (Δ) in the blood level of this substance which informs the secreting cell via the negative feedback loop to stop producing output. Reprinted with permission from Norris (1997).

regulators are secreted in a 24-h pattern, often with a single, repeatable daytime or nighttime peak. These are often called diurnal rhythms. Some rhythms that appear to be diurnal, however, may exhibit an endogenous period of slightly more than or less than 24 h under constant conditions. These daily rhythms are called circadian (*circa* = about; *dia* = a day). Under normal conditions, circadian rhythms are reset daily, usually by photoperiod, so that they appear to be diurnal. Other regulators may be secreted in patterns that show a much shorter periodicity than 24 h and are called ultradial. There also may be seasonal or annual patterns of regulator secretion, especially for reproductive hormones. These longer rhythms may be manifest in changes in diurnal rhythms as well. For example, the peak of a diurnal cycle may vary in amplitude and/or with timing of the daily peak according to the season. An excellent description of biological rhythms can be found in Binkley (1990) and Nelson (1999).

Types of Chemical Regulators

Regulators are defined by their origin and mode of transport to their target tissues, as described above. However, they also can be categorized according to their chemistry. Furthermore, a given chemical can function as more than one type of regulator. For example, the catecholamine epinephrine is a neurotransmitter in the central nervous system, but it is also a hormone secreted by the adrenal medulla. Somatostatin is a peptide neurohormone that affects pituitary secretion of growth hormone and is a paracrine regulator produced by D-cells of the pancreas that locally inhibits secretion of insulin by the pancreatic B-cells. In the next section, we examine some basic chemical classes of vertebrate regulators, how they are synthesized and released, how they are transported to their targets, and how they produce their effects. Many of these regulators or closely related molecules can be found among the invertebrates as well. The following classes of chemicals are discussed: catecholamines, indoleamines, peptides and proteins, steroids, and thyroid hormones.

Regulators That Bind to Cell Membrane Receptors

The cell membrane provides a lipid barrier containing embedded proteins, and it separates the contents of a cell from its aqueous environment. Catecholamines, indoleamines, peptides, and proteins are more or less compatible with an aqueous medium and do not readily enter cells. Consequently, these hormones all bind to receptors on the surface of their target cells, resulting in rapid cellular responses. Occupied receptors may directly alter ion channels, work through a G-protein–coupled second messenger system, or have innate enzymatic activity (e.g., tyrosine kinase activity; see table 1-3).

Table 1-3. Classification of receptor types.

Categories	Some Receptor Examples
I. Receptors coupled to G-proteins and using second messengers	
A. G_s protein that activates adenylyl cyclase to produce cAMP as a second messenger	Epinephrine, glucagon, ACTH, LH, FSH
B. G_i protein that inhibits cAMP production	Somatostatin, insulinlike growth factor-2, acetylcholine
C. G_q protein that activates phospholipase C to form inositol trisphosphate and diacylglycerol	Arginine vasopressin V_1-type receptors
II. Receptors with innate enzymatic activity	
A. Tyrosine kinases that phosphorylate tyrosine residues in themselves or in other enzymes	Insulin, many growth factors
B. Serine/threonine kinases	Inhibin, transforming growth factor-β
C. Guanylyl cyclase	Atrial naturiuretic peptide
III. Receptors coupled to ion channels	
A. Ligand gated	Glutamate receptors (e.g., N-methyl-D-aspartate receptor); nicotinic receptors
B. Second messenger gated	Voltage-regulated calcium channels; calcium channels associated with the smooth endoplasmic reticulum
IV. Ligand-activated transcription factors	Thyroid hormone, retinoic acid, estrogen, androgen, corticosteroid, progesterone, arylhydrocarbon

Types I–III are membrane receptors, whereas type IV are all intracellular receptors.

The Catecholamines

The catecholamines play important roles as neurotransmitters, neurohormones, and hormones in vertebrates. Relevant catecholamines are dopamine (DA), norepinephrine (NE; also called noradrenaline), and epinephrine (E; also called adrenaline). All of these compounds are synthesized from the amino acid tyrosine (figure 1-5) following decarboxylation to an amine and the addition of another hydroxyl group to the phenolic ring to produce a catechol. The first biologically active catecholamine produced from tyrosine is DA, an immediate precursor for the synthesis of NE. Similarly, NE is the precursor for the synthesis of E. The rate-limiting enzyme for catecholamine synthesis is tyrosine hydroxylase, and immunocytochemical identification of this enzyme is often used to identify catecholamine-synthesizing cells. Another important synthetic enzyme is phenylethanolamine-*N*-methyl transferase (PNMT), which converts NE to E. PNMT is an important marker for identifying cells capable

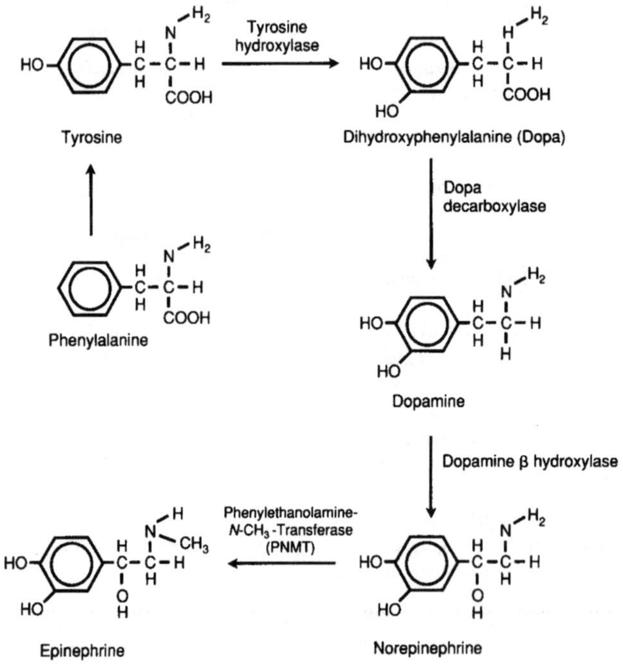

Figure 1-5. Synthesis of the catecholamine regulators. These regulators can be synthesized from the amino acids tyrosine or phenylalanine. Tyrosine hydroxylase is the rate-limiting enzyme for synthesis of all of the catecholamines and produces the precursor dopa. The enzymes necessary to synthesize dopamine, norepinephrine, and epinephrine also are indicated. The final product from a cell is determined by which enzymes are present. A given catecholamine may function as a hormone, a neurocrine, or a cytocrine depending on its cellular origin (endocrine cell, neuron, etc.) and where it is secreted (e.g., into blood or synapse). Reprinted with permission from Norris (1997).

of secreting E. After synthesis, catecholamines are stored in membrane-bound granules or vesicles within the cytoplasm until released.

Catecholamines may simply diffuse to their targets across a synaptic space and function as neurotransmitters or neuromodulators, diffuse through intercellular fluids as cytocrines, or they may be transported as hormones or neurohormones in the blood. In the latter case, these hormones bind reversibly to plasma proteins.

Each catecholamine binds to special receptors on target cells, although there is some overlap in their affinities for other catecholamine receptors. There are multiple types of receptors for DA (dopaminergic receptor), NE (α-adrenergic receptor), and E (β-adrenergic receptor) with a number of subtypes in each category. Degradation of catecholamines may involve different pathways in the brain than in peripheral tissues, producing different products for excretion.

The Indoleamines

The indoleamines are derived from the amino acid tryptophan. The double-ring structure of tryptophan is referred to as an indole, and decarboxylation of this amino acid yields an indoleamine. Serotonin or 5-hydroxytryptamine (5-HT) is the first biologically active indoleamine produced in this biochemical pathway, and it serves as a precursor for the production of melatonin in the pineal gland (figure 1-6). The rate-limiting enzyme for melatonin production is N-acetyltransferase (NAT). These hormones are stored in secretion granules in the cytoplasm of the secreting cell. 5-HT is as an important neurotransmitter in the brain, and, along with melatonin, it may be secreted by the pineal gland into the blood.

Indoleamines reach their targets in the same manner as catecholamines. Several receptor types and subtypes have been identified for 5-HT in the brain with somewhat differing modes of action. Melatonin receptors have not been studied to the same degree (see Norman and Litwack, 1998). Indoleamines are metabolized by the same enzymes responsible for catecholamine degradation.

Peptides and Proteins

Peptides and proteins are formed from amino acids through the classical mechanisms of polypeptide synthesis. A gene is represented by DNA in the cell nucleus. Through transcription, the nucleotides making up the DNA may be rewritten or transcribed into pieces of RNA that are then spliced to form a somewhat larger, single messenger RNA (mRNA) that leaves the nucleus and enters the cytoplasm. The nucleotide sequence of mRNA serves as a template to determine the sequence of amino acids in the resulting polypeptide. The number and sequence of amino acids gives each polypeptide its unique structure. Once in the cytoplasm, the mRNA interacts with cytoplasmic ribosomes of the rough endoplasmic reticulum (RER) to direct the synthesis of a polypeptide chain. Variations in splicing may result in multiple mRNAs, called splice variants, from the same gene, differing slightly in nucleotide sequences and producing variants of the basic polypeptide characterized as the product of that gene.

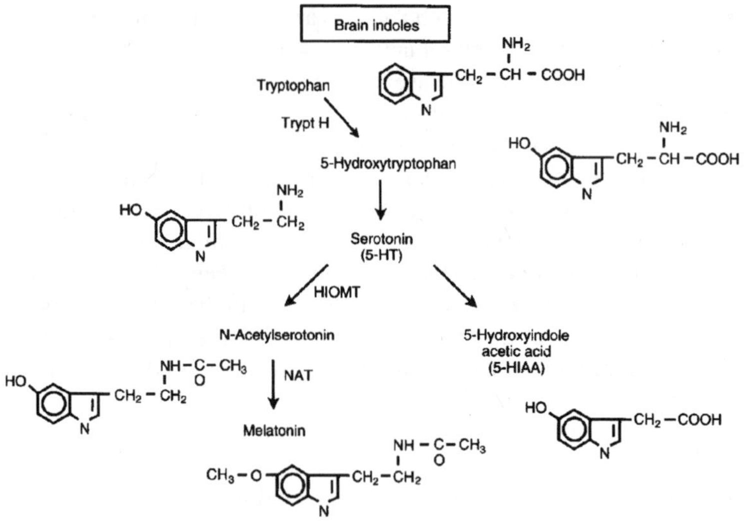

Figure 1-6. Synthesis of indoleamine regulators. Indoleamines are made from the amino acid tryptophan. Both serotonin and melatonin are important regulators. *N*-acetyl transferase (NAT) is the rate-limiting enzyme for the synthesis of melatonin from serotonin. Reprinted with permission from Norris (1997).

Peptide regulators are produced from polypeptides called preprohormones that are synthesized on ribosomes of the RER and transported to the Golgi apparatus for packaging into storage granules. A portion of the preprohormone (the signal peptide) is cleaved off as it passes into the RER to yield a smaller peptide called a prohormone. These prohormones are processed further by posttranslational enzymatic cleavage at the Golgi apparatus or in the storage granules to yield one or more inactive peptide fragments and the hormone. Both the inactive fragment and the hormone are released from the cell. The processing of the prohormone requires specific enzymes that hydrolyze the polypeptide between certain amino acids. Different cell types may process the same precursor but use different processing enzymes or convertases to produce different peptide products. For example, the polypeptide precursor known as proopiomelanocortin (POMC) may be processed in one pituitary gland cell type to produce the hormone corticotropin (ACTH) and in another cell type by different convertases to produce the hormone melanotropin (also called melanocyte- or melanophore-stimulating hormone) and β-endorphin.

Sometimes processing of a prohormone may produce several peptides, of which more than one may have biological activity. The resulting hormones and inactive fragments are retained in storage granules or vesicles until the cell receives a message to release the products. Transport to the targets may be by diffusion through extracellular fluids or via the blood or lymph. Hormones may travel as free peptides or may be loosely bound to plasma-binding proteins. The majority of neurotransmitters and neuromodulators in the brain as well as numerous neurohormones and hormones are peptides.

The final step in the synthesis of protein regulators usually involves the dimerization of two polypeptides to form the definitive protein containing two polypeptide subunits. Typically, each protein hormone is composed of unlike subunits and is called a heterodimer. In some cases, carbohydrates may be added to the subunits, usually in the Golgi apparatus. Thus, follicle-stimulating hormone (FSH) consists of a α-subunit and a β-subunit plus a number of carbohydrates, including sialic acid, attached to each subunit. Because of the carbohydrates such a hormone may be referred to as a glycoprotein. Growth hormone (GH), prolactin (PRL), and insulin usually are called protein hormones, although they are formed from a single polypeptide chain. In the case of GH and PRL, the polypeptide is folded through the formation of disulfide bridges, whereas the insulin molecule is first folded and then enzymatically cleaved to yield what appears to be a heterodimer (A and B chains) joined by disulfide bridges plus a free peptide (C-peptide). Most protein hormones travel free in the blood. Growth hormone travels in the blood bound to a circulating protein that probably is derived from a receptor protein.

Peptides and protein regulators may be degraded enzymatically to inactive products and/or amino acids by peptidases and proteases in the target cell or in nearby cells. They also may be inactivated by peptidases circulating in the blood. Both active hormones and inactive fragments may be excreted in the urine. Intact gonad-stimulating hormones (gonadotropins), for example, can be isolated from urine and are still biologically active when tested in animals.

Regulators That Enter Target Cells

Steroid hormones and thyroid hormones bind to intracellular receptors that function as transcription factors and stimulate or inhibit transcription of specific genes. This alters protein synthesis, bringing about changes in target cells. Because steroids are lipids, it was always assumed that they freely pass through cell membranes. Although thyroid hormones are derived from the amino acid tyrosine, they are not very soluble in aqueous medium and were believed to enter their target cells by diffusion. However, there is new evidence that suggests that steroids and thyroid hormones may be transported actively across cell membranes.

Because these intracellular receptors act at the nuclear level, they often are called nuclear receptors. In addition to thyroid hormone and steroid hormone receptors, there are numerous other nuclear receptors including the arylhydrocarbon receptor and the retinoic acid receptors (see table 1-3). The nuclear receptors have many common features and are distinct from the membrane-bound hormone receptors (see chapter 2).

Steroid Hormones

The precursor for all steroid hormones is the steroid cholesterol. Synthesis of cholesterol occurs primarily in the liver, where the rate-limiting enzyme for this pathway is hydroxymethylglutaryl-coenzyme A (HMG-CoA) reductase. Cholesterol consists of 27 carbons, of which 17 are formed into a peculiar 4-ring structure called the steroid nucleus (figure 1-7). Cholesterol is made from carbo-hydrate or triacylglycerols and is catabolized to a reduced number of carbons and modified to produce five classes of steroid hormones: androgens, corticosteroids, estrogens, progestogens, and members of the vitamin D complex (figure 1-8). The progestogens are C_{21} steroids that serve as precursors for androgens, cortico-steroids, and estrogens. The most common progestogen in vertebrates is proges-terone. Corticosteroids are also C_{21} steroids and are separable into two groups: the glucocorticoids (e.g., cortisol, corticosterone) and the mineralocorticoids (e.g., aldosterone). These terms are confusing when considering fishes, however, because cortisol usually functions as both a glucocorticoid (affecting carbo-hydrate formation and protein catabolism) and a mineralocorticoid (controlling Na^+/K^+ balance).

Progestogens give rise to androgens via a separate pathway. Androgens are C_{19} steroids responsible for stimulating development of male traits. The most common androgen among the vertebrates is testosterone, although others may be prominent in different species. Androgens are also precursors for the synthesis of estrogens, which are C_{18} compounds that stimulate development of female traits. Estradiol-17β is the most common vertebrate estrogen. The overall synthesis of corticosteroids and sex hormones is provided in figure 1-8, with some of the key enzymes indicated. Because of the precursor/product relationship of steroids, one must be cautious in interpretating effects produced by treatment with a steroid that can be readily metabolized to other active steroids (e.g., testosterone to estradiol; progesterone to either cortisol or testosterone; cortisol to aldoste-rone). Similarly, treatment with enzyme inhibitors may alter more than one steroidogenic pathway. For example, consider the impact of cyanoketone, an in-hibitor of 3β-hydroxysteroid dehydrogenase (HSD), on the scheme provided in figure 1-8.

The corticosteroids are synthesized only in the adrenal cortex or its homologue in nonmammals (e.g., the interrenal of bony fishes). The reproductive steroid hor-mones (progesterone, androgens, and estrogens) are formed in the gonads. How-ever, the adrenal cortex may synthesize androgens (especially androstenedione) as well. Furthermore, some peripheral tissues, including the liver and adipose tissue, can convert androgens to estrogens. Some steroid synthesis may occur in the brain, and these products are called neurosteroids. The conversion of cholesterol in the skin to vitamin D_3 involves opening up part of the steroid nucleus. The resulting compound is converted by sequential changes occurring in the liver and finally in the kidney to produce the active hormone 1,25-dihydroxycholecalciferol, which stimulates calcium uptake in the gut.

Steroids are transported in the blood mostly bound to specific plasma proteins. This binding prevents their rapid loss from the blood and helps maintain a high titer of circulating hormone. Sex steroid hormone-binding globulin (SHBG; also called testosterone- or androgen-binding protein) has a higher affinity for testosterone than estradiol, whereas a related androgen, androstenedione, and a related estrogen, es-trone, do not bind to SHBG and hence occur in the blood at very low concentra-tions. Corticosteroid-binding globulin (CBG) or transcortin has an equally high affinity for cortisol and corticosterone as well as for progesterone in most animals.

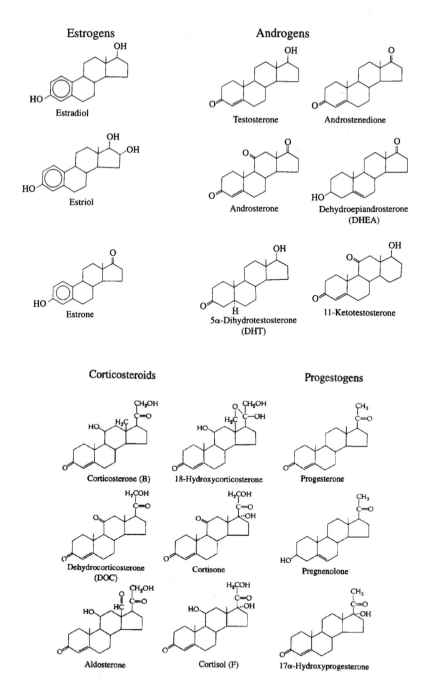

Figure 1-7. Some representative steroid hormones. Reprinted with permission from Norris (1997).

19

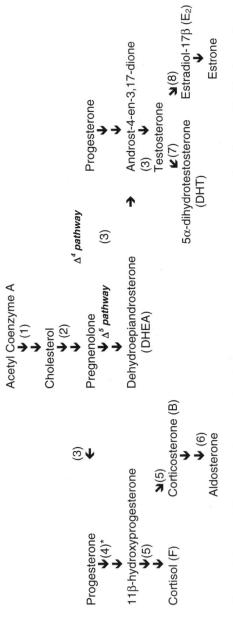

Figure 1-8. Simplified pattern for synthesis of androgens, estrogens, progesterone, and corticosteroids. Numbers refer to key enzymes in the process that can regulate the various pathways. Two arrows between compounds indicates that at least one intermediate step has been omitted from the diagram. (1) Hydroxymethylglutaryl-coenzyme A reductase, the rate-limiting enzyme for cholesterol synthesis; (2) P450 side-chain cleaving enzyme (P450$_{scc}$) converting the C_{27} skeleton of cholesterol to a C_{21} compound; (3) 3β-hydroxysteroid dehydrogenase; (4) P450$_{C21}$; (5) 11β-hydroxylase, P450$_{C11β}$; (6) aldosterone synthase, P450$_{C11AS}$; (7) 5α-reductase; (8) aromatase (P450$_{aro}$). *Only one of several paths that lead to the various corticosteroids from progesterone. See Norris (1997) or Norman and Litwack (1998) for more details of steroid syntheses.

In contrast, aldosterone does not bind well to CBG and is found primarily free, but at very low concentrations, in the blood.

Steroid hormones are metabolized primarily by conjugation with groups that interfere with interaction of the steroid to its binding proteins or receptors and render them more soluble in blood plasma. Consequently, they can be excreted readily in sweat, urine, and feces. Most steroid metabolism before excretion occurs in the liver, although some also occurs in target cells.

Some steroids may be metabolized further or transformed in the target cell cytoplasm before binding to a receptor. For example, in many brain cells, testosterone is first converted to an estrogen by a P450 aromatase enzyme before binding to an estrogen receptor and producing an effect. (P450 proteins are a large family of mitochondrial proteins that perform a variety of important functions.) In other target cells, testosterone may be converted to 5-dihydrotestosterone (DHT) by the enzyme 5α-reductase. DHT has a much stronger affinity for the androgen receptor than does testosterone.

Thyroid Hormones

Two thyroid hormones are formed by the thyroid gland from the amino acid tyrosine, but through an entirely different process than described for catecholamine synthesis. The thyroid gland is composed of multiple follicles, each consisting of a single epithelial cell layer surrounding a lumen filled with a proteinaceous fluid known as colloid. The base of each follicle cell is near a capillary that is part of a vascular network surrounding each follicle. The apex of each follicular cell contacts the colloid in the lumen. The presence of iodine atoms in thyroid hormones is unique among vertebrate hormones and is essential for their biological activity. Consequently, dietary deficiencies in inorganic iodide or disruption of iodide transport in an organism can lead to thyroid deficiencies.

The pathway for the synthesis of thyroid hormones and their structures are provided in figure 1-9. Synthesis begins in the follicular cell as tyrosine residues are incorporated into a large glycoprotein called thyroglobulin, which migrates to the apical end of the cell and enters the colloid by exocytosis. The phenolic ring of the tyrosines are iodinated by iodide absorbed from the blood, producing residues with first one and then two iodides per tyrosine and forming monoiodotyrosine (MIT) and diiodotyrosine (DIT) respectively. If enough iodide is available, all of the resulting iodinated tyrosines will be in the form of DIT. Iodide accumulates in the thyroid cell by an energy-requiring sodium–iodide co-transport pump in the basal membrane of the follicular cell, and iodination of tyrosine is accomplished by the enzyme thyroid peroxidase (TPO). Two DITs are coupled immediately by TPO to form tetraiodothyronine or thyroxine (T_4). Although initially T_4 was believed to be the active form of thyroid hormone, research has shown that it is actually a prohormone that, by the loss of an additional iodine atom from the outer phenolic ring, could be converted to triiodothyronine or T_3. Because the receptors in thyroid target cells have a higher affinity for T_3 than T_4, it is generally accepted that T_3 is the active form of the hormone. Nevertheless, both are generally referred to as thyroid hormones. Most of the T_3 in the circulation is produced from T_4 in the liver by a

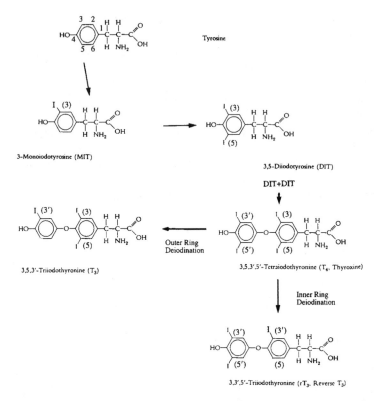

Figure 1-9. The synthesis of thyroid hormones. Thyroid hormones (T_3 and T_4) are made from tyrosine but via a very different process than for catecholamines. Tyrosine is first incorporated by thyroid cells into a large protein called thyroglobulin, iodinated, and then two iodinated tyrosines are coupled to form the hormones when the thyroglobulin is hydrolyzed enzymatically to release the hormones on demand (see chapter 4). Reprinted with permission from Norris (1997).

deiodinating enzyme, although some can be formed by the same process from T_4 in the thyroid gland before secretion into the blood.

In the blood, thyroid hormones are transported attached to binding proteins. The major binding protein is thyroid-binding globulin (TBG), which in mammals binds about 75% of the thyroid hormones. An additional 15% is transported attached to a prealbumin protein (transthyretin), with the remaining attached to serum albumin. TBG has the strongest affinity for T_4. About 99.9% of T_4 is bound to the thyroid-binding proteins, and only a small fraction is free in the blood. Only about 90% of the T_3 is bound, so that a greater proportion is available to enter target cells or to be degraded enzymatically (e.g., by liver deiodinases) and excreted.

Metabolism of thyroid hormones can take several routes. The most common pathway involves deiodination to biologically inactive forms by deiodinases in the liver and in target tissues. Some thyroid hormones are conjugated to sulfates or glucuronides like the steroids. Others are directly metabolized to inactive molecules such as tetraiodoacetic acid from T_4. As is the case for some of the steroids, T_4 may be converted by a selective deiodinase to T_3 in target cells before binding with thyroid receptors, supporting further the notion that T_4 is a sort of prohormone.

Mechanisms of Regulator Action

Regulators such as catecholamines, indoleamines, peptides, and proteins that bind to receptors on the cell surface generally do not pass across the cell membrane and therefore must generate intracellular second messengers that bring about the desired effect. If the regulator opens an ion channel, the second messengers might be ions entering the cell by diffusion and/or the change in the electrochemical gradient occurring across the cell membrane. In other cases, a new second messenger is synthesized at the cytoplasmic side of the membrane that then brings about changes within the target cell. These changes typically involve activation of enzymes that produce rapid effects. New protein synthesis may occur later through mechanisms that operate through gene transcription.

Steroid and thyroid hormones readily enter their target cells, where they bind to intracellular receptors in the cytoplasm or the nucleus. Their major effects are on protein synthesis via activation or repression of nuclear genes. In addition, however, some steroid hormones have been shown to produce rapid changes in target cells by activating cell membrane receptors in a manner similar to catecholamines and peptides. Details of the interactions of chemical regulators with their receptors and the mechanisms through which they produce effects on target cells is the subject of chapter 2.

Regulation of Development

The earliest stages of development are controlled by chemical regulators called inducers, secreted by embryonic cells. Later, as certain endocrine glands differentiate, hormones begin to influence development. The best known embryonic endocrine functions involve the thyroid (chapter 4), the adrenal cortex (chapter 5), and the gonads (chapter 6). For example, reproductive hormones are essential for early differentiation and development of sex accessory ducts in all tetrapods. Adrenal synthesis of androgens is essential for maintaining pregnancy in mammals, and fetal glucocorticoids are involved in the induction of labor. Thyroid hormones are important for postembryonic development in fishes, they initiate metamorphosis of larval fishes and amphibians, and they control early differentiation and functioning in the mammalian brain. Consequently, any interference with these systems by external chemicals could have profound effects.

Sex steroid hormones have two types of actions on development. They may produce an organizational effect such as the permanent designation of sex or a sexual trait. Once the trait is organized, it cannot change (i.e., it is determined). In other cases, the trait may be induced later in life by an activational action of the hormone.

This activational action may not be related to genetic sex but rather produces the same hormone-specific response in either sex. The involvement of steroids in sex determination and differentiation is discussed in chapter 6.

Vertebrate Nomenclature and Evolution

When reading the literature on vertebrates, one is easily distracted by the names used to refer to the various groups, in part because often they are complicated names that give no clues to the uninitiated as to what sort of vertebrates they are (e.g., chelonian, agnathan, dipnoan, acintopterygian). Additionally, different authors may use alternative names, creating further confusion. We provide a hierarchical listing of the major vertebrate groups with the more common synonyms and common names (table 1–4). Although this table provides a general gradation from the more primitive to the more advanced vertebrates, figure 1-10 provides a more accurate depiction of their evolutionary relationships.

Table 1-4. Vertebrate nomenclature and evolution.

Agnatha (jawless fishes)
  Cyclostomes
    Myxinoids (hagfishes)
    Petromyzontids (lampreys)
Gnathostomes (jawed vertebrates)
  Chondrichthyes (cartilaginous fishes)
    Elasmobranchs = selachians (sharks, rays, skates)
    Holocephalans (ratfishes)
  Osteichthyes (bony fishes)
    Sarcopterygians (lobe fin fishes)
      Crossopterygians (coelacanth, *Latimeria* spp)
      Dipnoi (lungfishes)
    Actinopterygians (ray fin fishes)
      Polypteriformes = Cladista (bichir)
      Chondrosteans (sturgeons, paddlefishes)
      Holosteans (bowfin, gars)
      Teleosts (most modern fishes)
Tetrapod gnathostomes
  Amphibia
    Anura (frogs, toads)
    Urodela = Caudata (salamanders, newts)
    Apoda = Gymnophiona (apodans: limbless)
  Reptiles
    Chelonia (turtles, tortoises)
    Rhynchocephalia (the tuatara)
    Squamata (lizards, snakes)
    Crocodilia (crocodiles, alligators, caiman)
  Aves (birds)
  Mammalia
    Prototheria = Protheria (egg-laying monotremes: platypus, echidnas)
    Metatheria (marsupials: kangaroos, wombats, possums, etc.)
    Eutheria (placental mammals)

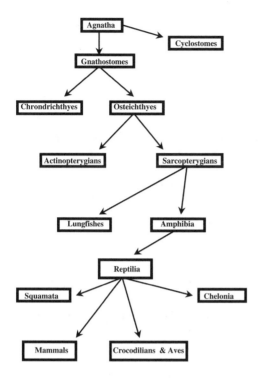

Figure 1.10. General evolutionary relationships of major vertebrate groups. See table 1-4 for an explanation of the group names.

The most primitive vertebrates were the jawless fishes (agnatha: *a* = without; *gnathos* = jaw), whose only living descendents are the cyclostomes (lampreys and hagfishes). However, the hagfishes are considered to be degenerate by many as evidenced by their anatomy. Many of their organ systems are simple and they lack vertebrae. Evolution of jaws is considered to be a major leap in vertebrate evolution that allowed fishes to become predatory and no longer linked to filtering the sea bottom for food. The early jawed fishes (gnathostomes: *stome* = mouth) gave rise to all of the modern bony fishes (Osteichthyes: *osteo* = bone; *ichthyos* = fish) as well as the cartilaginous fishes (Chondrichthyes: *chondro* = cartilage), the sharks and their relatives. The early osteichthyian fishes gave rise to both the modern ray-finned fishes (teleosts, etc.) and also the lobe-fin fishes (coelacanths and lungfishes). The group (crossopterygian fishes) giving rise to the latter two are also thought to be ancestral to the first tetrapod (four-footed) gnathostomous vertebrates, an amphibian.

Within the amphibians, frogs and toads (anurans) have diverged most from the ancestral salamander body plan of the amphibians that gave rise to the reptiles. There was a great radiation of reptiles almost simultaneously into many directions, leading to the modern turtles (chelonians), lizards and snakes (squamates), crocodilians and birds (archosaurs), and the mammals. The tuatara (Rhynchocephalia) of New Zealand represents the lone survivor of an early lizardlike group of reptiles. Among the mammals, we consider the egg-laying monotremes (echidnas, duckbilled platypus) to be the most primitive. The marsupials (kangaroos, koalas, opossum,

Tasmanian devil, etc.) represent an intermediate grade between the monotremes and the eutherian or placental mammals such as carnivores, rodents, cetaceans (e.g., whales, porpoises), and primates.

References

Di Giulio, R.T., Tillitt, D.E., 1999. Reproductive and Developmental Effects of Contaminants in Oviparous Vertebrates. SETAC Press, Pensacola, FL.

Kendall, R., Dickerson, R., Giesy, J., Suk, W., 1998. Principles and Processes for Evaluating Endocrine Disruption in Wildlife. SETAC Press, Pensacola, FL.

Norman, A.W., Litwack, G., 1998. Hormones, 2nd edition. Academic Press, San Diego, CA.

Norris, D.O., 1997. Vertebrate Endocrinology, 3rd edition. Academic Press, San Diego, CA.

For Further Reading

General Endocrinology

Becker, J.B., Breedlove, S.M., Crews, D., McCarthy, M.M., 2002. Behavioral Endocrinology, 2nd edition. The MIT Press, Cambridge, MA.

Bentley, P.J., 1998. Comparative Vertebrate Endocrinology, 3rd edition. Cambridge University Press, Cambridge.

Bolander, F.F., 1994. Molecular Endocrinology, 2nd edition. Academic Press, San Diego, CA.

Brown, R.E., 1994. An Introduction to Neuroendocrinology. Cambridge University Press, Cambridge.

Freedman, L.P., 1998. Molecular Biology of Steroid and Nuclear Hormone Receptors. Springer-Verlag, Berlin.

Hadley, M.E., 2000. Endocrinology, 5th edition. Prentice Hall, Upper Saddle River, NJ.

Nelson, R.J., 1995. An Introduction to Behavioral Endocrinology. Sinauer Associates, Sunderland, MA.

Weintraub, B.D. (Ed.), 1995. Molecular Endocrinology: Basic Concepts and Clinical Correlations. Raven Press, New York.

Vertebrate Biology

Butler, A.B., Hodos, W., 1996. Comparative Vertebrate Neuroanatomy: Evolution and Adaptation. Wiley-Liss, New York.

Duellman, W.E., Trueb, L., 1986. Biology of Amphibians. McGraw-Hill, New York.

Evans, D.H., 1997. The Physiology of Fishes, 2nd edition. CRC Press, Boca Raton, FL.

Feder, M.E., Burggren, W.W., 1992. Environmental Physiology of the Amphibians. University of Chicago Press, Chicago.

Groot, C., Margolis, L., 1998. Pacific Salmon Life Histories. UBC Press, Vancouver.

Helfman, G.S., Collette, B.B., Facey, D.E., 1997. The Diversity of Fishes. Blackwell Science, Malden, MA.

Kardong, K.V., 2002. Vertebrates: Comparative Anatomy, Function, Evolution, 3rd edition. McGraw Hill, Boston.

McNab, B.K., 2002. Physiological Ecology of Vertebrates. Cornell University Press, Ithaca, NY.

Pough, F.H., Heiser, J.B., McFarland, W.N., 1996. Vertebrate Life, 4th edition. Prentice Hall, Upper Saddle River, NJ.

Invertebrate Biology

Brusca, R.C., Brusca, G.J., Burness, T.J., 2003. Invertebrates, 2nd edition. Sinauer Associates, Sunderland, MA.

Nijhout, H.F., 1998. Insect Hormones. Princeton University Press., Princeton, NJ.

Penchenik, J.A., 1999. Biology of Invertebrates, 4th edition. McGraw-Hill, New York.

2

Hormone Action on Receptors

Miles Orchinik
Catherine Propper

The physiological actions of hormones are mediated by receptors, but the specificity of receptors is rarely absolute. Hormone receptors may bind endogenous compounds other than their principal cognate ligands and may also bind exogenous compounds. Therefore, a fraction of exogenous pharmaceuticals, as well as agricultural or industrial chemicals released into the environment, may bind to hormone receptors. A major subset of environmental endocrine-disrupting chemicals (EDCs) exert their actions by binding to hormone receptors. For example, agricultural compounds (insecticides, herbicides, and fungicides) and industrial compounds, such as dioxins (e.g., 2,3,7,8-tetrachlorodibenzo-p-dioxin; TCDD; see chapter 10), polychlorinated biphenyls (PCBs), and plastic derivatives (e.g., bisphenol A), can bind to endogenous hormone receptors. EDCs may also influence endocrine function through cellular mechanisms that do not directly involve EDC binding to the receptor for the hormone in question. In this chapter, we introduce some of the basic principles of receptor theory and practice necessary to understand and quantify the binding of endogenous or exogenous compounds to hormone receptors.

To evaluate the actions of EDCs, one needs an understanding of the basic theory and practice of ligand–receptor interactions. We have been accumulating data about the structure and function of hormone receptors and associated signaling molecules at an unprecedented rate. This vast new information base and its associated emergent technologies are tremendous resources that complement, but do not replace, the basic models of receptor theory and practice. Receptor pharmacology may seem complex at times, but a recent Receptor Biology Roundtable panel concluded that rigorous, basic receptor theory provides unifying concepts that are essential for analyzing EDCs (Limbird and Taylor, 1998). This chapter presents the mass action theory that describes ideal ligand–receptor interactions, several technical issues and limitations in radioligand binding studies, and then discusses several studies that investigated the binding of EDCs to steroid receptors. For more thorough and quantitative treatment of radioligand binding analysis, and particularly for nonequilibrium assays, which we do not discuss in detail, a number of excellent books are available (Yamamura et al., 1990; Kendall and Hill, 1995; Limbird, Keen, 1996, 1999; Picard, 1999). In addition, the following web sites are also excellent resources: GraphPad, Inc., makers of Prism, offers useful tools and links from their Data Analysis Resource Center (http://www.graphpad.com/welcome.htm) and "Theory and Practice of Receptor Characterization and Drug Analysis," an online course on receptor pharmacology taught by Richard B. Mailman and Jose Boyer (University of North Carolina; http://www.med.unc.edu/wrkunits/2depts/pharm/receptor/index.htm).

General Principles in Hormone–Receptor Binding

What Are Receptors?

The work of Ehrlich (1914) and Langley (1905) at the end of the 19th century led to the recognition that, in order for a compound to have a physiological effect, it must bind to a receptor; *corpora non agunt nisi fixata*: agents cannot act unless they bind. At the same time, everything binds to everything with low affinity. Therefore, one needs to carefully distinguish between binding to physiological receptors and nonspecific or artifactual binding. A receptor is generally a protein that selectively binds a chemical signal (ligand); the binding of endogenous ligand typically leads to a specific cellular response. The binding of an EDC to a physiological receptor may activate the receptor in a manner similar to an endogenous ligand, or it may interfere with the binding and actions of endogenous ligands without activating the receptor itself. In either case, one can only establish that a potential EDC is interfering with a physiological receptor system if the receptor of interest is clearly defined. A powerful tool in these investigations are radioligand binding studies. A number of criteria, listed below, are used to distinguish between artifactual binding sites and authentic physiological receptors, and many of these criteria can be assessed using radioligand binding studies.

1. Kinetic studies, done under nonequilibrium conditions, should demonstrate that the hormone–receptor interaction follows the law of mass action. Dissociation experiments should demonstrate that the binding of a radioligand to a receptor is reversible, although dissociation may only occur under certain assay conditions. Association rates derived from kinetic experiments should be consistent with the kinetics of physiological responses. In addition, the affinity estimates derived from the dissociation and association rate constants (k_{-1} and k_{+1}, respectively) should be similar to the equilibrium dissociation constant, K_d, derived under equilibrium conditions.
2. Equilibrium saturation binding studies should also indicate that the ligand–receptor interaction follows the law of mass action, and that binding sites are

saturable with a finite number of receptors. Even the most abundant hormone receptors are present only in very low concentrations—typically in the pico-mole/mg or femtomole/mg protein range.

3. The estimated affinity of a receptor for a given hormone should be reason-ably consistent with physiological responses and bioassays. Receptor affin-ity is usually expressed in terms of the equilibrium dissociation constant, K_d, the molar concentration of ligand that occupies half of the total receptor num-ber in a given preparation (B_{max}) at equilibrium. The smaller the K_d value, the higher the affinity of the receptor for that ligand. Hormone receptors have high affinity for their endogenous ligands, with K_d values typically between 0.01 and 10 nM. However, it is not uncommon to see a mismatch between the K_d for hormone binding and the concentration of hormone required to produce a half-maximal physiological response (ED_{50}; see below).

4. The receptor should bind a class of compounds that is consistent with physi-ological responses mediated by the given hormone. Steroid hormone recep-tors, for example, may display preference for specific stereoisomers, and this stereoselectivity should be consistent with physiological responses induced by specific isomers.

5. The distribution of binding sites should be consistent with physiology. Re-ceptors should be found in tissues or in cells that exhibit direct physiological responses to the ligand.

6. Control experiments should demonstrate that conditions that denature pro-teins, such as boiling and protease treatment, also inhibit binding and physi-ological responses. If receptor knockout models are available, specific binding and cellular responses should be greatly diminished in these animals relative to wild-types.

7. An authentic physiological receptor mediates a functional response, but ob-taining the type of data necessary to establish such a functional role is often arduous. Typically a variety of studies are required to determine if receptor activation by regulated release of hormone produces a meaningful response. Receptor isolation and reconstitution studies and gene cloning expression studies may be necessary to identify physiological receptors, but such stud-ies alone are not sufficient.

A note of caution: These criteria have been developed to establish the existence of physiological receptors for endogenous hormones or neurotransmitters, but EDCs may not be constrained by them. For example, an EDC, unlike the endogenous ligand, may bind irreversibly to a hormone receptor.

Dose Responses

An important criterion in confirming a suspected EDC can is the demonstration of a biological dose response to the compound within the range of concentrations found in the environment. One might expect the dose–response curve for a given com-pound to parallel the receptor occupancy curve for that compound in terms of shape, potency, and efficacy. However, pharmacodynamics and endocrinology indicate that biological dose responses are often quite different from receptor occupancy curves. Therefore, binding studies alone are not sufficient for screening potential EDCs.

Figure 2-1 displays an S-shaped receptor occupancy curve on a semilog plot, derived from an equilibrium saturation binding experiment representing a ligand and receptor following the law of mass action. In addition, four curves are shown representing hypothetical biological dose responses to four compounds that could have yielded the binding curve shown. In both ligand-binding and dose–response curves, several key parameters are represented: the maximal and minimal responses to the compound, the slope of the line between maximal and minimal response, and the dose of compound that produces a response midway between the maximal and minimal response. Therefore, it is important to use a range of concentrations to identify the maximal and minimal responses when estimating a compound's potency. The concentration that produces the half-maximal response in a dose–response study is the ED_{50} or EC_{50}; this is a measure of potency. In the simplest case (line C in figure 2-1), the biological dose response to compound C closely parallels the occupancy of its receptor, so the ED_{50} and K_d values are similar. In reality, dose–response curves often fall to the left of the receptor occupancy curve, as indicated by line A. The biological dose response to compound A is proportional to the amount of compound added, but the amount of compound required to produce its half-maximal effect is far less than the amount required to occupy half of its receptors. Comparing lines A and C, the efficacy is similar between compounds A and C, but the potency represented by line A is greater than line C. Line B might represent the dose response to a compound that acts as a partial agonist. Compound B has a potency similar to compound A; the ED_{50} for B is similar to the ED_{50} for A, even though B does not even produce 50% of A's maximal response. However, B is far less efficacious than A.

Therefore, to present data on EDCs in a meaningful way, one needs to determine the minimal and maximal response to a compound with a number of intermediate doses, analyze the shape of the curve by nonlinear regression, and then report both the compound's potency and efficacy.

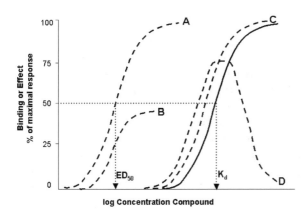

Figure 2-1. Hypothetical dose–response curves.

The mismatch in potency between the ED_{50} and K_d, as for compound A (figure 2-1), with the dose–response curve to the left of the fractional occupancy curve, is often explained by the concept of spare receptors. Many factors may account for this phenomenon, but functionally it means that activation of a small fraction of receptors may be sufficient to produce a maximal cellular response. A striking example of spare receptors is that occupancy of as few as 0.1% of estrogen receptors (ER) in prolactin-secreting cells derived from a pituitary tumor is sufficient to stimulate half-maximal cell proliferation (Chun et al., 1998). The ED_{50} for induction of prolactin gene expression by 17β-estradiol (E_2) via ER is several orders of magnitude greater than the ED_{50} for cell proliferation and similar to the K_d for E_2 binding to ER in this cell line. The implications is that concentrations of compounds, such as EDCs, may have important physiological consequences even if they only bind a small fraction of total receptor number, and that the same compound may produce different physiological responses with very different potency.

Unlike in vitro radioligand binding studies, where one assumes that ligand–receptor interactions are consistent with mass action principles, biological dose–response curves can assume almost any shape. For example, in figure 2-1, line D represents an inverted U-shaped dose–response curve. In contrast to toxicology, inverted U-shaped dose responses to hormones are common in endocrinology and should be expected in EDC research. Other types of atypical dose–response curves, with low doses producing responses not seen with higher doses, also appear in the EDC literature (Ramamoorthy et al., 1997; Park et al., 2001). The mechanisms producing atypical dose–response curves may be intracellular, or they may only occur in vivo. In the latter case, it may be that low doses of EDC produce receptor-mediated effects without activating the hepatic enzymes that metabolize the compound to inactive forms. At higher doses, hepatic metabolism of EDCs may reduce the plasma and tissue levels of the EDC below response threshold levels. At even higher doses, the metabolic enzymes may be overwhelmed.

A detailed discussion of the mechanisms underlying spare receptors or complex dose responses to hormones or EDCs is beyond the scope of this chapter, but their presence illustrates several important points for the study of endocrine disruption: (1) Reliance on K_d values derived from radioligand binding studies alone may lead to underestimation of the potency of exogenous compounds, particularly if activation of a small fraction of hormone receptors is sufficient to produce cellular responses; 2) potential actions of EDCs should be screened over a wide range of concentrations; and (3) it may be necessary to use several bioassays to discover responses induced by environmentally relevant concentrations of EDCs.

In addition, it should be remembered that many hormones are transported in the bloodstream bound to a binding globulin. In the case of steroids, 90–99% of the total circulating hormone is typically bound to binding globulin, meaning that only a small fraction of circulating hormone is biologically active. If an exogenous compound acts at receptors for such an endogenous hormone but does not bind to the hormone-binding globulin in plasma, then one may underestimate the potency of the exogenous compound, relative to the endogenous hormone, by 10- to 100-fold.

Types of Receptors and Responses

As implied in the discussion of dose–response curves, the biological result of a ligand–receptor interaction depends on many factors in addition to the concentration of ligand and the binding affinity of the compound. Cellular responses will vary depending on the types of receptors and transduction machinery present in a target cells and whether the ligand acts as a receptor agonist, antagonist, inverse agonist, partial agonist, or produces a mixed response. EDCs act as receptor agonists when they bind to the receptors, producing conformational changes in the receptor that lead to cellular responses, similar to the actions of the endogenous ligand. In contrast, EDCs act as receptor antagonists when they bind to the receptor but fail to produce a conformational change that leads to a cellular response. The cellular actions of receptor antagonists are therefore indirect, produced by blocking access of the endogenous hormone to the receptor binding site. Inverse agonists bind to receptors and produce cellular responses that are opposite to the response produced by an endogenous ligand.

Most hormonal effects on cells are mediated by hormone interaction with one or more of the following types of receptors: transmembrane receptor kinases, G-protein–coupled receptors, ligand-gated ion channels, or ligand-activated transcription factors. Much of the notoriety achieved by EDCs to date has been due to their demonstrable direct interaction with nuclear receptors belonging to the superfamily of ligand-activated transcription factors that includes intracellular steroid receptors. Steroid receptors appear to be particularly vulnerable to endocrine disruption. Ligands that bind to nuclear receptors, such as estrogenic EDCs, are powerful regulators of cell function because these receptors determine the rate of expression of specific genes. Other well-known examples of EDCs binding to nuclear receptors are dioxins and some PCBs that bind to the aromatic hydrocarbon receptor (AhR). Similar to steroid hormone receptors, the activated nuclear AhR binds to promoter regions of a number of genes. These target genes may directly alter endocrine function or may induce expression of enzymes that metabolize the phenolic compounds themselves.

EDCs may also bind to other types of receptors to produce endocrine disruption. For example, EDCs may bind to or alter the activity of steroid receptors found in plasma membranes rather than the cell nucleus or cytosol (Nadal et al., 2000). A number of pesticides (e.g., endosulfan and other organochlorine pesticides) bind with high affinity to vertebrate neurotransmitter receptors (Bloomquist, 1996), such as γ-aminobutyric acid (GABA)$_A$ receptors, and may alter endocrine function through this mechanism. In other words, an EDC does not need to bind to ER to be estrogenic, but it may disrupt normal endocrine function by acting on other signaling systems in the hypothalamus, pituitary, or peripheral targets. Some EDCs also bind to steroid binding proteins in plasma (Martin et al., 1995) and thereby alter the availability or action of steroid hormones. Although the plasma-binding proteins for steroids are not receptors in the conventional sense of the word, the interaction of ligands with these binding proteins follows the law of mass action, so the discussions of mass action are relevant for understanding this potential mechanism of EDC action. We should be open to the possibility that EDCs exert cellular effects by binding to other types of physiologically important molecules, such as enzymes, transport molecules, or DNA.

In general, hormone–receptor interaction involves a recognition step, a transduction mechanism, and an effector mechanism. The recognition site binds the chemical signal or ligand. The transduction mechanism transmits the signal from the cell surface, cytosolic, or nuclear receptor to an effector mechanism. The effector produces the cellular response—via enzymatic activity, ion channel kinetics, or gene transcription. These processes may involve one or several distinct molecules. In the case of G-protein–coupled receptors, the recognition site, the transduction mechanism, and the effector are found on three separate molecules. In the case of steroid receptors, one might argue that the receptor itself performs all three processes. EDCs may interact with any of these steps to disrupt endocrine function.

A major challenge in signal transduction, and therefore a challenge for studying EDCs, is the fact that a neurocrine, hormone, or cytocrine will likely bind to more than one type of receptor. Some of the receptors used by a hormone may be closely related, but other receptors used by the same hormone may represent an entirely different class of receptor with entirely different effector mechanisms. This diversity of receptor types means that one may need to examine multiple types of ligand–receptor interaction and effector outcomes in screening potential EDCs. For example, environmental estrogens may bind preferentially to one of two nuclear, ligand-activated transcription factors: estrogen receptor α (ERα) or ERβ. Because ERα and ERβ regulate the expression of different genes, and have quite different distributions, it may be important to examine multiple tissue types and regions. In addition, EDCs bind to estrogen receptors in cell-surface membranes (Watson et al., 1999; Nadal et al., 2000). Xenoestrogens may also bind directly to ion channels, in particularly the BK potassium channel (large conductance Ca^{2+}-activated or maxi-K channels) in cell membranes (Dick et al., 2001). Further, xenoestrogens may bind to sex-hormone–binding proteins in plasma and thereby alter the activity of circulating estrogens (Crain et al., 1998; Dechaud et al., 1999; Nagel et al., 1999). Therefore, screening for EDC action by assuming one mechanism of estrogen action may be insufficient or misleading. The diversity of receptors available to hormones makes the investigation of EDCs or suspected environmental toxins a daunting task, especially the interpretation of negative results based on the interaction of an EDC with one receptor subtype.

Radioligand Binding Studies

Equilibrium Properties of Hormone Receptors: Saturation

The simplest, most direct and powerful tool for studying the interaction of hormones or EDCs with receptors is the radioligand binding assay. In these studies, the hormone of interest, or a synthetic analog, is labeled with a radioactive isotope such as tritium ($[^3H]$) or iodine-125 ($[^{125}I]$), and the interaction of the radiolabeled ligand with responsive tissue is quantified. In 1971, Cuatrecasas labeled putative insulin receptors with ^{125}I-labeled insulin, and Goldstein et al. developed a theoretical and experimental strategy for detecting stereospecific opiate receptors. In these early studies, only 2% of the binding of the radioligand to tissue was specific; that is, only 2% of the total binding of radioligand could be displaced by unlabeled ligand (see discussion of nonspecific binding below). In 1973, Pert and Snyder and others used a filtration assay (see below) and radioligands with high specific activity to characterize endogenous opioid receptors; in these studies, the majority of radioligand binding was specific.

Equilibrium saturation binding experiments are the most familiar binding studies. From these experiments, we can obtain estimates of receptor density (B_{max}) and receptor affinity (K_d) for a given radioligand in a given tissue. Receptor density is typically expressed as the number of receptors sites per mg cytosolic or membrane protein, or the number of sites per cell. These experiments are performed by incubating the receptor preparation with a range of radioligand concentrations for a long enough period so that equilibrium is achieved. The amount of radioligand specifically bound to the receptors at equilibrium for each concentration of radioligand is determined. Mathematical analysis of the binding data involves several assumptions: (1) There are a finite number of binding sites; (2) the binding reflects a simple bimolecular reaction; and (3) specific binding is reversible.

In the simplest case, the interaction between ligand (L) and receptor (R) involves two simultaneous reactions: the formation of the ligand–receptor complex (LR) and the dissociation of the ligand from the ligand–receptor complex. This interaction is described by the law of mass action:

$$L + R \underset{k_{-1}}{\overset{k_{+1}}{\rightleftharpoons}} LR. \tag{2-1}$$

The rate equation for the formation of receptor–ligand complex is given by a second-order equation:

$$\frac{d[LR]}{dt} = k_{+1}[R][L], \tag{2-2}$$

where the rate of association of L and R depends upon the concentration of R and of L and the association rate constant (k_{+1}). The equation for the reverse reaction, the dissociation of L from the complex, is given by a first-order equation:

$$\frac{d[LR]}{dt} = k_{-1}[LR], \tag{2-3}$$

where rate of dissociation depends only upon the concentration of ligand–receptor complex and the dissociation rate constant (k_{-1}). When the binding reaction reaches equilibrium, the formation of RL equals the dissociation of LR into R and L, and therefore, at equilibrium:

$$k_{+1}[R][L] = k_{-1}[LR]. \tag{2-4}$$

The equilibrium dissociation constant (K_d) is, by definition, equal to:

$$K_d = \frac{k_{-1}}{k_{+1}} = \frac{[R][L]}{[LR]} \tag{2-5}$$

Because receptor density (B_{max}) equals the number of free receptors [R] plus the amount of ligand-occupied receptors [LR], $[R] = B_{max} - [LR]$. Rearranging:

$$K_d = \frac{[B_{max} - [LR]][L]}{[LR]} = \frac{B_{max}[L]}{[LR]} - \frac{[LR][L]}{[LR]}$$

$$K_d + [L] = \frac{B_{max}[L]}{[LR]}$$

and

$$[LR] = \frac{B_{max}[L]}{K_d + [L]}. \tag{2-6}$$

Equation 2-6 was derived by Hill in 1910, and is mathematically similar to the Michaelis-Menten equation for a rectangular hyperbola that describes the interaction between enzyme and substrate. As modified, it describes the interaction of ligand and receptor following the law of mass action in terms of receptor number and affinity. If a receptor preparation is incubated with an appropriate range of radioligand concentrations in an equilibrium saturation binding experiment, and the ligand and receptor follow the law of mass action, the data can be fit to equation 2.6 as shown in figure 2-2.

In an equilibrium saturation binding study, a receptor preparation is incubated with a range of radioligand concentrations to determine total binding. In parallel tubes, nonspecific binding (NSB) is determined by incubating the receptor preparation with radioligand and an excess of an unlabeled ligand that binds to the receptors of interest—usually at a concentration 200–500 times greater than the predicted K_d. Nonspecific binding, therefore, is the binding of radioligand to anything other than the receptor of interest. Specific binding to the receptor is estimated by subtracting NSB from the total binding at each concentration of radioligand. Shown in figure 2-2 are total binding, NSB, and specific binding. The points of particular interest are the K_d and the B_{max}. The K_d, indicated by the arrow pointing to the x-axis, is found on the steepest part of the slope and is the molar concentration of radioligand that occupies 50% of receptors. The B_{max} is estimated from the curve as the point at which receptors are fully occupied and is generally normalized to protein content or cell number.

The accurate determination of NSB is one of the key problems in radioligand binding assays. It is sufficiently problematic that experimenters and data analysis software, such as GraphPad Prism and LIGAND (Biosoft, Cambridge, UK), often circumvent the problem by estimating, rather than measuring, NSB. One can fit equilibrium saturation binding data to an equation that estimates the contribution of NSB to total binding as a component that increases linearly with radioligand concentration, c[L]:

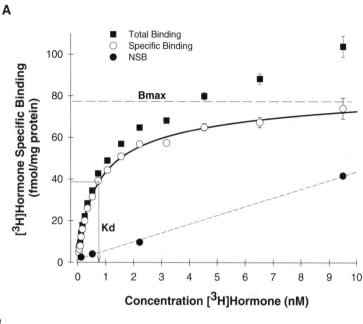

A

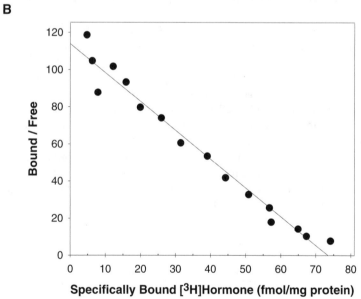

B

Figure 2-2. (A) Equilibrium saturation binding. (B) Scatchard-Rosenthal plot of data in Figure 2-1, curve A.

$$[LR] = \frac{B_{max}[L]}{K_d + [L]} + c[L]. \tag{2-7}$$

For the data in figure 2-2, a compromise approach was used, in which we measured NSB at several concentrations of radioligand; GraphPad Prism calculated a regression line for NSB based on these measurements and subtracted the estimated NSB from all determinations of total binding.

A common mistake leading to spurious parameter estimates in radioligand binding assays is the use of too high a concentration of unlabeled competitor to determine NSB. Equation 2-7 is based on the assumption that NSB increases linearly with radioligand concentration and is not saturable. In the range of radioligand concentrations usually used in saturation studies, this assumption holds. However, if one attempts to measure NSB by using a very high concentration of unlabeled ligand, a component of saturable NSB of radioligand may be displaced from non-receptor sites. In that case, specific binding estimates will be erroneous. To avoid this problem, one should try to use an unlabeled ligand that is structurally different from the radioligand but still binds to the receptor of interest. If a structurally different compound is unavailable, one should determine empirically, in a titration study, the concentration of unlabeled ligand that displaces 100% of radioligand binding to the specific receptor of interest but does not displace an additional component of NSB.

The data in figure 2-1 were fit to equation 2-6 using nonlinear regression analysis in GraphPad Prism, producing parameter estimates of K_d equal to 0.67 ± 0.03 and B_{max} equal to 73.6 ± 0.8 fmol/mg protein (table 2-1). Protein content in samples is typically derived from colorimetric assays in which sample reaction product is compared to a standard curve generated using known concentrations of protein. Most commonly used assays include the Bradford assay for cytosolic proteins (Bradford, 1976), modifications of the original assay for small sample sizes, Lowry assays for brain membrane proteins, or commercially available reagents.

Before the desktop computer and popular software packages, equilibrium saturation binding data were linearized and plotted according to the well-known Scatchard-Rosenthal equation. The Michaelis-Menten equation for a rectangular hyperbola can be rearranged to the following:

$$\frac{[LR]}{[L]} = -\frac{1}{K_d} \bullet [LR] + \frac{B_{max}}{K_d} \tag{2-8}$$

This can be plotted as [LR]/[L] versus [LR], the amount of radioligand bound/radioligand free versus the amount of radioligand bound. With ideal data and a simple bimolecular reaction following the law of mass action, the slope of the line is $-1/K_d$, the y-intercept is B_{max}/K_d, and the x-intercept is B_{max}. The data from figure 2-1A are replotted in figure 2-1B.

In a Scatchard-Rosenthal plot, the steeper the slope, the higher the affinity. In comparing two saturation experiments, if the lines on the Scatchard plot are different, but the x-intercept is the same, it is likely that the affinities differ, but the B_{max} values are the same between the samples. If lines fitting data from two experimen-

Table 2-1. Simplest case binding-parameter estimates derived by nonlinear regression analysis of data shown in figure 2-2, using GraphPad Prism.

Best Equation	Specific Binding Equation 1	vs.	Best-fit Values Equation 2	
		B_{max1}		31.9
Equation 1		K_{d1}		0.4207
Best-fit values		B_{max2}		42.42
B_{max}	73.57	K_{D2}		1
K_d	0.6702	Std. Error		
Std. Error		B_{max1}		187.1
B_{max}	0.7801	K_{d1}		1.035
K_d	0.02822	B_{max2}		186
95% Confidence intervals		K_{d2}		2.1
B_{max}	71.92 to 75.22	95% Confidence intervals		
K_d	0.6104 to 0.7300	B_{max1}		−369.3 to 433.1
Goodness of fit		K_{d1}		−1.800 to 2.641
Degrees of freedom	16	B_{max2}		−356.5 to 441.4
R	0.9967	K_{D2}		−3.504 to 5.504
Absolute sum of squares	33.78	Goodness of fit		
$Sy.x$	1.453	Degrees of freedom		14
		R		0.9969
		Absolute sum of squares		31.64
		$Sy.x$		1.503
		Comparison of fits		
		DFn, DFd		2, 14
		F		0.4723
		P value		0.6332
		Best fit fquation		Eq. 1
		Data		
		Number of X values		18
		Number of Y replicates		1
		Total number of values		18
		Number of missing values		0

tal groups are parallel, it is likely that the affinities are the same, but the B_{max} is different. The presence of a competitive inhibitor (a compound that binds to the same site as the radioligand; see below) in the saturation binding assay would alter the apparent K_d and produce a line with a different slope but with the same B_{max}. This is because increasing concentrations of radioligand displace the competitive inhibitor, eventually saturating the binding sites. In contrast, a noncompetitive or allosteric inhibitor would be expected to decrease the B_{max} estimate.

Because Scatchard-Rosenthal transformation may produce plots with more effective visual illustrations of treatment effects on K_d or B_{max}, it is good practice to replot data from equilibrium saturation binding assays. Another advantage to the Scatchard-Rosenthal replot is that a curvilinear, rather than a linear, Scatchard plot can indicate a number important features of the binding system in question. The assumption that a hormone binds to a single class of binding sites according to the law of mass action is often not met, so curvilinear Scatchard plots may also provide visual evidence for more complex binding kinetics, consistent with the existence of different affinity states of the receptor in question or binding to multiple types of receptors. More important, a curvilinear plot may indicate technical problems in the binding assay, such as an incorrect definition of NSB, nonequilibrium conditions, or degradation of ligand.

Nonlinear regression analysis, rather than Scatchard-Rosenthal analysis, should be used to estimate K_d and B_{max}, however, unless one is certain that the binding reaction is characterized by an ideal, simple bimolecular interaction following the law of mass action. This situation is rarely encountered. Analysis of equilibrium saturation binding data using a Scatchard-Rosenthal transformation violates most of the assumptions of linear regression analysis. Linear regression analysis assumes that (1) error is associated only with the dependent variable, (2) the distributions of x and y are independent, (3) errors are normally distributed, and (4) error variance is constant (Lutz and Kenakin, 1999). In a Scatchard analysis, the dependent variable [RL], the amount bound, appears in both the x-axis and y-axis, so any error in the measurement of the amount bound, x-axis, is passed on to the y-axis. This is not the case when fitting data to the equation for a rectangular hyperbola using in nonlinear regression analysis. In the Scatchard plot, variance is nonuniform, particularly at the intercepts. At low radioligand concentrations, binding values approach background, so that relatively small experimental errors in the low dose tubes are magnified by the linear transformation. At high radioligand doses, NSB is proportionally high and there is increased error associated with the determination of specific binding. In other words, the error associated with the dependent variable [LR] is dependent on the absolute value in violation of linear regression assumptions. At the same time, low radioligand concentrations are important in determining the K_d of a high-affinity interaction, and [LR] values obtained at high radioligand concentrations are particularly important in the determination of B_{max}. By using least squares methods of curve fitting untransformed binding data, nonlinear regression analysis avoids these pitfalls inherent in Scatchard linearization.

Equilibrium Properties of Hormone Receptors: Competition Studies

As discussed earlier, another critical step in identifying a physiological receptor or providing evidence that an exogenous compound interacts with an identified receptor is the demonstration of pharmacological specificity. One way to determine specificity (or to screen for EDCs) would be to perform equilibrium saturation analyses using a radiolabeled form of each compound of interest. However, relatively few candidate EDCs are available in radiolabeled form. Even if they were available in radiolabeled form, performing saturation analyses with each of these radioligands would be prohibitively expensive. Instead, one can demonstrate and characterize specificity by studying the ability of unlabeled compounds to inhibit the binding of a radioligand that binds to the known receptor. Assuming that the radioligand (L) and its inhibitor (I) bind to the same site on a receptor (competitive inhibition), and that the interactions of L and I with R follow the law of mass action, the reactions are described by equation 2-1 and by

$$[I]+[R]\underset{k_{-1i}}{\overset{k_{+1i}}{\rightleftharpoons}}[IR].\qquad(2\text{-}9)$$

According to the law of mass action, the rates of formation of bound radioligand [LR] and bound inhibitor [RI] are given by the following equations:

$$\frac{d[LR]}{dt}=k_{+1}[L]\big[B_{max}-[RI]\big]-k_{-1}[LR]\qquad(2\text{-}10)$$

$$\frac{d[IR]}{dt}=k_{+1i}[I]\big[B_{max}-[LR]-[RI]\big]-k_{-1i}[RI].\qquad(2\text{-}11)$$

At equilibrium, $d[L_1R]/dt = d[L_2R]/dt = 0$, and the following equation can be derived from this relationship:

$$[LR]=\frac{B_{max}[L]}{[L]+K_d\left(1+\dfrac{[I]}{K_{di}}\right)}.\qquad(2\text{-}12)$$

This is similar to the equation for the rectangular hyperbola describing saturation binding, where the amount of radioligand bound, [LR], is related to the B_{max} and K_d for radioligand binding to the receptor site, except that, in the presence of a competitor, the apparent K_d of the radioligand is increased by the factor $(1+[I]/K_{di})$. This reflects the fact that it requires additional radioligand to saturate binding sites when a competitive inhibitor is present. This apparent increase in the K_d in the presence of a competitor is due to a decrease in the association rate, without an effect on the dissociation rate. Therefore, if the K_d values obtained from equilibrium saturation experiments are greater than those published in the literature, one might suspect that a ligand, endogenous or otherwise, is present in the preparation.

One can characterize the binding of many unlabeled competitors at a receptor site using just one radioligand. These experiments are performed by incubating a receptor preparation with a single concentration of radioligand, less than the K_d, if practical, and a range of concentrations of inhibitor. A more general term for these types of studies is inhibition or titration experiments because, strictly speaking, an inhibition experiment is a competition experiment only if the inhibitor directly competes for the same binding site as the radioligand. The data from a competition experiment were plotted in a semilog plot in figure 2-3.

What should a competition curve look like? First consider equilibrium saturation binding using only radioligand. Rearranging the equation for a rectangular hyperbola to look at fractional occupancy,

$$\frac{[LR]}{B_{max}}=\frac{[L]}{[L]+K_d},\qquad(2\text{-}13)$$

one sees that the fraction of bound receptors is a function of amount of ligand present and the K_d for receptor–ligand interaction. If the ligand interacts with one class of receptor according to the law of mass action, at a radioligand concentration

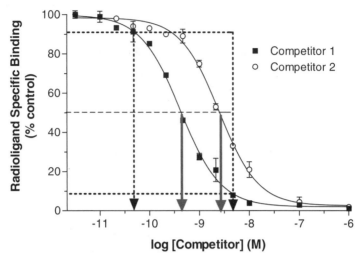

Figure 2-3. Inhibition of radioligand binding by two competitors with different affinities for the receptor in question.

one-tenth of the K_d, 1/11 of B_{max} will be occupied; at radioligand concentration equal to the K_d, half the receptors will be occupied; at radioligand concentration 10 times the K_d, 10/11 of B_{max} will be occupied.

Now consider the simplest competition experiment, in which the competitor is the unlabeled form of the radioligand. The mathematical relationship between ligand and receptor at equilibrium is essentially the same as for saturation binding. In a simple bimolecular reaction following mass action, the unlabeled ligand at a concentration of one-tenth the K_d will inhibit 9% of the specific binding of radioligand; at 10 times the K_d, the unlabeled ligand will inhibit 91% of specific binding of radioligand (dotted lines in figure 2-3 indicate these points for competitor 1). Therefore, a competition experiment should include concentrations of competitor at least an order of magnitude above and below the midpoint, and preferably two orders of magnitude above and below. If a competitor over this range does not produce a curve similar to the profile in figure 2-3, one should suspect that binding of the competitor to this site is not described by a simple bimolecular reaction.

The IC_{50} (or EC_{50}) is the concentration of inhibitor that produces half of the total displacement of radioligand that is maximally produced by that inhibitor. Similar to an ED_{50}, the IC_{50} is a measure of a compound's potency as a competitor for binding sites and is directly related to the affinity of the compound for the receptor. The efficacy of a compound in a competition experiment is a measure of the maximal displacement of the radioligand produced by the compound and is not directly related to its affinity for the receptor, nor is it necessarily related to its efficacy to produce cellular responses.

The arrows on the graph in figure 2-3 indicate the IC_{50} (or EC_{50}) values for the two competitors. The IC_{50} value determined in a competition experiment is not equivalent to the K_d for competitor binding to the receptor because the IC_{50} value depends not

Table 2-2. Parameter estimates provided by GraphPad Prism for displacement of radioligand binding by compounds with higher (competitor 1) and lower (competitor 2) affinity for the receptor labeled by radioligand (data shown in figure 2-3).

Parameter	Competitor 1	Competitor 2
Best-fit values		
Bottom	2.063	2.635
Top	99.79	98.38
Log EC$_{50}$	−9.375	−8.602
Hillslope	−1.088	−1.124
EC$_{50}$	4.22E-10	2.50E-09
K_i	1.41E-10	8.38E-10
Ligand (constant)	1.33	1.33
K_d (constant)	0.67	0.67
Std. Error		
Bottom	1.38	2.037
Top	1.544	1.312
Log EC$_{50}$	0.03011	0.0354
Hillslope	0.07304	0.09133
95% Confidence intervals		
Bottom	−1.058 to 5.184	−1.973 to 7.243
Top	96.29 to 103.3	95.42 to 101.3
Log EC$_{50}$	−9.443 to −9.307	−8.682 to −8.522
Hillslope	−1.253 to −0.9228	−1.331 to −0.9174
EC$_{50}$	3.608e-010 to 4.937e-010	2.080e-009 to 3.007e-009
K_i	1.209e-010 to 1.654e-010	6.967e-010 to 1.007e-009
Goodness of fit		
Degrees of freedom	9	9
R	0.9977	0.9962
Absolute sum of squares	45.67	65.94
S$y.x$	2.253	2.707
Data		
Number of X values	15	15
Number of Y replicates	3	3
Total number of values	13	13
Number of missing values	2	2

only on the affinity of the competitor for the receptor, but also on the affinity of the radioligand for receptor and the concentration of radioligand present in the incubation (for example, 1.33 nM for the data in figure 2-3 and table 2-2). The greater the concentration of radioligand used in the incubation reaction, the greater the amount of competitor required to displace 50% of radioligand binding, and the further the IC$_{50}$ value will be from the K_d value for the inhibitor (also known as the inhibition constant or K_i). IC$_{50}$ estimates are used to establish the rank order of potency for a series of compounds at a given receptor. It is important to remember that IC$_{50}$ values are not directly comparable between studies or even between experiments, unless the same radioligand and receptor preparation was used and the concentration of radioligand was identical in each assay. In addition, the IC$_{50}$ value, similar to an ED$_{50}$ value in a dose response, is only an estimate of a compound's potency, not its efficacy.

The affinity of an inhibitor for a receptor is expressed in absolute terms as the K_I, the equilibrium dissociation constant for the inhibitor. Under certain conditions, one can convert an empirically determined IC_{50} value to K_i using an equation worked out by Cheng and Prusoff (1973):

$$\frac{[LR]}{B_{max}} = \frac{[L]}{[L]+K_d},$$
(2-14)

As can be seen from this equation, if the radioligand concentration, $[L]$, is equal to the K_d for the radioligand, then the K_i value will be half of the IC_{50}. When radioligand is present at concentrations greater than the K_d, then the deviation of the IC_{50} from K_i is great. If the radioligand concentration is much lower than the K_d, the IC_{50} value will approximate the K_i. Table 2-2 is the output from a GraphPad Prism analysis in which K_i values were estimated for competitors 1 and 2 in an experiment where radioligand with a K_d for the receptor was present in the assay at 1.33 nM.

The conversion of IC_{50} to the K_i by to the Cheng-Prusoff equation (2-14) is only valid if certain assumptions are met in the binding reaction. The radioligand and the inhibitor must interact with a single population of binding sites, the binding reaction must be at equilibrium, and the concentration of receptors in the incubation must be much less than the K_d for the radioligand or the competitor. It is worth considering these factors when designing experiments, so that the conversion from IC_{50} to K_i can be made without violating the assumptions of the Cheng-Prusoff equation.

In table 2.2, the Hill coefficient (n) or slope factor was also calculated by GraphPad Prism as another means of testing whether competition data are fit by mass action describing a simple bimolecular reaction. Hill (1913) modified the equation for the rectangular hyperbola by expressing ligand concentration in exponential form:

$$[LR] = \frac{B_{max}[L]^n}{K_d+[L]^n}.$$
(2-15)

If $n = 1$, then the equation is equal to the rectangular hyperbola. In the above data set, competitor 1 and competitor 2 displaced radioligand binding with a Hill coefficient near unity, indicating a simple bimolecular reaction where the radioligand interacts with one population of non-interacting receptors. A Hill coefficient that deviates from unity ($n > 1.15$ or < 0.85) indicates that a more complex model of binding is necessary to account for the data (compare the two competition curves in figure 2-4 with the data shown in figure 2-3).

The displacement curve for competitor A is shallower than that for competitor B and the competition curves in figure 2-3. This is reflected in a Hill coefficient significantly < 1 (for this data set, $n = -0.5$; table 2-3). A negative Hillslope may indicate receptor heterogeneity (either multiple affinity states or receptor subtypes) or negative cooperativity. A Hill coefficient significantly > 1 may indicate positive cooperativity in the ligand–receptor interaction. Cooperativity in ligand binding to receptors is not an uncommon phenomenon and may be seen with estrogen receptors, for example (Brandt and Vickery, 1997).

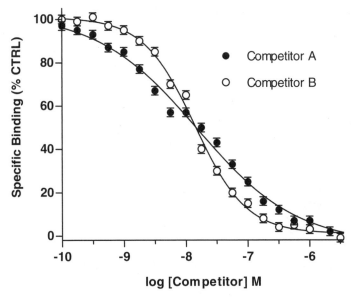

Figure 2-4. Comparison of simple (competitor B) and complex (competitor A) inhibition curves.

Further analysis of this data set is warranted because of the negative Hill coefficient. GraphPad Prism and LIGAND can analyze the competition curves simultaneously and test whether a more complex model of binding significantly improves the fit of data. For the data set shown in figure 2-4, the data for competitor B are best fit with a one-site model, whereas data for competitor B are best fit with a two-site model (table 2-4).

Remember that the addition of variables to a model will improve the goodness-of-fit of almost any data. Therefore, it is necessary to do some sort of statistical comparison between goodness-of-fits that takes into account the loss of degrees of freedom (number of data points minus the number of variables) in the more complex model. GraphPad Prism does this with an F-test that calculates a p-value indicating the probability that chance alone could account for the improvement of fit seen with a more complex model when the simpler model is really correct (see "comparison of fits" in tables 2-1 and 2-4). However, even with $p < .05$, parameter estimates derived from a more complex model may not necessarily make sense physiologically, so interpretation of output should always be based on sound biological principles.

A rigorous analysis of competition or saturation studies will also include estimates of the standard error of the computed parameters. Nonlinear regression analysis software will provide standard errors associated with estimates of potency (IC_{50} and K_d) and efficacy (difference between upper and lower plateaus of a binding curve or dose–response curve). Since the errors are a function of the number of data points, the distance of the points from the curve, and the overall shape of the curve, these values are critical in evaluating whether the parameter estimates reported are biologically meaningful. However, there are no strict guidelines for

Table 2–3. Analysis of data from competition experiment (figure 2-4).

Parameter	Competitor A	Competitor B
Best-fit values		
Bottom	−3.921	1.011
Top	103	100.9
Log EC_{50}	−7.822	−7.856
Hillslope	−0.5355	−1.043
EC_{50}	1.51E-08	1.39E-08
Std. error		
Bottom	2.268	1.189
Top	2.099	1.178
Log EC_{50}	0.05432	0.02612
Hillslope	0.03787	0.05923
95% Confidence intervals		
Bottom	−8.687 to 0.8444	−1.509 to 3.531
Top	98.57 to 107.4	98.37 to 103.4
Log EC_{50}	−7.936 to −7.708	−7.912 to −7.801
Hillslope	−0.615 to −0.456	−1.168 to −0.9171
EC_{50}	1.158e-008 to 1.959e-008	1.226e-008 to 1.582e-008
Goodness of fit		
Degrees of freedom	18	16
R	0.9959	0.9974
Absolute sum of squares	118.4	88.73
$Sy.x$	2.565	2.355
Data		
Number of X values	23	23
Number of Y replicates	3	3
Total number of values	22	20
Number of missing values	1	3

Assuming a simple, bimolecular reaction, GraphPad Prism output indicates that competitors A and B displace radioligand with similar EC_{50} values, but that hillslopes for the two competitors are different.

accepting or rejecting parameter estimates based on standard errors, and one needs to consider the biology of the system in question. GraphPad Prism provides 95% confidence intervals for key parameters (tables 2-1–2-4) based on error estimates, and the confidence intervals may be helpful in evaluating the data.

Applications

Binding Studies with ER

Numerous papers have used the titration assays described above to measure displacement of estradiol from ER by suspected EDCs (see table 2-5 for a small sample). The source of the ER varies considerably among studies, ranging from ERs derived from mouse uterine homogenates (Blizard et al., 2001) to frog hepatic cells (Kloas et al., 1999) to human ER transfected in yeast (Klotz et al., 1996). Radioligand binding assays were first used to study xenobiotic interaction with ER in 1974, when

Table 2-4. Analysis of data from competition experiment (figure 2-4): comparison of one-site versus two-site fits.

Parameter	Competitor A	Competitor B
Equation 1		
Best-fit values		
Bottom	6.438	0.4938
Top	92.74	101.4
Log EC_{50}	−7.817	−7.856
EC_{50}	1.53E-08	1.39E-08
Std. error		
Bottom	2.498	0.9848
Top	2.415	0.9739
Log EC_{50}	0.08082	0.02652
95% Confidence intervals		
Bottom	1.210 to 11.66	−1.584 to 2.572
Top	87.69 to 97.80	99.30 to 103.4
Log EC_{50}	−7.986 to −7.647	−7.912 to −7.800
EC_{50}	1.033e-008 to 2.252e-008	1.226e-008 to 1.586e-008
Goodness of fit		
Degrees of freedom	19	17
R	0.9738	0.9973
Absolute sum of squares	756.2	91.6
$Sy.x$	6.309	2.321
Equation 2		
Best-fit values		
Bottom	0.8377	0.8702
Top	99.68	100.6
Fraction1	0.4576	−0.2208
Log EC_{50_1}	−8.761	−8.386
Log EC_{50_2}	−7.103	−7.943
EC_{50_1}	1.73E-09	4.11E-09
EC_{50_2}	7.88E-08	1.14E-08
Std. error		
Bottom	0.971	1.144
Top	1.026	1.538
Fraction1	0.0311	1.994
Log EC_{50_1}	0.08606	2.114
Log EC_{50_2}	0.07181	0.3733
95% Confidence intervals		
Bottom	−1.211 to 2.886	−1.569 to 3.309
Top	97.52 to 101.8	97.31 to 103.9
Fraction1	0.3920 to 0.5232	−4.471 to 4.029
Log EC_{50_1}	−8.943 to −8.580	−12.89 to -3.881
Log EC_{50_2}	−7.255 to −6.952	−8.739 to -7.147
EC_{50_1}	1.140e-009 to 2.631e-009	1.285e-013 to 0.0001316
EC_{50_2}	5.559e-008 to 1.117e-007	1.826e-009 to 7.121e-008
Goodness of fit		
Degrees of freedom	17	15
R	0.9981	0.9974
Absolute sum of squares	54.08	86.16
$Sy.x$	1.784	2.397

(*continued*)

Table 2-4. Continued.

Parameter	Competitor A	Competitor B
Comparison of fits		
DFn, DFd	2, 17	2, 15
F	110.4	0.4729
P value	P<0.0001	0.6322
Best Fit Equation	Eq. 2	Eq. 1
Data		
Number of X values	23	23
Number of Y replicates	3	3
Total number of values	22	20
Number of missing values	1	3

Displacement of radioligand by competitor B is best described with a one-site model, whereas the goodness-of-fit for competitor A is significantly improved by addition of a second variable (site).

Nelson discovered that DDT, its analogs, and PCBs displaced radiolabeled 17β-estradiol from rat intracellular uterine ER. Since then, several alternative assays have been developed, such as fluorescently labeled ligands (Bolger et al., 1998), with the goal of determining which compounds bind to ER and the potency and efficacy of these compounds. Many papers report the binding affinity of suspected EDCs as the IC_{50} for displacing labeled estradiol from ER. Some studies compare the IC_{50} value for the competitor relative to the IC_{50} for 17β-estradiol and express this ratio as a relative binding affinity (RBA). Occasionally studies estimate RBAs based on the presumed maximal inhibition by one dose of competitor relative to displacement produced by unlabeled radioligand. In our opinion, one-point assays are not an acceptable substitute for titration studies across a range of concentrations that provide potency and efficacy estimates with standard errors.

A recent review of the literature on ER-specific xenobiotics (Bolger et al., 1998) found moderate agreement across studies in the reported IC_{50} or RBA values for xenobiotics. For example, the reported RBA for the methoxychlor metabolite, 2,2-bis(p-hydroxyphenyl)-1,1,1-trichloroethane (HPTE), ranged from 1.3 to 5.2 for the mouse ER. Using a fluorescent assay and human ER, the RBA for HPTE was 1.7 (Bolger et al., 1998). A more recent study reported an RBA for HPTE of 0.4 for binding specifically to ERα (Gaido et al., 1999). The RBAs reported for o,p'-DDT binding to ER have also been similar across a number of studies. For example, the RBA for o,p'-DDT binding to human ER transfected into a yeast system is 0.1 (estimated from data; Klotz et al., 1996); for mouse uterine receptor it is 0.3 (estimated from data; Shelby et al., 1996). However, the RBA for o,p'-DDT binding to a rat uterine leiomyoma cell line (0.00031) was orders of magnitude different (Hodges et al., 2000).

Differences in IC_{50} or RBA estimates may be explained by experimental variation across laboratories or by the use of different species or tissues as the source of ER (Matthews et al., 2000). The ligand binding domain of the human, rat, or mouse varies in amino acid sequence, and this may lead to observed changes in affinity for any given ligand. Such factors should be considered when making cross-study

Table 2-5. The agonistic and antagonistic properties of several EDCs known to bind directly to hormone receptors.

EDC	ER	AR	PR	20 Beta-S	GR	Sample Citations:
3-MeSO$_2$-CB149					0/–	Johansson et al., 1998
Alachlor	+/?		?/?			Vonier et al., 1996; Klotz et al., 1996
Arochlor	?/?					Nelson, 1974
Atrazine	?/?		?/?			Vonier et al., 1996
BBP	+/0	0/–				Bolger et al., 1998;
Bisphenol A	+/0	0/–				Bolger et al., 1998; Lutz and Kloas, 1999
Chlordane			?/?			Lundholm, 1988
Cyanazine	0/-					Tran et al., 1996
DDOH			0/–			Klotz et al, 1997
Dicofol			?/?			Vonier et al., 1996
Dieldrin	+/?	?/?				Soto et al., 1995; Danzo et al., 1997; Matthews et al., 2000
Endosulfan	+/?					Matthews et al., 2000
Endosulfan sulfate	?/?		?/?			Vonier et al., 1996
HPTE	+/–					Gaido et al., 1999; Shelby et al., 1996
Kepone	+/?		?/?	?/–		Bolger et al., 1998; Vonier et al., 1996; Dan and Thomas, 1999
M1		0/–				Kelce et al., 1994
M2		+/–				Kelce et al., 1994; Wong et al., 1995
methoxychlor	?/?					Nelson, 1974
Nonachlor	?/?					Klotz et al., 1996
Nonylphenol	+/0	+/0				White et al, 1994; Lutz and Kloas, 1999
o,p'-DDD	?/?		0/–	?/–		Klotz et al., 1996; Klotz et al., 1997; Dan and Thomas, 1999
o,p'-DDE	?/?		0/–			Nelson, 1974; Klotz et al., 1997
o,p'-DDT	+/-	0/–	0/–			Nelson, 1974; Edmunds et al., 1997;Klotz et al, 1997
Octylphenol	+/?					White et al, 1994; Lutz and Kloas, 1999
p,p'-DDD	?/?	?/–	0/–			Vonier et al., 1996; Klotz et al., 1997
p,p'-DDE	+/–	+/–				Kelce et al., 1995; Vinggaard et al., 1999; O'Connor et al., 1999
p,p'-DDT	?/?	?/–				Nelson, 1974
Procymidone		?/–				Hosokawa et al., 1993
Tetrachlorobiphenyl	+/?					Nesaretnam et al., 1996
Vinclozolin*	0/0	+/–				Wong et al., 1995

ER = estrogen receptor α or β; AR = androgen receptor; PR = progesterone receptor; 20 Beta-S = 17,20 Beta,21-trihydroxy-4-pregnen-3-one receptor in oocyte membranes; GR = Glucocorticoid receptor. The first mark represents agonistic properties, the second antagonistic ones. 0 = tested, no effect; + = agonistic activity; – = antagonistic activity; ? As of 2002, to our knowledge, binds to receptor but action untested.

*Action possibly through metabolism to M1 and M2 by cells used in transfection system.

comparisons (Andersen et al., 1999). Differences in IC$_{50}$ values between studies may also be due to differences in the amount of labeled ligand used in the competitive binding assays. As discussed earlier and illustrated by equation 2-14, the IC$_{50}$ is dependent on the concentration of and affinity of radioligand used, whereas the K_2 value is a property of the ligand–receptor interaction in that system corrected for amount of label in the assay. Unfortunately, many studies in the EDC literature are of limited utility because, for a number of reasons, they do not distinguish between potency and efficacy, they do not report parameter error estimates, and they compare IC_{50} values rather K_i values across experiments.

Nevertheless, the IC_{50} values reported for a number of potential EDCs are well below the value necessary to induce overt toxic symptoms in an organism and within the concentration of the contaminant found within the environment. For example, nonylphenol has been found in sewage and river systems at levels of 1–1000 ppb (http://www.eal.or.jp/DNN/199805/480.html; http://website.lineone.net/~mwarhurst/ summary.html; http://www.ecogent.ca/enviro/env_npe.htm) and in beluga whale fat at up to 1000 ppb (http://www.ecogent.ca/enviro/env_npe.htm). The IC_{50} for the displacement of labeled estradiol by nonylphenol in the MCF-7 breast cancer cell line is 7.2 μM (1500 ppb), and the maximal proliferative effect of nonylphenol in a bioassay of estrogenic compounds (MCF-7 cell proliferation; the e-screen assay) occurs at 1 μM (220 ppb; Soto et al., 1995). These data strongly suggest that environmental contaminants are present in concentrations high enough to have endocrine disrupting effects via ER.

Again, most studies have examined EDC interaction with ER, but estrogenic EDCs may also interact with membrane-bound ERs. A recent competitive binding study using horseradish peroxidase-bound estradiol mouse pancreatic islet cell membranes found that both bisphenol A and o,p'-DDT displaced estradiol from a plasma membrane receptor (Nadal et al., 2000).

Transfection Assays and ER

Recent technical advances in molecular biology allow one to screen ligands for endocrine action through transfection assays using cultured cells and reporter gene expression. For nuclear hormone receptors, these assays can be built in several different ways. First, the cell culture system used may be one that contains an endogenous specific steroid receptor. For example, an estrogen response element (ERE) from the *Xenopus* vitellogenin gene promoter region, linked to a luciferase reporter gene, has been transfected into rat pituitary tumor cells that express ER. Luciferase activity can then be monitored as an ER-mediated response after exposure to a number of environmental contaminants with suspected estrogenic activity (Edmunds et al., 1997). A number of other reporter protein systems are available for screening compounds that activate or inactivate ER. Other transfection systems for studying the actions of EDCs on hormone receptor activation use cell lines lacking endogenous receptors into which the receptor is transfected in addition to the ERE and reporter gene. For example, the ER gene, the ERE, and a reporter gene, such as β-galactosidase (which produces a color reaction upon addition of appropriate reagents), have been transfected into yeast cell lines used to screen EDCs (Klotz et al., 1996). HeLa cells have also been transfected with mouse ER and an ERE driving a chloramphenicol acetyl transferase (CAT) reporter gene to screen environmental contaminants for estrogenic activity (Shelby et al., 1996).

Results from transfection/reporter gene assays are typically reported as reporter gene activity at increasing concentrations of endogenous and exogenous ligand. Dose–response curves can be compared and an index of relative efficacy or potency developed. Most studies report the results of competitive binding assays within the transfection system to support the conclusion that the cellular responses seen are mediated by ligand activation of ER.

Transfection systems are appealing for several reasons. They do not involve radioactive waste, and they can be developed for high throughput systems for rapid analysis of many compounds. Further, they provide information about whether the compounds of interest have agonistic or antagonistic action at the specific receptor. Some disadvantages of transfection systems that may lead to spurious results are that the endogenous intracellular regulators of gene expression are likely to be somewhat different in a cell line than in vivo, and it is difficult to reproduce the naturally occurring stoichiometry of hormone, receptor, and response element in such systems.

Examples of Compounds That Bind to the Androgen Receptor

A number of environmental contaminants bind to the androgen receptor (AR) rather than, or in addition to, ER. Best studied are the fungicide vinclozolin and its metabolites and a metabolite of DDT (p,p'-DDE; see also chapter 12); all bind to AR and act as receptor antagonists. A number of competitive binding studies have used synthetic androgen [^{3}H]R1881 as radioligand. In such studies, using AR from rat ventral prostate cytosol, the IC_{50} for DDT metabolites ranged from 5 to 95 μM, with p,p'-DDE being the most potent (Kelce et al., 1995). Using a preparation from rat epididymis, vinclozolin and its metabolites M1 and M2 bound to cytosolic and nuclear AR; the IC_{50} values for M1 and M2 were 50 and 3 μmM, respectively (Kelce et al., 1994). Using human AR transiently transfected into monkey kidney COS cells, the IC_{50} values ranged from 10–20 μM for M1 to 500 nM for M2 (Wong et al., 1995).

As for ER, transfection systems have provided some insight into the character of EDCs that bind to the AR. Unlike most of the environmental contaminants that bind to ER, most EDCs studied with respect to the AR act as androgen antagonists. Using a yeast transfection system expressing human AR, p,p'-DDE, bisphenol A, and butyl benzylphthalate were all antiandrogenic (table 2-5), in that they inhibited dihydrotestosterone (DHT)-induced β-galactosidase expression (Sohoni and Sumpter, 1998). In this system, vinclozolin also had strong antagonistic activity. Similar results were reported for kidney CV1 cells expressing the human AR in which p,p'-DDE is the most potent antagonist, with o,p'-DDT producing substantial inhibition at higher doses (Kelce et al., 1995). The potency of the vinclozolin metabolites and vinclozolin to block DHT-induced expression of a luciferase reporter gene in monkey kidney CV1 cells transfected with a human AR closely paralleled the affinity estimates for receptor binding in monkey kidney COS cells transfected with the same human AR (Wong et al., 1995).

Several androgenic compounds also have been identified. Nonylphenol (see chapter 14) at high concentrations (about 80 μM) showed some androgenic activity in the yeast transfection system (Sohoni et al., 1998). This compound, well known for its estrogenic effects, does not exhibit any antiandrogen activity. Other compounds with antagonistic properties at AR are partial agonists at high concentrations.

For example, *p,p'*-DDE (Sohoni and Sumpter, 1998) and vinclozolin (Wong et al., 1995; Sohoni et al., 1998) are partial agonists. In these cases, the EDCs act as AR antagonists at lower doses but have androgenic properties at higher doses.

Examples of Compounds That Bind Progesterone Receptors

A few studies have investigated the interactions of EDCs on the cytosolic progesterone receptor (PR). The *o',p'* and *p', p'* isomers of DDE, as well as chlordane, and the PCB Arochlor 1242 (See also chapter 11) bind with moderate affinity to duck, chicken, and rabbit cytosolic progesterone receptors (Lundholm, 1988). Lundholm hypothesizes that these compounds are antagonistic at PR because egg-shell building is a progesterone-dependent event and these compounds induce egg-shell thinning. Using a crude cytosolic preparation from alligator oviductal tissue, 30 μM kepone, dicofol, endosulfan sulfate, cyanizine, and atrazine displaced the binding of radiolabeled progesterone receptor agonist [^{3}H]R5020, while DDT and its metabolites had no effect (Vonier et al., 1996). However, one is again reminded that single-point displacement assays are not a replacement for receptor characterization and competition studies using a range of unlabeled compounds. Using transfected human PR in yeast reporter assay, several PCBs bound to transfected PR and acted as receptor antagonists (Jin et al., 1997). Two forms of endosulfan and lindane also had antagonistic activity in this receptor but, surprisingly, they did not displace labeled progesterone from the transfected PR. This may reflect non-PR–mediated transcriptional activity of endosulfan and lindane, but it certainly emphasizes the importance of using a combination of binding and functional assays for EDCs.

Fish and amphibian oocytes contain a membrane steroid receptor that induces final meiotic maturation of the oocytes after binding of specific C_{21} steroids. Steroid action via these membrane receptors is independent of nuclear receptor-mediated changes in transcription. Das and Thomas (1999) investigated the activity of several pesticides at the 17,20β-21-trihydroxy-4-pregnen-3-one receptor (20β-S) from oocytes of spotted seatrout. They found that kepone and *o,p*-DDD both displaced radiolabeled 20β-S from membrane receptors and inhibited 20β-S-induced oocyte maturation. Because several steroid receptors in addition to a 20β-S receptor have been characterized in cell membranes, it is prudent to examine both intracellular and membrane-bound steroid receptors as targets for EDC action.

Lessons from the Ah Receptor

Planar aromatic hydrocarbons bind to the AhR, which then initiates cellular responses after translocation of the AhR–ligand complex to the nucleus (reviewed in Swanson and Bradfield, 1993; see also chapter 10). The prototypical ligand for AhR has been 2,3,7,8-tetrachlorodibenzo-*p*-dioxin (TCDD); studies using TCDD have revealed considerable complexity in the AhR. Like many steroid receptors, unliganded AhR is cytosolic, and it is translocated to the nucleus upon ligand binding (with the assistance of a protein called the Ah-R nuclear translocator), where it binds to a response element on DNA to alter gene transcription. The AhR exhibits two distinct binding affinities for a radiolabeled form of dioxin related to two different dissociation rates (Bradfield et al., 1988). It appears that cytosolic AhR bound to heat-shock proteins may have a lower affinity for ligand than the receptor free of heat-shock protein (Swanson et al., 1993).

Another assumption of binding studies is that all the ligand used in determining binding affinities is in solution. However, TCDD has a solubility factor of about 4×10^{-11} M, and many binding studies for TCDD to the AhR include values as high as 1×10^{-9} M. Furthermore, because many hydrocarbon compounds are hydrophobic, they may bind nonspecifically to other intracellular proteins, further decreasing the free form of the compound available to bind to the receptor. These problems make it difficult to determine overall binding kinetics of these compounds, and they interfere with the usefulness of binding studies in determining relative risk factors for exposure to these compounds.

TCDD has some antiestrogenic effects that are mediated via the AhR. However, unlike the ER-binding compounds discussed above, the actions of the TCDD–AhR complex are not through direct disruption of estrogen binding with its receptor. There appears to be cross-talk between the molecular pathways involving the bound AhR and the estrogen–ER complex (Safe et al., 1998). Some, though not all, estrogen-sensitive genes contain both an ERE and an inhibitory dioxin response element. When the activated AhR binds to its inhibitory dioxin response element, it antagonizes the action of the active ER.

Conclusion

The studies presented demonstrate that a strong understanding of the theory and methods behind ligand–receptor interactions and dose responses is necessary when investigating the effects of EDCs on endocrine physiology. In EDC studies, there is no substitute for rigorous investigation of one or more receptor systems, and no substitute for the use of appropriate receptor methodology and terminology by all investigators in the field. Whenever possible, the potencies and efficacies of compounds with error estimates should be presented, based on nonlinear regression analysis of wide range of doses of contaminant. Whenever possible, potencies should be reported as K_i values to allow for comparisons across studies and to facilitate more accurate comparisons with environmental levels of contaminants. The agonistic or antagonistic properties of the compounds should be determined using a range of bioassays. We also emphasize that radioligand binding studies alone are not sufficient for detection or characterization of EDCs. For example, arsenic or its metabolites may alter the binding of glucocorticoids to the glucocorticoid receptor through noncompetitive mechanisms under some assay conditions (Stancato et al., 1993), but they may also alter hormone signaling downstream of hormone binding. Arsenic or arsenite can alter the transcriptional activity of glucocorticoid receptors by altering molecular interactions of the glucocorticoid receptor in the cell nucleus (Huang et al., 2001; Kaltreider et al., 2001). This type of endocrine disruption, even though it is mediated by altered hormone receptor function, would not necessarily be detected by radioligand binding studies. Therefore, rigorous receptor binding studies are a critical component of EDC studies, but they need to be accompanied by an appreciation of receptor biology—the diversity of receptors and the inherent complexity in hormone signaling.

References

Andersen, H.R., Andersson, A.M., Arnold, S.F., Autrup, H., Barfoed, M., Beresford, N.A., Bjerregaard, P., Christiansen, L.B., Gissel, B., Hummel, R., Jorgensen, E.B., Korsgaard, B., Le Guevel, R., Leffers, H., McLachlan, J., Moller, A., Nielsen, J.B., Olea, N., Oles-Karasko, A., Pakdel, F., Pedersen, K.L., Perez, P., Skakkeboek, N.E., Sonnenschein, C., Soto, A.M., Sumpter, J. P., Thorpe, S. M., Grandjean, P., 1999. Comparison of short-term estrogenicity tests for identification of hormone-disrupting chemicals. Environ. Health Perspect. 107, Suppl 1, 89–108.

Blizard, D. Sueyoshi, T., Negishi, M., Dehal, S.S., Kupfer, D., 2001. Mechanism of induction of cytochrome p450 enzymes by the proestrogenic endocrine disruptor pesticide-methoxychlor: interactions of methoxychlor metabolites with the constitutive androstane receptor system. Drug Metab. Dispos. 29, 781–785.

Bloomquist, J.R., 1996. Ion channels as targets for insecticides. Annu. Rev. Entomol. 41, 163–190.

Bolger, R., Wiese, T.E., Ervin, K., Nestich, S., Checovich, W., 1998. Rapid screening of environmental chemicals for estrogen receptor binding capacity. Environ. Health Perspect. 106, 551–557.

Bradfield, C.A., Kende, A.S., Poland, A., 1988. Kinetic and equilibrium studies of Ah receptor-ligand binding: Use of [^{125}I]2-iodo-7,8-dibromodibenzo-p-dioxin. Mol. Pharmacol. 34, 229–237.

Bradford, M.M., 1976. A rapid and sensitive method for the quantification of microgram quantities of protein utilizing the principle of protein-dye binding. Anal. Biochem. 72, 248–254.

Brandt, M.E., Vickery, L.E., 1997. Cooperativity and dimerization of recombinant human estrogen receptor hormone-binding domain. J. Biol. Chem. 272, 4843–4849.

Cheng, Y., Prusoff, W.H., 1973. Relationship between the inhibition constant (K1) and the concentration of inhibitor which causes 50 per cent inhibition (I_{50}) of an enzymatic reaction. Biochem. Pharmacol. 22, 3099–3108.

Chun, T.Y., Gregg, D., Sarkar, D.K., Gorski, J., 1998. Differential regulation by estrogens of growth and prolactin synthesis in pituitary cells suggests that only a small pool of estrogen receptors is required for growth. Proc. Natl. Acad. Sci. USA 95, 2325–2330.

Crain, D.A., Noriega, N., Vonier, P.M., Arnold, S.F., McLachlan, J.A., Guillette, L.J., 1998. Cellular bioavailability of natural hormones and environmental contaminants as a function of serum and cytosolic binding factors. Toxicol. Ind. Health 14, 261–273.

Cuatrecasas, P., 1971. Insulin receptor interactions in adipose tissue cells: direct measurement and properties. Proc. Natl. Acad. Sci. USA 68, 1264–1268.

Danzo, B.J., (1997). Environmental xenobiotics may disrupt normal endocrine function by interfering with the binding of physiological ligands to steroid receptors and binding proteins. Environ. Health Perspect. 105, 294–301.

Das, S.,Thomas, P., 1999. Pesticides interfere with the nongenomic action of a progestogen on meiotic maturation by binding to its plasma membrane receptor on fish oocytes. Endocrinology 140, 1953–1956.

Dechaud, H., Ravard, C., Claustrat, F., de la Perriere, A.B., Pugeat, M., 1999. Xenoestrogen interaction with human sex hormone-binding globulin (hSHBG). Steroids 64, 328–334.

Dick, G.M., Rossow, C.F., Smirnov, S., Horowitz, B., Sanders, K.M., 2001. Tamoxifen activates smooth muscle BK channels through the regulatory beta 1 subunit. J. Biol. Chem. 276, 34594–34599.

Edmunds, J.S., Fairey, E.R., Ramsdell, J.S., 1997. A rapid and sensitive high throughput reporter gene assay for estrogenic effects of environmental contaminants. Neurotoxicology 18, 525–532.

Ehrlich, P., 1914. Chemotherapy. In: Himmelweit, F. (Ed.). The Collected Papers of Paul Ehrlich, vol. 3 (1960). Pergamon Press, London, pp. 505–518.

Gaido, K.W., Leonard, L.S., Maness, S.C., Hall, J.M., McDonnell, D.P., Saville, B., Safe, S., 1999. Differential interaction of the methoxychlor metabolite 2,2-bis-(p- hydroxyphenyl)-1,1,1–trichloroethane with estrogen receptors alpha and beta. Endocrinology 140, 5746–5753.

Goldstein, A., Lowney, L.I., Pal, B.K., 1971. Stereospecific and nonspecific interactions of the morphine congener levorphanol in subcellular fractions of mouse brain. Proc. Natl. Acad. Sci. USA 68, 1742–1747.

Hill, A.V., (1913). The Combinations of hemoglobin with oxygen and carbon monoxide, Biochem. J. 7: 471–480.

Hodges, L.C., Bergerson, J.S., Hunter, D.S., Walker, C.L., 2000. Estrogenic effects of organochlorine pesticides on uterine leiomyoma cells in vitro. Toxicol. Sci. 54, 355–364.

Hosokawa, S. et al. (1993). The affinity of procymidone to androgen receptor in rats and mice. J. Toxicol. Sci. 18, 83–93

Huang, C., Li, J., Ding, M., Costa, M., Castranova, V., Vallyathan, V., Ju, G., Shi, X., 2001. Transactivation of RARE and GRE in the cellular response to arsenic. Mol. Cell. Biochem. 222, 119–125.

Jin, L.,Tran, D.Q., Ide, C.F., McLachlan, J.A., Arnold, S.F., 1997. Several synthetic chemicals inhibit progesterone receptor-mediated transactivation in yeast. Biochem. Biophys. Res. Commun. 233, 139–146.

Johansson, M., Nilsson, S., and Lund, B.O., (1998). Interactions between methylsulfonyl PCBs and the glucocorticoid receptor. Environ. Health Perspect. 106, 769–772.

Kaltreider, R.C., Davis, A.M., Lariviere, J.P., Hamilton, J.W., 2001. Arsenic alters the function of the glucocorticoid receptor as a transcription factor. Environ. Health Perspect. 109, 245–251.

Keen, M. (Ed.), 1999. Receptor Binding Techniques, Methods in Molecular Biology. Humana Press, Totowa, NJ.

Kelce, W.R., Monosson, E., Gamcsik, M.P., Laws, S.C., Gray, L.E., 1994. Environmental hormone disruptors: Evidence that vinclozolin developmental toxicity is mediated by antiandrogenic metabolites. Toxicol. Appl. Pharmacol. 126, 276–285.

Kelce, W.R., Stone, C.R., Laws, S.C., Gray, L.E., Kemppainen, J.A., Wilson, E.M., 1995. Persistent DDT metabolite p,p'-DDE is a potent androgen receptor antagonist. Nature 375, 581–585.

Kendall, D.A., Hill, S.J. (Eds.), 1995. Signal Transduction Protocols, Methods in Molecular Biology. Humana Press, Totowa, NJ.

Kloas, W., Lutz, I., Einspanier, R., 1999. Amphibians as a model to study endocrine disruptors: II. Estrogenic activity of environmental chemicals in vitro and in vivo. Sci. Total Environ. 225, 59–68.

Klotz, D.M. et al. (1997). o,p'-DDT and its metabolites inhibit progesterone-dependent responses in yeast and human cells. Mol Cell Endocrinol 129, 63–71.

Klotz, D.M., Beckman, B.S., Hill, S.M., McLachlan, J.A., Walters, M.R., Arnold, S.F., 1996.

Identification of environmental chemicals with estrogenic activity using a combination of in vitro assays. Environ. Health Perspect. 104, 1084–1089.

Langley, J.N., (1905). On the reaction of cells and of nerve endings to certain poisons, chiefly as regards the reaction of striated muscle to nicotine and to curari. Journal of Physiology XXXIII, 374–413.

Limbird, L.E., 1996. Cell Surface Receptors: A Short Course on Theory and Methods. Kluwer Academic Publishers, Boston.

Limbird, L.E., Taylor, P., 1998. Endocrine disruptors signal the need for receptor models and mechanisms to inform policy. Cell 93, 157–163.

Lundholm, C.E., 1988. The effects of DDE, PCB and chlordane on the binding of progesterone to its cytoplasmic receptor in the eggshell gland mucosa of birds and the endometrium of mammalian uterus. Comp. Biochem. Physiol. C. 89, 361–368.

Lutz, M., Kenakin, T.P., 1999. Quantitative Molecular Pharmacology and Informatics in Drug Discovery. John Wiley and Sons, New York.

Martin, M.E., M. Haourigui, C. Pelissero, C. Benassayag, Nunez, E.A., 1995. Interactions between phytoestrogens and human sex steroid binding protein. Life Sci. 58, 429–436.

Matthews, J., Celius, T., Halgren, R., Zacharewski, T., 2000. Differential estrogen receptor binding of estrogenic substances: A species comparison. J. Steroid Biochem. Mol. Biol. 74, 223–234.

Nadal, A., Ropero, A.B., Laribi, O., Maillet, M., Fuentes, E., Soria, B., 2000. Nongenomic actions of estrogens and xenoestrogens by binding at a plasma membrane receptor unrelated to estrogen receptor alpha and estrogen receptor beta. Proc. Natl. Acad. Sci. USA 97, 11603–11608.

Nagel, S.C., vom Saal, F.S., Welshons, W.V., 1999. Developmental effects of estrogenic chemicals are predicted by an in vitro assay incorporating modification of cell uptake by serum. J. Steroid Biochem. Mol. Biol. 69, 343–357.

Nelson, J.A., 1974. Effects of dichlorodiphenyltrichloroethane (DDT) analogs and polychlorinated biphenyl (PCB) mixtures on 17beta-(3H)estradiol binding to rat uterine receptor. Biochem. Pharmacol. 23, 447–451.

Nesaretnam, K. et al. (1996). 3,4,3',4'-Tetrachlorobiphenyl acts as an estrogen in vitro and in vivo. Mol Endocrinol 10, 923–936.

O'Connor, J.C. et al. (1999). Detection of the environmental antiandrogen p,p-DDE in CD and long- evans rats using a tier I screening battery and a Hershberger assay. Toxicol. Sci. 51, 44–53.

Park, D., Hempleman, S.C., Propper, C.R., 2001. Endosulfan exposure disrupts pheromonal systems in the red-spotted newt: A mechanism for subtle effects of environmental chemicals. Environ. Health Perspect. 109, 669–673.

Pert, C.B., Snyder, S.H., 1973. Opiate receptor: Demonstration in nervous tissue. Science 179, 1011–1014.

Picard, D. (Ed.), 1999. Nuclear Receptors: A Practical Approach. Practical Approach Series 207. Oxford University Press, New York.

Ramamoorthy, K., Wang, F., Chen, I.C., Norris, J.D., McDonnell, D.P., Leonard, L.S., Gaido, K.W., Bocchinfuso, W.P., Korach, K.S., Safe, S., 1997. Estrogenic activity of a dieldrin/toxaphene mixture in the mouse uterus, MCF-7 human breast cancer cells, and yeast-based estrogen receptor assays: no apparent synergism. Endocrinology 138, 1520–1527.

Safe, S., Wang, F., Porter, W., Duan, R., McDougal, A., 1998. Ah receptor agonists as endocrine disruptors: Antiestrogenic activity and mechanisms. Toxicol. Lett. 102–103, 343–347.

Shelby, M.D., Newbold, R.R., Tully, D.B., Chae, K., Davis, V.L., 1996. Assessing environmental chemicals for estrogenicity using a combination of in vitro and in vivo assays. Environ. Health Perspect. 104, 1296–1300.

Sohoni, P., Sumpter, J.P., 1998. Several environmental oestrogens are also anti-androgens. J. Endocrinol. 158, 327–339.

Soto, A.M., Sonnenschein, C., Chung, K.L., Fernandez, M.F., Olea, N., Serrano, F.O., 1995. The E-screen assay as a tool to identify estrogens: An update on estrogenic environmental pollutants. Environ. Health Perspect. 103 (suppl. 7), 113–122.

Stancato, L.F., Hutchison, K.A., Chakraborti, P.K., Simons, S.S., Pratt, W.B., 1993. Differential effects of the reversible thiol-reactive agents arsenite and methyl methanethiosulfonate on steroid binding by the glucocorticoid receptor. Biochemistry 32, 3729–3736.

Swanson, H.I., Bradfield, C.A., 1993. The AH-receptor: genetics, structure and function. Pharmacogenetics 3, 213–230.

Tran, D.Q. et al. (1996). The inhibition of estrogen receptor-mediated responses by chloro-S-triazine-derived compounds is dependent on estradiol concentration in yeast. Biochem. Biophys. Res. Commun. 227, 140–146.

Vinggaard, A.M., Joergensen, E.C., and Larsen, J.C. (1999). Rapid and sensitive reporter gene assays for detection of antiandrogenic and estrogenic effects of environmental chemicals. Toxicol. Appl. Pharmacol. 155, 150–160.

Vonier, P.M., Crain, D.A., McLachlan, J.A., Guillette, L.J., Arnold, S.F., 1996. Interaction of environmental chemicals with the estrogen and progesterone receptors from the oviduct of the American alligator. Environ. Health Perspect. 104, 1318–1322.

Watson, C.S., Campbell, C.H., Gametchu, B., 1999. Membrane oestrogen receptors on rat pituitary tumour cells: Immuno- identification and responses to oestradiol and xeno-estrogens. Exp. Physiol. 84, 1013–1022.

White, R. et al. (1994). Environmentally persistent alkylphenolic compounds are estrogenic. Endocrinology 135, 175–182.

Wong, C., Kelce, W.R., Sar, M., Wilson, E.M., 1995. Androgen receptor antagonist versus agonist activities of the fungicide vinclozolin relative to hydroxyflutamide. J. Biol. Chem. 270, 19998–20003.

Yamamura, H.I., Enna, S.J., Kuhar, M.J., 1990. Methods in Neurotransmitter Receptor Analysis. Raven Press, New York.

3

The Hypothalamus–Pituitary Axis

James A. Carr
David O. Norris

For endocrine organs to respond appropriately to changes in an animal's environment, their activity must be coordinated with that of other effector systems controlled by the central nervous system. Because most endocrine organs have no way of detecting environmental changes on their own, the link between the nervous and endocrine systems is critical for the maintenance of homeostasis and coordination of complex physiological processes such as growth and reproduction. The brain communicates with the endocrine system via an intimate functional and anatomical linkage between the hypothalamus and pituitary gland (or hypophysis) known as the hypothalamus–pituitary (HP) axis. The HP axis consists of the pituitary gland and neurosecretory neurons in the hypothalamus of the brain that project to the pituitary. This account provides a synthesis of well known endocrine information based on the references and suggested readings at the end of this chapter. The first part of this chapter focuses on the mammalian HP axis. In the second part of the chapter comparative aspects of the HP axis are discussed.

Anatomy of the Mammalian Hypothalamus–Pituitary Axis

The Pituitary Gland

The pituitary gland is a small gland (usually weighing < 1 g in humans) located immediately below the hypothalamus, to which it is connected by the infundibular stalk (figure 3-1). The pituitary rests in a depression of the sphenoid bone called the sella turcica or Turkish saddle and is composed of five anatomically and functionally distinct regions. The adenohypophysis, consisting of the pars distalis (PD),

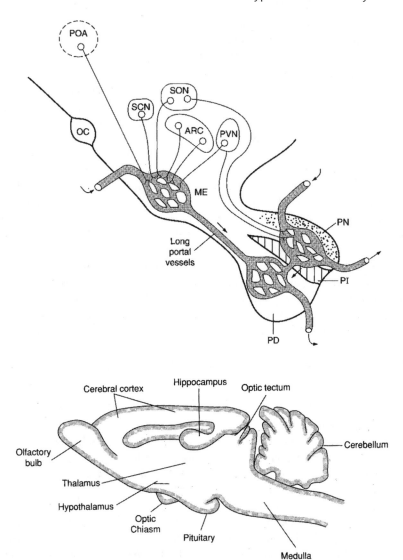

Figure 3-1. (Top) The hypothalamus–pituitary axis. Long portal veins connect capillaries in the median eminence (ME) to capillaries in the pars distalis (PD). See text for additional abbreviations. (Bottom) The mammalian brain. Both reprinted with permission from Norris (1997).

pars tuberalis (PT), and pars intermedia (PI), is derived from non-neural ectoderm. The neurohypophysis develops as an outgrowth of the brain and is composed of the median eminence (ME) and pars nervosa (PN) (figure 3-1).

The PD houses five phenotypically different cell types: corticotropes, which secrete the peptides corticotropin (ACTH), β-endorphin, and β-lipotropin (β-LPH); gonadotropes, which secrete luteinizing hormone (LH) and follicle-stimulating hormone (FSH); somatotropes, which secrete growth hormone (GH); lactotropes, which secrete prolactin (PRL); and thyrotropes, which secrete thyrotropin (TSH). In addition, nonendocrine cells called folliculostellate cells are found in the PD, where they seem to play a supportive role.

The PT is a narrow strip of glandular epithelium that develops as an outgrowth of the PD and extends along the ventral surface of the ME; its precise role is undefined, but there is evidence for its participation in the secretion of PRL from the PD. The PI contains a single type of endocrine cell (melanotropes) that secretes α-melanocyte-stimulating hormone (α-MSH). The gland is vestigial in adults of some species (including humans), whereas in other species, such as minks, it appears to play an important role in seasonal control of fur coloration. The PI plays a critical role in background color adaptation in amphibians and certain reptile and fish species. A second cell type is present in the pars intermedia of teleost fishes that secretes a hormone, somatolactin; this is involved with calcium regulation.

In contrast to the epithelial nature of the adenohypophysis, the neurohypophysis consists of neurosecretory nerve terminals and modified astroglial cells called pituicytes. The principal hormones of the PN are arginine vasopressin (AVP) and oxytocin (OXY). These peptides are produced in the hypothalamus and are axonally transported to the PN, where they are stored until released into local capillaries. Axon terminals in the ME release a number of peptides and amines into a specialized blood supply that travels to the PD. These hormones control the activity of cells in the PD and are discussed in detail later.

The Hypothalamus

The mammalian hypothalamus occupies a large portion of the brain just ventral to the thalamus, extending caudally to the mamillary body and laterally to the amygdala and the top of the third ventricle dorsally. Rostrally, the hypothalamus is continuous with the preoptic area (POA). Since the POA has no clear border with the hypothalamus and contains neurons that project to the pituitary gland, it is often considered part of the hypothalamus (figure 3-1).

The hypothalamus houses neurons that control adenohypophysial hormone secretion via the production and release of a number of peptide and nonpeptide hormones (hypothalamic releasing hormones, HRHs). With the exception of teleost fishes, delivery of hypothalamic hormones in vertebrates is vascular rather than synaptic; in other words, HRHs exit from nerve terminals in the vicinity of capillaries in the ME and are then transported via portal veins to sinusoidal capillaries in the PD (figure 3-1).

Hypothalamus–Pituitary Portal System

An important antecedent to the concept of HRHs was the demonstration that a specialized blood supply, called the hypothalamus-pituitary portal system, links the ME and PD. This specialized vascular delivery system transports HRHs released in the ME to tropic cells of the PD. The superior hypophysial artery perfuses an extensive capillary network in the ME that drains into a series of portal veins that feed low-pressure sinusoidal capillaries of the PD (figure 3-2). These capillaries allow for PD cells to be bathed in blood rich in hypothalamic hormones. Because the PN is perfused by an entirely separate blood supply, HRHs never reach this lobe. Although there is evidence for vessels linking the blood supply of the PN to the PD, evidence for a physiological interaction between these two pituitary lobes is equivocal. The PI, when present, is relatively avascular, and its secretion is controlled via direct innervation from the hypothalamus.

Development of the Hypothalamus–Pituitary Axis

The adenohypophysis forms from an ectodermal primordium called Rathke's pouch that develops at the base of the diencephalon (figure 3-3). Rathke's pouch is derived from the anterior neural ridge, a region of the neural plate that eventually gives rise to other non-neural structures such as the nasal cavity ectoderm and the olfactory placode. Adjacent regions of the neural plate become the most anterior neural structures including the ventral forebrain, hypothalamus, and neurohypophysis.

The five cell types of the PD develop in a well-defined temporal pattern that is coordinated by a cascade of transcription and differentiation factors. The first sign of adenohypophysial differentiation is the expression of the common glycoprotein

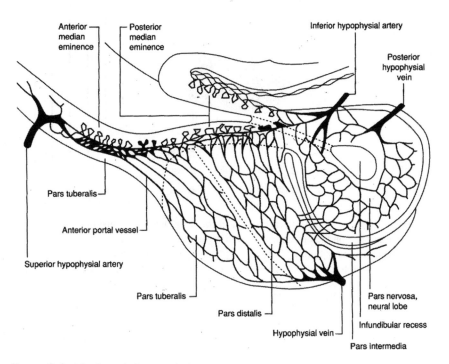

Figure 3-2. The hypothalamus–pituitary portal system of a cat. The long portal vessels carry blood from the capillaries in the median eminence to capillaries in the pars distalis. Modified from Matsumoto and Ishii (1992).

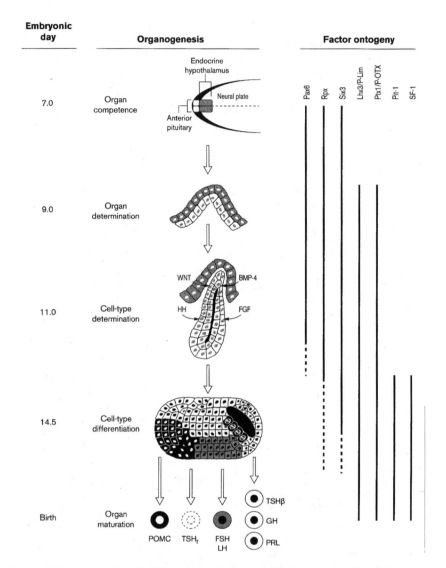

Figure 3-3. Pattern of cell differentiation in the developing pars distalis of the mouse. Different transcription factors are involved in organ determination, cell-type determination, and cell-type differentiation. By birth, five phenotypically-distinct cell types are present; corticotropes, lactotropes, somatotropes, gonadotropes, and thyrotropes. Rostral tip thyrotropes (TSHr) dissappear before birth. See text for abbreviations. Reprinted with permission from Treier and Rosenfield (1996).

α-subunit gene (GSU) in prospective gonadotropes and thyrotropes. This is followed by expression of the proopiomelanocortin (POMC) gene in corticotropes and by expression of the β-subunit of TSH in thyrotropes. Expression of GH and PRL genes in somatotropes and lactotropes, respectively, occurs shortly thereafter. The last cell types to appear are the gonadotropes, as determined by expression of genes encoding the unique β-subunits of LH and FSH (see below).

An important feature of adenohypophysial differentiation is that disruption of a single transcription factor early in the process can affect the production of multiple transcription factors downstream in the cascade pathway. For example, targeted disruption of thyroid-specific enhancer-binding protein (T/EBP), a transcription factor expressed in the ventral diencephalon as well as in the developing thyroid, can completely prevent the adenohypophysis from forming. T/EBP is required for the production of diffusable signals within the ventral diencephalon that promote differentiation of the adenohypophysis. Table 3-1 summarizes some of the other transcription and differentiation factors involved in pituitary and hypothalamus differentiation.

Pituitary Hormones

Adenohypophysial Hormones

The hormones of the adenohypophysis can be categorized based on chemical structure. Thyrotropin, LH, and FSH are glycoprotein hormones that each consist of two polypeptide subunits (total molecular weights about 30 kD), an α-subunit and a β-subunit. GH and PRL are polypeptide hormones ranging from 191 to 199 amino acids in length (about 20 kD), while ACTH, MSH, and the endorphins are peptides of 39 amino acids or less (table 3-2).

The glycoprotein hormones TSH, FSH, and LH contain a common α-subunit but different β-subunits (figure 3-4). It is the different β-subunits that confer unique activity to each hormone. Each subunit is proteolytically cleaved from a larger pro-subunit protein and glycosylated during posttranslational processing. Although certain cells in the PT express both the individual α- and β-subunits of TSH, they do not express the biologically active TSH heterodimer. These cells also lack certain features (transcription factor Pit-1 as well as receptors for thyroid hormone and TRH) required for physiological regulation of TSH secretion and do not appear to be involved in thyroid regulation.

Although structurally similar, FSH and LH generally control different gonadal events. In females, FSH promotes ovarian follicle growth, and LH induces ovulation. LH is named for its role in causing the formation (luteinization) of the corpus luteum from the remaining portion of the ovarian follicle after ovulation. Because both FSH and LH stimulate the gonads, they are also called gonadotropins (GTHs). Both GTHs are required for normal estradiol synthesis—LH stimulating the synthesis of androgens in thecal cells and FSH promoting aromatization of androgens to estrogen in the granulosa cell layer of the follicle. In males, FSH promotes estrogen synthesis that in turn stimulates spermatogenesis, while LH promotes

Table 3-1. Key transcription and differentiation factors involved in pituitary and hypothalamus development.

Factor	Abbreviation	Expression Site	Role in Pituitary/Hypothalamus Development
Brain-specific transcription factor-2	Brn-2	Developing hypothalamus	Required for arginine vasopressin, oxytocin and corticotropin-releasing hormone expression
Thyroid-specific enhancer-binding protein	T/EBP	Ventral diencephalon but not Rathke's pouch	Required for development of hypothalamus, pituitary gland and thyroid
Pituitary-specific transcription factor	Pit-1	Thyrotropes, somatotropes and lactotropes	Required for development of thyrotropes, somatotropes and lactotropes
LIM/homeobox gene-3	Lhx3	Rathke's pouch	Required for development of all pars distalis endocrine cell types except for corticotropes
Pituitary homeobox 1/pituitary OTX-related factor	Ptx1/P-OTX	Rathke's pouch/corticotropes	May influence proopiomelanocortin and other corticotrope gene expression
Steroidogenic factor-1	SF-1	Gonadotropes	Important for the development of gonadotrope responsiveness to gonadotropin-releasing hormone
Bone morphogenetic protein-4	BMP4	Ventral diencephalon	Chemical signal from diencephalon required for induction of Rathke's pouch
Fibroblast growth factor-8	Fgf8	Ventral diencephalon	Required for expression of Lhx3 and differentiation of pouch rudiment into a definitive pouch

Table 3-2. Mammalian adenohypophysial hormones and their receptors.

Peptide	Conventional Abbrevation	Structure	Receptor Name	Single Target Tissue	Signaling Mechanism
Thyrotropin	TSH	Heterodimeric glycoprotein	TSH-R	Thyroid follicular cells	Increase cAMP
Follicle-stimulating hormone	FSH	Heterodimeric glycoprotein	Follitropin receptor, FSH-R	Gonad	Increase cAMP
Luteinizing hormone	LH	Heterodimeric glycoprotein	Lutropin receptor, LH-R	Gonad	Increase cAMP
Corticotropin	ACTH	Peptide	Melanocortin (MC)-2 receptor	Adrenal cortex	Increase cAMP
β-Endorphin	None	Peptide	Mu opioid	Central nervous system	Reduce K^+ conductance
Melanotropin	αMSH	Peptide	MC-1	Melanocytes	Increase cAMP
Growth hormone	GH	Polypeptide	GH-R	Liver	JAK/STAT
Prolactin	PRL	Polypeptide	PRL-R	Mammary gland	JAK/STAT

JAK, janus kinase; STAT, signal transducer and activator of transcription.

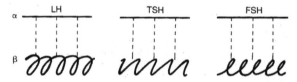

Figure 3-4. Generalized structure of pituitary glycoprotein hormones, luteinizing hormone (LH), thyroid-stimulating hormone (TSH), and follicle-stimulating hormone (FSH). The structure of the α subunit is shared by all three hormones, whereas the β subunit is unique and is responsible for the biological activity of the heterodimer. Reprinted with permission from Norris (1997).

spermiation and androgen synthesis in the Leydig cells of the testis. Thyrotropin acts on the thyroid gland to increase thyroidal iodide uptake and the synthesis and release of thyroid hormones. All of the glycoprotein hormones act via G-protein–coupled receptors linked to stimulation of intracellular cAMP.

GH and PRL are both polypeptides of similar size (191–199 amino acids; 22–23 kD) encoded by separate genes (figure 3-5). GH is important in longitudinal bone growth and as an anabolic hormone during development, with stimulatory effects on protein synthesis (especially in liver, spleen, kidney, thymus, and red blood cells) and lipid metabolism. Growth hormone has indirect and direct effects on peripheral tissues, with most of the indirect effects being mediated by insulinlike growth factor-I (IGF-I) of hepatic origins. Aside from stimulating milk production by the mammary glands, PRL has effects on growth, osmoregulation, parental behavior, the integument, and metabolism. Prolactin and GH target receptors link to the JAK/STAT) Janus Kinase) signal transduction pathway (table 3-2).

ACTH, β-endorphin, and β-LPH are peptides derived from POMC. ACTH is a 39-amino-acid peptide split from the middle region of POMC, whereas β-LPH is clipped from the C-terminus (figure 3-6). Further proteolytic cleavage of β-LPH results in β-endorphin, a 31-amino-acid peptide. All three peptides are co-stored in secretory granules and are co-released. ACTH acts on G_s-protein–coupled membrane melanocortin (MC-2) receptors to stimulate glucocorticoid synthesis and secretion from the adrenal cortex. These receptors belong to a family of five MC receptor types that carry out the actions of ACTH and α-MSH in the adrenal cortex, integument, and brain. β Endorphin (coined from "*endo*genous" and "*morphine*") is a potent analgesic peptide that acts on neuronal μ (morphine-selective) opioid receptors linked to inhibition of cAMP. Chemically related to β-endorphin, dynorphin and the enkephalins produced in the central nervous system have similar analgesic actions but through different receptors.

Melanotropes process POMC differently from corticotropes due to differential expression of two enzymes that process POMC (figure 3-6). In general, POMC undergoes more proteolytic cleavage in melanotropes, and the end products are subject to C-terminal amidation. The major end products of POMC processing in melanotropes are α-MSH and shorter and N-terminally acetylated forms of β-endorphin. The N-terminal 13 amino acids of ACTH are the source of α-MSH.

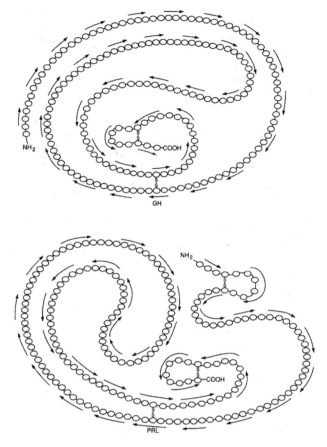

Figure 3-5. Comparison of growth hormone (GH; top) and prolactin (PRL; bottom) structures. Both are polypeptides of comparable size and sequence homology. Note the disulfide bonds (two in growth hormone, three in prolactin), which contribute to the three dimensional structure in each polypeptide. Given the similarity in structure it is not surprising that both hormones can exhibit similar effects when tested at high doses. Reprinted with permission from Norris (1997).

As a consequence, virtually all of the ACTH is cleaved to α-MSH and an inactive fragment called corticotropinlike peptide (CLIP) before secretion. The message sequence required for activation of MC-2 receptors is absent in the 13 amino acid sequence of α-MSH and, as a result, this peptide has little ability to stimulate glucocorticoid secretion. Its major role is the stimulation of melanin production in epidermal melanocytes via an interaction with the MC-1 receptor. Conversely, ACTH, which contains the entire α-MSH sequence, is effective at stimulating melanocytes and may do so when present in excessive amounts (e.g., in Cushing's syndrome).

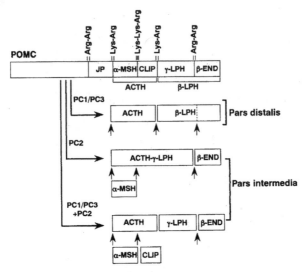

Figure 3-6. Posttranslational processing of proopiomelanocortin (POMC). Tissue-specific processing of POMC occurs via the differential expression and activity of proteolytic enzymes (prohormone convertase enzymes, PC). Proteolytic cleavage occurs between dibasic residue pairs and results in several biologically active peptides including corticotropin (ACTH)/lipotropin (LPH)/endorphin (END) in the pars distalis and endorphin and melanotropin (MSH) in the pars intermedia. From Norris (1997).

Neurohypophysial Nonapeptides

AVP and OXY are nonapeptides derived from similar but separate prohormones, each encoded by separate genes. They are produced by magnocellular neurons of the paraventricular (PVN) and supraoptic (SON) nuclei. Magnocellular neurons of the PVN and SON project to the PN, where AVP and OXY are stored until released in the vicinity of PN capillaries (figure 3-1). AVP is secreted in response to low blood volume or elevated blood osmolarity and acts on vasopressin-type 2 (V_2) receptors in the renal collecting ducts to increase water reabsorption. A second class of vasopressin receptors, the vasopressin type 1 (V_1) receptors, are linked to the inositoltrisphosphate (IP_3) hydrolysis pathway and mediate the vasomotor actions of AVP on vascular smooth muscle.

OXY stimulates milk release during lactation and is a powerful constrictor of reproductive smooth muscle in both males and females. It initiates and maintains uterine smooth muscle contraction during parturition and is responsible for contraction of the oviduct and vas deferens during sexual orgasm, contributing to ejaculation in males and intensifying the sensation of orgasm in both males and females. OXY is secreted as part of a reflex arc that begins with stimulation of sensory nerve ending in the nipples of the breast and the conduction of afferent neuronal signal to oxytocin neurons in the SON. Depolarization of these neurons results in OXY secretion from nerve terminals in the PN. OXY causes milk release by stimulating the contraction of myoepithelial cells surrounding the glandular alveoli in mammary tissue where milk is stored.

Hypothalamic Releasing Hormones (HRHs)

Hypothalamic HRHs hormones regulate the secretion of pituitary tropic hormones and are specifically named as releasing hormones or release-inhibiting hormones based on their function. Regulation of most tropic hormones involves only releasing hormones (FSH, LH, TSH, MSH) whereas PRL and GH have both releasing and release-inhibiting regulators. Corticotropin-releasing hormone (CRH), gonadotropin-releasing hormone (GnRH), growth hormone-releasing hormone (GHRH), GH release-inhibiting hormone (GH-RIH), and thyrotropin-releasing hormone (TRH) are all peptides. Prolactin release-inhibiting hormone has been identified as the catecholamine dopamine (Ben-Jonathan et al., 1989). Secretion of ACTH, TSH, and the gonadotropins (FSH, LH) is also regulated by negative feedback control (figure 3-7). All the HRHs act on G-protein–coupled receptors (table 3-3).

The first hypothalamic releasing factor to be purified and sequenced was TRH. This tripeptide is derived from proteolytic cleavage of a much larger precursor protein called proTRH that is synthesized by neurons in the PVN, the dorsomedial nucleus, and the SON. Despite multiple sites of synthesis, experimental evidence indicates that PVN-TRH cells are those that project to the ME and control the

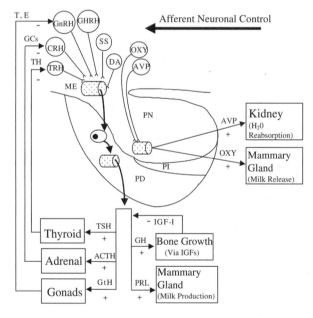

Figure 3-7. Pituitary hormone secretion and its regulation by the hypothalamus. Hypothalamic hormones are secreted in the portal vasculature and travel via portal vessels to the pars distalis (PD) where they regulate the secretion of TSH, ACTH, GtH, GH, and PRL. Oxytocin and vasopressin (AVP, also known as antidiuretic hormone) are synthesized in the hypothalamus and released into capillaries of the pars nervosa. Thyroid hormones (TH), glucocorticoids (GCs), and sex steroids inhibit their own secretion by acting at the level of the hypothalamus. See text for complete abbreviations.

Table 3-3. Mammalian hypothalamic hormones and their receptors.

Peptide	Abbreviation	Structure	Receptor Name	Receptor Location	Signaling Mechanism
Thyrotropin-releasing hormone	TRH	pGlu-His-Pro	TRHR	Thyrotropes, lactotropes	IP_3/DAG
Gonadotropin-releasing hormone	GnRH	pGlu-His-Trp-Ser-Tyr-Gly-Leu-Arg-Pro-Gly-NH_2	GnRHR	Gonadotropes	IP_3/DAG
Corticotropin-releasing hormone	CRH	Ser-Gln-Glu-Pro-Pro-Ile-Ser-Leu-Asp-Leu-Thr-Phe-His-Leu-Leu-Arg-Glu-Val-Leu-Glu-Met-Thr-Lys-Ala-Asp-Gln-Leu-Ala-Gln-Ala-His-Ser-Asn-Arg-Lys-Leu-Leu-Asp-Ile-Ala-NH_2	CRH-R_1	Corticotropes	Increase cAMP
Arginine vasopressin	AVP	Cys-Tyr-Phe-Gln-Asn-Cys-Pro-Arg-Gly-NH_2	V_{1b}	Corticotropes	IP_3/DAG
Growth hormone-releasing hormone	GHRH	Tyr-Ala-Asp-Ala-Ile-Phe-Thr-Asn-Ser-Tyr-Arg-Lys-Val-Leu-Gly-Gln-Leu-Ser-Ala-Arg-Lys-Leu-Leu-Gln-Asp-Ile-Met-Ser-Arg-Gln-Gln-Gly-Glu-Ser-Asn-Gln-Glu-Arg-Gly-Ala-Arg-Ala-Arg-Leu-NH_2	GHRH-R	Somatotropes	Increase cAMP
Pituitary adenylate cyclase activating peptide	PACAP	His-Ser-Asp-Gly-Ile-Phe-Thr-Asp-Ser-Tyr-Ser-Arg-Tyr-Arg-Lys-Gln-Met-Ala-Val-Lys-Lys-Tyr-Leu-Ala-Ala-Val-Leu-Gly-Lys-Arg-Tyr-Lys-Gln-Arg-Val-Lys-Asn-Lys-NH_2	PAC1, type 1 PACAP receptor or PVR1	Somatotropes	Increase cAMP IP_3/DAG
Somatostatin	SS	Ala-Gly-Cys-Lys-Asn-Phe-Phe-Trp-Lys-Thr-Phe-Thr-Ser-Cys	sst_1, sst_2, sst_3, sst_4, sst_5	Somatotropes, lactotropes	Decrease cAMP
Dopamine	DA	Catecholamine	D_2 receptor	Lactotropes, melanotropes	Decrease cAMP

DAG, diacylglycerol; IP3, inositol 1,4,5,-triphosphate.

pituitary thyroid axis. These neurons are richly endowed with thyroid hormone receptors and are inhibited by elevated circulating levels of thyroid hormone. Both thyrotropes and lactotropes possess TRH receptors. Although TRH stimulates PRL secretion under experimental conditions, its physiological role is unclear, as TRH levels in the portal circulation do not change during lactation.

Secretion of FSH and LH is controlled by the decapeptide GnRH first isolated from swine and sheep hypothalami in the early 1970s. Gonadotropin releasing hormone (GnRH) is synthesized from a larger prohormone in neurosecretory neurons of the medial preoptic nucleus and belongs to a family of GnRH peptides found throughout the vertebrates (see below). In both males and females, GnRH provides a constant, pulsate stimulation of gonadotropin (GTH) release. High circulating levels of androgen and estrogen inhibit these neurons. In addition, females have a specialized group of hypothalamic GnRH neurons that are stimulated by high blood estrogen levels. This so-called surge center is responsible for the surge in LH secretion that occurs during ovulation. A noteworthy feature of GnRH neurons is their extensive extrahypothalamic distribution, suggesting widespread involvement in central nervous system control of reproduction.

Mammalian CRH is a peptide consisting of 41 amino acids synthesized by neurosecretory neurons in the PVN and transported in the blood partially bound to a specific binding protein. This neuropeptide is the most important regulator of the pituitary adrenal axis and also mediates numerous other physiological events related to stress, including changes in cardiovascular function, energy metabolism, and behavior. The effects of CRH are carried out by two pharmacologically distinct membrane receptors, CRH receptors 1 and 2 (R1 and R2). The R1 is coupled to cAMP, is expressed by corticotropes, and is generally considered to be the receptor that mediates the effects of CRH on ACTH secretion. A peptide structurally related to CRH, urocortin, is synthesized within the PD as well as in several brain sites. Urocortin binds to R1 receptors and is capable of stimulating ACTH secretion, suggesting that this CRH-like peptide may have a role in local (paracrine) control of ACTH and possibly other PD hormones.

Under certain conditions, AVP potentiates CRH-induced secretion of ACTH. The AVP involved in ACTH secretion does not arise from magnocellular neurons, but from a subset of smaller neurons in the PVN that coexpress CRH and project to the ME. Approximately 50% of CRH neurons in the ME also contain AVP. Arginine vasopressin-containing CRH neurons are selectively activated during acute stress, suggesting that AVP may play a physiological role in stress-induced secretion of ACTH.

Dopamine acts as a neurotransmitter within afferent neuronal pathways controlling hypothalamic function as well as a hormone regulating PRL secretion. Dopamine is synthesized by neurosecretory neurons in the arcuate (Arc) nucleus and inhibits PRL secretion via dopamine D_2 G_i-coupled membrane receptors on lactotropes. An interesting feature of this control is that PRL secretion is under constant inhibitory regulation by DA. Elevations in PRL secretion usually involve release of DA inhibition either by inhibition of DA secretion at the point of release or inhibition of DA synthesis. Two additional hypothalamic peptides have been implicated in stimulating PRL release: vasoactive intestinal peptide (VIP) and TRH. Both are effective in a wide range of vertebrates, but only VIP levels are elevated in portal blood during suckling in mammals.

Growth hormone is released in a pulsatile manner that is controlled by two peptides with antagonistic effects on GH secretion; GH-RIH and GHRH. Pulsatile release of GH results from a reciprocal increase in GHRH secretion and decrease in GH-RIH secretion. Just the opposite pattern in GHRH and GH-RIH release causes the nadir or baseline in GH release. Growth hormone-releasing hormone is a peptide synthesized in the Arc nucleus that ranges from 44 amino acids (human, pig, goat, and sheep) to 43 and 42 amino acids in rat and mouse. Somatostatin is a peptide of 14 amino acids that is synthesized in the periventricular nucleus. Pituitary adenylate cyclase activating peptide (PACAP) is a 39-amino-acid peptide produced in the SON, and it stimulates GH secretion as well as the secretion of other pituitary hormones. In mammals, GHRH and PACAP are produced from separate prohormones encoded by separate genes. Although PACAP fulfills most requirements for hypothalamic releasing hormones, recent evidence for its production in gonadotropes raises the possibility of a paracrine role for this peptide in controlling GH secretion.

Hypothalamus–Pituitary Control of Gonadal, Thyroid, and Adrenal Hormone Secretion

The Hypothalamus–Pituitary–Gonadal Axis

Environmental cues such as photoperiod and temperature are used by many mammalian species to determine appropriate conditions for reproduction and offspring survival. The central nervous system integrates sensory information and regulates appropriate changes in GnRH secretion via a number of afferent neuronal pathways (figure 3-8). Hypothalamic control of the pituitary gonadal axis is required for the proper timing and development of secondary sex characteristics and in preparing the organism for reproduction. Feedback inhibition of GnRH and GTH secretion plays an important role in regulating the length of reproductive cycles.

The Hypothalamus–Pituitary–Thyroid Axis

In response to low body temperature, higher brain centers activate neuronal afferent pathways controlling TRH secretion from neurons in the PVN (figure 3-8). Elevations in TSH secretion lead to enhanced synthesis and secretion of thyroid hormones from the thyroid gland. In adults, thyroid hormones act on a wide variety of tissues (including the brain) to increase metabolic rate, oxygen consumption rate, and heat production (see chapter 4 for more details). During development, thyroid hormones play many critical roles, especially in the differentiation of the nervous system.

The Hypothalamus–Pituitary–Adrenal Axis

The HPA axis mediates an organism's ability to adapt to stressors in the environment. Changes in environmental conditions (heat, cold, salinity), perception of a

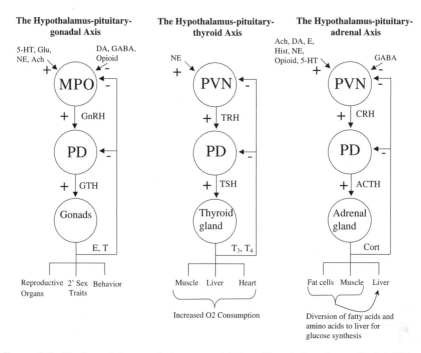

Figure 3-8. The hypothalamus–pituitary–gonadal, thyroid, and adrenal axes. Both inhibitory and stimulatory neurotransmitters regulate hypothalamic neurons producing gonadotropin-releasing hormone (GnRH), thyrotropin-releasing hormone (TRH), and corticotropin-releasing hormone (CRH). See text for other abbreviations.

physical or psychological threat, or alterations in homeostasis (e.g., low blood sugar) all lead to elevations in CRH secretion. Adrenal corticosteroids have important effects on metabolism that allow an organism to cope with stress (figure 3–8; see also chapter 5). An equally important action of corticosteroids is to turn off CRH secretion so that an individual's stress response does not go into overdrive. Excessive activation of the stress response can cause or exacerbate many diseases including cancer, diabetes, and cardiovascular disease.

Negative Feedback and Afferent Neuronal Regulation of Hypothalamic Hormones

Adrenal, gonadal, and thyroid hormones play an important role in regulating their own secretion via feedback inhibition of the hypothalamic neurons and tropic cells in the PD that control their secretion (figure 3-7). Hypothalamic hormone secretion is also controlled via afferent innervation from other brain areas as well as by hormonal feedback (figure 3-7). A number of neurotransmitters have been implicated in the afferent regulation of hypothalamic hormone secretion; a partial list is presented in table 3-4. Theoretically, disturbance of relevant neurotransmitter

Table 3-4. Afferent regulation of hypothalamic releasing hormones in mammals.

Hormone	Neurotransmitter								
	ACh	DA	E	GABA	Glu	Hist	NE[a]	Opioid	5-HT[a]
CRH	+	+	+	−		+	+	+	+
GnRH	+	−		+	+		+	−	+
GHRH		+		−		+	+(α_2)[b]	+	
Somatostatin	−	+		+			−(β_2)[c]	−	−
Dopamine		+		−	−	0	−	−	
TRH							+		+/−/0

+, stimulatory; −, inhibitory; 0, no effect; blank, not studied.

[a]Neurotransmitters known to be influenced by xenobiotics (see text for references).

[b]Mediated by α_2 adrenergic receptors.

[c]Mediated by β_2 adrenergic receptors.

ACh, acetylcholine; DA, dopamine; E, epinephrine; GABA, γ-amino butyric acid; Glu, glutamate; Hist, histamine; NE, norepinephrine; 5-HT, 5-hydroxytryptamine (serotonin). See text for hormone abbreviations.

processes at any level (synthesis, storage, secretion, postsynaptic action, turnover) could have potential effects on normal operation of the HP axis. In addition, disturbance of a single neurotransmitter system may have multiple effects on endocrine function, as virtually all of the neurotransmitter systems listed in table 3-4 control more than one endocrine pathway. For example, interfering with hypothalamic noradrenergic neurotransmission could potentially affect ACTH, TSH, GH, PRL, and gonadotropin secretion. Secretion of tropic hormones also can be influenced by feedback via hormones secreted by their target endocrine glands (e.g., thyroid hormones from the thyroid gland inhibit TSH release from the PD).

Anatomy of Hypothalamus–Pituitary Axis in Nonmammalian Vertebrates

The Pituitary Gland

Most features of the mammalian pituitary are recognizable in nonmammalian groups, with a few notable exceptions. Hagfishes are the most primitive chordate group to possess a recognizable adenohypophysis. The hagfish adenohypophysis does not contain clear divisions such as the PD and PI and is separated from the hypothalamus by connective tissue. In hagfish the PN develops as an extension from the wall of the third ventricle. There is no distinct ME, nor is there any evidence for direct innervation of adenohypophysial cells by hypothalamic neurons. In fact, the degree to which the hypothalamus regulates adenohypophysial function in this group is largely unknown. There is some evidence that hypothalamic hormones may be able to diffuse from the third ventricle to the adenhypophysis, leading to the suggestion of a "diffusional median eminence" in this primitive group of fish. It is important to point out that the primitive condition of the HP axis in this group is not considered to be an ancestral trait for all craniates. The lack of a vascular or neuronal mechanism for delivery of hypothalamic hormones may actually represent a loss of the ancestral condition.

In contrast to hagfish, lampreys resemble more recently evolved vertebrates in having a well-developed adenohypophysis clearly partitioned into a PD and PI. In lamprey and bony fishes the PD is separated into rostral and caudal (also called proximal PD) divisions that differ in the cell types present (figure 3-9). Tropic cells may be organized into discrete areas of the PD (figure 3-9). Elasmobranchs possess an additional pituitary lobe, the saccus vasculosus, that appears to be composed of epithelial-like ependymal cells but has no known endocrine function.

The adenohypophysis of amphibians and reptiles resembles that of mammals, with a well-developed PT, a feature found only in tetrapods. In contrast to mammals, where tropic cells are homogeneously distributed throughout the PD, amphibians may demonstrate regional organization of tropic cell types within the PD in a fashion similar to that of lampreys and bony fishes. A recognizable PI never forms in birds, although melanotropelike cells may be found in the PD.

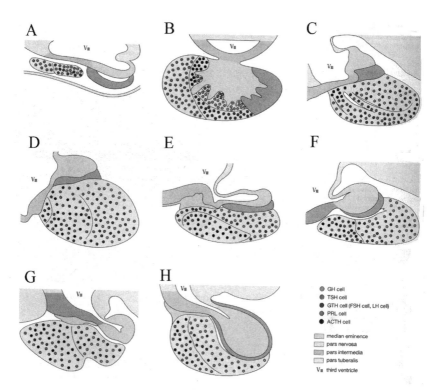

Figure 3-9. Anatomical distribution of cellular phenotypes in the pars distalis of various vertebrates. A, lamprey, *Lampetra japonica*; B, teleost, *Orizias latipes*; C, newt, *Triturus pyrrhogaster*; D, toad, *Bufo japonicus*; E, tortoise, *Geoclemys reevesii*; F, lizard, *Takydromus tachydromoides*; G, Japanese quail, *Coturnix coturnix japonicus*; H, cat. Modified from Matsumoto and Ishii (1992). See text for abbreviations.

In lampreys, elasmobranchs, chondrostean and holostean fishes, and tetrapods the neurohypophysis is composed of a ME and PN (figure 3-9). In teleosts, PD cells are directly innervated by hypothalamic neurosecretory neurons and an ME is not present. Hypothalamic hormones in these fishes are released in the vicinity of tropic cells and do not require a portal vasculature system for delivery.

The Hypothalamus

A striking feature of hypothalamic organization in fishes and amphibians is the limited migration of neurons outside the periventricular cell layer and less division of hypothalamic areas into clearly defined nuclear groups (figure 3-10). Furthermore,

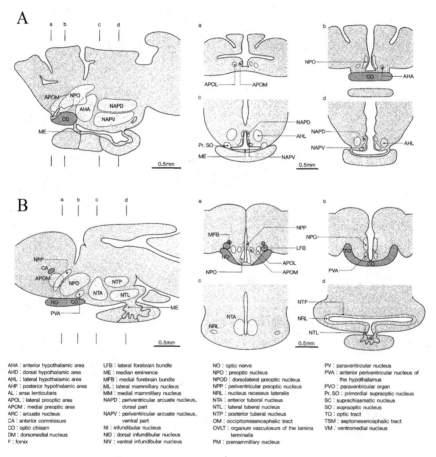

AHA : anterior hypothalamic area	LFB : lateral forebrain bundle	NO : optic nerve	PV : paraventricular nucleus
AHD : dorsal hypothalamic area	ME : median eminence	NPO : preoptic nucleus	PVA : anterior periventricular nucleus of
AHL : lateral hypothalamic area	MFB : medial forebrain bundle	NPOD : dorsolateral preoptic nucleus	the hypothalamus
AHP : posterior hypothalamic area	ML : lateral mammillary nucleus	NPP : periventricular preoptic nucleus	PVO : paraventricular organ
AL : ansa lenticularis	MM : medial mammillary nucleus	NRL : nucleus recessus lateralis	Pr. SO : primordial supraoptic nucleus
APOL : lateral preoptic area	NAPD : periventricular arcuate nucleus,	NTA : anterior tuberal nucleus	SC : suprachiasmatic nucleus
APOM : medial preoptic area	dorsal part	NTL : lateral tuberal nucleus	SO : supraoptic nucleus
ARC : arcuate nucleus	NAPV : periventricular arcuate nucleus,	NTP : posterior tuberal nucleus	TO : optic tract
CA : anterior commissure	ventral part	OM : occipitomesencephalic tract	TSM : septomesencephalic tract
CO : optic chiasm	NI : infundibular nucleus	OVLT : organum vasculosum of the lamina	VM : ventromedial nucleus
DM : dorsomedial nucleus	NID : dorsal infundibular nucleus	terminalis	
F : fornix	NIV : ventral infundibular nucleus	PM : premammillary nucleus	

Figure 3-10. Anatomical organization of the hypothalamus and nomenclature of hypothalamic nuclei in representative vertebrate species. A, lamprey, *Entosphenus japonica*; B, teleost fish (eel), *Anguilla japonica*; C, American bullfrog, *Rana catesbeiana*; D, snake, *Elaphe conspicillata*; E, Japanese quail, *Coturnix coturnix japonicus*. Reprinted with permission from Matsumoto and Ishii (1992).

the assignment of hypothalamic neurosecretory neurons are not organized into clearly defined nuclei (figure 3-10). As a consequence, there are generally fewer hypothalamic cell groups that project to the pituitary compared to mammals, and individual nuclei may contain many different hypothalamic releasing hormones. For example, GnRH, TRH, CRH, somatostatin, POMC-derived peptides, and AVP have all been localized to neurons of the preoptic nucleus in amphibians.

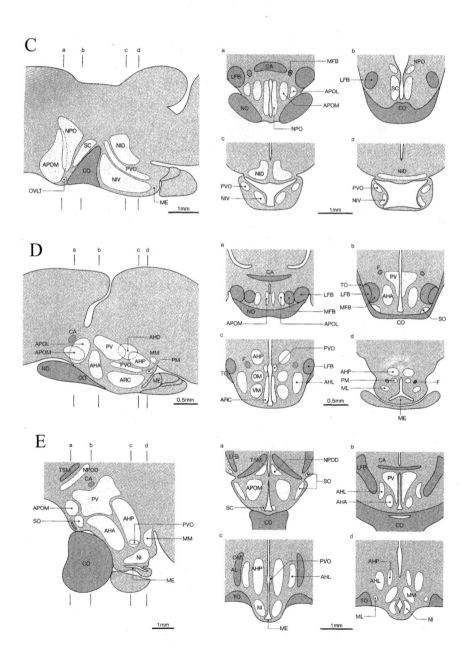

Comparative Aspects of Hypothalamic Releasing Hormones

Neurons immunoreactive to TRH have been identified in the hypothalami of lamprey, bony fishes, amphibians, reptiles, and birds. Despite the presence of a TRH-like peptide in the hypothalamus, mammalian TRH does not stimulate the pituitary thyroid axis of fishes or larval amphibians. The reptilian pituitary gland releases TSH in response to TRH as well as to CRH, sauvagine (a CRH-like peptide first isolated from amphibians), and mammalian GHRH and GnRH. CRH also appears to be the physiological stimulator of both ACTH and TSH release in larval amphibians. Obviously, there have been major changes in the structure and pattern of expression of the TRH receptor during vertebrate evolution.

Ten GnRH molecules have been sequenced or cloned from vertebrates and two from tunicates. They have been named according to the taxa where they were first found, but most have much wider distributions among the vertebrates. Hence mammalian GnRH (mGnRH) is found in amphibians and chicken-II (cGnRH-II) is found from fishes to mammals. Recently, chicken-II has been renamed as GnRH-II. All of these decapeptides share greater than 50% homology in their amino acid sequence (figure 3-11). Four of the vertebrate forms are found in tetrapods: mGnRH, guinea pig GnRH (gpGnRH), chicken GnRH-I; and chicken GnRH-II (cGnRH-II). The remaining are restricted to fishes (figure 3-12) and include lamprey GnRH-I (lGnRH-I) and III (lGnRH-III), dogfish GnRH (dfGnRH), salmon GnRH (sGnRH), seabream GnRH, (sbGnRH), and catfish GnRH (cfGnRH).

Most vertebrates possess more than one form of GnRH. Lamprey GnRH-I and III are restricted to this group, GnRH is found only in certain elasmobranchs. All

```
            1  2  3  4  5  6  7  8  9  10

tGnRH-I      pGlu-His-Trp-Ser-Asp-Tyr-Phe-Lys-Pro-Gly-NH2

tGnRH-II     pGlu-His-Trp-Ser-Leu-Cys-His-Ala-Pro-Gly-NH2

lGnRH- I     pGlu-His-Tyr-Ser-Leu-Glu-Trp-Lys-Pro-Gly-NH2

lGnRH- III   pGlu-His-Trp-Ser-His-Asp-Trp-Lys-Pro-Gly-NH2

dfGnRH       pGlu-His-Trp-Ser-His-Gly-Trp-Leu-Pro-Gly-NH2

cfGnRH       pGlu-His-Trp-Ser-His-Gly-Leu-Asn-Pro-Gly-NH2

sbGnRH       pGlu-His-Trp-Ser-Tyr-Gly-Leu-Ser-Pro-Gly-NH2

sGnRH  pGlu-His-Trp-Ser-Tyr-Gly-Trp-Leu-Pro-Gly-NH2

cGnRH-I      pGlu-His-Trp-Ser-Tyr-Gly-Leu-Gln-Pro-Gly-NH2

cGnRH-II     pGlu-His-Trp-Ser-His-Gly-Trp-Tyr-Pro-Gly-NH2

mGnRH        pGlu-His-Trp-Ser-Tyr-Gly-Leu-Arg-Pro-Gly-NH2

gpGnRH       pGlu-Tyr-Trp-Ser-Tyr-Gly-Val-Arg-Pro-Gly-NH2
```

Figure 3-11. Comparision of mammalian and nonmammalian gonadotropin-releasing hormone (GnRH) sequences. The C-terminus dipeptide is conserved in all forms, including tunicate. The N-terminus dipeptide is conserved in all but guinea pigs. Abbreviations: t = tunicate, l = lamprey, df = dogfish, cf = catfish, sb = seabream, s = salmon, c = chicken, m = mammalian, gp = guinea pig.

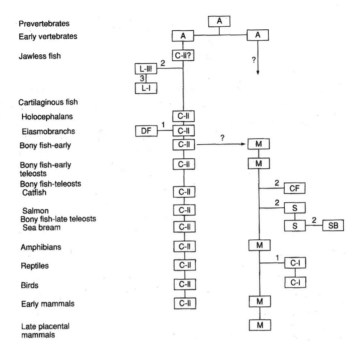

Figure 3-12. Phylogenetic distribution of gonadotropin-releasing hormone (GnRH)-like peptides in vertebrates. In this scheme the gene encoding an ancestral GnRH peptide (A) has undergone several duplication events, resulting in cGnRH II (C-II) and mGnRH. Reprinted with permission from Sherwood et al. (1994).

elasmobranchs and bony fishes examined so far have GnRH-II, a characteristic shared with tetrapods. The second form of GnRH found in bony fishes differs depending on the species; mGnRH is found in lobe-finned and more primitive ray-finned fishes and sGnRH is found in most teleosts. In an evolutionary sense, teleosts have replaced mGnRH with sGnRH and other specialized forms of the peptide. This notion is supported by the fact that eels (*Anguilla* sp.), primitive teleosts, retain cGnRH-II and mGnRH but lack sGnRH. The unique GnRH forms found in catfish and seabream appear to be restricted to these groups. Nucleotide base-substitutions in the mGnRH gene leading to a single amino-acid switch at position 8 resulted in the GnRH-I form found throughout reptiles. Birds, which evolved from the crocodilian lineage, also have the chicken GnRH-I form as well as chicken GnRH-II. Because of their wide distribution among vertebrates some investigators have adopted GnRH-I to represent mGnRH and GnRH-II for cGnRH-I.

GnRH neurons that project to the pituitary gland have been identified in all jawed fishes with the exception of the elasmobranchs. In elasmobranchs, GnRH is released into the systemic blood circulation from nerve terminals at the base of the telencephalon.

In some species expressing multiple forms of GnRH, it is clear the two forms of GnRH carry on different functions. In amphibians, for example, mGnRH controls

GTH secretion, while the second form acts as a neuromodulator regulating neural pathways. These observations cannot easily be extrapolated to all species, as in fish two forms of GnRH are released in the vicinity of gonadotropes and appear to bind to the same receptor. This does not necessarily rule out the possibility of different actions on gonadotropes. Studies in goldfish indicate that two GnRH peptides bind to one gonadotrope receptor but activate different intracellular signaling pathways.

Although nonmammalian GnRH receptors have not been cloned, it appears that there has been some coevolution of receptor and peptide, as endogenous forms of the peptide seem to be the most potent in stimulating GTH secretion in those species that have been studied. Mammalian GnRH is the most potent stimulator of FSH and LH release in mammals. In goldfish, GnRH-II and sGnRH, the endogenous forms in this species, are equally potent, while other nonendogenous forms (mGnRH, GnRH, and GnRH-1) are active but are significantly less potent.

The structure of CRH has been highly conserved in vertebrate evolution; teleost CRH differs from the rat/human CRH by only 3–10 amino acids, depending on the species. Amino acid residues 9–21 are required for activation of ACTH secretion and are conserved in all vertebrate CRHs, suggesting that the role of this peptide in controlling ACTH secretion has been conserved (figure 3-13). This is borne out by studies showing that mammalian CRH stimulates ACTH secretion in every vertebrate class so far examined. CRH also stimulates TRH secretion in fishes, amphibians, reptiles, and birds.

Urocortin and CRH are part of a larger family of neuropeptides that includes the urotensins and sauvagine (figure 3-13). As in mammals, multiple CRH-like peptides appear to be the rule of thumb in nonmammalian species. Urotensin is a peptide that is structurally related to urocortin and was first isolated from the urophysis (*uro* =

Primary Structure	Species	Taxon
10 20 30 40		
Urotensin-I/Sauvagine/Urocortin Series		
.DDPPLSIDLTFHLLRTLLE LARTQSQRER AEQNRIIFDS V	rat/sheep UCN	Eutheria
.EDLPLSIDLTFHLLRTLLE LARTQSQRER AEQNRIILNA V	hamster UCN	
.DNPSLSIDLTFHLLRTLLE LARTQSQRER AEQNRIIFDS V	human UCN	
.QGPPISIDLSLELLRKMIE IEKQEKEKQQ AANMPLLLDT I	frog SVG	Anura
PAETPNSLDLTFHLLREMIE IAKHENQQMQ RDSNRRIMDT I	shark UI	Chondrichthyes
NDDPPISIDLTFHLLRNMIE MARNENQREQ AGLNRKYLDE V	carp UI	Cypriniformes
NDDPPISIDLTFHLLRNMIE MARIENEREQ AGLNRKYLDE V	sucker UI	
NDDPPISIDLTFHLLRNMIE MARIESQKEQ AELNRKYLDE V	trout UI	Salmoniformes
SEEPPMSIDLTFHMLRNMIH RAKMEGEREQ ALINRNLMDE V	sole UI	Pleuronectiformes
SEEPPMSIDLTFHMLRNMIH RAKMEGEREQ ALINRNLLDE V	maggy sole UI	
SEDPPMSIDLTFHMLRNMIH MAKMEGEREQ BQINRNLLDE V	European flounder UI	
Corticotropin-releasing factor Series		
SEEPPISLDLTFHLLREVLE MARAEQLAQQ AHSNRRLMEI I	rodent, human CRF	Eutheria
SEEPPISLDLTFHLLREVLE MARAENLAQQ AHSNRRLMEI I	rat testes CRF	
SEEPPISLDLTFHLLREVLE MARAEQLAQQ AHSNRRLMEI F	porcine major CRF	
SEEPPISLDLTFHLLREVLE MARAEQLAQQ AHSNRRLMEN F	porcine minor CRF	
SQEPPISLDLTFHLLREVLE MTKADQLAQQ AHSNRRLLDI A	ovine/caprine CRF	
SQEPPISLDLTFHLLREVLE MTKADQLAQQ AHSNRRLLDI A	bovine CRF	
AEEPPISLDLTFHLLREVLE MARAEQIAQQ AHSNRRLMDI I	Xenopus CRF	Anura
SEEPPISLDLTFHLLREVLE MARAEQLAQQ AHSNRMMEI F	sucker CRF1	Cypriniformes
SEEPPISLDLTFHLLREVLE MARAEQLVQQ AHSNRMMEI F	sucker CRF2	
SDDPPISLDLTFHMLRQMME MSRAEQLQQQ AHSNRRMMEI F	sockeye salmon CRF	Salmoniformes

Figure 3-13. A comparison of the primary amino acid sequences for urotensins-I (UI) and corticotropin-releasing hormone/corticotropin-releasing factor (CRH/CRF) peptides in vertebrates. Reprinted with permission from Lovejoy and Balment (1999).

tail, *physis* = growth), a neurohypophysial-like structure in the spinal cord of fishes. The urophysis may serve as a pituitary-independent regulator of adrenocortical tissue, as urotensin is a potent releaser of cortisol synthesis in bony fishes (figure 3-14). Urotensin may be involved in controlling ACTH secretion, as it is produced in hypothalamic neurosecretory neurons as well as in the urophysis. Sauvagine is found in the brains of only anuran amphibians, where it is located in the cells of the preoptic nucleus that project to the neurohypophysis. Its function is unknown, although it can influence release ACTH under experimental conditions.

The inhibitory control of GH secretion by GH-RIH has been evolutionarily conserved; the tetradecapeptide sequence of GH-RIH, with the same amino acid sequence, is found in representatives from every vertebrate class, suggesting that GH-RIH is a phylogenetically ancient molecule. Fishes and amphibians appear to possess a second GH-RIH gene (preprosomatostatin-II) that encodes for a variant (Tyr^7, Gly^{10}) GH-RIH of undetermined physiological significance. Although GH secretion appears to be under dual inhibitory and stimulatory regulation in all vertebrates, the nature of the stimulatory hormone varies among species. Sequences for GHRH and PACAP are known for a number of teleost species. Contrary to the situation in mammals, where separate genes encode GHRH and PACAP, teleosts possess a single gene for a prohormone that encodes both peptides, phylogenetic evidence that synthesis of these peptides is coordinated, at least at the level of gene expression. The ability of these neuropeptides to stimulate GH secretion varies between species and depends on the nutritional and behavioral (stress vs. nonstress) status of the organism. In eels, for example, PACAP but not GHRH is capable of stimulating GH secretion, but to a lesser degree than CRH.

Neurohypophysial Nonapeptides

In general, two nonapeptides structurally related to AVP and OXY are present in the nonmammalian PN (table 3-5). Chemically, AVP is a basic molecule due to the presence of the amino acid arginine, whereas OXY is a neutral molecule. In tetrapods and lungfishes, the basic homologous peptide to AVP is arginine vasotocin (AVT), and the peptide homologous to OXY is mesotocin (MST), whereas AVT and isotocin (IST) predominate in teleosts (figure 3-15). The primitive lamprey,

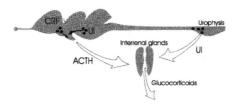

Figure 3-14. Anatomical relationship between brain corticotropin-releasing factor and urotensin I (UI) peptides and the urohypophyis, which also secretes UI in fishes. Both cortocotropin (ACTH) and UI are believed to play a role in adrenal corticosteroid secretion in fishes. Reprinted with permission from Lovejoy and Balment (1999).

Table 3-5. Phylogenetic distribution of oxytocin- and vasopressin-related neuropeptides in vertebrates.

	Amino acid									Occurrence
	1	2	3	4	5	6	7	8	9	
Vasotocin/vasopressin family										
Vasotocin[a]	Cys	Tyr	Ile	Gln	Asn	Cys	Pro	Arg	Gly(NH$_2$)	All nonmammalian vertebrates
Vasopressin	—	—	Phe	—	—	—	—	—	—	Mammals
Lysopressin	—	—	Phe	—	—	—	—	Lys	—	Pig, macropodids, didelphids, peramelids
Phenypressin	—	Phe	Phe	—	—	—	—	—	—	Macropodids
Oxytocinlike hormone family										
Oxytocin	—	—	—	—	—	—	—	Leu	—	Ratfish, placental mammals
Mesotocin	—	—	—	—	—	—	—	Ile	—	Lungfishes, nonmammalian tetrapods, marsupials
Isotocin	—	—	—	Ser	—	—	—	Ile	—	Bony fishes
Glumitocin	—	—	—	Ser	—	—	—	Gln	—	Rays
Aspargtocin	—	—	—	Asn	—	—	—	Leu	⎫	Spiny dogfish
Valitocin	—	—	—	—	—	—	—	Val	⎬	
Asvatocin	—	—	—	Asn	—	—	—	Val	⎫	Spotted dogfish
Phasvatocin	—	Phe	—	Asn	—	—	—	Val	⎬	

From Acher (1996).

[a]Residues identical with those of vasotocin are indicated by dashes.

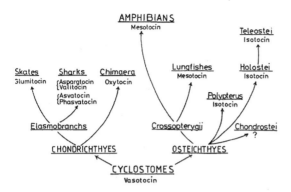

Figure 3-15. Phylogenetic distribution of oxytocinlike peptides in anamniotes. Reprinted with permission from Acher (1996).

however, secretes only AVT and does not produce a neutral nonapeptide. In contrast to the condition in bony fishes, there has been a striking radiation in the genes encoding OXY-like peptides in cartilaginous fishes (table 3-5). Although AVT plays an important osmoregulatory role in tetrapods, the physiological significance of the many OXY-like forms in bony and cartilaginous fishes remains unclear. AVT is an important stimulator of reproductive smooth muscle and has been implicated in oviposition and birth by amphibians and reptiles as well as spawning and birth in fishes.

Pituitary Hormones

Adenohypophysial Hormones

Although structurally related counterparts for all of the adenohypophysial hormones have been identified in fishes, amphibians, and reptiles, there are important differences in the action of these hormones among vertebrate groups. Functional constraints appear to have limited variation in the structures of some hormones, while other tropics have entirely different actions compared to their role in mammals. Because tropic and hypothalamic hormone receptors may be located in unexpected organs or cell types, a priori assumptions that similarities in hormone structure carry over to similarities in function can be misleading. Knowledge of hormone structure as well as receptor function and location are important when predicting the effects of hormones in nonmammals based on mammalian data. Only a handful of pituitary hormone receptors have been sequenced in nonmammals, so our knowledge of these receptors is based largely on cross-phyla comparisons of target tissue responsiveness and pharmacological binding studies.

Immunohistochemical studies using antibodies against the mammalian tropic hormones indicate the presence of corticotropes, thyrotropes, lactotropes, somatotropes, and gonadotropes in the adenohypophysis of all vertebrate classes (figure 3-9). Corticotropin appears to function in all vertebrates as the major hormone regulating steroid synthesis and secretion from adrenocortical tissue. The ACTH prohormone has been sequenced in lamprey, lungfish, and several amphibian species. In contrast to bony fish and tetrapods, lampreys possess two separate genes that encode a POMC-like prohormone. Mammalian TSH stimulates thyroid hormone secretion in every craniate group that has been tested except for hagfish. The fact that gonadotropins are capable of stimulating thyroid hormone secretion in some species but not in others suggest a great deal of interclass variation in the structure of the TSH receptor.

The diversity of reproductive modes in vertebrates make generalizations about GTH function difficult, except to say that gonadal function is under pituitary control. Pituitary GTH cells have been described from lampreys through birds. Lampreys and elasmobranchs appear to have a single LH-like GTH, while teleosts clearly have two, GTH-I and GTH-II. Teleost GTHs are heterodimeric glycoproteins produced in different cells of the pituitary and generally have different effects on gonadal function. In salmon, GTH-I, but not GTH-II, stimulates ovarian vitellogenin uptake, whereas oocyte maturation is primarily controlled by GTH-II. Both GTHs stimulate testicular and ovarian steroid synthesis. In salmon, GTH-I and GTH-II may have different roles in regulating testicular function, as GTH-I is the major GTH circulating in males during spermatogenesis, whereas GTH-II predominates during spermiation. Although amphibians and birds are similar to mammals in having distinct FSH- and LH-like GTHs, squamate reptiles have lost the ability to synthesis an LH-like GTH; an FSH-like hormone carries out all GTH-related function in this group.

It is impossible to generalize about the varied functions of PRL in vertebrates. Although we normally associate PRL with reproduction and lactation in mammals, it is a vitally important osmoregulatory hormone in freshwater fish and acts as an antimetamorphic hormone in amphibians. Although piscine PRL is not effective in avian and mammalian bioassays, mammalian PRL is capable of activating piscine PRL receptors. The evolution of PRL receptors in phylogentically "new" tissues (mammary gland, crop sac) appears to have coincided with the selection for PRL molecules that contain additional message sequences for receptor activation. Mammalian GH stimulates growth in those nonmammalian species that have been studied, and nonmammalian GHs are generally effective in mammalian bioassays. A clear understanding of the physiological role of GH in nonmammals is clouded by the fact that PRL as well as other hormones (thyroid hormones, corticosteroids) affect many of the same aspects of growth as GH in these animals. Although teleost somatolactin shows a significant amount of sequence homology with GH and PRL (hence its name), its physiological role remains unclear.

Hypothalamus–Pituitary Control of Gonadal, Thyroid, and Adrenal Activity in Nonmammals

As discussed above, most of the mammalian pituitary and hypothalamic hormones have structural counterparts in nonmammalian vertebrates, suggesting a remarkable conservation of hormone structure during the course of vertebrate evolution. Afferent neuronal activity and hormonal feedback appear to regulate the gonadal, thyroid, and adrenal (or interrenal) axes as in mammals (figure 3-8). However, important differences emerge when examining the control of pituitary hormone secretion. Many tropic cell types in nonmammals possess receptors for multiple HRHs. As a result, HRHs sometimes show less fidelity to a specific tropic hormone cell type than is generally the case in mammals. Unfortunately, exactly how this translates to differences in hypothalamic control remain unclear. It is one thing to demonstrate that a peptide of hypothalamic origin stimulates tropic hormone secretion and quite another to show its involvement in endogenous hypothalamic control. At present, the best example of plasticity in hypothalamic control occurs in the regulation of TSH secretion in fishes and amphibians. In these organisms, CRH controls both the pituitary–thyroid and pituitary–adrenal axes.

References

Acher, R., 1996. Molecular evolution of fish neurohypophysial hormones: Neutral and selective evolutionary mechanisms. Gen. Comp. Endocrinol. 102, 157–172.

Ben-Jonathan, N., Arbogast, L.A., Hyde, J.F., 1989. Neuroendocrine regulation of prolactin release. Prog. Neurobiol. 33, 399–447.

Lovejoy, D.A., Balment, R.J., 1999. Evolution and physiology of the corticotropin-releasing factor (CRF) family of neuropeptides in vertebrates. Gen. Comp. Endocrinol. 15, 1–22.

Matsumoto, A. and Ishii, S. (Eds.), 1992. Atlas of Endocrine Organs. Springer-Verlag, New York.

Norman, A.W. and Litwack 1998. Hormones, 2nd Ed, Academic Press, San Diego.

Norris, D.O., 1997. Vertebrate Endocrinology. Academic Press, San Diego, CA.

Sherwood, N.M., Parker, D.B., McRory, J.E., Lescheid, D.W., 1994. Molecular evolution of growth hormone-releasing hormone and gonadotropin-releasing hormone. In: Sherwood, N.M., Hew, C.L., (Eds.), Fish Physiology, vol. XIII. Academic Press, San Diego, CA, pp. 3–66.

Treier, M., Rosenfield, M.G., 1996. The hypothalamic-pituitary axis: Co-development of two organs. Curr. Opin. Cell Biol. 8, 833–843.

Selected Readings

Argente, J., Chowen, J.A., 1994. Neuroendocrinology of growth hormone secretion. Growth Gen. Horm. 10, 1–5.

Barraclough, C.A., 1992. Neural control of the synthesis and release of luteinizing hormone-releasing hormone. Ciba Foundation Symp. 168, 233–251.

Bertherat, J., Bluet-Pajot, M.T., Epelbaum, J., 1995. Neuroendocrine regulation of growth hormone. Eur. J. Endocrinol. 132, 12–24.

Calogero, A.E., 1995. Neurotransmitter regulation of the hypothalamic corticotropin-releasing hormone neuron. Ann. N.Y. Acad. Sci. 771, 31–40.

Corpas, E., Harman, M., Blackman, M.R., 1993. Human growth hormone and aging. Endocr. Rev. 14, 20–39.

Licht, P., 1979. Reproductive endocrinology of reptiles and amphibians: Gonadotropins. Annu. Rev. Physiol. 41, 337–351.

Lin, X.W., Otto, C.J., Peter, R.E., 1998. Evolution of neuroendocrine peptide systems: Gonadotropin-releasing hormone and somatostatin. Comp. Biochem. Physiol. C Pharmacol. Toxicol. Endocrinol. 119, 375–388.

Peter, R.E., 1986. Vertebrate neurohormonal systems. In: Vertebrate Endocrinology: Fundamentals and Biomedical Implications, vol. I. Morphological Considerations (Pang, P.K.T., Schreibman, M.P., eds.). Academic Press, New York,. pp. 57–102.

Peter, R.E., Yu, K-L., Marchant, T.A., Rosenblum, P.M., 1990. Direct neural regulation of the teleost adenohypophysis. J. Exp. Zool. 4, 84–89.

Schwanzel-Fukuda, M., Jorgenson, K.L., Bergen, H.T., Weesner, G.D., Pfaff, D.W., 1992. Biology of normal luteinizing hormone-releasing hormone neurons during and after their migration from olfactory placode. Endocr. Rev. 13, 623–634.

Sherwood, N.M., Lovejoy, D.A., Coe, I.R., 1993. Origin of mammalian gonadotropin-releasing hormones. Endocr. Rev. 14, 241–254.

Silva, J.D.B., Nunez, M.T., 1996. Facilitory role of serotonin (5-HT) in the control of thyrotropin-releasing hormone/thyrotropin (TRH/TSH) secretion in rats. Braz. J. Med. Biol. Res. 29, 677–683.

Wittkowski, W., Bockman, J., Kreutz, M.R., Bockers, T.M., 1999. Cell and molecular biology of the pars tuberalis of the pituitary. Intl. Rev. Cytol. 185, 157–194.

4

The Thyroid Gland

James A. Carr
David O. Norris

A ll craniates produce iodine-containing hormones derived from the amino acid tyrosine. These hormones, collectively called thyroid hormones, have a constellation of direct and indirect actions on energy metabolism, reproduction, and development. Cells that accumulate inorganic iodide (I^-) and synthesize thyroid hormones are arranged into follicles that, in tetrapods and some fishes, form a discrete endocrine organ, the thyroid gland.

The Mammalian Thyroid Gland

The mammalian thyroid gland consists of two lobes connected by a narrow isthmus covering the ventral surface of the larynx (figure 4-1). The gland is composed of a single layer of epithelial cells, called thyrocytes, that possess numerous microvilli, a feature shared with other absorptive epithelia. Thyrocytes are joined at their apical ends by tight junctions to form hollow, soccer ball-shaped structures called follicles (figure 4-1). The hollow cavity of each follicle is filled with a protein-rich fluid called colloid that serves as the storage site for thyroid hormones. The thyroid gland is unique among the endocrine glands in that it stores secretory products extracellularly. Both thyrocyte size and colloid amount vary proportionately with the degree of stimulation by thyrotropin (TSH), a feature that can be exploited for histological assessment of thyroid function. The thyroid gland also contains cells that secrete the calcium regulatory hormone calcitonin. These cells, called parafollicular cells, are located in the interfollicular connective tissue.

Figure 4-1. The thyroid gland in humans. This butterfly-shaped gland is located in the throat region. The gland consists of many hollow follicles that are lined by epithelial cells that synthesize thyroid hormone. Thyroid hormone is stored in the colloid that fills the lumen of each follicle. Modified from Norman and Litwack (1998).

Thyroid Hormone Structure

The structures of thyroid hormones and related compounds are shown in figure 4-2. Thyroid hormones are composed of two tyrosine residues covalently bound to iodine. In the coupling process, the outer iodotyrosine molecule loses its alanine side chain and couples directly with the phenolic ring of the inner iodotyrosine molecule via an ether linkage, producing a molecule called a thyronine. The ether bond makes these hormones relatively hydrophobic. All of the carbons are numbered to aid with nomenclature; carbons on the outer ring structure are given a prime desig-

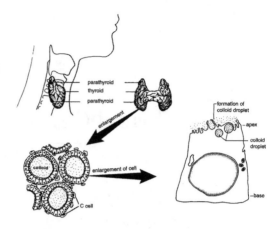

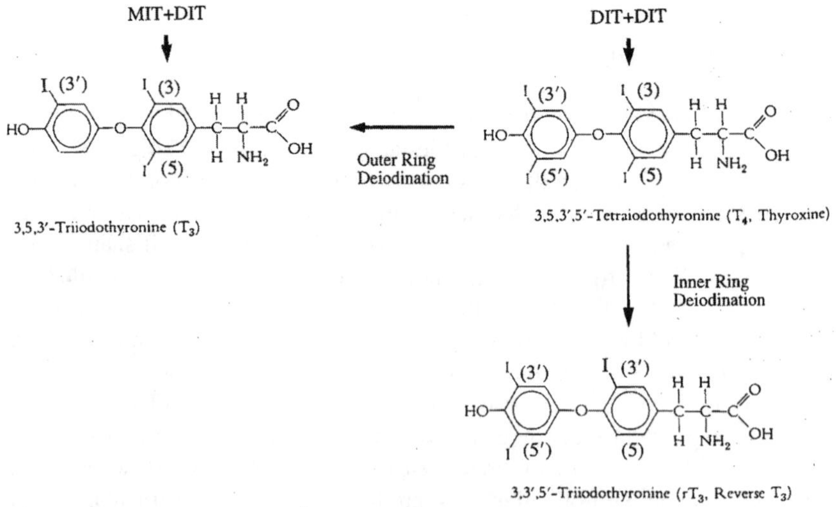

Figure 4-2. The structures of vertebrate thyroid hormones and reverse T_3. Reprinted with permission from Norris (1997).

nation. For example, tetraiodothyronine (T_4) has iodine molecules attached at carbons 3, 5, 3', and 5', while 3, 5, 3'-triodothyronine (T_3) lacks an iodine on the 5' carbon. Reverse T_3 (rT_3), a biologically inactive metabolite of T_4, is iodinated on the 3, 3', and 5' carbons. Given the large atomic mass of iodine, it is not surprising that the location and number of bound iodine atoms plays a critical role in the biological activity of these hormones; T_3 has 10-fold greater affinity for thyroid hormone receptors than T_4.

The Hypothalamus–Pituitary–Thyroid Axis

As discussed in chapter 3, synthesis and secretion of thyroid hormones is regulated by TSH produced in the pars distalis (see figure 3-8). TSH secretion, in turn, is regulated by the hypothalamic hormone thyrotropin-releasing hormone (TRH). The principal stimulus for TRH secretion in adult mammals is low environmental temperature. TSH stimulates iodide accumulation by the thyroid, synthesis of thyroglobulin and thyroid hormones, and the release of thyroid hormones into the circulation. Hypothalamic control of the hypothalamus–pituitary–thyroid HPT axis ensures that thyroid hormone secretion is coordinated with other events in the homeostatic regulation of body temperature. During development, thyroid hormones are important for many critical events, including proper development of the brain.

Thyroid Hormone Synthesis, Storage, and Secretion

Dietary intake of I^- is required for thyroid hormone synthesis. In the course of evolution, the thyroid gland has developed the ability to provide T_4 and T_3 at relatively constant rates despite intermittent I^- intake. T_4 is stored in a precursor form in the follicular colloid, creating a pool of readily available T_4 and T_3. Thyroid hormone synthesis and secretion can be divided into four distinct stages: (1) I^- uptake, (2) thyroglobulin synthesis, (3) iodination and coupling of iodinated tyrosines, and (4) endocytosis of thyroglobulin and thyroid hormone release.

Iodine Uptake

Thyrocytes have the capacity to selectively concentrate I^- from extracellular fluid containing relatively low I^- concentrations (10^{-8} M–10^{-7} M). Although other tissues (gastric mucosa, salivary glands, mammary glands, and choroid plexus) share the capacity to concentrate I^-, the thyroid is the only tissue to use I^- in hormone synthesis. Iodide is transported into thyrocytes by an integral plasma membrane glycoprotein, called the thyroid Na^+/I^- symporter (NIS). The genes responsible for both rat and human NIS have been cloned, and they share 84% sequence homology. The NIS cotransports Na^+ and I^- across the basolateral plasma membrane of thyrocytes at an $Na^+:I^-$ ratio of 2:1. This process is driven by the Na^+/K^+ ATPase transport protein, which maintains a "downhill" diffusion gradient for Na^+ entry into thyrocytes. For this reason thyroidal I^- uptake is sensitive to oubain, a poison that selectively disrupts Na^+/K^+ ATPase activity.

Thyrocytes must concentrate I^- from extracellular fluid rich in Cl^-. Not surprisingly, ion selectively plays a critical role in NIS function. The rank potency for anion inhibition of I^- transport by NIS is $TcO_4^- \geq ClO_4^- > ReO_4^- > SCN^- > BF_4^- > I^- > NO_3^- > Br^- > Cl^-$. This selectivity profile has been exploited clinically in the use of perchlorate (ClO_4^-, usually as the $KClO_4$ salt) to treat hypersecretion of thyroid hormones. Perchlorate has been introduced into surface and groundwater supplies as a by-product of rocket fuels at various sites around the United States. Because of its inhibitory effects on thyroid function, perchlorate is an important EDC and is especially critical to humans when it occurs in domestic water supplies.

Thyrocytes trap I^- to a much greater degree than other tissues expressing the NIS gene due to thyroid-specific transcription factors that promote NIS gene transcription. Thyroid transcription factor-1 (TTF-1) activates the NIS gene promoter in thyrocytes. The NIS gene promoter also contains a unique TSH response element, suggesting a possible interaction between TSH and TTF-1 in regulating thyroidal I^- uptake.

I^- must move across the apical plasma membrane into the lumen of the thyroid follicle for it to be incorporated into thyroid hormones (figure 4-3). Efflux of I^- into colloid is mediated in part by a TSH-responsive I^- channel in the apical plasma membrane of thyrocytes. A recently identified protein called the human apical iodide transporter (Rodriguez et al., 2002) may mediate iodide transport through the apical plasma membrane of thyrocytes. In addition, pendrin, an 86-kDa protein encoded by the Pendred syndrome gene (PDS), may act as an apical I^- transporter. Individuals with Pendred syndrome display mutations in the PDS gene, develop goiters (enlarged thyroid glands), and exhibit defects in I^- incorporation into thyroglobulin (see below). It is unlikely that pendrin is responsible for all apical I^- efflux, as expression of the PDS gene is not affected by TSH.

Thyroglobulin Synthesis

Thyroglobulin (TG) is a 660-kDa glycosylated protein synthesized by thyrocytes. Thyroglobulin serves as the substrate for synthesis of thyroid hormones, which are eventually released from TG by proteolysis. Thyroglobulin also serves as a storage form for thyroid hormones in colloid, representing up to 20–30% of the total colloidal protein in thyroid follicles. TG is synthesized in a monomeric 330-kDa form, each monomer containing approximately 67 tyrosine residues. Dimers of two identical TG chains are glycosylated and transported across the apical plasma membrane into the follicular lumen by an unidentified receptor or binding protein.

Iodination and Coupling of Iodinated Tyrosines

Inorganic iodine must be converted to a higher oxidation state before incorporation into TG. Oxidation of I^- is carried out by the enzyme thyroid peroxidase (TPO), located on the outside of the apical membrane of the thyrocyte. This enzyme also carries out the incorporation of oxidized I^- into tyrosine residues of TG and cou-

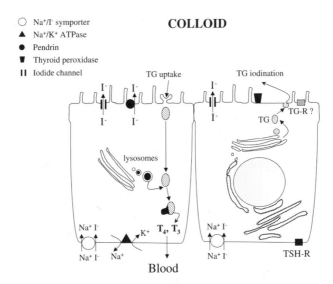

Figure 4-3. Two thyroid follicle cells (connected by tight or occluding junctions) and the various steps involved in thyroid hormone synthesis and secretion. Iodide is taken up across the basal plasma membrane via sodium-dependent co-transport, a step that can be blocked by the perchlorate anion. Sodium-dependent iodide transport is driven by a sodium gradient established by Na^+/K^+-ATPase pumps in the basolateral plasma membrane. Iodination of thyroglobulin (TG) occurs in the extracellular colloid in proximity to the apical plasma membrane and involves the enzyme thyroid peroxidase. The antithyroid drugs propylthiouracil and methimazole act by inhibiting the activity of this enzyme. Iodide may move across the apical plasma membrane due to the action of iodide transporter (pendrin) or through iodide channels. Iodinated thyroglobulin is taken up by the process of endocytosis; this process may involve specific thyroglobulin receptors. Thyroid hormones are enzymatically liberated from thyroglobulin and pass across the basolateral plasma membrane into the blood circulation. TSH-R, thyroid-stimulating hormone receptor.

pling of iodinated tyrosines to form thyroid hormones. TPO belongs to a class of mammalian peroxidase enzymes that include myeloperoxidase, lactoperoxidase, eosinophil peroxidase, and salivary peroxidase. Iodination of TG is also believed to free TG from binding proteins in the apical plasma membrane, thereby allowing iodinated TG to freely associate with follicular colloid.

Endocytosis of Thyroglobulin and Release of TH

Unbound and iodinated TG is taken up by pinocytosis across the apical plasma membrane upon stimulation of thyrocytes by TSH. Pinocytotic vesicles containing TG fuse with lysosomes where TG is cleaved proteolytically, and T_4 is released into the cytoplasm. Mono- and di-iodotyrosines formed during the proteolytic attack on TG are recovered by a specific transporter in lysosomal membranes, deiodinated by an iodotyrosine deiodinase in the cytosol, and recycled as inorganic iodine and tyrosine. Some T_4 is also deiodinated to form T_3.

Thyroglobulin Suppression of TSH Actions

Although the classic view of TG function is as a passive support matrix for synthesis of thyroid hormones, recent evidence supports a dynamic role for TG in regulating several aspects of TSH-induced gene expression. TG suppresses TSH-induced NIS expression and I^- transport as well as expression of the TPO, TG, and TSH receptor genes. TG does this by suppressing the expression of thyroid-specific transcription factors (TTF-1, TTF-2) that mediate TSH actions. The effects of TG on thyroid-specific gene expression may be mediated by a TG-receptor in the apical plasma membrane of thyrocytes. Thus, TG may play an important in negative feedback regulation of TSH actions in the thyroid.

Transport of Thyroid Hormones

T_4, the major form of thyroid hormone secreted by the thyroid gland, is transported in blood bound to three different classes of binding protein that each exhibit different affinities and capacities for T_4. Of the three binding proteins, thyroxine binding globulin (TBG) has the greatest affinity for T_4 and appears to be found only in mammals. Transthyretin (TTR), found in most vertebrates, shows a moderate affinity for T_4 but a greater capacity for T_4 binding than TBG. Finally, some T_4 is weakly bound to plasma albumin, which acts as a low-affinity, high-capacity transporter of T_4 in blood. Virtually all (> 99%) T_4 circulating in blood is bound to these plasma proteins, whereas only about 90% of the circulating T_3 is bound, making it easier for T_3 to enter target cells than T_4.

Metabolism of Thyroid Hormones

Approximately 50% of the circulating T_4 is converted to rT_3 in the liver by the process of deiodination. Deiodination is an important mechanism for both activating and deactivating thyroid hormones and is discussed further in a later section. Another important mechanism for thyroid hormone metabolism involves conjugation (via the phenolic hydroxyl group) with sulfate or glucuronic acid and rapid excretion in bile. The alanine side chain of either thyroid hormone may be converted to acetic acid, and the resultant molecules are excreted in the urine.

Disorders Associated with Abnormal Thyroid Hormone Synthesis

The inability to synthesize normal quantities of thyroid hormone is generally referred to as hypothyroidism. Two forms of hypothyroidism are commonly recognized. Congenital hypothyroidism is associated with a defective thyroid gland due to abnormal development of thyroid tissue or defects in thyroid hormone synthesis. Congenital hypothyroidism is usually detected shortly after birth and is treated by thyroid hormone administration. Endemic hypothyroidism results from inadequate dietary I$^-$ and is usually treated by dietary supplementation, generally in the form of iodized table salt. Endemic hypothyroidism is prevalent in underdeveloped countries with iodide-deficient soil. Mountainous areas of the United States as well as the lands surrounding the Great Lakes have iodine-deficient soils, but use of iodized salt protects people in these areas from endemic hypothyroidism. Natural animal populations may not fare so well, as exemplified by the presence of goitrous thyroids in Great Lakes fishes.

Insufficient thyroid hormone secretion during development can lead to abnormal growth of the nervous system and mental retardation. Endemic cretinism is a particularly severe form of mental retardation associated with I$^-$ deficiency during prenatal life. The neurological defects associated with endemic cretinism are generally more severe than observed in congenital hypothyroidism, presumably because the effects are manifested at earlier stages of embryogenesis.

Certain disorders result in an enlargement of the thyroid gland called a goiter. Goiters most commonly result from one or more defects in the thyroid hormone biosynthetic/secretion pathway that result in lower circulating T$_4$ levels. As a result, negative feedback on TSH secretion is reduced and TSH levels are elevated, leading to enlargement of the thyroid gland. Less common are goiters that form in hyperthyroid individuals. In these cases goiter formation is associated with elevated levels of thyroid hormone and reduced levels of TSH in the bloodstream. A special form of goiter may arise during pregnancy as a result of increased plasma TBG in the maternal circulation. This reduces circulating levels of free or unbound thyroid hormone, thus leading to elevated TSH secretion and an enlarged thyroid that returns to normal size after delivery.

Deiodination of Thyroid Hormones

Virtually all of the known physiological effects of thyroid hormones are carried out by T$_3$. Some of the T$_4$ is deiodinated to T$_3$ before leaving the thyroid gland. Approximately 40% of T$_4$ is converted in the liver by deiodinase enzymes to T$_3$, accounting for most of the T$_3$ in the circulation. After release from the thyroid gland, or liver T$_3$ and T$_4$ are transported via binding proteins in the blood circulation to the liver and to various target tissues. Also, T$_4$ commonly is converted to T$_3$ in the cytosol of target tissues before entering the nucleus and binding to thyroid receptors. The conversion of a hormone to a more active form at the site of action is known also for some steroid hormones in certain target tissues and ensures precise, tissue specific control of hormone action.

There are three major deiodinase enzymes in mammals, all encoded by separate genes (table 4-1). Two isozymes, 5'-deiodinases type I (5'DI) and type II (5'DII), convert T$_4$ to T$_3$. 5'DI, which is also capable of converting T$_4$ to rT$_3$, is

Table 4-1. Characteristics of mammalian deiodinase enzymes.

Property	Type I 5'-deiodinase	Type II 5'-deiodinase	Type III 5-deiodinase
Substrate preference	$rT_3 > T_4 > T_3$; inner or outer ring	$T_4 > rT_3$; outer ring	T_3(sulfate) > T_4; inner ring
Inhibitors	Thiouracil, iopanoic acid, iodoacetate, flavonoids	Thiouracils, iopanoate, iodoacetate, flavonoids	Thiouracil, iopanoate, flavonoids
Molecular weight	55 kDa	200 kDa	Unknown
Tissue distribution	Liver, thyroid, kidney, pituitary, CNS	CNS, pituitary, brown adipose tissue	Virtually every tissue except for liver, kidney, thyroid and pituitary
Subcellular distribution	Liver: RER, SER; kidney: basolateral plasma membrane		

Modified from Kohrle (1999). RER, rough endoplasmic reticulum; SER, smooth endoplasmic reticulum.

primarily within the liver, although it is also found in kidney, thyroid, and pituitary gland. Most of the T_3 produced in the liver enters the blood circulation. 5'DII is primarily in the central nervous system, placenta, skin, and in brown adipose tissue in rodents. This enzyme may also be responsible for supplying T_3 during fetal development, especially for the important development effects of T_3 on the fetal brain. The third deiodinase enzyme, 5-deiodinase type III (5D), removes iodine exclusively from the 5 (or 3) position of the inner phenolic ring resulting in rT_3 or diiodothyronine. These products lack affinity for nuclear thyroid hormone receptors and are therefore biologically inactive. 5D may suppress the biological activity of thyroid hormones by limiting availability of the T_4 prohormone. This enzyme is expressed in tissues that exhibit limited thermogenic responsiveness to T_3 (skin, placenta, and brain) but is absent from other important thyroid hormone target tissues such as liver, kidney, thyroid, and pituitary as well as from thyroid follicular cells.

All of the deiodinase enzymes are integral membrane proteins with different subcellular locations. In the liver, 5'DI is restricted to the smooth and rough endoplasmic reticulum, whereas in the kidney it is found in association with the basolateral plasma membrane of cells in the proximal tubule of the renal nephron. Thyrocyte 5'DI is stimulated by TSH via the cAMP/protein kinase A pathway, whereas liver, kidney, and pituitary 5'DI are stimulated by both T_4 and T_3. Deiodinase activity in these tissues is therefore sensitive to circulating T_4 and T_3 levels; hypothyroidism depresses activity in all tissues except for thyroid, where 5'DI is stimulated as a result of elevated TSH secretion produced by interruption of normal negative feedback mechanisms. Unlike 5'DI, the activity if 5'DII is suppressed by elevated levels of circulating T_4 but is relatively insensitive to circulating T_3 titers. 5'DII activity is stimulated by glucocorticoids and elevated cAMP levels.

Thyroid Hormone Action

Thyroid hormones act on virtually every cell in the body to influence gene expression. They do this by interacting with nuclear receptors that bind directly with elements in DNA called thyroid hormone response elements (TRE).

Thyroid Hormone Transport into Target Cells

Despite their hydrophobic core structure, thyroid hormones do not enter target cells by passive diffusion as once thought. There is now overwhelming evidence that thyroid hormones enter target cells by carrier-mediated or facilitated transport. Several transporter proteins, including the Na+/taurocholate cotransporting polypeptide (Ntcp, rat; NTCP, human) and members of the organic anion transporter protein (OATP) family, are capable of transporting thyroid hormones (table 4-2). It is important to note that none of these proteins can truly be called a "thyroid hormone transporter" because they all have broad substrate preferences. Ntcp was first identified as the protein responsible for Na+-dependent uptake of bile salts in liver, whereas the OATP family of proteins transports a wide range of amphipathic molecules (bile salts, unconjugated and conjugated steroids, cardiac glycosides, mycotoxins, leukotrienes) by Na+-independent mechanisms in liver, kidney, and the central nervous system.

Thyroid Hormone Receptors and Mechanism of Action

Thyroid hormone receptors (TR) belong to a large family of hormone-regulated (i.e., ligand-activated) transcription factors that includes receptors for sex steroid hormones, vitamin D, prostaglandins, retinoids, and fatty acids. Sequencing of the genes that encode TR has revealed two TR genes, α and β, that each encode unique receptors. Each TR (as well as other nuclear hormone receptors) has several

Table 4-2. Characteristics of plasma membrane proteins that transport thyroid hormones.

Transporter	Tissue Location	K_m (μM)		References
		T_3	T_4	
LST-1	Liver	2.7	3.0	Abe et al., 1999
Ntcp	Liver	+	+	Friesma et al., 1999
OATP1	Liver, kidney	+	+	Friesma et al., 1999
OATP2	CNS (hippocampus, cerebellum, choroid plexus), retina, liver	5.9	6.5	Abe et al., 1998
OATP3	Kidney, retina	7.3	4.9	Abe et al., 1998

K_m, Michaelis-Menten constant (substrate concentration at which initial reaction rate is half maximal); +, indicates uptake but K_m not determined. LST-1, human liver-specific organic anion transporter; Ntcp, Na+/taurocholate cotransporting polypeptide; OATP, (Na+-independent) organic anion transporting polypeptide.

functional domains: the A/B domain at the N terminus is responsible for transcription activation; the C domain is responsible for DNA binding; the D domain is believed to function as a molecular hinge; and the E domain at the C terminus is important for ligand binding (figure 4-4). The two TR genes share 70–80% homology in the C-E domains but no homology in the A/B domain. Each gene is also capable of producing two isoforms, TRα1 and TRα2 from the TRα gene and TRβ1 and TRβ2 from the TRβ gene. These isoforms result from differential processing of the TR mRNAs before their translation. The function of the TRα2 form is not known because it does not bind either T_3 or T_4. The other TR isoforms may mediate different physiological functions of thyroid hormones.

TRs differ from steroid receptors (see chapters 2 and 6) in that they normally exist bound to DNA within the nucleus, even in the absence of T_3. An interesting feature of this interaction is that unoccupied TRs are not inert; in fact, in the ab-

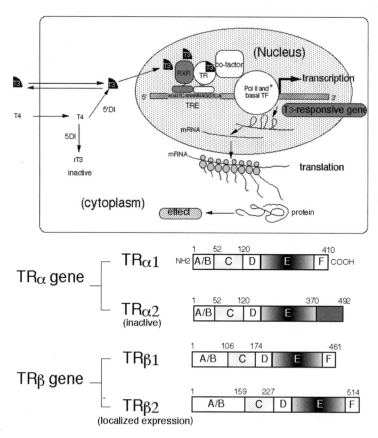

Figure 4-4. (Top) The intracellular events involved in thyroid hormone receptor signaling. (Bottom) Functional domains of thyroid hormone receptors. The D-E-F domain is the hormone binding domain. The C domain in the mid-region of the receptor serves as the DNA binding domain and is highly conserved. See text for abbreviations. Both panels reprinted with permission from Murata (1998).

sence of T_3, TRs may suppress transcription of the same genes that become activated upon T_3 binding. This complex regulation of gene transcription has to do with the dimerization state of the TR and the influence of several coregulatory proteins called coactivators and corepressors. Most TR exists as a heterodimer with the retinoid-X receptor (figure 4-5). TR can also associate with DNA as a homodimer (TR–TR) or as a monomer. Binding of T_3 stimulates dissociation of one or more corepressor proteins from the TR and promotes the association of coactivator proteins with the T_3–TR complex (figure 4-5). Formation of the coactivator complex facilitates histone acetyltransferase activity (HAT) within the chromatin structure, thereby allowing greater accessibility of DNA to TR and other transcription factors. Unoccupied TR allows for the recruitment of corepressor proteins that pro-

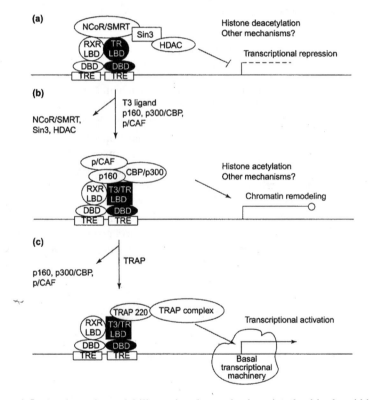

Figure 4-5. A schematic model illustrating the mechanisms involved in thyroid hormone receptor (TR) activation and transcription. In the absence of T3 (ligand), transcription is repressed due to the association of the retinuic acid receptor (RXR)-TR heterodimer and its association with a corepressor complex consisting of NCoR/SMRT, Sin3, and HDAC. When T3 binds to the TR, the receptor undergoes a conformational change and associates with a coactivator complex consisting of several proteins including p/CAF and CBP/p300, preparing the chromatin for transcriptional activity. The T3/TR/RXR complex then associates with thyroid hormone-associated protein (TRAP), which activates transcription. See text for complete abbreviations. Reprinted with permission from Wu and Koenig (2000).

mote histone deacetylation and a tighter chromatin structure that prevents DNA interaction with transcription factors (figure 4-5).

Coactivators interact with a 9-amino-acid sequence in the C termini of TRs called the activator function 2 (AF-2) domain. This domain is critical for ligand-dependent gene activation by many nuclear receptors. Several coactivator and corepressor proteins have been identified (table 4-3). Steroid coactivator-1 belongs to a family of 160-kD$_a$ proteins (p160 family) that act as nuclear receptor coactivators. All contain the amino acid sequence LXXLL (where L is the amino acid leucine and X is any amino acid), a region that mediates binding to nuclear receptors. These coactivator proteins require T_3 binding and an AF-2 domain on the TR in order to function. Other coactivator proteins are listed in table 4-3. SRC-1 and other co-activator proteins possess intrinsic HAT activity. Corepressors suppress T_3/TR-mediated gene transcription indirectly by interacting with a protein called Sin3A that activates histone deacetylase activity (figure 4-5).

Thyroid Hormone Actions on Heat Production and Energy Metabolism

The ability of thyroid hormones to increase metabolic rate in birds and mammals has been known for more than 100 years. Hypothyroid mammals undergo a 20–30% reduction in basal metabolic rate (BMR), whereas hyperthyroidism is associated with an approximately 50% increase in BMR. Thyroid hormones increase oxygen consumption in virtually every organ in the body (with the exception of the brain, spleen, and testes) by influencing cellular respiration via a number of long- and short-term actions on mitochondrial energy transfer. Long-term effects involve alterations in genomic gene transcription and synthesis of mitochondrial proteins involved in the respiratory pathway as well as possible effects on mitochondrial membrane lipid composition. Short-term effects on mitochondrial function may be mediated via direct action of T_3 on mitochondrial TRs, although the evidence for a direct action of T_3 on mitochondria

Table 4-3. Coactivators and corepressors of T_3/TR-mediated gene transcription.

Name	Abbreviation
Coactivators	
CREB-binding protein	CBP
p300/CREB binding protein (CBP)-associated factor	PCAF
PPAR gamma coactivator 1	PGC-1
Steroid receptor coactivator-1	SRC-1
Transcription intermediary factor 1	TIF1
Thyroid hormone receptor and Rb interacting protein of 230 kD	TRIP230
Corepressors	
Nuclear corepressor	N-COR
Silencing mediator of retinoid and thyroid receptors	SMRT

CREB, cyclic adenosine monophosphate (cAMP) response element binding protein; PPAR, peroxisome proliferator-activated receptor; Rb, retinoblastoma protein.

is controversial. High concentrations of T_3 are required for direct action on mitochondria, and the effects are not specific. For example, 3,5-diiodothyronine, which is virtually devoid of thyroid hormone action in vivo, mimics the direct actions of T_3 on mitochondria. What is relatively certain is that T_3 does not increase BMR by simply uncoupling mitochondrial ATP production, as once thought, as the concentrations required for this effect exceed physiological levels of the hormone.

Thyroid hormone-dependent elevations in metabolic activity are fueled by enhanced production and utilization of energy substrates such as fatty acids. Thyroid hormones also influence metabolism by potentiating the actions of other metabolic-stimulating hormones such as GH, epinephrine, and glucagon. They stimulate hepatic fatty acid synthesis and have a permissive role in enhancing catecholamine-induced lipolysis. Fatty acids constitute the principal energy source used in hyperthyroid individuals. The accelerated lipogenesis–lipolysis cycle stimulated by thyroid hormones appears, at first glance, to serve no other role than to maintain steady-state levels of fatty acids (the so-called futile cycling of fatty acids). However, this recycling of energy substrate is energetically expensive and may contribute up to 15% of the total heat production accounted for by thyroid hormone activity. Thyroid hormones influence carbohydrate metabolism as well, as increased production and utilization of glucose occurs in hyperthyroidism. Thyroid hormones may influence gluconeogenesis via receptor-dependent effects on the expression of genes encoding gluconeogenic enzymes such as phosphoenolpyruvate carboxykinase.

Within a range of environmental temperatures called the thermoneutral range, homeotherms maintain a constant body temperature without adjustments to BMR. Maintenance of a constant body temperature below the thermoneutral range requires additional heat production (nonshivering thermiogenesis). In small mammals and neonatal humans, the principal site for nonshivering thermiogeneis is brown adipose tissue (BAT). Thyroid hormones are important regulators of BAT thermiogenesis; nuclear TRs are more numerous in BAT than any other tissue except for the pituitary. BAT expresses the deiodinase enzyme 5'DII, which rapidly converts T_4 to T_3. Thyroid hormones are important for the expression of a unique enzyme in BAT, uncoupling protein (UCP) or thermogenin, a nuclear-encoded mitochondrial protein that uncouples ATP synthesis and respiration, producing heat. T_3 stimulates BAT heat production by acting on nuclear TRs to increase transcription of the UCP gene. The importance of BAT for cold adaptation in adults is not clear. The recent discovery of UCP in other tissues suggests that T_3-mediated changes in UCP gene expression may be a common form of thyroid hormone action.

Thyroid Hormones and Nervous System Development

Hypothyroidism during development can result in severe neurological deformations. Although maternal thyroid hormones may cross the placenta and enter the fetal circulation, they cannot replace deficits in fetal thyroid hormone production. Fetal thyroid hormone synthesis is required for normal neuronal growth and differentiation, synthesis of neurotransmitters and synaptic machinery, maturation of neuronal metabolic pathways, normal glial development, and production of myelin. A summary of thyroid hormone actions on the developing CNS is presented in figure 4-6.

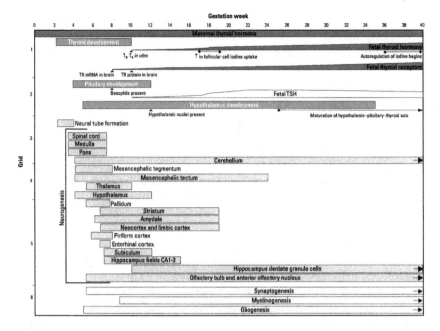

Figure 4-6. The temporal relationship between thyroid function and brain development in humans. Reprinted with permission from Howdeshell (2002).

The actions of thyroid hormone on the central nervous system appear to take place at key times during development that differ depending on the species. In rats, which have an underdeveloped nervous system at birth, the critical period for thyroid hormone action is during the early postnatal period. In humans the critical period is during late intrauterine life. Hypothyroidism during these critical periods of central nervous system development can have long-lasting deleterious effects on learning and memory, auditory function, and neuroendocrine control of growth and reproduction. Congenital hypothyroidism may also be a cause of certain forms of epilepsy and dwarfism (table 4-4). These effects probably result from deficiencies in the development of one or more key brain areas. Brain areas most sensitive to neo- or prenatal hypothyroidism include the cerebellum, auditory neocortex, striatum, olfactory bulbs, and hippocampus, a brain area that is critical for short-term memory processing.

It is generally assumed that thyroid hormone actions on the developing central nervous system are mediated by nuclear TRs. TRα1 is the most abundant TR in fetal brain, accounting for more than 60–70% of total nuclear radiolabeled T_3 binding. Roughly 30–40% of the remainder is bound to TRβ1 and TRβ2. Despite the fact that TRβ1 accounts for only a small percentage of total T_3 binding in the fetal brain, a great deal of attention has been directed at this receptor because its appearance during development coincides with a prenatal surge in fetal T_3 levels.

Table 4-4. Thyroid hormone actions on the developing nervous system.

Cellular actions
    Increase synaptogenesis and neurotransmitter synthesis
    Increase neurite outgrowth
    Required for normal cell migration
    Required for Purkinje cell differentiation
    Required for normal oligodendrocyte differentiation and myelin synthesis
Abnormalities associated with neo- or prenatal hypothyroidism
    Mental retardation
    Cretinism
    Hearing loss
    Increased audiogenic seizure susceptibility, lower threshold for electroconvulsive shock

Comparative Endocrinology of the Thyroid Gland

All adult craniates possess thyroid follicles, although the degree of follicular organization varies. The anatomical location of the thyroid gland in various vertebrates is shown in figure 4-7. In hagfishes, lampreys, and most teleosts, follicular tissue is not organized into a discrete thyroid gland. Rather, thyroid follicles are scattered throughout the pharyngeal area. In adult hagfishes and lampreys, thyroid follicles are loosely scattered in the connective tissue lining the ventral midline of the body cavity, rostral to the ventral aorta. Adult lampreys possess thyroid follicles that lack colloid; thyroid hormones are stored intracellularly. Larval lampreys (ammocoetes) do not possess thyroid follicles but have an endostyle, a structure used for filter feeding located ventral to the mouth (figure 4-8). The endostyle is capable of trapping I^- and synthesizing thyroid hormones. Although the structure of the endostyle is quite different from that of the thyroid follicle, this primitive organ probably represents the ancestral condition of the thyroid gland, as the only other organisms in which it is found are cephalo- and urochordates. At metamorphosis, the lamprey endostyle forms the adult's thyroid follicles.

Elasmobranchs have compact thyroid embedded in the connective tissue of the lower jaw, whereas in most bony fishes follicles are found scattered in the connective tissue surrounding the ventral aorta. Exceptions to this general rule are the Hawaiian parrotfish (*Scarus dubius*) and certain species of tuna. In these fishes, thyroid follicles exist in compact lobes surrounding the branchial arteries. In amphibians, thyroid tissue forms a paired gland that attaches to the hyoid cartilage. A similar pattern is seen in birds, where the paired thyroid glands attach to the carotid arteries. In reptiles, a single compact gland is found attached to the trachea.

The nonmammalian thyroid gland does not have parafollicular cells. In all vertebrates except for mammals, these calcitonin-producing cells are localized to a structure called the ultimobranchial gland. This gland has a follicular architecture that resembles that of the thyroid gland. The anatomical location of the ultimobranchial gland in various vertebrates is shown in figure 4-9.

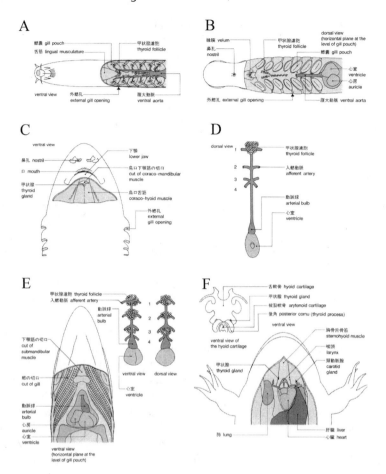

Figure 4-7. Anatomical location of the thyroid gland in representative vertebrates. A, hagfish, *Eptatretus burgeri*; B, lamprey, *Lampetra japonica*; C, shark, *Triakis scyllium*; D, eel, *Anguilla japonica*; E, salmon, *Onchorhynchus masou masou*; F, bullfrog, *Rana catesbeiana*; G, bullfrog tadpole; H, tortoise, *Clemmys japonica*; I, lizard, *Takydromus tacchydromoides*; J, snake, *Rhabdophis tigrinus tigrinus*; K, Japanese quail; *Coturnix coturnix japonians*; L, rat. Reprinted with permission from Matsumoto and Ishii (1992).

Comparative Aspects of Thyroid Hormone Synthesis, Transport, and Deiodination

Thyroid hormone synthesis in most species is thought to occur much as it does in mammals, with the exception of lampreys. In these organisms thyroid hormones are produced and stored intracellularly. The enzymatic machinery required to iodinate protein-bound tyrosine appears to be phylogenetically ancient, as TPOs are present in the urochordate endostyle. TG structure is conserved in those vertebrates in which the protein has been purified or the gene for TG cloned. Nonmammalian thyroid hormones

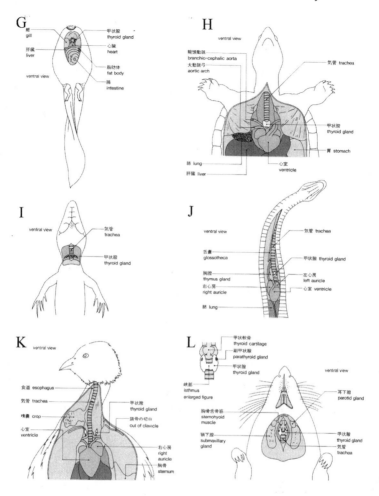

are structurally identical to their mammalian counterparts. As a result, techniques (such as radioimmunoassay and high-performance liquid chromatography) for measuring thyroid hormones in mammalian species can be applied successfully to nonmammals.

As in mammals, in other animals thyroid hormones are transported in the blood attached to transport proteins. Transthyretin mRNAs have been sequenced in several teleost, amphibian, reptilian, and avian species, suggesting that a transthyretinlike binding protein has been conserved evolutionarily. Both T_3 and T_4 are present in the circulation of all craniates studied so far. Deiodination is presumed to be an important means of hormone activation and deactivation, as in mammals; homologs of the mammalian 5'DI, 5'DII, and 5D are present in nonmammalian tetrapods and teleosts. In contrast to rats, 5'DI activity in at least some teleost species is relatively insensitive to propylthiouracil, a potent inhibitor of the mammalian enzyme. Although no deiodinase enzyme has yet been cloned in cartilaginous or jawless fishes, pharmacological and physiological studies indicate that elasmobranchs and lampreys possess

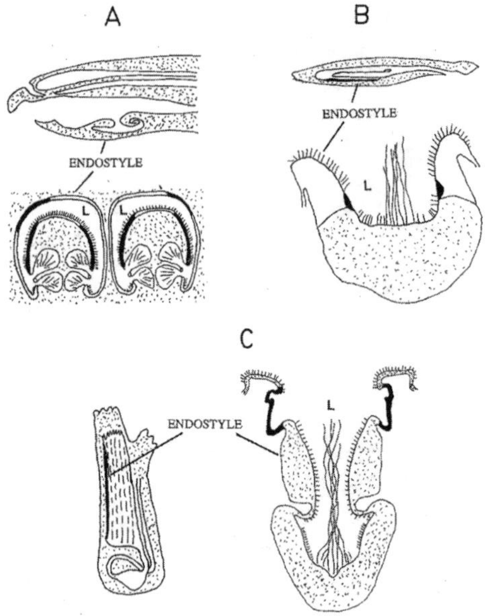

A   B

C

Figure 4-8. Structure of the endostyle in the ammocoete larva of a
lamprey (A), amphioxus, a cephalo-
chordate (B), and a tunicate (C).
Reprinted with permission from Eales
(1997).

an outer-ring deiodinase capable of converting T_4 to T_3. In addition, lampreys possess at least one intestinal inner-ring deiodinase that may be responsible for the drop in circulating T_4 levels that accompanies metamorphosis in these animals. Lampreys are the only vertebrates exhibiting a distinct metamorphosis that is prevented by thyroid hormones rather than induced by them.

The function of the hypothalamo–hypophysial–thyroid axis has been essentially conserved in the sense that thyroid hormone secretion is ultimately controlled by TSH in all vertebrates except cyclostomes. Surgical removal of the pituitary gland (and presumably the source of TSH) has no effect on thyroid function in these jawless fishes. Furthermore, lampreys are the only vertebrates that do not respond to mammalian TSH. There are more dramatic species differences with respect to hypothalamic control of TSH. For example, dopamine may be more important than TRH in the control of TSH secretion in teleost fishes. In teleosts, larval amphibians, and hatchling birds, hypothalamic corticotropin-releasing hormone (CRH) may also be an important stimulator of pituitary TSH secretion.

Thyroid Hormone Action in Nonmammals

Attempts to influence oxygen consumption with thyroid hormones in anamniotes have met with mixed results, and there is no general agreement as to the role of the thyroid in regulating energy metabolism in these organisms. Thyroid hormones are important for maintaining standard metabolic rate in reptiles, and they may be involved in temperature acclimation as well. Thyroid hormones generally have the same effects on energy metabolism in birds as they do in mammals, increasing BMR and tissue oxygen consumption.

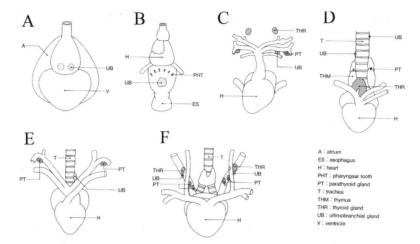

Figure 4-9. Anatomical location of ultimobranchial glands in various vertebrates. A, sting-ray; B, goldfish; C, newt; D, rat snake; E, lizard; F, starling. Reprinted with permission from Matsumoto and Ishii (1992).

Thyroid hormones have an evolutionarily ancient role in coordinating developmental events. Maternal thyroid hormones are deposited into oocytes before ovulation in bony fishes, amphibians, and bird species, suggesting actions on early embryonic development. Thyroid hormones are important for fin and limb development in postembryonic bony fishes and amphibians, respectively, and play an especially critical role in anuran and teleost metamorphosis (see below). Thyroid hormones are required for normal somatic growth in reptiles and birds and appear to be critical for the development of endothermy in altricial bird species. There is no known role for thyroid hormones in elasmobranch development.

Thyroid hormone secretion is an important seasonal cue for annual reproductive, migratory, and molting events in many species and has been especially well documented in seasonally breeding birds. Through a complex interaction with photoperiod and temperature, thyroid hormones influence seasonal gonadal growth, body weight, fat deposition, and postnuptial molt. Thyroid hormones also have been implicated in seasonal molting in certain mammalian, reptilian, and amphibian species. Thyroid hormones appear to have especially important roles in the reproductive cycles of seasonally breeding fish. (The role of the thyroid in reproduction is considered in chapter 6.)

Metamorphosis

As part of their normal life-history pattern, some fishes and most amphibian species undergo a series of dramatic morphological, physiological, and biochemical changes known as metamorphosis. In most cases, metamorphosis is associated with a transition from one ecological niche to another—for example, from an aquatic larval form to a semi- or fully terrestrial form (amphibians) or from a freshwater to a marine environment (smoltification in salmon). Thyroid hormones are required for metamorphosis in amphibians and fishes but appear to play an inhibitory role in lamprey metamorphosis.

Amphibian Metamorphosis: A Model for Studying Developmental Aspects of Thyroid Hormone Action

The importance of thyroid hormones in amphibian development is unparalleled in any other vertebrate group. In most amphibian species, thyroid hormones initiate and maintain the massive reprogramming of gene expression that accompanies the transition from aquatic to terrestrial environments. These actions are particularly dramatic during anuran metamorphosis, where thyroid hormones are required for resorption of the tail, reorganization of the gastronintestinal tract, limb development, restructuring of the cranial skeleton, and a switch in hepatic nitrogen metabolism from ammonia to urea production (figure 4-10).

The amphibian TRs that have been cloned thus far are similar in structure and function to their mammalian counterparts. Consequently, there is every reason to believe that receptor-mediated gene transcription is the major form of thyroid hormone action during metamorphosis. Several theories have been proposed to explain how thyroid hormones act on a limited variety of receptors to produce such varied and dramatic tissue reorganization (figure 4-11). In the most commonly accepted scenario, thyroid hormones directly influence the expression of immediate early response genes whose products, in turn, affect the expression of late response genes,

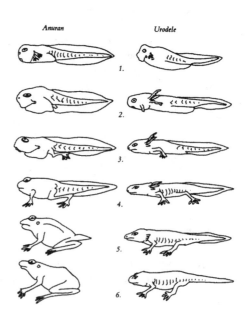

Figure 4-10. A comparison of the morphological changes that occur during metamorphosis of an anuran (left panel) and a urodele. Reprinted with permission from Norris (1997).

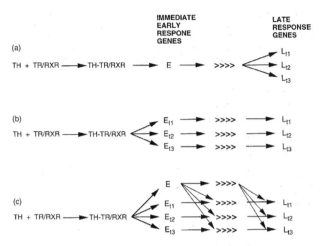

Figure 4-11. Schematic view of gene cascades that may be initiated by the triiodothyronine/thyroid hormone receptor/retinoic acid receptor (T3/TR/RXR) complex during amphibian metamorphosis. Genes (E, early response; L, late response) with numbered subscripts refer to tissue-specific genes. For example, Et1, Et2, and Et3 are genes that become activated in three different tissues, whereas gene "E" is ubiquitously expressed. Reprinted with permission from Shi (1999).

resulting in a gene regulation cascade (figure 4-12). Early thyroid hormone response genes are those responding (altered mRNA levels) to thyroid hormone within 24 h of treatment, whereas late-response genes may not show evidence of enhanced transcription for up to 1 month after treatment. Many of the early thyroid hormone response genes encode proteins that act as transcription factors (including the TRβ gene) by either repressing or activating the expression of late-response genes.

A number of hormones are capable of either facilitating or inhibiting the action of thyroid hormones on amphibian metamorphosis. Corticosteroids (see chapter 5) generally accelerate thyroid hormone-induced metamorphosis, although the effects are stage and tissue dependent. Corticosteroid levels in blood are elevated during metamorphosis, and receptors for these steroids are present in a number of larval organs, including tail and liver. The pituitary hormone prolactin (PRL) is a potent inhibitor of thyroid hormone-induced metamorphosis, although its physiological role is unclear because PRL and PRL receptor gene expression are both upregulated during metamorphosis.

Some urodele species never undergo metamorphosis and are capable of breeding as larvae. These organisms exhibit a profound retention of larval characteristics generally referred to as neoteny. Two forms of amphibian neoteny are recognized. In facultative neoteny, larvae undergo spontaneous metamorphosis, but only under favorable environmental conditions. Species exhibiting this form of neoteny include the tiger salamander (*Ambystoma tigrinum*) and the Mexican axolotl (*Ambystoma mexicanum*). Other species never undergo spontaneous metamorphosis under natural conditions, a condition referred to as obligate neoteny. The mudpuppy *Necturus*

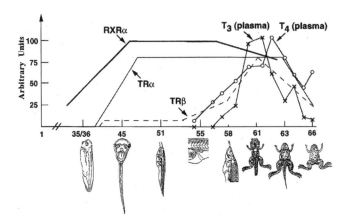

Figure 4-12. Temporal relationships among developmental stage, thyroid hormone receptor (TR) expression, and plasma levels of T3 and T4. RXR, retinoid X receptor. Reprinted with permission from Shi (1999).

maculosus does not respond to thyroid hormone treatment, possibly due to a lack of TRβ expression. Neoteny has not been observed in any anuran species.

Thyroid hormone secretion is elevated during amphibian metamorphosis, correlating with forelimb emergence (anurans only), tail (or tail-fin) resorption, and expression of at least one isoform of TR, TRβ (figure 4-11). Deiodinase activity shifts from reliance on D'II to D'I, resulting in more circulating T_3. Early attempts to induce thyroid hormone secretion and metamorphosis with TRH were unsuccessful, suggesting that control of the pituitary–thyroid axis in larval amphibians was very different from what occurs in mammals. Work by Denver and colleagues (see Shi, 1999) has shown that CRH, and not TRH, is the principal hypothalamic cue triggering the pituitary-thyroid axis during metamorphosis. An interesting feature of this control is that exposure to certain stressors may trigger metamorphosis, presumably through activation of the pituitary–thyroid axis (see chapter 3). Many other environmental factors have been implicated in the control of metamorphosis, including light, temperature, diet, iodide availability, and crowding. Whether any or all of these factors influence metamorphosis through the pituitary–thyroid axis is likely but unknown.

References

Abe, T., Kakyo, M., Sakagami, H., Tokui, T., Nishio, T., Tanemoto, M., Nomura, H., Hebert, S.C., Matsuno, S., Kondo, H., Yawo, H., 1998. Molecular characterization and tissue distribution of a new organic anion transporter subtype (oatp3) that transports thyroid hormone and taurocholate and comparison with oatp2. J. Biol. Chem. 273, 22395–22401.

Abe, T., Kakyo, M., Tokui, T., Nakagomi, R., Nishio, T., Nakai, D., Nomura, H., Unno, M., Suzuki, M., Naitoh, T., Matsuno, S., Yawo, H., 1999. Identification of a novel gene family encoding human liver specific organic anion transporter LST-1. J. Biol. Chem. 274, 17159–17163.

Bolander, F.F., 1989. Molecular Endocrinology Academic Press, San Diego, CA.

Eales, J.G., 1997. Iodine metabolism and thyroid-related functions in organisms lacking thyroid follicles: Are thyroid hormones also vitamins? Proc. Soc. Exp. Biol. Med. 214, 302–317.

Friesema, E.C., Docter, R., Moerings, E.P., Stieger, B., Hagenbuch, B., Meier, P.J., Krenning, E.P., Hennemann, G., Visser, T.J., 1999. Identification of thyroid hormone transporters. Biochem. Biophys. Res. Commun. 254, 497–501.

Howdeshell, K.L., 2002. A model for development of the brain as a construct of the thyroid system. Environ. Health Perspect. 110: 337–348.

Köhrle, J., 1999. Local activation and inactivation of thyroid hormones: the deiodinase family. Mol. Cell. Endocrinol. 151, 103–119.

Matsumoto, A., Ishii, S. (Eds.), 1992. Atlas of Endoerine Organs. Springer Verlag, New York.

Murata,Y., 1998. Multiple isoforms of thyroid hormone receptor: An analysis of their relative contribution in mediating thyroid hormone action. Nagoya J. Med. Sci. 61, 103–115.

Norman, A.W., Litwack, G., 1998. Hormones, 2nd edition. Academic Press, San Diego, CA.

Norris, D.O., 1997. Vertebrate Endocrinology. Academic Press, San Diego, CA.

Rodriguez, A.M., Perron, B., Lacroix, L., Caillou, B., Leblanc, G., Schlumberger, M., Bidart, J.M., Pourcher, T., 2002. Identification and characterization of a putative human iodide transporter located at the apical membrane of thyrocytes. J. Clin. Endocrinol. Metab. 87, 3500–3503.

Shi, Y.-B., 1999. Amphibian Metamorphosis. From Morphology to Molecular Biology. Wiley-Liss, New York.

Wu, Y., Koenig, R.J., 2000. Gene regulation by thyroid hormone. Trends Endocrinol. Metab. 6, 207–211.

Selected Readings

Apriletti, J.W., Ribeiro, R.C.J., Wagner, R.L., Feng, W., Webb, P., Kushner, P.J., West, B.L., Nilsson, S., Scanlan, T.S., Fletterick, R.J., Baxter, J.D., 1998. Molecular and structural biology of thyroid hormone receptors. Clin. Exp. Pharmacol. Physiol. 25, S2–S11.

Christ, M., Haseroth, K., Falkenstein, E., Wehling, M., 1999. Nongenomic steroid actions: Fact or fantasy? Vit. Horm. 57, 325–373.

Freake, H.C., Oppenheimer, J.H., 1995. Thermogenesis and thyroid function. Annu. Rev. Nutr. 15, 263–291.

Goglia, F., Moreno, M., Lanni, A., 1999. Action of thyroid hormones at the cellular level: The mitochondrial target. FEBS Lett. 452, 115–120.

Koenig, R.J., 1998. Thyroid hormone receptor coactivators and corepressors. Thyroid 8, 703–713.

Nilsson, M., 1999. Molecular and cellular mechanisms of transepithelial iodide transport in the thyroid. BioFactors 10, 277–285.

Oppenheimer, J.H., Schwartz, H.L., 1997. Molecular basis of thyroid hormone-dependent brain development. Endocr. Rev. 18, 462–475.

Rodriguez-Pena, A., 1999. Oligodendrocyte development and thyroid hormone. J. Neurobiol. 40, 497–512.

Royaux, I.E., Suzuki, K., Mori, A., Katoh, R., Everett, L.A., Kohn, L.D., Green, E.D., 2000. Pendrin, the protein encoded by the Pendred syndrome gene (PDS), is an apical porter of iodide in the thyroid and is regulated by thyroglobulin in FRTL-5 cells. Endocrinology 141, 839–845.

Silva, J.E., 1995. Thyroid hormone control of thermiogenesis and energy balance. Thyroid 5, 481–492.

Suzuki, K., Mori, A., Lavaroni, S., Ullianich, L., Miyagi, E., Saito, J., Nakazato, M., Pietrarelli, M., Shafran, N., grassadonia, A., Kim, W.B., Consiglio, E., Formisano, S., Kohn, L.D., 1999. Thyroglobulin regulates follicular function and heterogeneity by suppressing thyroid-specific gene expression. Biochimie 81, 329–340.

Suzuki, K., Mori, A., Saito, J., Moriyama, E., Ullianich, L., Kohn, L.D., 1999. Follicular thyroglobulin suppresses iodide uptake by suppressing expression of the sodium/iodide symporter gene. Endocrinology 140, 5422–5430.

Taurog, A., 1999. Molecular evolution of thyroid peroxidases. Biochimie 81, 557–562.

Venkatesh, S.G., Deshpande, V., 1999. A comparative review of the structure and biosynthesis of thyroglobulin. Comp. Biochem. Physiol. Part C 122, 13–20.

Wolff, J., 1998. Perchlorate and the thyroid gland. Pharmacol. Rev. 50, 89–105.

5

The Adrenal Glands

James A. Carr
David O. Norris

The adrenal glands play a vital role in maintaining homeostasis and coping with stress. These paired organs lie on the superior aspect of each kidney, surrounded by adipose tissue (figure 5-1). Each gland consists of an outer cortical region and an inner medullary region that differ in embryonic origin and function. Ceolomic mesoderm in the urogenital ridge gives rise to the cortex, whereas the medulla is composed of cells that migrate from the neural crest, the same tissue that gives rise to sympathetic neurons and melanocytes.

The adult cortex, part of the hypothalamus–pituitary–adrenal (HPA) axis (see chapter 3), is organized into three major functional zones: the zona glomerulosa (ZG), the zona fasciculata (ZF), and the zona reticularis (ZR; figure 5-2). Cells in each zone synthesize and secrete steroid hormones (corticosteroids) important for regulating electrolyte balance (the mineralocorticoid hormones) and carbohydrate metabolism (the glucocorticoid hormones). In mammals, the ZG is the primary source of the mineralocorticoid hormone aldosterone (ALDO); the ZF secretes the glucocorticoid hormones cortisol and corticosterone. The ZR is responsible for producing the weak-acting androgens dihydroepiandrosterone (DHEA) and androstenedione (ASD).

The fetal adrenal is essential for maintenance of pregnancy and for important changes related to birth. It consists mainly of a special fetal zone that synthesizes both adrenal androgens and glucocorticoids. The fetal zone regressess soon after birth, and the adrenal gland remains relatively inactive, growing slowly until it reaches adult size at puberty.

Although derived from neural crest cells, adrenomedullary cells do not resemble neurons and do not normally exhibit dendrites or axons. However, they do share a number of functional similarities with other neural crest-derived cells related to the

111

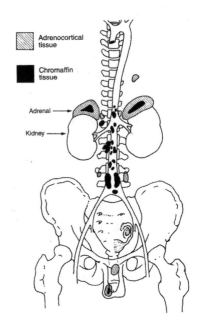

Figure 5-1. The location of the adrenal glands and extra-adrenal chromaffin tissue in humans. Reprinted with permission from Norris (1997).

use of tyrosine for biosynthesis of catecholamines (sympathetic neurons) and the pigment melanin (melanocytes; figure 5-3).

The adrenal cortex and medulla are functionally linked by a specialized vascular supply. The adrenal gland is well vascularized, receiving a disproportionate share of total cardiac output despite its small size (approximately 0.02% of total body weight). The main blood supply to the gland travels centripetally from the outer connective tissue capsule through sinusoids in the cortex and into the medulla, supplying blood that is rich in corticosteroids to cells of the medulla. Exposure of medullary cells to glucocorticoids is required for normal expression of phenylethanolamine-N-methyltransferase (PNMT), the enzyme responsible for the final step of epinephrine biosynthesis (see chapter 1).

Adrenal Steroidogenesis

Corticosteroids are synthesized from cholesterol that is derived from (in decreasing order of importance) receptor-mediated uptake of lipoproteins, intracellular stores of esterified cholesterol or de novo synthesis from acetate. As with gonadal steroids, corticosteroids are not stored but are synthesized upon demand in response to regulatory hormones (corticotropin, ACTH or angiotensin II). Corticosteroids readily diffuse out of adrenocortical cells after synthesis.

Many of the steroidogenic enzymes belong to a large family of P450 enzymes localized to the mitochondria and smooth endoplasmic reticulum. These enzymes

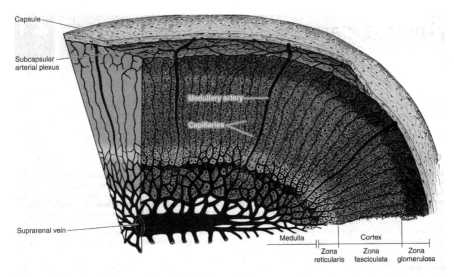

Figure 5-2. Architecture of and blood flow in the mammalian adrenal gland. Reprinted with permission from Junqueira et al. (1998).

are named for the carbon position modified by their action, whereas the genes encoding these enzymes are given the prefix *CYP*. For example, P450 21-hydroxylase (P450$_{C21}$) catalyzes the addition of a hydroxyl group to C-21 of progesterone, creating deoxycorticosterone. The corresponding gene directing the synthesis of P450$_{C21}$ is *CYP21A2*. The nomenclature of steroidogenic enzymes is presented in table 5-1.

The first step in corticosteroid synthesis is the transfer of cholesterol across the intermembrane space to the inner mitochondrial membrane, where P450$_{SCC}$ removes the 6-carbon side chain of cholesterol, forming pregnenolone (SCC = side-chain cleaving). Transfer of cholesterol across the aqueous environment of the mitochondrial intermembrane space is carried out by the steroidogenic acute regulatory (StAR) protein (figure 5-4). Although it is generally agreed that P450$_{SCC}$ is the rate-limiting enzyme in corticosteroid synthesis, transfer of cholesterol by the StAR protein is the true rate-limiting step in the acute phase of this process.

In adrenocortical cells, the fate of pregnenolone is determined by the differential expression of two enzymes in the smooth endoplasmic reticulum that compete for pregnenolone as a substrate. Pregnenolone is directed down the Δ^5 pathway by 3β-hydroxysteroid dehydrogenase (3β-HSD) and down the Δ^4 pathway by P450$_{C17}$ (figure 5-5). In the ZG and the ZF, relatively high levels of 3β-HSD and a lack of P450$_{C17}$ result in pregnenolone being converted to progesterone. Progesterone serves as a precursor for ALDO in the ZG and cortisol or corticosterone, in the ZF. In the ZR, low levels or absence of 3β-HSD and expression of P450$_{C17}$ result in most of the pregnenolone being converted to DHEA. Some species, such as rats, lack adrenal

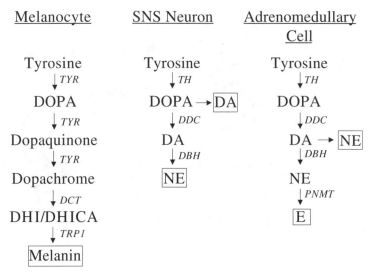

Figure 5-3. Comparison of biosynthetic fate of tyrosine in adrenomedullary cells and related cell types derived from neural crest. Biosynthetic end-products are shown in boxes and enzymes are shown in italics. Melanocytes use tyrosine as a biosynthetic precursor for melanin. Both sympathetic nervous system (SNS) neurons and adrenomedullary cells use tyrosine as a biosynthetic precursor for catecholamine synthesis. Adrenomedullary cells have an additional enzyme, the glucocorticoid-dependent enzyme phenylethanolamine-N-methyl transferase (PNMT), that carries out the conversion of norepinephrine (NE) to epinephrine (E). DBH, dopamine β-hydroxylase; DCT, dopachrome tautomerase; DDC, DOPA decarboxylase; DHI/DHICA, 5,6-dihydroxyindole-2-carboxylic acid; DOPA, 3,4-dihydroxyphenylalanine; TH, tyrosine hydroxylase; TRP1, tyrosinase-related protein 1; TYR, tyrosinase.

expression of $P450_{C17}$ altogether. In these species, there are no appreciable levels of adrenal androgens circulating in the blood.

Differences Between Adrenal and Gonadal Steroid Synthesis

Although the initial formation of pregnenolone from cholesterol occurs in adrenocortical cells much as it does in gonadal tissue, corticosteroids differ significantly from gonadal steroids with respect to other aspects of biosynthesis, structure, and biological activities. The adrenal cortex has enzymes ($P450_{C21}$, $P450_{C11}$ and $P450_{aldo}$) required for synthesis of C_{21} corticosteroids—enzymes not found in gonadal steroidogenic tissue (figure 5-5). $P450_{C21}$ is restricted to the ZG and ZF, where it catalyzes the addition of a hydroxyl group to C-21 of progesterone, creating deoxycorticosterone. Deoxycorticosterone, in turn, acts as a substrate for $P450_{aldo}$ in the ZG and $P450_{C11}$ in the ZF. These enzymes are 93% identical, and both carry out 11β-hydroxylation of deoxycorticosterone. $P450_{aldo}$ has the additional ability to carry out 18-hydroxylation and 18-oxidase steps required for the synthesis of ALDO from 11-deoxycorticosterone in the ZG.

Table 5-I. Nomenclature and intracellular location of steroidogenic enzymes.

Enzyme	Abbreviation	Gene[a]	Tissue Location	Intracellular Location	Enzyme Substrate
Cholesterol side-chain cleavage enzyme	P450$_{scc}$	CYP11A	Gonads, adrenals	Mitochondria	Cholesterol
3β-Hydroxysteroid dehydrogenase Δ5-Δ4 isomerase	3β-HSD	HSD3B[b]	Gonads, adrenals	SER	Pregnenolone, DHEA
P450 17α-hydroxylase/17-20 lyase	P450$_{C17}$	CYP17	Gonads, adrenals	SER	Pregnenolone, 17α-hydroxypregnenolone, 17α-hydroxyprogesterone
P450 21-hydroxylase	P450$_{C21}$	CYP2 1A2[c]	Adrenals	SER	Progesterone, 17α-hydroxyprogesterone
P450 11β-hydroxylase	P450$_{C11}$	CYP11B1	Adrenals	Mitochondria	Deoxycorticosterone
Aldosterone synthase	P450$_{aldo}$	CYP11B2	Adrenals	Mitochondria	Deoxycorticosterone, corticosterone
5α-reductase	5αR	SRD5A[b]	Many androgen target tissues, including gonads	Cytoplasm	Testosterone
Aromatase	P450$_{aro}$	CYP19	Gonads	SER	Testosterone
17β-hydroxysteroid dehydrogenase	17β-HSD	HSD17B[b]	Gonads	Mitochondria	Androstenedione, estrone, estradiol

SER, smooth endoplasmic reticulum.
[a]Human (Jene Mappmg Workshop (HUMW)- approved symbols.
[b]Mulitple isoforms exist.
[c]CYP21A1 encodes a nonfunctional enzyme (White et al., 1988).

115

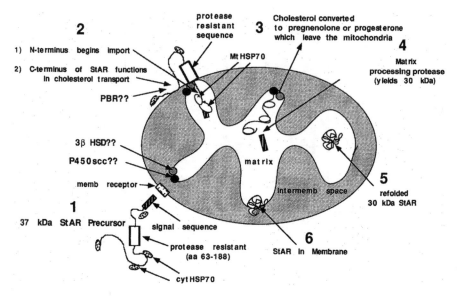

Figure 5-4. Sequence of events involved in steroidogenic acute regulatory protein (StAR)-assisted cholesterol transport movement to the inner mitochondrial membrane where side-chain cleavage of cholesterol occurs. Reprinted with permission from Stocco (2000).

End products of pregnenolone metabolism in the ZR are C_{19} androgenic steroids, primarily DHEA and ASD. Although the basic pattern of steroid synthesis is similar to that in the testis, ZR cells lack the enzyme 17β-hydroxysteroid dehydrogenase 17β-HSD required to convert ASD to testosterone. Both ASD and DHEA are weak androgens, having only a fraction of the potency of testosterone in bioassays. The ZR also contains an enzyme for sulfating DHEA. DHEA sulfate is the major form of DHEA circulating in plasma.

Species Differences in Cholesterol Delivery and Corticosteroid Synthesis

Although the majority (90% or more in rats) of cholesterol needed for corticosteroid synthesis is derived from plasma lipoproteins, the lipoprotein pathway used is species dependent. In humans, cholesterol destined for adrenocortical cells is carried on low-density lipoproteins (LDLs). LDLs dock with receptors on the surface of adrenocortical cells and are quickly internalized. In contrast, rodents transport cholesterol bound to high-density lipoproteins (HDL), and the scavenger receptor-BI (SR-BI) mediates the uptake of cholesterol from HDLs. Although the LDL receptor is the principal means of cholesterol uptake by adrenocortical cells in humans, the SR-BI pathway may play an important backup role when there are defects in the LDL receptor pathway.

In humans, pigs, sheep, and cows, $P450_{C17}$ in the ZF diverts some of the pregnenolone and progesterone substrate away from corticosterone synthesis and to-

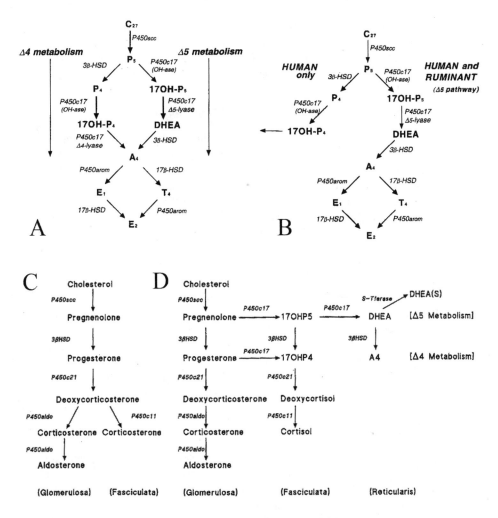

Figure 5-5. A schematic representation of (A,B) the Δ 4 and Δ 5 steroidogenic pathways in humans and other mammals and (C,D) tissue-specific pathways of steroid biosynthesis in the adrenal gland. Reprinted with permission from Conley and Bird (1997).

ward the synthesis of cortisol (figure 5-5). These species appear to retain the ability to produce corticosterone upon inhibition of the P450$_{C17}$ pathway.

In humans and rats, 11β-hydroxylation is carried out by P450$_{aldo}$ in the ZG and by P450$_{C11}$ in the ZF. In certain mammalian species, including cows, sheep, and pigs, a related enzyme encoded by the gene *CYP11B* carries out the last three steps in ALDO synthesis. Exactly how production of glucocorticoids and mineralcorticoids is regionalized is not known, as this enzyme is not restricted to the ZG but is expressed throughout the adrenal cortex. One hypothesis is that levels of adrenal corticosteroids in the blood flowing through the adrenal cortex produce local negative feedback on enzyme activity. Hence, blood flowing from the ZG is rich in ALDO and inhibits conversion of corticosterone to ALDO in the ZF, whereas blood reaching the ZR has high levels of ALDO as well as corticosterone and/or cortisol and drives the ZR cells to synthesize androgens (ASD and DHEA).

Pharmacological Manipulation of Adrenal Steroidogenesis

Synthesis of glucocorticoids can be inhibited or blocked by the drug metyrapone, which inhibits the activity of $P450_{C11}$. Cyanoketone inhibits 3β-HSD activity by occupying the enzyme's binding site for pregnenolone and preventing the synthesis of progesterone, androgens, and estrogens by the gonads. The role of $P450_{C17}$ in adrenal and gonadal androgen synthesis has made it a therapeutic target for intervention in prostate cancer and has fueled the development of a number of $P450_{C17}$ inhibitors. These compounds include both steroid (abiraterone) and nonsteroid (YM-116) inhibitors of $P450_{C17}$.

An important feature of steroidogenic processing common to both gonadal and adrenocortical tissue is that many of the enzymes involved can act on more than one substrate. For example, 3β-HSD acts on pregnenolone, 17α-hydroxy pregnenolone, and DHEA. $P450_{C17}$ can act on both pregnenolone and progesterone, and $P450_{aldo}$ can act on both deoxycorticosterone and corticosterone. This is an important consideration when manipulating steroid hormone synthesis. Alterations in the activity or expression of one enzyme may effect processing at many points in the steroidogenic pathway. Furthermore, the experimental addition of one steroid may influence the levels of others by acting as a substrate for their synthesis.

Control of ALDO Synthesis

Secretion of ALDO is controlled by the renin–angiotensin system. Angiotensin II (ANG II) and extracellular K^+ are the major factors regulating ALDO synthesis and secretion from the ZG. Although both stimulate ALDO secretion independently, their individual potencies depend on one another. For example, when circulating ANG II levels are low, K^+ has less of a stimulatory effect. Alternatively, angiotensin is less effective under conditions of K^+ depletion. Therefore, steady-state levels of plasma ALDO generally reflect the combined effects of K^+ and ANG II.

Angiotensin II (ANG II) is an octapeptide derived from proteolytic cleavage of angiotensin I (ANG I). In response to low blood pressure or low Na^+ levels, specialized secretory cells in the afferent arteriole of the renal nephron release renin, an enzyme that catalyzes proteolytic cleavage of ANG I from its precursor, angiotensinogen. This substrate (sometimes called renin substrate) that is present in circulating blood is a product of the liver. ANG I is converted to ANG II through the action of angiotensin converting enzyme (ACE). Vascular endothelial cells in the lungs and peripheral blood vessels are the major source of ACE.

ANG II stimulates ALDO secretion via an interaction with G-protein–coupled receptors (AT1 receptors) linked to the inositol trisphosphate (IP_3) hydrolysis signaling system. These receptors are upregulated by Na^+ depletion and downregulated by excessive Na^+ loading. Activation of AT1 receptors in ZG cells causes membrane depolarization and elevated intracellular Ca^{2+} levels. Steroidogenic factor-1 (SF-1) and cAMP response element-like binding sites on the *CYP11B2* gene promoter may mediate ANG II- and K^+-induced ALDO synthesis. Atrial natriuretic peptide (ANP), originally discovered in cardiac muscle cells, affects natriuresis and blood pressure in part by inhibiting ANG II and K^+ stimulation of ALDO secretion.

Control of Glucocorticoid Synthesis

The principal stimulus for glucocorticoid synthesis is ACTH secreted from the pars distalis (PD). ACTH acts on melanocortin-2 receptors to stimulate intracellular cAMP accumulation, initiating cholesterol transport to the mitochondria and synthesis of glucocorticoids in ZF cells. Although ACTH-mediated elevations in cAMP are the trigger for StAR synthesis, the *StAR* gene promoter does not appear to have a cAMP response element. The *StAR* gene contains a binding domain for the transcription factor Ad4BP. This transcription factor can promote transcription of the *StAR* gene and may mediate cAMP-induced increases in *StAR* gene transcription. The orphan nuclear receptor DAX-1 (dosage sensitive sex reversal adrenal hypoplasia congenital, critical region on the X chromosome gene-1) has been implicated in the inhibition of mediated adrenal steroidogenesis. ACTH-induced transcription of *CYP11B1* requires activator protein-1 transcription factors (dimers of the Jun and Fos family of proteins). ACTH enhances the expression of Jun and Fos mRNAs in the ZF.

In the rat, a nocturnally active animal, basal (unstimulated) glucocorticoid secretion displays a 24-h rythmicity, with peak levels during the evening and lowest levels early in the morning. In contrast, glucocorticoid levels in humans peak in the morning, and lowest levels occur in the evening. Daily patterns in glucocorticoid secretion are an important factor to consider when investigating effects on adrenal glucocorticoid secretion.

Transport of Glucocorticoids in Blood

Most glucocorticoids in plasma (90% or greater) are bound with high-affinity (apparent K_d of 10–100 nM) to a transport glycoprotein called corticosteroid-binding globulin (CBG; transcortin) that is synthesized in the liver. Plasma albumins also bind glucocorticoids but with a much lower affinity. CBGs dramatically prolong the plasma clearance of glucocorticoids. ALDO, which does not bind to CBGs, is cleared from plasma 10 times more quickly than glucocorticoids. Consequently, plasma ALDO concentrations are generally 200- to 2000-fold lower than plasma glucocorticoid levels. During human pregnancy, maternal plasma glucocorticoids increase threefold because of increased CBG synthesis.

Biological Actions of Corticosteroids

Lack of mineralocorticoids leads to death resulting from decreased blood volume, decreased cardiac output, and decreased blood pressure. ALDO plays an essential role in the maintenance of blood-volume blood pressure by regulating electrolyte transport in the kidney. The main action of ALDO is to increase Na$^+$ and water reabsorption in the kidney, leading to increased K$^+$ excretion, reduced Na$^+$ excretion, and expansion of blood and extracellular fluid volume. These effects involve ALDO binding to an intracellular receptor in kidney epithelial cells of the proximal and distal tubules and transcription of amiloride-sensitive epithelial sodium channels and possibly other proteins involved in electrolyte transport.

Cortisol and corticosterone, the two principal glucocorticoid hormones in mammals, elevate blood glucose concentrations (hyperglycemia) and increase glycogen stores in liver via a number of direct and indirect mechanisms (figure 5-6). Glucocorticoids stimulate the synthesis of glucose from amino acids and/or lipids (gluconeogenesis) in liver and decrease uptake and utilization of glucose by muscle and adipose tissues. Glucocorticoids increase protein catabolism in skeletal muscle and increase the level of amino acids available for gluconeogenesis. Furthermore, they increase the transcription of enzymes required for deamination of amino acids in liver and depress the uptake of amino acids by other tissues except for liver, where amino acid transport is actually enhanced. Glucocorticoids also increase the level of phosphoenol pyruvate carboxykinase, a key enzyme for gluconeogenesis. As a result, there is a net increase in the delivery of amino acids to hepatic cells for deamination and conversion to glucose. Glucocorticoids also reduce glucose uptake by adipose tissue and facilitate (GH)-induced lipolysis, thereby contributing to elevated bloods levels of nonesterified fatty acids. Thus, glucocorticoids shift the entire

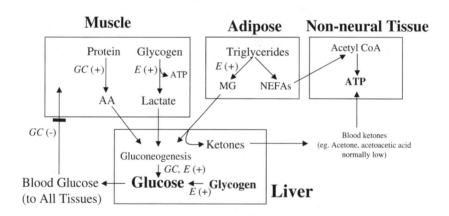

Epinephrine (E):
1. Promotes glycogenolysis.
2. Promotes gluconeogenesis in liver.
3. Promotes triglyceride breakdown
   in adipose tissue.
4. Elevates blood glucose.

Glucocorticoids (GC):
1. Promote protein catabolism.
2. Inhibit glucose uptake by muscle.
3. Promote gluconeogenesis in liver
   via stimulation of PEPCK
4. Permissive effect on E action
   in adipose tissue.
5. Elevate blood glucose.

Figure 5-6. Effects of epinephrine and glucocorticoids on energy metabolism. AA, amino acids; ATP, adenosine triphosphate; MG, monoacylglyceride; NEFA, nonesterified fatty acids; PEPCK, phosphoenolpyruvate carboxykinase.

energy balance of an organism to one that conserves carbohydrate energy stores at the expense of utilizing mostly proteins (about 80%) as well as fats (about 20%) for energy sources.

Effects of Glucocorticoids on the Inflammation/Immune Response

Glucocorticoids interfere with fibroblast proliferation, collagen synthesis and degradation, deposition of connective tissue ground substances, formation of blood vessels (angiogenesis), wound contraction, and reepithelialization. Glucocorticoids inhibit inflammation by enhancing transcription of anti-inflammatory proteins (lipocortin-1, interleukin-10, interleukin-1 receptor antagonist, and neutral endopeptidase) while simultaneously inhibiting the transcription of inflammation-response proteins (cytokines, adhesion molecules, receptors). The anti-inflammatory actions of glucocorticoids represent a form of negative feedback control, as a number of cytokines (e.g., interleukins) indirectly enhance activity of the HPA axis.

Mechanisms of Corticosteroid Receptor Action

Corticosteroids are lipophilic and readily pass through the plasma membrane of target cells, where they interact with intracellular receptors. These receptors belong to a large family of nuclear hormone receptors that include receptors for androgens, estrogens, thyroid hormones, and retinoic acid as well as a number of "orphan receptors" such as the aryl hydrocarbon receptor (AhR) that binds dioxins (see chapter 2). Glucocorticoids act on a 94-kDa cytosolic receptor protein called the type II glucocorticoid receptor or simply the glucocorticoid receptor (GR). This receptor differs from some other members of the family (e.g., estrogen receptor, thyroid hormone receptor) in that when unoccupied it resides in the cytosol. The unbound GR belongs to a protein complex that also includes at least two different heat shock proteins (hsp90 and hsp70) as well as a number of less well-characterized proteins. Binding of glucocorticoids to the GR induces a conformational change in the GR complex that causes dissociation of the GR/hsp complex and translocation of the bound GR to the nucleus. Once inside the nucleus, there are two modes of action: After forming dimers, the occupied GR dimer can bind to the glucocorticoid response elements in the promoter region of glucocorticoid-sensitive genes, or it can interact directly with other transcription factors. The first mechanism usually leads to enhanced transcription. The second mechanism usually results in direct interference with the activity of other transcription factors and subsequent inhibition of gene transcription. This latter mechanism is believed to underlie the anti-inflammatory actions of glucocorticoids (see above).

Mineralocorticoid receptors (MR = type I glucoccorticoid receptors) are structurally related to GRs and will bind both glucocorticoids and ALDO with high affinity. The fact that plasma glucocorticoid titers are 200–2000 times greater than ALDO levels has raised speculation as to the physiological significance of these receptors. Two theories have been proposed to explain ALDO–receptor interactions. First, CBGs produced locally in the kidney may effectively lower available glucocorticoid concentrations, thereby allowing ALDO access to its receptor. Second, the enzyme 11β-HSD may provide "on-site" protection of MR by metabolizing glucocorticoids to inactive forms. However, MRs in the hippocampus of the brain do bind glucocorticoids and appear to be directly involved in glucocorticoid inhibition of corticotropin-releasing hormone (negative feedback).

Some tissues may express plasma membrane GRs. A G-protein–coupled corticosterone receptor has been identified in the central nervous system of the newt *Taricha granulosa*. Plasma membrane GRs may mediate some of the rapid behavioral actions of glucocorticoids.

Adrenomedullary Catecholamine Synthesis

Adrenomedullary cells produce and secrete catecholamines into the blood, hormones derived from the amino acid tyrosine. Based on their secretary phenotype, these cells can be divided into two types: those that produce norepinephrine and those that possess the additional biosynthetic enzyme PNMT required for conversion of norepinephrine to epinephrine. Expression of PNMT depends on glucocorticoids from the adrenal cortex and distinguishes adrenomedullary cells from sympathetic neurons, which lack this enzyme. The catecholamine biosynthetic pathway is outlined in figure 5-3.

Biological Actions of Adrenomedullary Catecholamines

Adrenal catecholamines (especially epinephrine) have a number of important actions concerned with preparing an organism for confrontational situations, the fight-or-flight response. Anticipation or anxiety results primarily in release of norepinephrine whereas "emergency" situations or emotional stimuli evoke epinephrine release. The actions of these hormones include increasing heart rate, blood pressure, reducing gastrointestinal motility (except for the large intestine), and decreasing the activity of salivary glands. Catecholamines have vasodilatory effects on some vascular beds and vasoconstrictive actions on others. The vasomotor actions depend on the type of receptor present; activation of α_2-adrenergic receptors results in relaxation of vascular smooth muscle, whereas activation of α_1-adrenergic receptors causes their constriction. Adrenal catecholamines also have both indirect and direct actions on energy metabolism (figure 5-6). Direct effects of epinephrine include direct glycogen breakdown and indirect stimulation of gluconeogenesis (via enhanced lipolysis) in liver. Indirect effects include inhibition of insulin secretion via activation of α_2-adrenergic receptors on pancreatic islet cells. The net result of these actions is pronounced hyperglycemia.

Developmental Effects of Corticosteroids

At birth the human adrenal gland is 20 times larger than the adult gland when differences in body size are considered. Instead of three steroidogenic layers characteristic of the adult cortex, the fetal adrenal is composed of a small, outer, definitive zone composed of proliferative stem cells; an inner fetal zone that produces DHEA; and a transitional zone, sandwiched between the outer and inner zones, that is the source of fetal cortisol synthesis. The outer definitive zone is composed of tightly packed small and highly proliferative epithelial cells that compose 15% or less of the entire fetal gland. These cells do not exhibit a steroidogenic phenotype until just before to birth and synthesize both cortisol and ALDO.

The inner fetal zone is responsible for the majority of steroid production during fetal life, accounting for 80–85% of the total organ weight. This zone expresses only minimal amounts of 3β-HSD but does have enzymes necessary for DHEA and DHEA sulfate (DHEAS) synthesis. In some species, DHEAS is involved in the timing of parturition by acting as a precursor for placental estrogen synthesis during gestation. Fetal corticosteroids secreted late in gestation facilitate conversion of androgens to estrogens. In humans, placental CRH may directly influence contractions of uterine smooth muscle and thereby influence birth. Glucocorticoids are required for proper maturation of the fetal lung and the development of the pulmonary surfactant system. Glucocorticoids available to the developing fetus include those produced by the maternal adrenal as well as cortisol derived from conversion of cortisone in amniotic membranes and lung fibroblasts. Low or nonexistent expression of 3β-HSD in the fetal adrenal suggest that this organ is not an important source of glucocorticoids until after birth.

Stress and the Adrenal Glands

In 1936, Hans Selye described a syndrome in laboratory rats after exposure to a number of different noxious stimuli (see Selye, 1976). Administration of many different foreign agents (tissue extracts, toxic chemicals [the "pharmacology of dirt" quipped an unimpressed critic of Selye's early work]) or changes in an animal's physical environment (heat, cold, exercise) all produced the same effects: atrophy of the lymphatic organs, enlargement of the adrenal glands, and ulceration of the intestinal lining (figure 5-7). Selye termed this the "general adaptation syndrome." According to Selye, animals respond to physical or psychological challenges in three distinct stages: the alarm reaction, the stage of resistance, and, after prolonged exposure, the stage of exhaustion (which he incorrectly interpreted as exhaustion of the HPA axis). Selye is generally credited as the first to use the engineering term stress to describe this physiological state, and he described causative agents as stressors.

Modern molecular, physiological, and biochemical methodologies have revealed a great deal about the mechanisms underlying stress and adaptation. Within seconds after exposure to a stressor, epinephrine is secreted for a short period of time from adrenal medulla cells. Secretion of epinephrine is part of an immediate, coordinated response by the sympathetic nervous system (SNS) that is responsible for most of the early cardiovascular, respiratory, and gastrointestinal changes that accompany stress (table 5-2). Activation of the SNS is followed within minutes by elevation in plasma ACTH and a subsequent rise in plasma glucocorticoids, which has prolonged metabolic effects. After the initial exposure to a stressor, however,

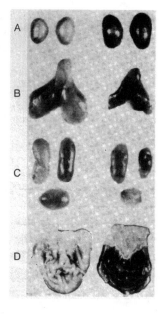

Figure 5-7. The appearance of the adrenal glands, lymphoid organs, and stomach in normal (left) and stressed (right) rats. A, adrenals; B, thymus; C, lymph nodes; D, stomach. Note the marked enlargement of the adrenal glands and the atrophy of the lymphoid organs. The stomach of the stressed rat exhibits characteristic ulcerations. Selye referred to this suite of stress-related changes as the "triad." Reprinted with permission from Selye (1976).

only glucocorticoid secretion is elevated so long as the stressor is present. Once the stressor is removed, plasma glucocorticoids may return to baseline or pre-stress levels within 24 h or may require as long as a week, depending on the intensity of the stress, regardless of the type of stressor involved.

Stress and Energy Metabolism

Stress is associated with almost immediate elevations in blood glucose, unbound fatty acids, and lipoproteins (LDLs, HDL/cholesterol) resulting from the actions of epinephrine (Figure 5-6). Elevated blood glucose is maintained during stress through the actions of glucocorticoids, largely through gluconeogenesis from amino acids (figure 5-6). The resultant elevations in blood glucose are used almost exclusively by the nervous system because glucocorticoids block glucose uptake in other tissues, especially skeletal muscle. Prolonged stress leads to many deleterious effects in energy metabolism, including muscle protein deficiency due to glucocorticoid-induced protein catabolism and utilization of amino acids for energy production rather than for new protein synthesis.

Stress and Immunity

Stress has both suppressive and enhancing effects on immune function that depend on the timing and duration of the stressor. Chronic stress leads to decreased circulating levels of lymphocytes, reduced antibody production, and alterations in normal cytokine communication. On the other hand, certain types of acute stress seem to prepare the immune system for an impending emergency, just as stress does for

Table 5-2. The general adaptation syndrome.

Alarm phase
   Elevated epinephrine and glucocorticoid secretion
   Mobilization of stored glucose and elevated blood glucose
   Inhibition of insulin secretion
   Mobilization of lipids
   Increased heart rate, cardiac output, blood pressure, and respiratory rate
Adaptation or resistance phase
   Elevated glucocorticoid secretion
   Increased hepatic gluconeogenesis and conservation of stored glucose
   Prevention of glucose uptake by muscle and adipose tissue
   Increased mobilization of fat and protein reserves; elevated blood levels of glycerol, fatty acids,
     and amino acids
   Increased secretion of aldosterone; maintenance of fluid volume and electrolyte balance
Exhaustion phase
   Exhaustion of lipid reserves
   Adrenal exhaustion; inability to synthesize corticosteroids
   Failure to maintain electrolytes
   Leads to death if stressor not eliminated

other effector systems. Acute stress leads to recruitment of lymphocytes to the largest organ in the body, the skin (hence the rapid decrease in circulating lymphocytes) and enhances skin delayed-type hypersensitivity reactions, which that provide resistance to foreign pathogens.

The effects of stress on immune function involve complex interactions between stress hormones (glucocorticoids, epinephrine), the SNS, and cytokines. Glucocorticoids and epinephrine have direct receptor-mediated effects on lymphocytes. Noradrenergic SNS nerves directly innervate immune lymphoid organs such as the spleen and influence immune function via direct synaptic contacts with lymphocytes. The "hardwiring" of immune organs by the CNS appears to be part of a complex regulatory scheme that involves reciprocal control of the SNS and HPA axis by lymphocyte-generated cytokines such as the interleukins.

Stress and Reproduction

Reproduction is not critical for survival in an emergency. Stress, either acute or chronic, leads to depressed levels of circulating gonadotropin and sex steroid hormones. Consequently, virtually every process influenced by reproductive hormones, from courtship behavior to fertilization, is inhibited. Stress-induced inhibition of reproductive hormone secretion appears to take place "upstream" of gonadal hormone secretion at the level of the hypothalamus. Endorphins are normally released during stress and have important analgesic effects within the CNS during stress. However, endorphins also inhibit secretion of gonadotropin-releasing hormone (GnRH) from hypothalamic neurons during stress, thereby reducing overall activity of the hypothalamus–pituitary–gonadal (HPG) axis (see Chapters 3 and 6).

Stress and Behavior

Although commands for carrying out specific behaviors reside within the central nervous system, the endocrine status of an organism plays a critical role in modulating the effectiveness of various stimuli to elicit different types of behavior. This is particularly true during stress. Stress hormones not only prepare an organism's internal environment for confrontation, but they also have profound impacts on how behavioral stimuli are interpreted. For example, chronic stress reduces feeding behavior and causes loss of body weight, leading to a condition called stress-induced anorexia. Stress suppresses behaviors that might place an animal at risk of predation or harm (such as exploratory behavior) and enhances behaviors that facilitate escape from a threatening stimulus (such as fear-induced startle behavior). The hormonal control of these behaviors is complex and not well understood, but it certainly involves an interaction of stress hormones with numerous different neurotransmitter and neuromodulatory pathways that are activated during stress. For example, administration of CRH increases locomotor activity and exploratory behavior in otherwise unstressed animals. This profile of behavioral effects changes to one of behavioral suppression during a stressful situation.

Stress and Growth

Stress affects growth in a number of ways that involves both direct effects of stress hormones on peripheral tissues as well as secondary effects on the secretion of GH. In general, chronic stress is detrimental to longitudinal bone growth. Part of this effect is mediated by glucocorticoids, which interfere directly with bone growth by blocking intestinal Ca^{2+} absorption and promoting bone resorption. Prolonged exposure to glucocorticoids inhibits GH secretion indirectly by enhancing the secretion of somatostatin (see chapter 3). Stress can also suppress neonatal development in humans and other mammals. Mother–infant separation results in increased activity of the HPA axis, reductions in circulating GH, and depressed activity of the enzyme ornithine decarboxylase, a sensitive marker enzyme for growth and differentiation. Curiously, acute stress in humans leads to an initial elevation in GH secretion. It has been suggested that GH may be involved in energy metabolism during the early stages of stress in humans.

Stress and Aging

Normally, secretion of glucocorticoids is regulated by negative feedback on the HPA axis; elevated levels of glucocorticoids inhibit CRH and ACTH secretion. This ability declines with increasing age in mammals. Loss of glucocorticoid-sensing neurons in the hippocampus leads to excessive glucocorticoid secretion under resting conditions and the inability to shut off the stress response, exacerbating the deleterious effects of stress. Excessive secretion of glucocorticoids can accelerate cardiovascular disease, cancer, type 1 or type 2 diabetes, and a host of other disorders. The fact that glucocorticoids may actually cause hippocampal damage has led some to speculate that stress accelerates the aging process of the brain over the life of an individual. Exposure of neonatal animals to stress can affect the response of the HPA axis to stressors when these animals become adults.

Comparative Anatomy of the Adrenal Gland

There are major differences in the organization and location of the adrenal glands throughout the vertebrates (figure 5-8). In general, craniates that do not produce an amniotic egg (anamniotes; includes hagfishes, lampreys, elasmobranchs, and bony fishes as well as amphibians) tend to have clumps or cords of mixed adrenocortical and adrenomedullary tissue that intermingle to varying degrees with kidney tissue. There often are no distinct boundaries between adrenocortical and adrenomedullary tissues, and there is no zonation within the adrenocortical tissue in these organisms. These homologs of the mammalian adrenal glands are sometimes referred to as interrenal glands because of their location. In lampreys, presumptive masses of adrenocortical and adrenomedullary tissues exist within adipose tissue attached to

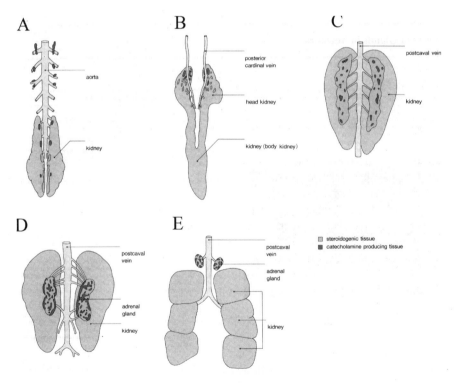

Figure 5-8. Comparative anatomy of the adrenal glands in various vertebrate groups. A, chondricthyes (*Raja maculata*); B, teleost (*Chelidotrigla kumu*); C, amphibian (*Rana catesbeiana*); D, reptile (*Psuedemys troostii*); E, bird (*Columba livia*). Reprinted with permission from Kawamura and Kikuyama (1992).

the cardinal veins. In elasmobranchs, adrenocortical and adrenomedullary tissues are completely isolated from one another. Adrenocortical cells form discrete masses of cells encapsulated with connective tissue within the posterior kidney, while adrenomedullary cells form clumps along the anterior kidney and aorta. Adreno-cortical tissue in teleosts exists as isolated clumps scattered along the posterior cardinal vein and as aggregated masses in the anteriormost portion of the kidney (the so-called head kidney). Although the head kidney forms the anterior portion of the mesonephric kidney in the group, it is cytologically quite different from kidney tissue and is composed of hematopoietic lymphoid tissue.

In urodele amphibians, mixed adrenocortical and adrenomedullary tissues appear as scattered clumps along the ventral aspect of the kidney, whereas anurans have relatively compact aggregates of adrenocortical and adrenomedullary tissue on the anterior ventromedial region of each kidney (figure 5-8). Reptiles and birds have encapsulated adrenal glands anatomically separated from the kidney, resembling the situation in mammals. However, as with other nonmammals, adrenocortical and adrenomedullary tissues are not segregated into discrete cortical and medullary regions, and there is no distinct functional zonation of adrenocortical tissues.

Despite differences in organization and location, the cytology of adrenocortical and adrenomedullary tissue is remarkably conserved throughout vertebrates. Adrenocortical cells in nonmammalian vertebrates have well-developed smooth endoplasmic reticulum, specialized mitochondria, and lipid granules. The adreno-cortical tissue of some anuran amphibians (i.e., ranid frogs) possesses a cell type unique to this group, the Stilling or summer cell. These cells have cytoplasmic granules and resemble mast cells, although their precise function is unknown.

Comparative Aspects of Adrenal Steroidogenesis

There is a great deal of variation in the major circulating corticosteroid hormones in craniates. This variation generally reflects differences in the expression of genes that encode enzymes in the corticosteroidogenic pathway. It is generally assumed that the initial stages in steroidogenesis are identical to those in mammals.

Despite the presence of circulating corticosteroids in lampreys and hagfishes, there is still confusion as to the site of corticosteroid production in cyclostomes. Presumptive adrenocortical cells do not express 3β-HSD do not synthesize corticosteroids under experimental conditions. Cortisol, corticosterone, and 11-deoxycorticosterone are the principal circulating corticosteroids in cyclostomes (table 5-3). The adrenocortical tissue of sharks and rays possess a unique enzyme, 1α-hydroxylase, that is responsible for the synthesis of 1α-hydroxycorticosterone (1αHC), a corticosteroid found only in this group of elasmobranchs. Elasmobranchs also have detectable circulating levels of cortisol, corticosterone, and 11-deoxycorticosterone. Cortisol appears to be the major circulating glucocorticoid in ratfish. In all bony fishes except for the lungfishes, cortisol is the predominant circulating adrenocorticosteroid, although lesser amounts of corticosterone can be detected. In lungfish, both cortisol and corticosterone, and to a lesser degree deoxycorticosterone and ALDO, are secreted. In amphibians, reptiles, and birds, corticosterone and ALDO are the principal corticosteroids circulating in blood.

Table 5-3. Circulating corticosteroids (mg/100 ml) in representative craniate species.

Species	1α-HC	DOC	11-DOC	Cortisol	Corticosterone	Aldosterone
Hagfish						
Myxine glutinosa				7	2	
Eptatretus stouli			2.7	0.8		
Lampreys						
Petromyzon marinus				0.5	0.2	
Cartilaginous fishes (sharks and rays)						
Raja laevis	94	42	3		16	
Squalus acanthias	280		16		250	
Bony fishes						
Ray-finned						
Chondrostean						
Acipenser oxyrhynchus		0.08	0.07	0.18	0.07	
Holosteans						
Amia calva				0.76		
Teleosteans						
Salmo trutta				2.4	0.022	
Carassius auratus		0.8		44	7.2	0.11
Clupea harengus			0.03	75	0.06	
Lobe-finned						
Lepidosiren paradoxa			0.03	6	0.16	0.58
Neoceratodus fosteri			0.88	0.28		0.03
Amphibians						
Rana catesbeiana					8.8	1.8–50.2
Reptiles						
Crocodylus nioticus					0.5	0.008
Tiliqua rugosa					2	0.015
Birds						
Meleagris gallopavo					2.3	0.010

1α-HC, 1α-hydroxycorticosterone; DOC, deoxycorticosterone; 11-DOC, 11-deoxycorticosterone.

Effects of Corticosteroids in Nonmammals

The terms mineralocorticoid and glucocorticoid hormone are functional definitions based largely on the effects of corticosteroids in mammals and do not necessarily apply to nonmammalian species. Differences in corticosteroid function have to do with differences in the structure, location, affinity, and number of corticosteroid receptors, and not with the structure of the corticosteroid hormones themselves. The structures of cortisol, corticosterone, ALDO, and other major corticosteroids appear to be identical in all vertebrates.

Osmoregulatory Effects

ALDO, the major mineralocorticoid in mammals, is not found in cartilaginous fishes and most bony fishes (table 5-3). In both cartilaginous and bony fishes, a single corticosteroid appears to carry out gluco- and mineralocorticoid actions. In sharks and rays, 1α-hydroxycorticosterone is the important osmoregulatory corticosteroid. In these organisms, body fluids are isotonic with that of seawater, largely due to elevated blood levels of urea. 1α-hydroxycorticosterone may play a role and in maintaining elevated levels of urea and trimethylamine oxide, a solute that counteracts the deleterious effects of elevated intracellular urea. 1α-hydroxycorticosterone also prevents water loss under conditions of limited urea synthesis by reducing Na⁺ excretion from the gills, kidney, and rectal gland, a unique salt-excreting structure in these organisms. In bony fishes, except for lungfishes, cortisol is the major osmoregulatory corticosteroid. This hormone is especially important in seawater adaptation, promoting the differentiation of chloride cells in the branchial epithelium and stimulating gill Na^+,K^+-ATPase activity and salt secretion. In salmonid species undergoing parr/smolt transformation, an elevation in plasma cortisol coincides with the acquisition of seawater tolerance. Lungfish adrenocortical tissues synthesize ALDO and deoxycorticosterone, but the importance of either hormone in lungfish osmoregulation has yet to be established.

Both ALDO and corticosterone have osmoregulatory effects in nonmammalian terapods. In amphibians, corticosterone and ALDO are both effective in stimulating epithelial sodium transport and maintaining blood Na⁺ titers in hypophysectomized animals, largely through actions on the urinary bladder epithelium. The toad urinary bladder has been used for many years as a model system for studying the molecular aspects of ALDO action on target cells. ALDO and corticosterone reduce sodium excretion by the nasal salt glands in reptiles and also have been implicated in the control of avian nasal gland secretion.

Effects on Carbohydrate Metabolism

Administration of ACTH, cortisol, or corticosterone elevates blood glucose levels in every vertebrate group so far examined. In sharks and rays, 1α-hydroxycorticosterone appears to be the major glucocorticoidlike hormone. The effects of cortisol on carbohydrate metabolism in teleosts are complex and depend on the dose administered, length of treatment, and whether the animal is hypophysectomized. In general, acute treatment with cortisol elevates blood glucose, liver glycogen, and triglycerides and stimulates hepatic gluconeogenesis. Chronic treatment with high doses of cortisol also leads to hyperglycemia but causes depletion of liver glycogen. Corticosterone is the primary corticosteroid influencing carbohydrate metabolism in amphibians. ALDO elevates blood glucose in one species of frog (*Rana temporaria*) but has no effect in other species. Although cortisol is not a major end product of adrenocortical steroid metabolism in adult amphibians, it is detected in larval amphibians and may be involved in regulating blood glucose levels during metamorphosis.

Adrenocortical Regulation in Nonmammals

ACTH bioactivity and its precursor protein POMC (see chapter 3) have been demonstrated in pituitaries from representatives of all vertebrate classes as well as in many invertebrates. Molecular studies indicate little variation in the amino acid sequence of ACTH across vertebrates groups. Mammalian ACTH stimulates adrenocortical secretion in every vertebrate group so far examined, although the types of steroids released may differ among species. ACTH stimulates serum cortisol and corticosterone levels in hagfishes and lampreys. In elasmobranchs, ACTH stimulates the secretion of 1α-hydroxycorticosterone while hypophysectomy reduces plasma levels of this steroid. Secretion of cortisol is clearly controlled by hypophysial ACTH secretion in bony fishes. In addition, teleosts appear to have an adrenocortical regulatory pathway independent of the HPA axis. In these fishes, products first discovered in the caudal neurosecretory system regulate cortisol secretion. This system consists of neurosecretory neurons and a neurohemal organ called the urophysis, in which two peptides are stored: urotensin I and urotensin II. Although urotensin I and II are structurally similar to CRH peptides, their effects on cortisol secretion appear to be directly on adrenocortical tissue rather than at the hypothalamus of pituitary. In nonmammalian tetrapods, ACTH stimulates both ALDO and corticosterone secretion.

All nonmammalian vertebrates appear to have a renin–angiotensin system, and the structure of ANG-II is evolutionarily conserved (table 5-4). However, the ability of ANG-II to stimulate adrenocortical hormone secretion differs greatly among nonmammalian vertebrates. Even within a single vertebrate class, there may be species differences in responsiveness to ANG-II. ANP also is found in nonmammalian vertebrates but it does not always inhibit mineralocorticoid secretion as in mammals. For example, ANP stimulates cortisol secretion in teleost fishes. In birds, the effects of atrial natriuretic peptide are species dependent; this peptide stimulates ALDO secretion in some species (e.g., turkey, *Meleagris gallopavo*) and inhibits it in others (e.g., duck, *Anas platyrhynchos*).

Daily as well as seasonal patterns in corticosteroid secretion have been documented for many nonmammalian species, especially with respect to seasonal changes in energy metabolism. Corticosterone may play a role in the acquisition of energy stores before seasonal migration events and in conservation of energy stores during seasonal periods of fasting.

Table 5-4. The amino acid sequences of angiotensin I in various vertebrate groups.

	Amino Acid Sequence									
	1	2	3	4	5	6	7	8	9	10
Common Structure	—	Arg	Val	Tyr	—	His	Pro	Phe	—	Leu
Human, pig, rabbit, rat, dog, horse	Asp				Ile				His	
Bovine, turtle	Asp				Val				His	
Fowl	Asp				Val				Ser	
Snake	Asp/Asn				Vat				Tyr	
Bullfrog	Asp				Val				Asn	
Goosefish	Asn				Val				His	
Salmon	Asn/Asp				Val				Asn	
Eel	Asp/Asn				Val				Gly	
Elasmobranch	Asn		Pro		Ile				Gln	

The enzyme renin cleaves amino acids 9–10 of angiotensin I to produce the octapeptide angiotensin II. From Hazon et al. (1999).

Comparative Aspects of Adrenocortical Hormones and Stress

Activation of the HPA axis appears to be a universal trait of the response to noxious environmental stimuli in nonmammalian and mammalian vertebrates. The principle corticosteroid released during stress in elasmobranchs is 1α-HC. Cortisol is released during stress in bony fishes, whereas nonmammalian tetrapods secrete corticosterone. In all vertebrates corticosteroids and adrenomedullary catecholamines are believed to regulate the same types of metabolic changes that accompany stress in mammals (see above). Stress is also accompanied by alterations in electrolyte levels in bony fishes, which is not surprising given the dual mineralocorticoid/glucocorticoid for cortisol in this group.

Stress can be a developmental cue in amphibians, enhancing or inhibiting certain aspects of growth and metamorphosis. Stress plays a critical developmental role in certain desert-dwelling toads, accelerating metamorphosis when ephemeral ponds begin to dry up. Crowding stress can accelerate certain aspects of metamorphosis (tail regression) while inhibiting others (hind limb growth). As in mammals, stress can also influence behavior. Stress decreases feeding and reduces weight gain in most nonmammals, a phenomenon that probably has a mechanism similar to stress-induced anorexia in mammals. Stress plays an important role in conspecific social interactions in many fishes, reptiles, and birds. In certain species, subordinate social structure is associated with elevated adrenocortical and adrenomedullary activity.

Stress is almost always associated with decreased reproductive capacity in nonmammals. Acute stressors can cause rapid declines in blood levels of reproductive hormones, and chronic stress can reduce fertility and offspring survival.

Although relatively rare in natural animal populations, chronic and severe stress can lead to symptoms of the general adaptation syndrome exhaustion phase as described by Selye (1976). After spawning, male salmon exhibit peptic ulcers, depressed immune function, and enlarged adrenal tissue as well as increased neurodegeneration, presumably due to a lack of negative feedback control of the HPA axis and subsequent oversecretion of cortisol. In certain species of marsupial mice and squirrels, all the males in the population die immediately after their first mating, whereas the females survive. During the mating period, males exhibits elevated plasma cortisol levels as well as immunosuppression and peptic ulcers.

Finally, it is important to use appropriate endpoints when evaluating stress in nonmammals. Adrenal size is difficult to measure in fishes and amphibians because of the diffuse nature of the gland. In these animals, other markers of adrenocortical and adrenomedullary activity (e.g., measurement of plasma corticosteroids and/or ACTH, measurement of circulating catecholamines, qualitative assessment of steroidogenic enzyme activity) must be substituted for measurements of adrenal size. Stress testing can be used effectively to assess the status of the HPA axis. In addition, endpoints other than size of the thymus are often used to gauge immune responsiveness, a practical necessity in species (e.g., lampreys) that lack a well-defined thymus gland, although many of these endpoints (e.g., a reduction in circulating lymphatic levels) are still open to multiple interpretations.

References

Babu, P.S., Bavers, D.L., Beuschlein, F., Shah, S., Jeffs, B., Jameson, J.L., Hammer, G.D., 2002. Interaction between Dax-1 and steroidogenic factor-1 in vivo: Increased adrenal responsiveness to ACTH in the absence of Dax-1. Endocrinology 143, 665–673.

Conley, A.J., Bird, L.M., 1997. The role of cytochrome P450 17α-hydroxylase and 3β-hydroxysteroid dehydrogenase in the integration of gonadal and adrenal steroidogenesis via the Δ5 and Δ4 pathways of steroidogenesis in mammals. Biol. Reprod. 56, 789–799.

Hazon, N., Tierney, M.L., Takei, Y., 1999. Renin-angiotensin system in elasmobranch fish: A review. J. Exp. Zool. 284, 526–534.

Jaffe, R.B, Mesiano, S., Smith, R, Coulter, C.L., Spencer, S.J., Chakravorty, A., 1998. The regulation of fetal adrenal development in human pregnancy. Endoc. Res. 24, 919–926.

Junqueira, L.J., Carneiro, J., Kelley, R.O., 1998. Basic Histology. Appleton & Lange, Stanford, CT.

Kawamura, K., Kikuyama, S., 1992. Adrenal gland. In: Matsumoto, A., Ishii, S. (Eds.), Atlas of Endocrine Organs. Springer-Verlag, Berlin.

McKewn, B., 2002. Sex, stress and the hippocampus: allostasis, allostatic load and the aging process. Neurobiol. Aging 23, 921–939.

Müller, J., 1998. Regulation of aldosterone biosynthesis: The end of the road? Clin. Exp. Pharmacol. Physiol. 25(suppl.), S79–S85.

Mulrow, P.J., 1999. Angiotensin II and aldosterone regulation. Reg. Peptides 80, 27–32.

Sapolsky, R.M., 1994. Why Zebras Don't Get Ulcers. W.H. Freeman, New York.

Seckl, J.R., Walker, B.R., 2001. 11β-hydroxysteroid dehydrogenase type 1—a tissue-specific amplifier of glucocorticoid action. Endocrinology 142, 1371–1376.

Selye, H., 1936. The Stress of Life. McGraw-Hill, New York.

Stewart, P.M., Krozowski, Z.S., 1999. 11β-hydroxysteroid dehydrogenase. Vit. Horm. 57, 249–324.

Stocco, D.M., 2000. The role of the StAR protein in steroidogenesis: challenges for the future. J. Endocrinol. 164, 247–253.

Stocco, D.M., 2001. Tracking the role of a StAR in the sky of the new millennium. Mol. Endocrinol. 15, 1245–1254.

Vinson, G.P., Whitehouse, B., Hinson, J., 1992. The Adrenal Cortex. Prentice Hall, Englewood Cliffs, NJ.

White, P.C., Vitek, A., Dupont, B., New, M.I., 1988. Characterization of frequent deletions causing steroid 21-hydroxylase deficiency. Proc. Natl. Acad. Sci. USA 85, 4436–4440.

6

The Gonads

James A. Carr
David O. Norris

The reproductive system includes not only the hypothalamus–pituitary–
gonadal (HPG) axis affecting the production of gametes and various
sex accessory structures, but also includes the modulating roles of both the hypothalamus–pituitary–thyroid (HPT) axis and the hypothalamus–pituitary–adrenal
(HPA) axis. No other endocrine system is so strongly influenced by selective processes of the environment as the HPG axis. Thus, many reproductive adaptations
have evolved independently in diverse groups of animals. For example, we find
viviparous cartilaginous and bony fishes, amphibians, and reptiles as well as some
ovoviviparous species (eggs retained until hatching within the maternal body) exhibiting specializations of the female reproductive system for transferring nutrients
from the mother to the developing offspring. Similarly, egg laying occurs in all vertebrate groups, including the protherian mammals. Viviparity, then, is a strategy to
solve certain environmental problems or restraints that has evolved numerous times
independently as a modification of the basic habit of producing yolky eggs. In this
chapter, we focus first on the roles of pituitary and gonadal hormones in mammals
and then compare the mammalian system to other vertebrate groups. Finally, we
discuss briefly roles of the HPT and HPA axes as they affect reproduction.

Some Generalizations

Hormones

In all vertebrates, reproduction is under primary control by a hypothalamic
gonadotropin-releasing hormone (GnRH), which activates the pituitary gland to

secrete two separate gonadotropins (possibly just one gonadotropin in squamate reptiles). The amino acid composition of this decapeptide varies in the different groups, and two forms of GnRH are present in most vertebrates, one that functions as a neurohormone affecting the pituitary and one that acts principally as a neurotransmitter or neuromodulator in the central nervous system (see chapter 3). Additional hypothalamic factors (e.g., neuropeptide Y) influence the pituitary as well. One of the gonadotropins (follicle-stimulating hormone, FSH) regulates gamete production, and the second (usually called luteinizing hormone; LH) controls gamete release. Both gonadotropins may be involved in the regulation of gonadal steroid secretion, usually involving a two-cell model for synthesis of androgen (stimulated by LH) and its conversion to estrogen (stimulated by FSH). The principal androgen in vertebrates is usually testosterone and the principal estrogen is 17β estradiol (E_2), and all vertebrates secrete progesterone, although it is not always the dominant progestogen (table 6-1). For example, an important progestogen in fishes is $17\alpha,20\beta$-dihydroxyprogesterone (DHP). These steroids are secreted into the blood and may have local paracrine effects in the gonads as well. In addition to secretion of steroids, gonads produce a number of other regulators that may have local effects or be involved in selective feedback on gonadotropin secretion (e.g., inhibin, GnRH).

Gonads

The testes are organized in one of two anatomical patterns. There is a cystic pattern of organization in fishes and amphibians where the spermatogenetic cells occur in cysts composed of masses of like cells. In some species, the cysts are organized into lobules, and a testis may consist of several lobules. Spermatogenesis, the process whereby diploid spermatogonia undergo meiotic and mitotic cell division to form haploid spermatozoa, is synchronized within a cyst so that all cells are in the same stage at the same time (first all are primary spermatogonia, then all synchro-

Table 6–1. Prinicipal gonadal steroids secreted by various vertebrates.

Taxon	Androgens	Estrogens	Progestogens
Agnathan fishes			
Hagfishes	T, 6βHT, 7αHT	?	?
Lampreys	T, 15αHT, 15βHT	E_2, $15\alpha E_2$, $15\beta E_2$	?
Chondrichthyes	T>DHT, 11-KT, A,11-KA	E_2	P_4>DHP
Teleosts	11-KT>DHT, 11βHT	E_2	DHP>P_4
Amphibians			
Urodeles	T, DHT, 11βHT, 11-KT	E_2	P_4
Anurans	T, DHT	E_2	P_4
Reptiles	T, DHT	E_2	P_4
Birds	T, DHT	E_2	P_4
Mammals	T, DHT	E_2	P_4

A = androstenedione; DHP = $17\alpha,21\beta$-dihydroxyprogesterone; DHT, 5α-dihydrotestosterone; HT = hydroxytestosterone; K = keto; P_4 = progesterone, T= testosterone. Based on Norris (1997); Selcer and Clemens (1998); Sower and Gorbman (1998); Manire et al. (1999); Rasmussen et al. (1999).

nously become primary spermatocytes, etc., and final all undergo spermiogenesis at the same time). Furthermore, all of the cysts in a lobule are usually at the same stage. The steroidogenic interstitial cells are found within the connective tissue covering of each cyst and are called lobule-boundary cells. The organizational pattern is modified in anuran amphibians, where each testis is organized into distinct tubules, but the synchronous cystic pattern of spermatogenesis persists within each tubule. In contrast to the anamniotes (fishes, amphibians), the testes of amniotes (reptiles, birds, mammals) are all organized into distinct seminiferous tubules with separate steroidogenic Leydig (interstitial) cells within the connective tissue regions between the tubules. All stages of spermatogenesis may be present at the same time in the testes of amniotes.

The ovaries of all vertebrates are organized in a similar pattern. At some point in ovarian development, primary oocytes become invested with a single layer of stromal cells within the ovary to form a follicle. In fishes and amphibians, follicle growth typically consists of marked enlargement of the oocyte within the follicle and a minimal proliferation of the surrounding follicle cells so that the follicle wall consists of an inner layer of cells, the granulosa, and an outer layer of cells, the theca. In amniotes, the follicle wall develops into an inner granulosa of multiple layers separated from the thecal layer by a distinct basal lamina. The outermost multiple layer of follicle cells is the theca, which in reptiles, birds, and mammals may be separable into an outer connective-tissue like layer, the theca externa, and an inner steroidogenic layer, the theca interna. The steroidogenic roles of these cell types are described below.

Sex Accessory Structures

The sex accessory ducts of vertebrates have similar embryonic origins in all but the bony fishes, with the müllerian ducts giving rise to oviducts and/or uteri and the Wolffian ducts functioning as paired sperm ducts, the vas deferens. The wolffian ducts also act as urinary ducts in elasmobranch fishes, teleosts, and amphibians, and so are retained in females as well as in males. In amniotes (reptiles, birds, and mammals), the evolution of the paired metanephric kidneys, each with a new urinary duct, the ureter, results in the wolffian duct becoming exclusively a sperm duct in males and degenerating in females.

In teleosts, sperm ducts and oviducts may develop as unique structures. In some species, gametes are extruded through temporary openings or pores in the body wall and gonaducts are lacking. Numerous sex accessory structures and pigmentation patterns have been described in all vertebrates that are dependent on estrogens or androgens for their formation and/or maintenance.

Development of the Gonads and Sex Accessory Structures

There are three mechanisms known for controlling sex determination in vertebrates. Genetic sex determination frequently includes the designation of one pair of heteromorphic (unlike) sex chromosomes, usually designated as XY or ZW. Species where the heterogametic sex (produces two kinds of gametes) is male are designated as XY (male) and XX (female). In species with ZZ/ZW type of sex determination, the female sex is heterogametic. In either case, there appears to be one or more genes associated with the Y or Z chromosome that determine the sex of the individual. Thus, the homogametic (one kind of gamete) condition is often considered the default sex. Mammals have the XX/XY type of sex determination, with females being homogamtetic, whereas birds have the ZZ/ZW type, with the male is homogametic. In mammals, genes on the Y chromosome (e.g., *SrY*) result in production of androgens, which initiate a change from the basic female default phenotype to that of a male. Recent studies in mammals also indicate that estrogen secretion as well as the absence of androgen may be necessary for normal gonadal differentiation in females, so it is not simply the absence of androgen that makes a female.

In birds, the *cSox9* gene is expressed in ZZ individuals rather than in ZW individuals and confers maleness. Reptiles, amphibians, and fishes generally do not have heteromorphic pairs of chromosomes, and the importance of genetic determination of sex is not clear. Some studies show that anuran amphibians tend to exhibit the XX/XY type, whereas caudate amphibians appear to rely on a ZZ/ZW pattern. In salmonid fishes, males have been shown to be the heterogametic sex through studies of the progeny of sex-reversed individuals mated with normal fish. Several other teleosts also exhibit XX/XY type of sex determination, but ZZ/ZW also occurs (see Devlin and Nagahama, 2002). Recently, a sex-determining gene was reported to determine male sex in two species of medaka (Matsuda et al., 2002). This *dmY* gene is a member of the *dmrt1* gene family, which has been shown to be involved in testis determination and/or differentiation in fishes, amphibians, birds, and mammals. However, the *dmY* gene appears to be unique to two species of medaka and is not present in other teleosts (Y. Nagahama, personal communication).

Temperature-dependent sex differentiation has been well documented in some teleostean fishes and numerous reptiles, especially among the turtles and crocodilians. In these cases, incubation temperature determines the sex of the offspring. In one species, incubation may produce all females at high temperatures, all males at low temperatures, and normal sex ratios at intermediate temperatures. This pattern may be completely reversed in another species. Thus, selection of a nest site can have profound effects on the sex of the offspring. Estrogens appear to be important mediators of the temperature effect in both turtles (e.g., Belair et al., 2001) and alligators (e.g., Gabriel et al., 2001).

Unique modes of sex determination are found among sequential hermaphroditic teleosts, including behavioral sex determination. In certain species, there is a social system where one or a few males dominate harems of females, preventing other males from ready access to mates. Some of these species are diandric, having two male phenotypes with different mating strategies. Removal of the dominant male or a reduction in the ratio of dominant males to females quickly triggers the largest female to exhibit male behaviors, with a slower conversion of the ovary into a functional testis. Estrogens appear to be involved in the sex change of males to females in protandrous teleost species, which begin life as males and later transform into females (e.g., Lee et al., 2000).

Sex accessory duct development in tetrapods is determined by the early secretions of the gonads (figure 6-1). Testes secreting androgens and müllerian-inhibiting hormone (MIH; also called anti-müllerian hormone) cause degeneration of the müllerian ducts, and androgens stimulate development of the wolffian duct into a vas deferens. In females, estrogen from the ovary both protects the müllerian duct and stimulates its development into an oviduct and/or uterus. Estrogens are antagonistic to development of the vas deferens.

Control of Gonadal Steroidogenesis

Gonadal steroids are synthesized in the same manner as adrenal steroids from cholesterol (see chapter 5). In most cases, the Δ^4-pathway is used to synthesize progesterone, androgens, and estrogens from pregnenolone. The critical enzymes for these syntheses are listed in chapter 5 (table 5-1). In the testis, most of the progesterone is converted first to androstenedione (a weak androgen), which is the immediate precursor for synthesis 11-ketoandrostenedione (11-KA) and for the principal vertebrate androgen, testosterone. Depending on the species, some of the testosterone may be converted to other androgens such as 11-ketotestosterone (11-KT; e.g., in many fishes) or dihydrotestosterone (DHT; e.g., in fishes and tetrapods). In the ovary of tetrapods, androstenedione and testosterone are mostly converted to E_2, which may serve as the precursor for synthesis of other less potent estrogens. In fishes and amphibians, there may be significant secretion of testosterone in females, which in some cases exceeds the levels for E_2. The adrenal androgen dehydroepiandro-

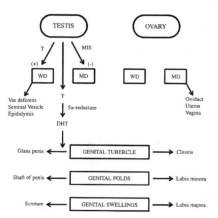

Figure 6-1. Patterns of development for primary sexual characters. In males, the testes secrete both testosterone (T) that stimulates differentiation of the wolffian ducts (WD) and müllerian-inhibiting substance (MIS) that causes regression of the müllerian ducts. Dihydrotestosterone (DHT) is necessary for the genital tubule, folds, and swellings to differentiate in the male direction. Development of the female structures (and degeneration of the WD in female amniotes) requires not only the absence of androgens but the presence of estradiol. Reprinted with permission from Norris (1997).

sterone (DHEA) can be converted readily to androstenedione by either testes or ovaries and under some circumstances can be a significant source of testosterone and/or E_2. In pregnant mammals, DHEA and androstenedione synthesized by the fetal adrenal are converted to E_2 by the placenta.

Synthesis of gonadal steroids in mammals is controlled primarily by the sequential actions of the gonadotropins. LH stimulates the synthesis of testosterone in the Leydig cells of the testis or in the thecal cells of the ovarian follicles as well as by the thecal interstitial ovarian cells. Under the influence of FSH, which stimulates production of the aromatase enzyme ($P450_{aro}$), granulosa cells in the ovarian follicle convert androgens into estrogens. In males, Sertoli cells produce aromatase and convert some testosterone into E_2. Two exceptions to this two-cell pattern of steroidogenesis apparently occur in the ovaries of birds and reptiles. In birds, the granulosa cells produce progesterone, which is converted to androgens by the adjacent thecal cells (theca interna), whereas conversion of the androgens to estrogens takes place in the outermost thecal cells (theca externa). Furthermore, the thecal layer can synthesize DHEA via the Δ^5 pathway. In reptiles, steroidogenesis does not take place in the thecal layer.

Control of Gamete Production

The formation of active gametes is under the control of an FSH-like gonadotropin (GTH-I in teleosts). FSH stimulates the proliferation of spermatogonia in males and the growth of follicles in females. Androgens produced locally under the influence of LH are generally considered responsible for the initiation of meiosis in males. Release of gametes from the testes (spermiation) or the ovaries (ovulation) is stimulated by LH (GTH-II in teleosts). In many species, LH also stimulates the postovulatory follicle or corpus luteum to secrete progesterone.

Roles for Estrogens in Males and Androgens in Females

Traditionally, estrogens have been considered female hormones, but they are also synthesized in males and have numerous roles (see review by Staub and DeBeer, 1997). For example, estrogens are involved in the regulation of seasonal cycles of androgen secretion, spermatogenesis, masculinization of the brain and HPA axis, and can influence behavior. Although high levels of androgens, androgen receptors, and androgen-metabolizing enzymes have been described in females of all vertebrate groups, the biological role of androgens (except as requisite precursors for estrogen synthesis) has not been examined thoroughly. Possibly the most thorough study of the importance of androgens in female development and reproductive biology has been that of the spotted hyena (see Glickman et al., 1987). Thus, estrogenic, antiestrogenic, androgenic, and antiandrogenic endocrine-disrupting compounds (EDCs) could have important effects in the heterologous sex, and future studies should not focus on the effects of an EDC in members of only one sex.

Reproduction in Fishes

The Agnathans

The hagfishes are marine agnathans (jawless fishes) whose reproductive endocrinology is poorly known, as is their reproductive biology in general (see Sower, 1998; Gorbman, 1997). The protogenous (female first) sequential hermaphrodite, *Eptatretus stouti*, has been studied more than any other hagfish, but even for this fish little is really known. Hagfishes lack gonaducts, and there is little evidence of secondary sexual characters. The HPG axis of hagfishes is poorly developed, but they do produce two forms of GnRH (see chapter 3). However, there is little evidence that the adenohypophysis is functional. Hypophysectomy seems to have no effect on the gonads. Although arginine vasotocin (AVT) is present in hagfishes, no reproductive role has been demonstrated for this peptide. Estrogens are present, as are putative estrogen receptors (ERs), and administration of estrogen stimulates synthesis of a vitellogenin (Vtg) by the hagfish liver. Vtg is a precursor protein sequestered from the blood by the ovary and used as a source for the synthesis of yolk proteins.

Lampreys are semelparous (spawn once and die) agnathans that live in freshwater or marine habitats, but all species spawn in fresh water. Much of their biology and reproductive habits are well known, especially for the parasitic species (see Hardisty and Potter, 1971; Sower, 1998). Like hagfishes, lampreys lack gonaducts. Secondary sexual characters consist of swelling of the cloacal labia and appearance of an erectile cloacal structure in males. The lamprey HPG axis is characterized by two distinct GnRH forms (see chapter 3), although gonadotropins have not been identified. E_2 and progesterone are associated with reproduction in both males and females, whereas testosterone levels are either very low or undetectable in males. Estrogen receptors are found in both ovaries and testes, but androgen receptors apparently are lacking. There are unique derivatives of E_2 and testosterone in hagfishes (e.g., 6β- and 7α-hydroxytestosterone) and lampreys (e.g., 15α- and 15β-hydroxytestosterone) that may be of some importance in reproduction (see Selcer and Clemons, 1998; Sower and Gorbman, 1998).

The Chondrichthyes

The elasmobranch fishes (sharks, rays, and skates) are mostly live bearing, with about 42% (including all skates) exhibiting oviparity. Of the viviparous species, only about 9% are truly placental (sharks); the remainder use stored yolk or yolk and uterine secretions. The chimaeras or ratfishes (holocephalans) are all oviparous, but their reproductive endocrinology has not been well studied.

In general, the reproductive endocrinology of cartilaginous fishes is typical of other vertebrates. Gonadal gametogenesis and steroidogenesis are regulated by pituitary gonadotropins. Males produce primarily testosterone as well as some 11-KT, 11-KA, and DHT. In at least one species, the bonnethead shark (*Sphyrna tiburo*), there is elevated secretion of DHP during early annual testicular development. Plasma levels of testosterone are considerably greater than E_2 in females, although the pattern of secretion for these two hormones is similar. Females also secrete small amounts of DHT, 11-KT, and 11-KA as well. DHP increases during gonadal maturation in females, and 11-KA is highest during pregnancy in the placental bonnethead shark. The role of the other androgens in females is not clear. A major role for ovarian E_2 in elasmobranchs is the stimulation of the liver to produce Vtg. The postovulatory follicle or corpus luteum (postovulatory follicle) secretes progesterone, although secretion of progesterone often occurs just before ovulation in at least some species. E_2 also stimulates the shell gland, which synthesizes and secretes the egg capsule used by oviparous species into which the eggs are deposited. Thinner capsules are produced around the eggs of viviparous species. In viviparous species, E_2 produces a local effect by stimulating ovarian production of relaxin, an insulinlike peptide that reduces muscular contractions in the pregnant uterus. Mammalian relaxin similarly quiets uterine muscle contractions.

The Bony Fishes: Sarcopterygians

The living lungfishes are restricted to Africa (four species of *Protopterus*), South America (*Lepidosiren paradoxa*), and Australia (*Neoceratodus fosteri*). All are oviparous, and *Neoceratodus* is considered the most primitive species. Treatment of *Neoceratodus* with mammalian GnRH stimulates secretion of E_2 in females and testosterone in males. These fish apparently do not secrete either 11-KT or DHT.

Coelacanths, *Latimeria chalumnae* and *L. menadosensis*, are living representatives of a group of fishes thought to have gone extinct 66 million years ago (e.g., Munsick et al., 1991). In marked contrast to the living lungfishes, coelacanths are live-bearing (ovoviviparous). However, because of its limited distribution, secretive behavior, and its rare and endangered status, virtually nothing is known about the endocrine control of reproduction. Morphological studies suggest that the pattern of live bearing is similar to that described for nonplacental elasmobranchs with nutrition based on stored yolk, possible eating of supernumerary eggs, and possibly including limited involvement of some uterine secretions to supplement the yolk of the eggs. Although the yolk sac of each embryo makes contact with the "uterus" in which development occurs, there is little evidence of nutritional transfer between parent and offspring. In spite of our ignorance about their reproduction, we do know that adipose and liver tissues of living coelacanths are contaminated with polychlorinated biphenyls (PCBs) (see chapter 11) as well as DDT and its metabolites (see chapter 12; Hale et al., 1991).

The Bony Fishes: Teleosts

Life-history patterns of the teleostean fishes are the most varied, exhibiting virtually every reproductive adaptation observed in any other vertebrate. Teleosts may be dioecious or hermaphroditic, and several patterns of sex changing fishes have been described. Some species are diandric, and the two forms of males may have alternative reproductive strategies. Most species are iteroparous, breeding repeatedly, often within a single reproductive season. A few species are semelparous, and the fish die after a single spawning (e.g., Pacific salmon, genus *Oncorhynchus*). Teleosts generally exhibit genetic sex determination, although temperature-dependent sex determination has been described for a few species. Among some tropical reef fishes, behavioral sex determination has been described. Natural sex reversals have been described in teleosts, and some species are functional hermaphrodites.

Reproductive maturation and breeding are usually controlled by environmental factors such as photoperiod, temperature, tidal rhythms, wet and dry seasons, and so on, operating through the neuroendocrine axis. Most teleosts have hollow ovaries, and numerous cases of viviparity occur among the teleosts with the eggs retained in the ovary and the young being reared within the ovarian cavity. Eggs of oviparous species may be deposited in nests, attached to vegetation or other objects, or simply released into the water where external fertilization occurs. Varying degrees of parental care are exhibited by teleosts, varying from none to aggressive defense of a nest or young. Some fish will brood their fertilized eggs in their mouth, and the young fish may seek shelter there even after hatching.

In response to environmental cues, GnRH is secreted from the hypothalamus and stimulates release of gonadotropins from the pituitary gland. GTH-I (see chapter 3) is responsible for gonadal growth, gametogenesis, and steroidogenesis, resulting in elevated plasma levels of steroids (males: mainly 11-KT, testosterone; females: mainly E_2). Oviposition and birth in viviparous fishes is stimulated by arginine vasofocin released from the pars nervosa. In fishes, AVT seems to perform those functions served by both arginine vasopressin (AVP) and oxytocin in mammals.

Gonadal steroids are responsible for stimulating sex accessory characters and controlling reproductive behaviors. Many of these behaviors appear to be mediated by E_2 or testosterone converted to E_2, as the teleost brain contains much higher levels of aromatase than do brains of other vertebrates. Spermiation and ovulation are caused by GTH-II from the pituitary, which induces the formation of a progesterone derivative (commonly DHP) that mediates these processes and which may have additional behavioral effects. DHP produced in the ovary is a reproductive pheromone affecting the males in some species. Apparently, in the Indian catfish (*Heteropneustes fossilis*), a corticosteroid released from the adrenocortical (interrenal) tissue stimulates ovulation and circulating cortisol is elevated during vitellogenesis.

In females, E_2 normally stimulates the liver to synthesize vitellogenin. Although males naturally do not secrete vitellogenin, exposure to natural or synthetic estrogens will induce the appearance of vitellogenin in the blood.

Prolactin from the pituitary gland may influence parental behavior in teleosts. In the discus fish, *Symphysodon aequifasciata*, prolactin stimulates secretion of mucus by skin glands. This mucus is fed upon by hatchlings.

Reproduction in Amphibians

Amphibians may have the most varied life-history patterns of any vertebrate group, in part because in many ways they represent the transitional state between aquatic and terrestrial vertebrates (figure 6-2). Some species are totally aquatic, but many

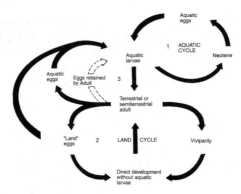

Figure 6-2. Summary of amphibian life-history patterns. Three basic patterns occur: (1) a totally aquatic cycle in which larvae become sexually mature (called neotenes or paedomorphs); (2) a totally terrestrial (land) cycle with direct development to miniature adults and no aquatic phase; (3) an aquatic–land cycle (amphibious).with aquatic egg and larval phases followed by metamorphosis to a terrestrial adult phase. The white arrow indicates a transitional phase where eggs are retained on land, but when the larvae hatch, they are deposited in the water either by chance or by parental behavior. Reprinted with permission from Norris (1997).

species only have aquatic larvae that undergo a series of complex biochemical, physiological, and anatomical changes (metamorphosis; see chapter 4) into a terrestrial or semiterrestrial adult; some species have abandoned the aquatic habitat and are completely terrestrial. In addition, there are species that lie among these three groupings as well as species that in some places may be totally aquatic (become sexually mature as larvae) and at other sites have aquatic larvae and metamorphosed terrestrial adults. Amphibians appear to have only genetic sex determination, although this issue has not been investigated extensively.

Most urodeles (salamanders and newts), most anurans (frogs, toads) and some apodans (caecelians) are oviparous. Most anuran and urodele hatchlings undergo a prolonged period of postembryonic growth and development as free-swimming larvae. The transition from larval to adult body form is called metamorphosis. Some species (such as the desert-dwelling species of the genus *Spea*) have a very short larval period of 2–3 weeks while, others (such as the American bullfrog, *Rana catesbeiana*) may remain as larvae for several years. In certain species, the larval period is entirely absent, and the hatchlings emerge from the eggs as miniature adults. This developmental strategy, called direct development, is most often seen in anurans that lay their eggs on land (such as members of the genus *Eleutherodactylus*).

Viviparity is most common among apodans and occurs in a few frogs and salamanders. For example, *Salamandra salamandra* and *S. atra* give birth to larvae or fully metamorphosed offspring, respectively, after relatively long gestational periods (4 years for *S. atra*). The young at birth of the truly viviparous species are much larger than the original egg and may collectively exceed the weight of the mother, which is evidence for maternal transfer of nutrients to the young during gestation. Most of the viviparous frogs actually can be classified as ovoviviparous with eggs being retained during develop in the oviducts, in dermal pouches or actually embedded in the skin, in the vocal sacs off the mouth, and even in the stomach. Still others, such as the male midwife toad, carry the eggs around with them, periodically moistening them in water to prevent desiccation.

Gonadal development is stimulated by FSH from the pituitary in response to GnRH. Androgen secretion is stimulated by LH and conversion of androgens to estrogens by FSH. Spermiation and ovulation are also regulated by LH. In the case of ovulation, LH stimulates the synthesis of progesterone, which in turn stimulates germinal vesicle breakdown and ovulation. Progesterone also increases the sensitivity of the oviduct to AVT which controls oviductal contractions and hence oviposition. AVT is also responsible for sperm release and spermatophore production in urodeles (see below).

Fertilization is internal in all but the most primitive urodeles and in all of the apodans, but most anuran amphibians practice external fertilization. In urodeles, internal fertilization is accomplished through the production of a gelatinous spermatophore by the male, which is deposited by the male on the substrate. After elaborate courtship behaviors, which involve pheromonal regulation of female behavior, the female is encouraged to pick off the sperm packet at the tip of the spermatophore with her cloaca. The sperm are then stored in a specialized region of the cloaca (spermatheca) until the eggs are ovulated. Apodans have a special organ, the phallodeum, that is used to transfer sperm directly into the female cloaca.

In males, the major androgens are testosterone and DHT, with testosterone plasma levels usually exceeding those of DHT, but there are exceptions in some species (e.g., in the bullfrog, *Rana catesbeiana*). In urodeles, androgens are synthesized by lobule boundary cells (as in teleost fishes), but in anurans androgens are synthesized by interstitial Leydig cells. Androgens stimulate sex accessory structures including the vas deferens, cloacal glands of male salamanders, nuptial pads in newts or thumb pads in frogs that are used for clasping the female during mating (amplexus), and possibly the production of pheromones used in breeding. Androgens also stimulate development of glands in the skin.

The role of androgens in male mating behavior may differ among species. In some species, the elevation in androgens does not occur at the time of mating (dissociated pattern), whereas in others androgens are positively associated with mating behavior. In the rough-skinned newt, *Taricha granulosa*, testosterone plays a permissive role in maintaining the responsiveness of brain regions to the stimulatory actions of GnRH and AVT resulting in clasping of the female, an early behavior in a complex mating ritual. Stress or the administration of corticosterone blocks this response.

The amphibian ovary is hollow like that of teleosts. The ovarian follicle consists of a thin granulosa layer and a thin thecal layer. Follicle growth is initiated by FSH and steroidogenesis of testosterone and E_2 is regulated by LH and FSH, respectively, as in mammals. E_2 from the ovary stimulates Vtg secretion by the liver, which is essential for yolking of the oocytes. Development of the oviducts is also stimulated by estrogens as well as the development of the pouch from egg brooding in the marsupial frog, *Gastrotheca riobambae*. The synthesis of progesterone by the ovary can be initiated by LH and the action of progesterone locally on the follicle is to stimulate final oocyte maturation (germinal vesicle breakdown) and ovulation. Short-lived corpora lutea usually develop from ovulated follicles and may persist in viviparous species throughout gestation. Corpora lutea be important in maintaining gestation through continued secretion of progesterone.

An unusual feature of bufonid anurans is the presence of an ovarylike structure called the Bidder's organ that appears as a small mass of oocytes before the differentiation of gonads in both sexes. The paired Bidder's organs are located just anterior to where the gonads develop. In females, the Bidder's organ becomes incorporated into the ovary, but in males it persists and one is usually fused to the anterior end of each testis. The male Bidder's organ undergoes a gonadotropin-dependent seasonal development that begins when the testicular cycle begins, but development of the oocytes stops just before yolking while the testes continue to develop. The significance of Bidder's organ is unknown.

Pheromones have been implicated in the reproductive activities of a variety of salamander species, but their possible roles in anurans and apodans are unknown. Territorial scent marking by cloacal gland secretions has been reported in plethodontids, and many male urodeles use glandular secretions in courtship to influence female behavior. These glands appear to be androgen dependent.

Reproduction in Reptiles

Reproductive patterns in reptiles range from oviparity to ovoviviparity to true viviparity with a functional placental arrangement. Fertilization is always internal, however, even in oviparous species. Turtles, the most primitive of living reptiles, and crocodilians, the closest living reptilians to birds, are strictly oviparous, but ovoviviparity and viviparity have evolved independently many times among the lizards and snakes (squamates). Parental behavior is lacking in turtles after digging a nest and laying the eggs, whereas crocodilians show complicated parental care before and after the offspring hatch in the nest. Some squamates exhibit parental care, but most do not. Sex is determined by incubation temperature in the nests of turtles and crocodilians as well as in a few squamates. This effect is mediated by gonadal steroids.

The reptilian testis is organized into spermatogenetic or seminiferous tubules as in birds and mammals. Androgen secretion is controlled by LH, except in squamates, which appear to have only a single FSH-like gonadotropin that regulates both gametogenesis and steroidogenesis. The principal androgens in reptiles are testosterone and DHT. These androgens are responsible for male-specific behaviors and stimulate sex accessory structures including the sexual segment of the kidney which stores sperm. Androgens may control aggressive behavior as well. Fertilization is always internal, even in the oviparous species, and males are equipped with one or two copulatory organs for transferring sperm to the cloaca of a receptive female.

The ovaries of turtles and crocodilians are organized like those of birds, with a single cuboidal layer of steroidogenic granulosa cells surrounded by a connective-tissuelike theca. Unlike other amniotes, the reptilian thecal layer lacks 3β-hydroxysteroid dehydrogenase (3β-HSD) and does not appear to be steroidogenic (see chapter 5). The squamate follicle originally has three cell types in the follicle wall, but by the time vitellogenesis begins, the follicle wall resembles that of turtles and crocodilians. Females secrete both testosterone and E_2, and in many species the plasma levels of testosterone exceed those of E_2. Vitellogenin production by the liver is stimulated by $\dot{E}_2$. Progesterone secreted by the ovary is elevated in gravid lizards and snakes, whereas in turtles the postovulatory levels of progesterone are much lower than for preovulatory females, reflecting different reproductive roles. Live-bearing squamates maintain elevated levels of progesterone throughout pregnancy. Detailed studies have documented that E_2 regulates sexual receptivity in both female lizards and snakes. Oviposition or birth appears to be controlled by AVT from the pars nervosa, which stimulates oviductal and/ or uterine contractions.

Although visual cues are important in reptilian reproduction, pheromones have been implicated in reproductive behaviors of all major groups of reptiles, but they have been studied in most detail in squamates. The secretions of androgen-dependent femoral glands that occur along the inner thighs of many lizards are used in both courtship and territorial displays. The skin of emerging female red-sided garter snakes secretes a pheromone that signals readiness to mate. Female garter snakes can also secrete another chemical after mating to repel male attention.

Reproduction in Birds

All birds are oviparous, and fertilization is always internal. There are no dramatic differences among their reproductive modes. Eggs are typically deposited in a nest that is constructed and protected by one or both parents. Even after hatching, there is considerable care of the young by the parents. Birds use their high body temperatures to warm the eggs in the nest and hasten development. Birds are often classified as precocial or altricial depending on how independent or dependent on the parent they are at hatching, respectively. For example, precocial birds, in marked contrast to altricial birds, have well-developed feathers and are capable of thermoregulating and feeding themselves soon after hatching.

The sexes are always separate, and some species show marked sexual dimorphism, but others exhibit little or none. Among the dimorphic species, it is usually the male that exhibits colorful plumage and the female that exhibits more drab plumage, but these roles are reversed in a few species. Sex determination is of the ZZ/ ZW type in birds, with the males being the homogametic sex. Sex reversal has not been described in birds. Reproduction is tightly bound to environmental cues such as availability of nest materials, food, and/or water and photoperiod and/or temperature conditions.

The anatomy of most birds has been greatly modified for flight, and these modifications have their counterparts in the reproductive system. Associated with the ability to fly, many species are migratory and breeding sites may be separated from non-breeding sites by great distances. Most female birds have only one functional ovary and one oviduct in adults. The second gonad usually remains undeveloped as a rudimentary ovotestis or testislike structure. During the nonbreeding season the gonads of migratory species are extremely small and grow markedly after arrival at the breeding site. The testes of passerine birds, for example may increase 500× within a few weeks of arriving at the breeding site.

The endocrine regulation of reproduction in birds is much like that of other vertebrates with GnRH stimulating LH and FSH secretion and similar feedback loops. In males, the testes, which are active only during the breeding season, consist of seminiferous tubules containing spermatogenetic tissue and Sertoli cells. Leydig cells located between the tubules are responsible for androgen synthesis. The Sertoli cells also are steroidogenic. Plasma testosterone and DHT are both secreted by the gonads. At the beginning of the breeding season, there is a sharp increase in testosterone, and elevated androgen levels are responsible for male–male aggression and territorial behavior as well as for singing. Androgens are responsible for the seasonal proliferation of neurons in the sexually dimorphic nucleus of the preoptic area that controls singing. The sensitivity of this brain region to androgens is abolished by estrogen treatment of eggs, although the action of testosterone in the adult bird depends on conversion of testosterone to E_2 by brain aromatase. Elevated androgens are responsible for nest-building behavior, too. During the period of egg incubation, androgen levels fall, with an associated decrease in these behaviors. Castration abolishes all of these behaviors, whereas androgen therapy accentuates or prolongs them.

In female birds, the ovary produces considerable amounts of testosterone as well as E_2. In some species (e.g., white crowned sparrows) testosterone levels are higher in the female, whereas in other species (e.g., king penguins) the reverse is true. Both E_2 and testosterone peak about the same time just before egg laying and incubation. A major role for E_2 is the stimulation of vitellogenin production by the liver. Recent studies of kestrels have demonstrated deposition of increasing amounts of androgens in successive eggs laid by a female, and it is suggested that these androgens increase growth so that individuals hatching later are at less of a disadvantage with respect to earlier-hatching siblings.

In some birds, E_2 together with prolactin causes defeathering and an increase in the vasculature of a ventral region of skin (called an apterium or incubation patch) that allows for more efficient heat transfer from parent to egg. In pigeons, prolactin also stimulates proliferation of the epithelium of the crop sac of the parent, which then sloughs off and is regurgitated to feed the young birds.

Estrogens and androgens have early developmental roles in sex determination and differentiation. In juvenile songbirds, estrogens are important in learning the species' characteristic song. Treatment of juvenile male songbirds with tamoxifen, an estrogen receptor antagonist, blocks song learning.

Other Hormones Affecting Reproduction

The cooperative role for thyroid hormones on reproduction is well known in mammals. In mammals, thyroid hormone deficiencies can delay sexual maturation, and spermatogenesis may not occur. Ovarian weight is reduced in hypothyroid females, and ovarian cycles become irregular. There are reports of a positive relationship between thyroid function and successful reproduction in some fish and reptilian species, but these conclusions are based more on correlational observations than on experiments that demonstrate cause–effect relationships. In zebrafish, severe disruption of thyroid function with ammonium perchlorate has no affect on reproduction (Patiño et al., 2003). Experiments using thyroidectomy or treatment with antithyroid substances support an important role for thyroid hormones in enhancing reproduction of amphibians, although field experiments have not shown a cause–effect relationship in natural populations. In birds, the picture may be more complicated. Thyroid function is positively associated with reproduction in birds, as supported by studies of domestic species using thyroidectomy and antithyroid drugs. However, in all wild species studied, thyroid hormones appear to antagonize reproduction, and thyroidectomy may yield precocial sexual development.

Stress, and hence corticosteroids, are well known for their ability to shut down reproduction in many different species. Simply capturing an animal and holding it captive for a few hours may markedly depress circulating gonadal steroid levels, which may have drastic consequences for reproduction. There are a few exceptions to the rule concerning negative impacts of stress and corticosteroids on reproduction. The males of marsupial mice and both males and females of Pacific salmon undergo sexual development and breed when corticosteroid levels are extremely high. However, all of these animals die after spawning only once. Another exception, noted earlier, is the apparent positive involvement of a corticosteroid in the Indian catfish.

Considerable evidence indicates that melatonin secreted by the pineal gland may inhibit reproductive activity at the hypothalamic level. Long photoperiod has stimulatory effects on reproduction by the inhibitory action of light on melatonin secretion. Clinical studies have linked many cases of sexual precocity in humans to pineal dysfunction, especially in males. Melatonin treatment always produces inhibition of reproductive functions in amphibians, reptiles, and birds, whereas in teleosts and mammals the effect depends on the species examined. However, even in these latter groups, the predominant case is for inhibition by melatonin. In mammals, melatonin inhibits cells in the pars tuberalis that normally secretes a paracrine factor (tuberalin) that stimulates prolactin release from lactoropes in the nearby pars distalis of the pituitary (Lincoln, 2000).

References

Belair, B., Richard-Mercer, N., Pieau, C., Dorizzi, M., 2001. Sex reversal and aromatase in the European pond turtle: Treatment with Letrozole after the thermosensitive period for sex determination. J. Exp. Zool. 290, 490–497.

Devlin, R.H., Nagahama, Y., 2002. Sex determination and sex differentiation in fish: An overview of genetic, physiological, and environmental influences. Aquaculture 208, 191–364.

Gabriel, W.N., Blumberg, B., Sutton, S., Place, A.R., Lance, V.A., 2001. Alligator aromatase cDNA sequence and its expression in embryos at male and female incubation temperatures. J. Exp. Zool. 290, 439–448.

Gorbman, A., 1997. Hagfish development. Zool. Sci. 14, 375–390.

Hale, R.C., Greaves, J., Gunderson, J.L., Mothershead, R.F., II, 1991. Occurrence of orga-nochlorine contaminants in tissues of the coelacanth *Latimeria chalumnae*. Environ. Biol. Fish. 32, 361–367.

Hardisty, M.W., Potter, I.C., 1971. The general biology of adult lampreys. In: Hardisty, M.W., Potter, I.C. (Eds.), The Biology of Lampreys. vol. 1. Academic Press, London, pp. 127–206.

Lincoln, G., 2000. Melatonin modulation of prolactin and gonadotrophin secretion. Sys-tems ancient and modern. In: Olcese, J. (Ed.), Melatonin after Four Decades. Kluwer Academic/Plenum Publishers, New York, pp. 137–152.

Munsick, J.A., Bruton, M.N., Balon, E.K. (Eds.) 1991. The Biology of *Latimeria chalumnae* and Evolution of Coelacanths. Kluwer Academic Publishers, Dordrecht, the Nether-lands.

Patiño, R., Wainscott, M.R., Cruz-Li, E.I., Balakrishnan, S., McMurry, C., Blazer, V.S., Anderson, T.A., 2003. Effects of ammonium perchlorate on the reproductive perfor-mance and thyroid follicle histology of zebrafish. Environ. Toxicol. Chem. 22, 1115–1121.

Selcer, K.W., Clemens, J.W., 1998. Androgens, subavian species. In: Knobil, E., Neill, J.D. (Eds.), Encyclopedia of Reproduction, vol. 1. Academic Press, San Diego, CA, pp. 207–213.

Sower, S.A., 1998. Brain and pituitary hormones of lampreys, recent findings and their evolutionary significance. Am. Zool. 38, 15–38.

Sower, S.A., Gorbman, A., 1998. Agnatha. In: Knobil, E., Neill, J.D. (Eds.), Encyclopedia of Reproduction, vol. 1. Academic Press, San Diego, pp. 83–90.

Staub, N.L., DeBeer M., 1997. The role of androgens in female vertebrates. Gen. Comp. Endocrinol. 108, 1–24.

For Further Readings

General Vertebrate Reproduction

Bentley, P.J., 1998. Comparative Vertebrate Endocrinology, 3rd edition. Cambridge Uni-versity Press.

Gorbman, A., Sower, S.A., 2003. Evolution of the role of GnRH in animal (metazoan) bi-ology. Gen. Comp. Endocrinol. 134, 207–213.

Jones, R.E., Baxter, D.C., 1991. Gestation, with emphasis on corpus luteum biology, pla-centation, and parturition. In: Pang, P.T.K., Schriebman, M. (Eds.), Vertebrate Endo-crinology: Fundamentals and Biomedical Implications, vol. 4, pp. 205–302. Academic Press, NY.

Knobil, E., Neill, J.D., 1999. The Encyclopedia of Reproduction. Academic Press, New York.

Lange, I.G., Hartel, A., Meyer, H.H.D., 2003. Evolution of oestrogen function in vertebrates. Steroid Biochem. Mol. Biol. 1773, 1–8.

Naz, R.K. 1999. Endocrine Disruptors. Effects on Male and Female Reproductive Systems. CRC Press, Boca Raton, FL.

Norris, D.O., 1997. Vertebrate Endocrinology, 3rd edition. Academic Press, New York.

Norris, D.O., Jones, R.E., 1987. Hormones and Reproduction in Fishes, Amphibians and Reptiles. Plenum Press, New York.

Van Tienhoven, A., 1983. Reproductive Physiology of Vertebrates, 2nd edition. Cornell University Press, Ithaca, NY.

Fishes

Callard, G.V., 1988. Reproductive physiology. Part B. The Male. In: Shuttleworth, T.J. (Ed.), Physiology of Elasmobranch Fishes. Springer-Verlag, Berlin, pp. 292–317.

Callard, I.P., Klosterman, L., 1988. Reproductive physiology. Part A. The female. In: Shuttleworth, T.J. (ed.), Physiology of Elasmobranch Fishes. Springer-Verlag, Berlin, pp. 277–291.

Goetz, F.W., Thomas P., 1995. Proceedings of the Fifth International Symposium on the Reproductive Physiology of Fish. FishSymp 95, The University of Texas, Austin TX.

Groot, C., Margolis, L., 1991. Pacific Salmon Life Histories. UBC Press, Vancouver.

Hoar, W.S., Randall, D.J., Donaldson, E.M., 1983. Fish Physiology, vol. IX. Reproduction. Academic Press, New York.

Koob, T.J., Callard, I.P., 1999. Reproductive endocrinology of female elasmobranchs: Lessons from the little skate (*Raja erinacea*) and spiny dogfish (*Squalus acanthias*). J. Exp. Zool. 284, 557–574.

Lee, Y.-H., Lee, Y.L., Yeuh, W.-S., Tacon, P., Du, J.-L., Chang, C.-N., Jeng, S.-R., Tanaka, H., Chang, C.-F. 2000. Profiles of gonadal development, sex steroids, aromatase, and gonadotropin II in the controlled sex change of protandrous black porgy, *Acanthopagrus schegeli* Bleeker. Gen. Comp. Endocrinol. 119, 111–120.

Lethimonier, C., Madigou, T., Muñoz-Cueto, J.-A., Lareyre, J.-J., Kah, O., 2004. Evolutionary aspects of GnRHs, GnRH neuronal systems and GnRH receptors in teleost fish. Gen. Comp. Endocrinol. 135, 1–16.

Mainre, C.A., Rasmussen, L.E.L., Gross, T.S., 1999. Serum steroid hormones including 11-ketotestosterone, 11-ketoandreostenedione, and dihydrohyprogesterone in juvenile and adult bonnethead sharks, *Sphyrna tiburo*. J. Exp. Zool. 284, 595–603.

Matsuda, M., Nagahama, Y., Shinomiya, A., Sato, T, Matsuda, C., Kobayashi, T., Morrey, C.E., Shibata, N., Asakawa, S., Shimizu, N., Hori, H., Hamaguchi, S. and Sakaizumi, M., 2002. *DMY* is a Y-specific DM-domain gene required for male development in the medaka fish. Nature 417, 559–563.

Rasmussen, L.E.L., Hess, D.L., Luer, C.A. 1999. Alterations in serum steroid concentrations in the clearnose skate, *Raja eglanteria*: Correlations with season and reproductive status. J. Exp. Zool. 284, 575–585.

Scott, A.P., 1987. Reproductive endocrinology of fish. In: Chester Jones, I., Ingelton, P.M., Phillips, J.G. (Eds.), Fundamentals of Comparative Endocrinology. Plenum Press, New York, pp. 223–256.

Sorenson, P.W. et al. 1995. Sulfated 17,20-dihydroxy-4-pregnen-3-one functions as a potent and specific olfactory stimulant with pheromonal actions in the goldfish. Gen. Comp. Endocrinol. 100, 128–142.

Warner, R.R., Swearer, S.E., 1991. Social control of sex change in the bluehead wrasse, *Thalassoma bifasciatum* (Pices: Labridae). Biol. Bull. 181, 199–204.

Wourms, J.P., Atz, J.W., Stribling, M.D., 1991. Viviparity and the maternal-embryonic relationship in the coelacanth *Latimeria chalumnae*. Environ. Biol. Fishes 32, 225–248.

Amphibians

Duellman, W.E., Trueb, L., 1985. Biology of Amphibians. McGraw Hill, New York.

Houck, L.D., Woodley, S.K., 1994. Field studies of steroid hormones and male reproductive behaviour in amphibians. In: Heatwole, H. (Ed.), Amphibian Biology, vol. 2. Social Behaviour. Surrey Beatty and Sons, Chipping Norton, NSW, Australia.

Moore, F.L., 1987. Reproductive biology of amphibians. In: Chester Jones, I., Ingelton, P.M., Phillips, J.G. (Eds.), Fundamentals of Comparative Endocrinology. Plenum Press, New York, pp. 207–221.

Pancak-Roessler, M.K., Norris, D.O., 1991. The effects of orchidectomy and gonadotropins on steroidogenesis and oogenesis in Bidder's organs of the toad *Bufo woodhousii*. J. Exp. Zool. 260, 323–336.

Taylor, D.H., Guttman, S.I.. 1977. The Reproductive Biology of Amphibians. Plenum Press, New York.

Wake, M.H., 1985. Oviduct structure and function in nonmammalian vertebrates. In: Duncker, H.R., Fleischer, G. (Eds.), Functional Morphology in Vertebrates. Gustav Fischer Verlag, Stuttgart, pp. 427–435.

Reptiles

Guillette, L.J., Jr., 1993. The evolution of viviparity in lizards. Bioscience 43, 742–751.

Jones, R.E. et al., 1991. Loss of nesting behavior and the evolution of viviparity in reptiles. Ethology 88, 331–341.

Owens, D.W., Morris, Y.A., 1985. The comparative endocrinology of sea turtles. Copeia 1985, 723–735.

Birds

Balthazart, J., Ball, G.F., 1995. Sexual differentiation of brain and behavior in birds. Trends Endocrinol. Metab. 6, 21–29.

Mammals

Adashi, E.Y., Leung, P.C.K., 1993. The Ovary. Raven Press, New York.

Burger, H., de Kretser, D., 1989. The Testis, 2nd edition. Comprehensive Endocrinology revised series. Raven Press, New York.

de Kretser, D., 1993. Molecular Biology of the Male Reproductive System. Academic Press, New York.

Findlay, J.K., 1994. Molecular Biology of the Female Reproductive System. Academic Press, New York.

Fraser, H.M., Lunn, S.F., 1993. Does inhibin have an endocrine function during the menstrual cycle? Trends Endocrinol. Metab. 4, 187–194.

Glickman, S.E., Frank, L.G., Davidson, J.M., Smith, E.R., Siiteri, P.K., 1987. Androstenedione may organize or activate sex reversed traits in female spotted hyenas. Proc. Natl. Acad. Sci. USA 84, 3444–3447.

Jones, R.E., 1997. Human Reproductive Biology, 2nd edition. Academic Press, New York.

Kierszenbaum, A.L., 1994. Mammalian spermatogenesis *in vivo* and *in vitro*: A partnership of spermatogenic and somatic cell lineages. Endocr. Rev. 15, 116–134.

Knobil, E., Neill, J.D., 1994. The Physiology of Reproduction, 2nd edition, vols. 1 and 2. Raven Press, New York.

Lee, M.M., Donahoe, P.K., 1993. Mullerian inhibiting substance: A gonad hormone with multiple functions. Endocr. Rev. 14, 152–164.

McLachlan, R.I., Wrefortd, N.G., Robertson, D.M., de Kretser, D.M., 1995. Hormonal control of spermatogenesis. Trends Endocrinol. Metab. 6, 95–100.

Part II

Toxicology and Risk Assessment

Introduction to Part II

Ernest E Smith

James A. Carr

David O. Norris

The purpose of part II (chapters 7–9) is to provide a brief introduction to the science of toxicology and to emphasize the need for stronger interactions between endocrinologists and toxicologists. A recurrent theme throughout these chapters is the importance of integrating toxicology and endocrinology to minimize logical interpretation from a false premise. The collection and interpretation of endocrine disruption data and their ecological importance require integration of the fields of endocrinology and toxicology, as well as integration of several major interdependent academic fields of study. It is obvious that interactions leading to an interdisciplinary team of toxicologists and endocrinologists are highly favorable for experimental design, data collection, and interpretation of data that are ultimately used to identify potential endocrine-disrupting chemicals (EDCs).

It is easy to see how chemicals that may alter endocrine function fall within the traditional toxicology paradigm. Similarly, it is quite clear that assessing endocrine-specific endpoints presents many new and different challenges. Many chemicals disrupt endocrine function in seemingly subtle ways that may not lead to obvious or immediate adverse effects but over time may adversely affect the population with respect to evolutionary fitness. In addition, exposure to chemical agents that may disrupt endocrine function at different stages of development further complicates interpretation and application of data pertinent to risk assessment. Assessment and interpretation become even more difficult when experimental designs are complicated by in utero, lactational, or lifetime exposure. Knowledge of exposure, toxicity, and dose–response relationships is required to understand the impact of potential EDCs because timing and duration play a major role in the toxicity of certain toxicants.

The manifestation of physiological effects caused by a toxicant depends on the magnitude of exposure, so dose–response relationships can indicate the characteristics of exposure and the degree to which a potential EDC might affect endocrine function. However, dose–response relationships for potential EDCs present some unique problems, as many EDCs may have effects at concentrations well below those that produce toxic effects (e.g., Welshons et al., 2003). For example, as described in chapter 2, the apparent dissociation constant (the ligand concentration required to bind one-half of the receptor molecules) for many hormone receptors is in the low nanomolar range. Thus, some have argued that testing contaminants at high concentrations may saturate closely related receptors and produce responses that are not mediated by the receptor in question. Toxicologists and endocrinologists should work together to identify and eventually integrate low-dose response endpoints into conventional risk assessment approaches.

The concept of hormesis illustrates a difference between traditional toxicology and the emerging field of endocrine disruption. This concept has emerged from the molecular biology and chemical toxicology literature and can be generally defined as a dose–response relationship in which there are inverse effects at low and high doses, resulting in a non-monotonic dose–response curve (Calabrese and Baldwin, 1999, 2003). Consideration of nonlinear dose–response relationships for EDCs are not surprising, given that some hormones can act on multiple receptor types (and subtypes) with potentially different intracellular signaling pathways. Thus, at low concentrations a ligand may bind with high affinity to receptor A and produce a stimulatory effect, whereas at higher concentrations the EDC may also bind with low-affinity to receptor B and produce an inhibitory effect. A similar observation might result from changes in metabolic processing of the chemical at the higher dose.

Some have argued that nonlinear dose–response curves challenge the traditional methods for determining no-observed-adverse-effect levels (NOAELs), defined as the greatest dose or concentration that can be administered to an organism without resulting in an adverse effect (see chapter 7). Following this line of reasoning, if a NOAEL is defined from a linear dose–response relationship as the lowest (or threshold) concentration that produces an adverse effect, then accepting a threshold effect concentration may overlook effects that are not necessarily adverse but occur at lower concentrations of the EDC. Although there are reports of non-monotonic dose–response effects for natural ligands and EDCs, the utility of using hormetic responses for assessing risk of a particular EDC (see chapter 8) has generated considerable controversy because it is often difficult to extrapolate from highly controlled ligand–receptor interactions to a response determined at the organism, population, and ecosystem levels. For example, an estrogenic effect at the whole-animal level may be caused by multiple mechanisms including direct action on estrogen receptors, induction of aromatase, alterations in the affinity of plasma sex hormone binding proteins for estradiol, or changes in the metabolism of estradiol. Nevertheless, the observed effects are real, regardless of understanding the pathway.

Quantification of EDC effects and relationships require clearly defined and validated biomarkers of effect and biomarkers of exposure, as well as knowledge about the life-history stage that is most sensitive to a contaminant. For example, there is a substantial amount of evidence that the sensitivity of an individual to an endogenous or exogenous gonadal steroid agonist or antagonist depends on the life-stage of that individual. Specifically, the gametes, embryonic stages, larval and fetal stages, juveniles, and adults represent different windows of sensitivity and thus in part determine impacts of EDCs (see Benson et al., 1999). Biomarkers are biochemical, cytological, histological, physiological (see Anderson et al., 1997) or behavioral indicators of exposure to EDCs.

Dose–response relationships, expression of biomarkers of exposure, and induction of biomarkers of effect are strongly influenced by routes of exposure. Although there are widely varying scenarios for EDC exposure, they generally can be divided into four categories: acute, subacute, subchronic, and chronic. Emphasis on routes of exposure will have to be reconsidered in evaluating EDCs in multiple species. An approximate descending order of effectiveness for the routes of exposure in mammals would be inhalation, intraperitoneal, subcutaneous, intramuscular, intradermal, oral, and dermal. However, there are major species differences in routes of exposure when one studies aquatic vertebrates, particularly during embryonic development, when the embryo is generally only protected by a thin gelatinous matrix. Dickerson and Smith provide a detailed description of the impact of routes and length of exposure for EDCs in chapter 7.

The results of single-species toxicity tests are often used to predict effects on the ecosystem. Multispecies tests for assessing ecosystem-level impacts of EDCs are not currently available, and in many cases the impact of EDCs at the ecosystem level cannot be extrapolated with any level of confidence. Moreover, demonstrating a causal link between a laboratory (or field) observation and environmental factors that cause disease or reduced fitness at the population level is not a traditional aspect of endocrine research.

The recognition of experimental limitations, such as that of studying a single species, must include the fact that animals are being exposed to mixtures of EDCs with similar biological acitivity. Some of these EDCs may be working through the same mechanism (e.g., two chemicals binding to the same receptor) or through different mechanisms (e.g., one chemical affecting synthesis of a hormone and a second chemical affecting its metabolism and inactivation). Hence there may be additive effects when an animal is exposed to a mixture of chemicals (see Ramamoorthy et al., 1997; Rajapaske et al., 2002; Silva et al., 2002; Tollefsen, 2002). Furthermore, a mixture of EDCs with different actions (estrogenic and antithyroid activity) may produce unique effects not seen with either EDC alone.

Knowledge of basic ecological relationships as well as the ecology of individual species (i.e., population studies) throughout their life cycle is critical to an understanding of the impacts of EDCs. Because of their unpredictable nature, communities and ecosystems are more difficult to examine (see van Straalen, 2003). Many of the same ecological principles apply when examining terrestrial and aquatic biomes, yet each has unique features that relate to studies of toxic substances and EDCs. Freshwater ecosystems, for example, can be separated into static systems (lakes, ponds) and flowing systems (rivers, streams), and each has unique subregions with peculiar types of organisms and relationships. A large lake may be segregated into distinct habitats containing specific inhabitants such as found in the littoral zone (shallow areas near shore), the limnetic zone (open water), and the benthic zone (bottom), whereas a small stream may be separated into riffles and pools (Kalff, 2001). Similarly, headwaters are very different environments from those downstream (Gomi et al., 2002).

Bioaccumulation refers to the uptake, storage, and accumulation of organic and inorganic contaminants by organisms living in any habitat (see Streit, 1998; Facemire, 2000; Kime, 2001). Bioaccumulation may be a consequence of bioconcentration resulting from absorption through gills, skin, or lungs after exposure to contaminated water, air, or soil. It can also result from dietary sources (biomagnification), and thus the ingestion of EDCs or toxicants can be biomagnified by several million-fold as they are transferred through the food chain or food web to the top predators. Bioaccumulation can result in deleterious effects on organisms even when environmental levels are very low (Streit, 1998; Facemire, 2000; Kime, 2001).

Effects of EDCs observed first at the molecular and cellular level may have effects at the organismal or population levels. For example, killifish, *Fundulus heteroclitus*, chronically exposed to an environment containing mixed contaminants exhibit altered behaviors that could affect their evolutionary fitness (Weis et al., 2003). Environmental stress of this nature can lead to reductions in population size and even extinction (see Hoffman and Hercus, 2000). How do we use such information in risk assessment?

To fully understand the risks posed by EDCs, a fundamental knowledge and application of basic ecotoxicological principles, as well as standard means for establishing causality are required. Recently, the International Program on Chemical Safety (Damstra, 2002) proposed criteria for assessing EDCs based on well-established criteria for identifying disease-causing agents. These include the temporality of the association, the strength of the association, the consistency of the effect, the biological plausibility of the effect, and evidence of recovery from the effect. These criteria have been used for years to assess health risks to humans and have recently been incorporated into ecological risk assessment methods (see chapter 9).

References

Anderson, M.J., Barron, M.G., Diamond, S.A., Lipton, A., Zelikoff, J.T., 1997. Biomarker selection for restoration monitoring of fishery resources. In: Dwyer, F.J., Doane, T.R., Hinman, M.L. (Eds.), Environmental Toxicology and Risk Assessment: Modeling and Risk Assessment. American Society for Testing Materials, pp. 333–359.

Benson, W.H., Aum, L., deFur, P., Gooch, J., Mihaich, E.M., Tyler, C., 1999. Reproductive and developmental effects of contaminants in oviparous vertebrates: Workshop summary, conclusions, and recommendations. In: Di Giulio, R.T., Tillitt, D.E. (Eds.), Reproductive and Developmental Effects of Contaminants in Oviparous Vertebrates. SETAC Press, Pensacola, FL, pp. 403–416.

Calabrese, E.J., Baldwin, L.A., 1999. Reevaluation of the fundamental dose-response relationship. Bioscience 49, 725–732.

Calabrese, E.J., Baldwin, L.A., 2003. Toxicology rethinks its central belief. Nature 421, 691–692.

Damstra, T., Barlow, S., Bergman, A., Kavlock, R., Van Der Kraak, G. (Eds.), 2002. Global Assessment of the State-of-the-Science of Endocrine Disruptors. World Health Organization, International Programme on Chemical Safety, Geneva.

Facemire, C.F., 2000. Bioaccumulation, storage, and mobilization of endocrine-altering contaminants. In: Guillette, L., Jr., Crain, D.A. (Eds.), Environmental Endocrine Disrupters. Taylor and Francis, London, pp. 52–81.

Giesy, J.P., 2001. Hormesis—Does it have relevance at the population, community or ecosystem levels of organization? Hum. Exp. Toxicol. 20, 517–520.

Gomi, T., Sidle, R.C., Richardson, J.S., 2002. Understanding processes and downstream linkages of headwater systems. Bioscience 52, 905–916.

Hill, A.B., 1965. The environment and disease: Association or causation? Proc. R. Soc. Med. 58, 295–300.

Hoffman, A.A., Hercus, M.J., 2000. Environmental stress as an evolutionary force. Bioscience 50, 217–226.

Kalff, J., 2001. Limnology. Prentice Hall, Englewood Cliffs, NJ.

Kendall, R.J., Anderson, T.A., Baker, R.J., Bens, C.M., Carr, J.A., Chiodo, L.A., Cobb, G.P. III, Dickerson, R.L., Dixon, K.R., Frame, L.T., Hooper, M.J., Martin, C.F., McMurry, S.T., Patino, R., Smith, E.E., Theodorakis, C.W., 2001. Environmental toxicology. In: Klaassen, C. (Ed.), Casarett & Doull's Toxicology: The Basic Science of Poisons, 6th edition. McGrawHill, New York.

Kime, D.E., 2001. Endocrine Disruption in Fish. Kluwer Academic Publishers, Boston.

Ramamoorthy, K., Vyhlidal, C., Chen, I.-C., McDonnell, D.P., Leonard, L.S., Gaido, K.W., 1997. Additive estrogenic activities of a binary mixture of 2',4',6'-trichloro- and 2',3',4',5'-tetrachloro-4-biphenylol. Toxicol. Appl. Pharmacol. 147, 93–100.

Rajapaske, N., Silva, E., Kortenkamp, A., 2002. Combining xenoestrogens at levels below individual no-observed-effect concentrations dramatically enhances steroid hormone action. Environ. Health Perspect. 110, 917–921.

Silva, E., Rajapaske, N., Kortenkamp, A., 2002. Something for "nothing"—Eight weak estrogenic chemicals combined a concentrations below NOECs produce significant mixture effects. Environ. Sci. Technol. 36, 1751–1756.

Streit, B., 1998. Bioaccumulation of contaminants in fish. In: Braunbeck, T., Hinton, D.E., Streit, B. (Eds.), Fish Ecotoxicology. Birkhäuser Verlag, Basel, pp. 353–387.

Tollefsen, K.-E., 2002. Interaction of estrogen mimics, singly and in combination, with plasma sex steroid-binding proteins in rainbow trout (*Oncorhynchus mykiss*). Aquat. Toxicol. 56, 215–225.

U.S. EPA, 1992. Framework for Ecological Risk Assessment. EPA/630/R-92/001. Office of Research and Development, U.S. Environmental Protection Agency, Washington, DC.

van Straalen, N.M., 2003. Exotoxicology becomes stress ecology. Environ. Sci. Technol. 325, 324A–330A.

Weis, J.S., Zhou, T., Santiago-Bass, C., Weis, P., 2001. Effects of contaminants on behavior: Biochemical mechanisms and ecological consequences. Bioscience 51, 209–217.

Welshons, W.V., Thayer, K.A., Judy, B.M., Taylor, J.A., Curran, E.M., vom Saal, F.S., 2003. Large effects from small exposures. I. Mechanisms for endocrine disrupting chemicals with estrogenic activity. Environ. Health Perspect. 111, 994–1006.

7

Introduction to the Science of Toxicology

Richard L. Dickerson
Ernest E. Smith

Toxicology is not a single discipline of science but a synthesis of the disciplines of anatomy, biochemistry, biology, chemistry, physiology, physics, and modeling. Toxicology is a science that was borrowed from the ancient poisoners but has expanded and advanced to the study of molecular biology using toxicants as tools (Gallo, 1995). Toxicology is concerned with adverse effects of physical and chemical agents (toxicants) on living organisms. These agents include drugs, industrial chemicals, plant and animal poison or venoms, light, noise, and radiation. Effects elicited by exposure to these agents can range from alteration of function at the molecular level to organismal death. Moreover, in some cases, the effects of the agent can be passed on to offspring.

Natural toxicants are referred to as toxins. These include the venom of animals, poisonous plants and animals, and the endotoxins and exotoxins of various prokaryotes. Classification of other agents is more complex, and several systems have been proposed. Examples of toxicant classification systems include source (animal, manufacturing, plant), structure (metal, polychlorinated dibenzo-*p*-dioxin, steroid), target organs (liver, kidney, central nervous system, gonads), use (pesticides, solvents, food additives), and effects (cancer, immunosuppression, feminization, masculinization). Other classifications include physical state, mechanism of action, and degree of toxicity.

Many classes of toxicants can act as endocrine-disrupting chemicals (EDCs). These include compounds that mimic natural hormones and thus can bind to their receptors, resulting in either activation or inhibition of function. These compounds, which may be endogenous or exogenous, can increase or decrease the concentration of an endogenous compound by altering the activity of its biosynthetic enzymes or by altering its metabolism and excretion as well as plasma binding. Certain toxicants can induce apoptosis of gonadal cells, resulting in diminution of function. Others can influence the migration of germ cells to the primordial gonads during development. Still others may affect the circadian clock, thus altering normal release patterns of hormones. Known EDCs include metals (Cd, Hg, etc.; see chapter), combustion products (polynuclear aromatic hydrocarbons [PAHs], polychlorinated dibenzo-*p*-dioxins and furans [PCDDs and PCDFs]; see chapters 10, 11), pesticides (chlordane, DDT); see chapters 12, 16 and plant and fungal compounds (coumestrol, zearalenone; see chapter 15). This chapter is intended to give a broad view of the general principles of toxicology.

Exposure

For a substance to adversely affect an organism, it must enter the organism with sufficient magnitude and duration at the target site. In general, the amount of compound applied to the animal is called the external dose; the amount inside the organism is called the internal dose, and the amount that reaches the target organ is called the target organ dose. In some cases, the compound itself is not toxic and must be transformed into a bioactive compound by the action of the organism's own biotransformation enzymes. An example is the pesticide methoxychlor, which must be demethoxylated (*O*-dealkylation resulting in a hydroxyl group on the phenyl ring instead of a methoxy group, a reaction catalyzed by a cytochrome P450) to bind to the estrogen receptor (ER). The toxicity of a specific chemical is based on duration and frequency of exposure, magnitude of exposure, and route of administration, as well as the sex, strain, and species of recipient. The biological half-life of a toxicant, defined as the time for the plasma concentration to decrease to one-half of its initial value, depends on the rates of absorption, metabolism, and excretion for that compound.

The most common routes of exposure for environmental toxicants are ingestion, dermal absorption, and inhalation. Some toxicants and many drugs are injected into the subcutaneous space, muscle, veins and arteries, the peritoneal cavity, or into the cerebrospinal fluid. In general, ingestion and dermal absorption are slow compared to inhalation or intravenous administration. Ingestion is hindered by the rate of passage of the compound from the oral cavity to the area of the gastrointestinal tract from which it is absorbed. This subject is further explored in the section on absorption. Similarly, dermal absorption is diffusion dependent. The compound may have to pass through multiple tissue layers before it reaches systemic circulation. Due to these differences in the rate by which a compound reaches systemic circulation, a dose of compound that is lethal by inhalation or injection may be only damaging by ingestion or dermal absorption. For EDCs, ingestion and dermal absorption predominate as routes of exposure.

The other important factor in exposure is time of exposure, and this includes both duration and frequency. The exposure of animals to xenobiotics can be divided into four categories by length of exposure: acute (less than 24 h); subacute (continuous or repeated exposure for less than 1 month); subchronic (continuous or repeated exposure for 1–3 months) and chronic (continuous or repeated exposure for greater than 3 months) (Eaton and Klaassen, 2001). The effects elicited by acute exposure often differ from those resulting from chronic exposure. For example, acute exposure to benzene results in typical, ethanol-like central nervous system depression (primarily due to benzene itself), whereas repeated exposure may lead to aplastic anemia or lymphocytic leukemia (due to bioactivation of benzene to benzoquinone). Intermittent, long-term exposure may result in a combination of acute and chronic effects.

The frequency of repeated exposure is also an important consideration. All compounds that an individual is exposed to are biotransformed and excreted. The rate of biotransformation/excretion compared to the frequency of administration determines whether a compound will bioaccumulate over time or will ebb and fall. Generally, if the time required to eliminate 50% of the compound (biological half-life) exceeds the dosing interval, the compound may be expected to bioaccumulate in the organism, leading to an increase in toxicity.

Bioaccumulation, Bioconcentration, and Biomagnification

Depending on the nature of a compound, it may be refractory to metabolic inactivation and excretion. Many compounds are lipophilic and stable in the environment, including DDE [1,1-dichloro-2,2-bis(4-chlorphenyl)ethene], DDT, the polychlorinated biphenyls (PCBs), PCDDs, PCDFs, and chlorinated cyclodienes (dieldrin, aldrin and heptachlor) and camphors (toxaphene). For terrestrial animals, accumulation and storage of these compounds in fat and other lipid-rich tissue is referred to as bioaccumulation. The source of these compounds is predominantly from the food chain, but dermal absorption may be significant for some compounds. Inhalation of these compounds is uncommon due to their low volatility, but inhalation may occur in dusty environments, as these compounds adhere to dust particles. This time-dependent increase in concentration of a toxicant may result in deleterious effects and/or death. A similar concept is used for aquatic organisms. The ratio between the concentration of a toxicant in water to that in a target organism is referred to as the bioconcentration factor. This ratio can be predicted by examining the partition coefficient, K_{ow} (octanol-water partition coefficient), for the compound between water and an organic solvent. Basically, the accumulation rate of a compound in fish is described by the following equations:

$$dC_F/dT = k_1 C_w - k_2 C_F$$

and

$$1 / k_2 = t_w K_{ow} + t_o.$$

In these equations, C_W is the concentration in the water, C_F is the concentration in the fish, T is time, and k_1 and k_2 are the uptake and clearance rate constants. A more detailed explanation can be found in *Multimedia Environmental Models* (Mackay, 1991). Oil and octanol are frequently used. For example, the K_{ow} for DDE is approximately 300. Rainbow trout exposed to 1 ppt DDE for 242 days had a tissue concentration of 172 ppb, a bioconcentration factor of 1.8×10^5 (Matsumura et al.,

1975; Hamelink, 1977). There is evidence that the bioconcentration factor increases with the organism's position in the food chain. A classic example concerns PCB residues in the Great Lakes.

Polychlorinated biphenyls were produced until 1977 for use in transformers, capacitors, heat transfer fluids, printer's inks, and many other products. The only U.S. producer was Monsanto Chemical Company of St. Louis, Missouri, who marketed the Aroclor series of PCBs. Their attractive industrial properties of high dielectric constant, thermal stability, low volatility, and minimal chemical reactivity resulted in wide use but also caused environmental persistence. These compounds are extremely lipophilic (among the highest known values) and are readily stored in fat. Midges bioconcentrate PCBs from the water and sediment and are a food source for gizzard shad, a fish species exposed to PCBs in the water as well. The shad were then consumed by trout and other salmonids, which were in turn consumed by gulls, ospreys, and bald eagles. This trophic concentration of PCBs and other toxicants is referred to as biomagnification. The trout and other salmonids developed blue sac disease and goiter, while the piscivirous birds developed brain asymmetry, craniofacial deformities, and reproductive failure. The ultimate bioconcentration is four orders of magnitude, depending on the exact species.

Timing of Exposure

There is a substantial amount of evidence that sensitivity of an individual to gonadal steroids depends on the life stage of that individual. Specifically, the fetus appears to be the most sensitive life stage for lasting impacts of gonadal steroids or agonists/antagonists in mammals (Birnbaum, 1994; Blanchard and Hannigan, 1994; Ojasoo et al., 1992). This is because of two important features that generally, but not always, distinguish developmental toxicology from other endpoints: sensitivity (temporal and dose) and irreversibility. Temporal sensitivity is conferred by the developmental sequence; a toxicant may act only at one stage lasting a brief time (hours to weeks), while other stages show no response. A compound that mimics a natural hormone may produce widely differing effects if administered before or after the hormone is naturally present. Alternatively, a compound may have little effect at environmentally relevant concentrations on a postpubescent animal but may prevent normal development if exposure occurs during embryonic development or puberty. This may vary within a species according to target tissues or may exhibit species-specific characteristics not observed in other organisms. Research with PCBs and dioxin in mammals has shown that gestational exposure is more critical than lactation exposure in eliciting developmental effects (Bjerke and Peterson, 1994; Bjerke et al., 1994a,b). Moreover, in extrapolating from rodent data to human data, it must be remembered that the beginning of the second trimester of human fetal development is equivalent to the first few days of postnatal development in the rat. Therefore, if multigenerational dose–response assessment is not performed, the following critical developmental stages may require evaluation: gametogenesis, early gestation, late gestation, lactation, puberty, spermatogenesis, oogenesis, mating behavior, and copulation. Similar consideration of relevant developmental stages are equally important for invertebrates and nonmammalian vertebrates. In assessing effects of a potential EDC, it is important to emphasize that the timing of exposure is everything.

Dose sensitivity to many compounds, including EDCs, is generally higher in fetal and perinatal individuals than in adults. However, in some cases, the presence of fetal serum binding proteins may result in lower sensitivity to these compounds. For example, the ability of α-fetoprotein to bind 17β-estradiol (E_2) appears to protect fetal male rats from maternal estrogen (Herve et al., 1990). Without this protection, feminization would occur. Thus, agents that bind to or inhibit the formation of α-fetoprotein can result in endocrine disruption. Although this is a common mechanism in rodents, other species have developed other mechanisms to protect developing embryos from maternal or fetal hormones. The sex steroid-binding globulins, cortisol-binding globulin, and other plasma proteins such as albumin can also bind both endogenous hormones and mimetics. These are found in many vertebrates.

Often, a rapidly developing organism does not repair damage effectively as an adult does; consequently, chemicals that are reversible toxicants in adults may be irreversible toxicants during early development. An EPA workshop identified development of reproductive capability as the highest research priority in consideration of the features discussed above (USEPA, 1995). Therefore, any testing protocol must take effects on reproductive capability into consideration.

Assessment of Toxicity

As humans, we like to classify things in terms of relative importance, and toxicants are no exception. It is our desire to know whether a compound is of low toxicant potency, like oxygen, or of high toxic potency, like botulinum toxin, as well as its source. Typically, the toxicity of a compound is determined by testing in organisms. The simplest measurement is the amount of compound required to kill 50% of the exposed subjects within a given time period. This test is performed by subjecting organisms to a graded exposure of a given toxicant and measuring mortality. The dose required to kill 50% of the organisms is called the lethal dose 50% or LD_{50}. When exposure is to an environmental concentration of a toxicant, such as found in the atmosphere or in lake water, it is referred to as the lethal concentration 50% or LC_{50}. In both of these tests, it is common to specify a time in which the lethality occurred, such as 96 h or 30 days. Table 7-1 lists the toxicities of some well-known toxicants. Data used to generate such information are usually obtained from a dose–response curve (see below), which plots the number of organisms that died versus the amount of toxicant administered. These data also can be used to generate other relative measures of toxicity such as LD_{10} or LC_{90}, which are the doses required to kill 10% of organisms and the concentration required to kill 90% of the organisms exposed, respectively.

Nonlethal endpoints are also used in relative toxicity studies. For example, one might examine the dose-dependent decrease in cholinesterase activity versus an applied dose of a certain organophosphorus insecticide. In this case, toxicologists use the term effective dose or concentration required to elicit an effect. Thus, an

Table 7-1. LD_{50} values for well-known toxicants in the rat.

Substance	LD_{50} (mg/kg)
Ethanol	10,000
Morphine sulfate	900
Phenobarbital	150
Picrotoxin	5
Nicotine	1
Hemocholinium-3	0.2
Tetrodotoxin	0.1
2,3,7,8-tetrachlorodibenzo-p-dioxin	0.001
Botulinum toxin	0.00001

Adapted from Eaton and Klaassen (2001).

ED_{50} dose would be one that is required to obtain a 50th percentile response. The dose–response curves thus determined are then used to generate other reference points for judging toxicity. One of these is the no-observed-effect level, or NOEL. This is the highest dose or concentration that can be administered to an organism without an observable effect. A similar endpoint is the no-observed-adverse-effect level, or NOAEL. This is the highest dose or concentration that can be administered to an organism without an adverse effect.

Other useful indices that are commonly used by regulatory toxicologists in determining the safety of a compound are the allowable daily intake (ADI), maximum tolerated dose (MTD), margin of safety (MOS), and the therapeutic index (TI). The ADI is defined as amount of a compound that can be absorbed on a daily basis by an organism over its life span without causing adverse effects. The MTD is the largest dose that can be given to an organism through the course of a chronic exposure without resulting in overt toxic responses such as weight loss, lethargy, or other behavioral changes. The TI is defined as the ratio between the lethal dose and effective dose to achieve a beneficial response, both at the 50th percentile. The MOS is defined as the ratio of the amount of compound required to kill 1% of the population to the effective dose for 99% of the population, a much more restrictive standard. Other related terms include the margin of exposure (MOE) and the benchmark dose (BMD). The MOE is the ratio of the NOAEL, as determined in animals in units of mg/kg/day, to the anticipated daily exposure in humans in the same units. No factors are used, but a MOE less than 100 is generally considered sufficient to warrant further evaluation. The BMD approach is considered by many to be more precise. In this approach, the dose–response curve is modeled and upper and lower confidence limits set. The lower confidence bound for a given dose, at a specified response level, is calculated and denoted as the benchmark response. This is used to determine a corresponding BMD. The BMD is then used to calculate a reference dose (RfD) (Kimmel and Gaylor, 1988; Kavlock et al., 1995).

Another aspect of toxicology that may be used therapeutically is selective toxicity. This occurs when a compound is more toxic to one form of life than another, even though the life forms are living in close proximity. Selective toxicity is the basis for pesticides and antibiotics. The injured form is usually referred to as the uneconomic form, whereas the protected form is the economic form. These forms may be parasite and host organisms or a cancer in an organism. The chemical agents may be selective for several reasons. First, the compound may be equitoxic but may accumulate in the uneconomic cells. Second, the drug or toxicant may react with a cytological or biochemical feature that is present in uneconomic cells or organisms but absent in economic cells or organisms.

Mechanisms of Toxicology: Toxicodynamics

Toxicants can interact with organisms to cause damage by a number of mechanisms, depending on the species or target tissue. These mechanisms may be complex or simple. For example, the *Diffenbachia* toxin oxalic acid causes damage by blocking renal tubules as it precipitates due to the change in pH. Ethylene glycol first must be converted by alcohol dehydrogenase in the liver to the aldehyde and then to oxalic acid before it reaches the kidney. Basically, a typical mechanism begins with delivery of the toxicant to the organism. For simple toxicants such as oxalic acid, this is sufficient for a toxic response. Other toxicants require delivery followed by interaction with target molecules, resulting in cellular dysfunction, injury, and disrepair. The study of the interactions of a toxicant with target molecules is called toxicodynamics. The most common mechanisms of toxicity are discussed below.

Oxidative Stress

During normal aerobic metabolism, oxygen radicals are formed and play a role in oxygen toxicity. Radiation also generates reactive oxygen radicals. These reactive oxygen radicals include the hydroxyl radical, perhydroxyl radical, superoxide anion, and hydrogen peroxide. Other compounds can form free radicals as well, which can react with molecular oxygen or water to form reactive oxygen radicals. By definition, a free radical is a molecule or fragment of a molecule that contains one or more unpaired electrons in its outer electron shell or orbital. Free radicals are formed by reduction–oxidation (redox) reactions or by fission of a covalent bond.

A number of electrophilic xenobiotic compounds can accept electrons from reducing agents, such as reductases, to form radicals. Examples include paraquat, diquat, doxorubicin, and nitrofurantoin (Kappus, 1986). These xenobiotic radicals then transfer an electron to molecular oxygen, forming superoxide anions and regenerating the parent compound, which is then free to accept yet another electron. This process is referred to as redox cycling, and through this process, one xenobiotic molecule can generate many superoxide anion radicals.

Nucleophilic xenobiotics, such as phenols, catechols, hydroquinones, thiols, and similar compounds, lose an electron to peroxidases and thus contribute to the formation of free radicals. Hydroquinones and catechols undergo stepwise one-electron oxidations to semiquinones and quinones that are quite reactive themselves as electron acceptors and electrophiles. Polynuclear aromatic hydrocarbons can be converted to radical cations or dihydrodiol epoxides by sequential one-electron oxidations catalyzed by cytochromes P450 or peroxidases.

Carbon tetrachloride can be induced to fission by a one-electron transfer catalyzed by cytochrome P4502E1. This results in the formation of a trichloromethyl free radical and an inactivated P4502E1 because the chlorine radical forms a covalent bond with the reactive center of the enzyme, thus acting as a "suicide substrate" (Parkinson, 2001). The trichloromethyl free radical then can react with molecular oxygen to form a trichloromethylperoxy radical, which is extremely reactive. The hydroxyl radical can be formed from hydrogen peroxide by homolytic fission by a similar process.

Many cells of an organism commonly produce oxygen radicals (superoxide) to achieve biological function. An example of this is the respiratory burst in phagocytic white blood cells used to kill engulfed microorganisms. However, these cells contain superoxide dismutase and catalase or glutathione peroxidase, which protects them from oxygen radicals. Superoxide dismutase converts the superoxide anion into hydrogen peroxide. Both catalase and glutathione peroxidase are then able to convert hydrogen peroxide into water and molecular oxygen. If these detoxification pathways are not present, free radicals can damage cells by causing lipid peroxidation or by forming covalent bonds with proteins or nucleic acids.

DNA and Protein Adduct Formation
by Electrophilic Metabolites

Electrophiles are molecules that have one or more electron-deficient atoms, resulting in a partial or full positive charge. These molecules seek to achieve neutrality by obtaining an electron from another molecule. Thus, an electrophile reacts with electron-rich molecules, referred to as nucleophiles, by sharing electron pairs. Such nucleophiles include thiol groups of proteins and the imino groups of nucleic acids. The formation of a covalent bond between the electrophile and DNA or a protein is called adduct formation. These adducts may alter or eliminate the function of the molecule to which they are attached. Electrophiles are formed from a variety of parent compounds, usually by a cytochrome P450-mediated oxidation. These compounds are identified as a principal cause of chemical carcinogenesis (Miller, 1970). A notable example is formation of the 7,8-dihydrodiol-9,10-epoxide of benzo[a]pyrene. These reactive metabolites readily form covalent bonds with nucleic acids and proteins if not detoxified by conjugation with glutathione, catalyzed by glutathione-S-transferase (Ketterer, 1988). Both electrophiles and nucleophiles can be classified on a scale of soft to hard, depending on the magnitude of positive or negative charge present.

Enzymatic Alteration of Cellular Macromolecules

Some toxins act enzymatically on cellular proteins. Examples include ricin, diphtheria toxin, cholera toxin, and a number of snake venoms. Ricin, isolated from the castor bean (*Ricinus commonus*), causes the hydrolytic fragmentation of subunits making up the eukaryotic ribosome, thus interrupting protein synthesis (Norton, 1995). A number of snake venoms, including the crotalids (rattlesnakes), contain hyaluronidases, transaminases, phospholipase, phosphodiesterase, endonucleases, and cholinesterase (Gomez and Dart, 1995). These enzymes result in severe necrosis and alteration of cellular function.

Alteration of Gene Expression

Toxicants can alter gene expression by inducing or inhibiting transcription, stabilizing or destabilizing messenger RNA, and stabilizing or destabilizing the gene product. In general, transcription is controlled by formation of a transcriptional complex around the gene promoter. This may be facilitated in some cases by the binding of transacting elements to specific sequences in the enhancer region of the gene. Xenobiotic compounds may interfere with the promoter of the gene itself or with transcription factors or other transacting factors.

In general, two types of transcription factors are recognized: ligand activated and signal activated. Another superfamily of nuclear transcriptional factors is represented by the steroid-thyroid hormone receptors. Steroid and thyroid hormone receptors are examples of ligand-activated transcription factors. A number of xenobiotic compounds, both natural and anthropogenic, possess affinity for these receptors, resulting in inappropriate activation or inhibition of function. Examples include diethylstilbestrol (DES), DDE, DDT, methoxychlor, and a number of compounds used in the plastics industry (Arnold et al., 1996; Vonier et al., 1996; Foster et al., 1983). These compounds can alter both development and the ability to reproduce (Jobling et al., 1995; Kelce et al., 1995; McMaster et al., 1996). For example, E_2 binds to the ER, transforming it into a form capable of binding the estrogen response element, which activates the progesterone receptor gene, among others. A number of xenobiotics, including o,p'-DDT and DES, also can bind to the ER and initiate transcription and thus cause inappropriate mitogenesis.

A number of signal-activated transcription factors are known. Most of these involve phosphorylation of a transcription factor, resulting in its activation. Perhaps the best understood of these is activation protein-1 (AP-1), which is involved in many toxicant-induced alterations in gene expression (Angel and Karin, 1991). AP-1 is a family of dimeric proteins composed of subunits from the Fos and Jun families (c-*fos* and c-*jun*). AP-1 initiates gene transcription through the tetradecanoylphorbol acetate response element. This pathway involves activation of protein kinase C by exogenous or endogenous factors, which in turn activate AP-1. Perturbation of signaling pathways with resulting dysregulation of gene expression is thought to be one of the pathways leading to apoptosis (see below).

Cell Death

Cells must constantly synthesize endogenous compounds, assemble these into complexes, membranes and organelles, generate energy, and maintain their intracellular environment by movement of ions and molecules across their membranes. Interference with any of these functions can result in cell death. Cellular toxicants can be classified as compounds interfering with mitochondrial ATP synthesis, agents causing elevated cytosolic Ca^{2+}, or agents that block essential transport into or from the cell.

Xenobiotics can disrupt ATP formation by disrupting any part of oxidative phosphorylation, including inhibiting hydrogen delivery to the electron transport chain, inhibiting electron transport, inhibiting oxygen delivery to the electron transport chain, or inhibiting ADP phosphorylation. Examples of xenobiotics that interfere with hydrogen delivery include iodoacetate (glycolysis), coenzyme A depletors (gluconeogenesis), 4-pentenoic acid (fatty acid oxidation), arsenite (pyruvate dehydrogenase), fluoroacetate (aconitase), DCVC S-(1,2-dichlorovinyl)-1-cysteine (isocitrate dehydrogenase), and malonate (succinate dehydrogenase). Inhibitors of electron transport include rotenone (NADH-coenzyme Q reductase), antimycin A (coenzyme Q-cytochrome c reductase), and cyanide and hydrogen sulfide (cytochrome oxidase). Agents that inhibit oxygen delivery to the electron transport chain include central nervous system depressants, convulsants, ergot alkaloids, and carbon monoxide. Agents that affect ADP phosphorylation include DDT, pentachlorophenol, and chlordecone.

Exposure to a number of xenobiotic compounds causes a sustained elevation of intracellular Ca^{2+}. In a typical cell, there is a 10,000–fold greater concentration of Ca^{2+} in the extracellular fluid, as compared to cytosol, that is maintained by Ca^{2+} ATPases. Toxicants can cause an increase in cytosolic Ca^{2+} by two general mechanisms: causing Ca^{2+} influx into the cytoplasm and preventing Ca^{2+} export from the cytoplasm. These agents may work by inhibiting the function of Ca^{2+} ATPases, opening holes or pores in the plasma membrane, or by affecting voltage-gated or ligand-gated ion channels. Examples of agents working on ligand-gated ion channels include capsaicin, resiniferatoxin, glutamate, kainate, and domoate. The hydroxyl free radical and maitotoxin directly affect voltage-gated channels. Some agents (including maitotoxin, amphoceterin B, chlordecone, alkyl mercury and tin compounds, detergents, phospholipases, carbon tetrachloride, and phalloidan) can create "pores" or disrupt membrane integrity. Toxicants affecting Ca^{2+}ATPases include acetaminophen, dichloroethylene, carbon tetrachloride, chloroform, and metal ions such as vanadate, cystamine, diquat, and diamide. Alterations in cytosolic Ca^{2+} levels result in increased mitochondrial uptake of Ca^{2+} and impaired ATP production, microfilamental dissociation causing plasma membrane blebbing, and activation of proteases, phospholipases, and nucleases (Nicotera et al., 1992).

Agents that cause lipid peroxidation and thus disrupt membranes, agents that disrupt the cytoskeleton, and agents that disrupt protein or DNA synthesis can also result in cell death. Such agents include chlorinated solvents, phalloidan, actinomycin-D, cycloheximide, and ricin. The mechanisms of actual death may be a combination of Ca^{2+} influx and structural damage.

Apoptosis and Necrosis

Apoptosis is the directed deletion of damaged or unnecessary cells. An estimated 60–70 billion cells die each day in a human, mostly from apoptosis. This is a necessary function; insufficient cell death can lead to cancer and autoimmune disease, whereas excessive cell death can lead to neurodegenerative diseases such as Parkinson's or immune dysfunction such as AIDS. Apoptosis differs from necrosis in several ways but primarily in terms of morphology and mechanism. Apoptosis is an orderly, active process requiring gene activation, whereas necrosis is generally passive and disorderly from the standpoint of the cell (Bursch et al., 1992). In apoptosis, the cell shrinks, chromatin and cytoplasmic contents condense, and the cell breaks apart into membrane-surrounded apoptotic fragments, which are phagocytized by other cells without causing inflammation. Necrosis, in contrast, begins with the swelling of both the cell and its organelles, followed by lysis of the plasma membrane. This results in release of cellular debris, triggering an inflammatory response, which may result in damage to surrounding cells.

Apoptosis may be triggered by a number of factors. These include damage to the cell, damage to DNA, external signals such as glucocorticoids, or removal of signals that suppress apoptosis, such as placental uterotrophic hormones. Signal-induced apoptosis is a normal feature of development. However, xenobiotic-induced apoptosis may lead to the removal of normal cells needed to maintain function. Moreover, a number of compounds or agents may induce apoptosis at low doses, yet cause necrosis at higher exposures. Indeed, a continuum seems to exist, beginning with disrepair of DNA, progressing to apoptosis, and ending with necrosis. Examples of agents that trigger apoptosis include ionizing radiation and antitumor (chemotherapeutic) drugs.

Apoptosis involves a series of biochemical events that begin with detachment of chromosomes from the nuclear scaffold, resulting in chromatin condensation (Corcoran et al., 1994). This is followed by endonuclease-mediated hydrolysis of DNA in regions between histones resulting in DNA laddering. This may be facilitated by depletion of polyamines and the increase of intracellular Ca^{2+}. Transglutaminase synthesis is induced during the same time, resulting in cross-linking of ϵ-(γ-glutamyl) lysine, which may facilitate formation of membrane-enclosed apoptosis bodies. Activation of protein kinase A promotes apoptosis, whereas activation of protein kinase C inhibits the process. Apoptotic cells increase the formation of transforming growth factor β (TGF-β), a hormonal factor that inhibits cell division and increases apoptosis by acting through an autocrine loop. In addition, Nagata and Goldstein (1995) reported that cytotoxic T-lymphocytes produce a factor called fas ligand, which binds to the fas receptor, resulting in apoptosis of liver, heart, and lung cells.

Genetic control of apoptosis begins with activation of the c-*myc* gene, which takes the cell from G_0 to G_1. The tumor-suppressor gene *p53* blocks the progression from G_1 to S, perhaps to give time for DNA repair to occur. However, if this does not happen, the *bax* gene is activated, resulting in the initiation of apoptosis. The action of bax protein is opposed by bcl-2 protein, presumably due to the formation of a bax–bcl-2 heterodimer (Oltavi et al., 1993). Bcl-2 protein is located on the outer of the two mitochondrial membranes, as are at least 13 other members of the *Bcl-2* gene family. Of these proteins, five appear to be antiapoptotic and nine are proapoptotic. The interaction of these proteins in the control of apoptosis is still unclear. However, cytochrome C is released from between the mitochondrial membranes into the cytosol, where it activates the caspase family of cell death proteases. Bax, a member of the proapoptotic bcl-2 family (Jurgenmeier et al., 1998), mediates this release. The released cytochrome C induces formation of a complex that includes apoptosis protease activating factor 1 (apaf-1) and procaspase 9. This complex then causes procaspase 9 to cleave itself, thereby producing casp 9. Casp 9 then cleaves procaspase 3 into casp 3, the putative ultimate protease in apoptosis (Hakem et al., 1998).

Alteration of Cellular Maintenance

Damaged cells are either repaired or replaced by proliferation of stem cells or dedifferentiated parenchymal cells. Replacement involves the triggering of mitosis by factors such as hepatic growth factor and transforming factor α. Mitosis also can be observed after treatment with agents such as carbon tetrachloride (Noji et al., 1990; Lindroos et al., 1991). Such processes may be hindered by the development of inflammation. Circulating cells of the immune system, predominantly monocytes and granulocytes, are attracted to sites of cellular injury due to release of chemotactic agents by injured cells. These leukocytes undergo a respiratory burst, releasing hydroxyl radicals, superoxide anion radicals, and hydrolytic enzymes. The primary purpose of these radicals and enzymes is to eliminate microorganisms, but they may also injure nearby cells.

Failure to repair cells or tissues leads to necrosis at the cellular or tissue level. If severe enough, necrosis will lead to death of the organism. Necrosis can be circumvented by two mechanisms designed to limit the extent of the damage: apoptosis and proliferation. Apoptosis results in the tidy removal of a cell and does not cause release of chemotactic agents that may result in inflammation. Proliferation of cells within a tissue results in more than simple replacement of damaged cells; it appears to limit the toxic response. For example, pretreatment of rats with chlordecone, an insecticide which also inhibits proliferation, results in necrosis at much lower doses when rats are subsequently treated with carbon tetrachloride (Calabrese et al., 1993).

Necrosis, if not lethal, may result in fibrosis. Fibrosis is the deposition of extensive amounts of extracellular matrix of atypical composition and is one common form of tissue disrepair. Normally, cellular injury is followed by proliferation and production of extracellular matrix that ceases once wound healing has occurred and tissue remodeling has been completed. However, fibrosis occurs when cytokine-mediated overproduction of extracellular matrix is not turned off during wound healing. Examples include hepatic cirrhosis following ethanol consumption or exposure to carbon tetrachloride. Fibrosis in the lungs may result from a number of compounds including asbestos, oxygen, and bleomycin. Fibrosis is undesirable because it constricts parenchymal cells and vessels, isolates endothelial cells from parenchymal cells resulting in malnutrition, and stiffens the tissue, resulting in diminution of function.

Another form of tissue disrepair is carcinogenesis. Chemical and viral carcinogenesis result from alteration of the host genome by chemicals or alteration of the cell by the viral genome. In the case of chemical carcinogenesis, failure of a number of protective and repair mechanisms must occur. In the case of some chemicals, bioactivation of the compound from a procarcinogen to an ultimate carcinogen must also occur. Other steps in carcinogenesis are failure of DNA repair, failure to terminate proliferation, and failure to induce apoptosis. The occurrence of cancer often requires more than one insult to the genome of the cell. For example, the progression of a polyp to colon cancer requires mutation or deletion in five genes

(Kinzler and Vogelstein, 1996). These genes are a relatively small set of proto-oncogenes and tumor-suppressor genes and are affected by many carcinogens. The protooncogenes are essential for many signal transduction pathways, and alteration of a number of these protooncogenes and tumor-suppressor genes can result in uncontrolled cellular proliferation (Barrett, 1992).

EDCs may interact with multiple targets. There is evidence for EDCs acting at every level of hormone synthesis, secretion, transport, site of action, and metabolism, as well as causing reproductive cancers. EDCs may act at the level of hormone receptor, or they may disrupt the enzymes that synthesize or metabolize the endogenous hormones (see chapter 1). In addition, there may be multiple levels of response. There may be cascade responses in which a xenobiotic compound blocks negative feedback at the level of hypothalamic or pituitary receptors, resulting in abnormally high levels of circulating hormones (see chapter 3). In other cases, synergism or antagonism may exist between endogenous and exogenous ligands. A xenobiotic compound may exert effects at the receptor level through multiple mechanisms. These include differential effects at multiple receptor types or direct effects on intracellular signaling pathways, thereby directly influencing hormone action at the target tissue. Alternatively, EDCs can affect hormone-binding proteins in blood, thereby disrupting hormone transport. Xenobiotic compounds may act on the endocrine system by affecting transcription and signal transduction, and they can act through receptor-mediated or non–receptor-mediated mechanisms. For example, genistein has been shown to be a weak ER agonist. However, it also modulates the activity of tyrosine kinases and DNA topoisomerases (Piontek et al., 1993; Makela et al., 1994, 1995; Okajima et al., 1994; Olsen et al., 1994; Whitten et al., 1995). These examples point to a number of challenges in the selection of both doses and endpoints to be used in the development of dose–response assessments of potential EDCs. Dose selection should include doses below the apparent NOAEL to determine if a U-shaped curve is present. Often, the same response may result from two different exposures to the same toxicant. However, one may on the upward of the curve and the other on the downward portion of the curve.

If a compound acts through a mechanism other than receptor-mediated agonism or antagonism, it may still adversely alter endogenous levels of a critical hormone by inducing or inhibiting biosynthetic or metabolic enzyme activities (see chapter 2). These changes can alter endogenous hormone levels at the active site to cause endocrine disruption. For example, some phytoestrogens can interact with the 17β-dehydrogenase that regulates E_2 and estrone levels, suggesting they can modulate overall estrogen levels in addition to acting as a ligand for the ER. Also, perchlorate competitively inhibits thyroidal iodide uptake, thereby disrupting thyroid hormone synthesis.

It is postulated that certain EDCs disrupt stimulus-secretion coupling in endocrine cells, thereby disrupting normal hormone secretion. It has been known for many years that Cd^{2+} is a non-selective Ca^{2+} blocker that can disrupt Ca^{2+}-dependent exocytosis in hypothalamic neurosecretory neurons and pituitary endocrine cells, for example (Childs et al., 1987).

A number of compounds can cause apparent sex reversal, sex behavior alterations, and/or genital malformations in various species. For example, antiandrogens such as *p,p'*-DDE and 2,3,7,8-tetrachlorodibenzo-p-dioxin (TCDD) cause feminization of mammals, whereas estrogens cause feminization in a number of reptile species (Mably et al., 1992a,b,c; Guillette et al., 1994, 1996; Kelce et al., 1995). In contrast, many chlorinated compounds such as the polychlorinated biphenyls, alkanes, and dioxins cause masculinization of a number of bird species. These alterations affect apparent sex, genital differentiation, and mating behavior to the point of reproductive failure (see chapter 10, 11, 12).

Absorption, Distribution, Metabolism, and Excretion of Toxicants-Toxicokinetics

Absorption

In general, toxicants must pass through several cell layers in multicellular organisms before reaching their target. For mammals, these cell layers include the cornified squamous epithelium of the skin, the pneumocytes of the lung, the endothelial cells of the capillaries, and the cells lining the gastrointestinal (GI) tract. These cells may facilitate or hinder the passage of toxicants. The processes by which a toxicant enters an organism, reaches systemic circulation, is delivered to the target tissue, and is metabolized and excreted is referred to as toxicokinetics. With a thorough knowledge of physiology and biochemistry in the target species as well as the physicochemical properties of the toxicant, researchers can use physiologically based toxicokinetic (PBTK) models to predict distribution in an organism.

Toxicants may cross cell membranes by several mechanisms, which may be passive means such as diffusion and filtration (concentration is the driving force, and rates are predicted by Fick's laws) or active (energy-dependent) means such as active transport, facilitated diffusion, phagocytosis, and pinocytosis. The process by which substances cross membranes and enter the systemic circulation is absorption and is the focus of the remainder of this section.

Absorption of toxicants from the GI tract can occur in any location from the oral mucosa to the rectum. For diffusion, which is the predominant mechanism for most toxicants, the ionization state determines the ease of crossing the plasma membrane. A nonionized molecule enters the plasma membrane much more readily than does an ionized molecule. The ionization state of many molecules depends on the pH of the fluid within which they are dissolved. The pH varies from an average of 2 in the stomach to an average of 6 in the intestine. Weak acids are nonionized at pHs below their pKa and ionized at pHs above their pKa. Weak bases, on the other hand, are ionized at pHs below their pKa and nonionized at pHs above their pKa. Knowledge of the acid-base properties of the toxicant can enable the toxicologist to predict the part of the GI tract in which a toxicant will be absorbed. The transport of other toxicants can be predicted by their resemblance to endogenous molecules for which a transporter is present (e.g., pyrimidine transport system, iron transport system, calcium transport system, peptide transport system). Examples include transport of lead by the calcium transporter and transport of cobalt, thallium, and manganese by the iron transporter system.

Absorption of toxicants by skin and lung is somewhat simpler because neither organ exhibits substantial variation in pH. With the exception of compounds that are absorbed by active (energy-requiring) transport, most substances cross cellular membranes by diffusion. Examples include organic and chlorinated solvents, gases, and volatile anesthetics.

Distribution

After absorption, a toxicant is translocated throughout the compartments of the body by a process referred to as distribution. Once absorption takes place, distribution is quite rapid, being dependent on rate of perfusion of the tissue in which absorption is occurring and that of the target tissue. The water in an organism can be divided into three compartments: intracellular water, interstitial water, and plasma. The concentration in plasma depends on the volume of distribution of the toxicant. The volume of distribution depends on binding to protein and/or storage in bone, fat, or liver. Plasma proteins do not bind some xenobiotics, but others, such as warfarin, may be 90% bound to plasma proteins such as albumin. A toxicant that is not readily soluble in plasma membrane phospholipids will have restricted distribution with a higher plasma concentration.

Some organs may have a partial barrier to passage of some classes of compounds that is typically formed by tight junctions between the cells. Others have special transporters, such as the multidrug resistance proteins (MDR), that move chemicals into the blood. Examples of such barriers include the blood–brain barrier and the blood–testis barrier. Both serve to isolate sensitive tissues from contaminants, primarily by decreasing or eliminating pores through which contaminants can diffuse by establishing tight junctions between capillary endothelial cells, surrounding the capillaries by astroglial or other support cells, reducing the concentration of contaminants in the interstital fluid, and actively removing contaminants (MDR ATP-dependent transporter).

Metabolism

As discussed earlier, toxicants can be divided into two classes, lipophilic (hydrophobic) and hydrophilic (lipophobic), depending on their solubility in lipids or water. Because of the lipoidal nature of the plasma membrane, lipid-soluble compounds readily enter cells, but hydrophilic compounds have difficulty entering cells unless transported. One mechanism to eliminate toxicants is to render them water soluble and get them into the interstitial or plasma compartments, where they reach the bile or urine. Most cells can enzymatically alter xenobiotic compounds to add or expose functional groups, which can be used by conjugation enzymes to attach polar molecules, thus increasing the water solubility of the parent compound. These enzymes, which are referred to as drug metabolizing enzymes or xenobiotic metabolizing enzymes, fall into two broad classes: phase I (redox enzymes) and phase II (conjugation enzymes).

The purpose of phase I enzymes is twofold: to increase the water solubility of a compound and to add a functional group to serve as a "handle" for conjugation enzymes. The primary classes of enzymes that perform these functions are the cytochromes P450, the flavin-containing mixed-function oxidases, the esterases, alcohol dehydrogenase, epoxide hydralase, and rhodanase. Of these enzymes, the cytochromes P450 have the greatest capacity, broadest substrate specificity, and widest tissue distribution, although the highest concentration of these enzymes are found in the liver, lung, and kidney. Of the many cytochromes P450, relatively few are involved in xenobiotic oxidation reactions; the remainder perform oxidations on endogenous molecules to allow their excretion. The cytochromes P450 are transmembrane proteins, embedded in the smooth endoplasmic reticulum (SER) with the active site facing the lumen of the SER tubule. A number of P450 molecules surround a molecule of cytochrome P450 reductase. The enzymatic reaction uses a heme iron unit of the P450 molecule to accept and donate electrons. Cytochromes P450 can add hydroxyl groups, create epoxides and arene oxides across double bonds, dealkylate at carbon, oxygen, nitrogen, and sulfur groups, and perform a variety of dehalogenations. The cytochrome P450 genes that are responsible for the majority of xenobiotic transformation include *cyp1A1*, *cyp1A2*, *cyp1B1*, *cyp2A6*, *cyp2C*, *cyp2D6*, *cyp2E1*, and *cyp3A*. Of these, the most abundant form of the enzymes in human liver is CYP4503A, while in rat liver it is CYP4502C. Moreover, CYP4502D6 and CYP4503A metabolize the greatest number of pharmaceutical compounds, whereas CYP4501A1/2 metabolize the widest variety of polynuclear aromatic hydrocarbons and aromatic amines.

The purpose of phase II drug-metabolizing enzymes is to add a charged group (very polar) to a functional group, most commonly a hydroxyl or epoxide group formed by phase I metabolism. These groups may be original to the molecule or added through the action of a phase I drug-metabolizing enzyme. Types of conjugation include acetylation, methylation, or the addition of glucuronic acid, glucose, glutathione, or sulfate. The enzymes responsible for these conjugations also have wide tissue distribution but are most abundant in the liver and kidney. The substrates used by the conjugation enzymes are high energy. Examples include UDP-glucuronic acid (glucuronosyl transferase), phosphoadenosyl phosphorylsulfate (PAPS; sulfotransferase), and acetyl CoA (*N*-acetyltransferase). Of the conjugation enzymes, glucuronosyl transferase and glutathione-*S*-transferase have the highest capacity in terms of enzyme turnover rate, availability, and sufficiency of substrate. The addition of these charged groups, followed by their removal from the cell, ensures that the metabolite is water soluble and is hindered from being reabsorbed by other cells.

Bioactivation versus Detoxification

The process of metabolism is designed to render toxicants or endogenous lipid-soluble compounds water soluble and thus aid in the excretion of these compounds. However, occasionally either phase I or phase II reactions alter the compound into a form more toxic than the parent compound. This process of metabolically activating a compound to a more toxic form is called bioactivation.

An example of bioactivation is the metabolism of benzo[*a*]pyrene. This compound consists of five benzene rings fused together. It is found in smoke, soot, and some cooked foods. Potts demonstrated that soot was responsible for scrotal cancer in chimney sweeps, and it was later shown that the carcinogenic agent in soot was primarily benzo[*a*]pyrene (see Lawley, 1994). The parent compound is incapable of forming either DNA or protein adducts; thus, it must be altered to allow it to be active toward these biochemical macromolecules. Miller and Miller (1947) demonstrated that a number of chemicals must be bioactivated into electrophiles in order to become ultimate carcinogens. In the case of benzo[*a*]pyrene, two metabolic reactions are required, both catalyzed by CYP4501A1. In the first step, benzo[*a*]pyrene may be hydroxylated or an arene oxide formed in a number of sites on the parent compound. The majority of these products are not carcinogenic. However, one of two possible 7,8-dihydrodiol metabolites is more carcinogenic than the parent compound yet still unable to bind to macromolecules. After a second round of cytochrome P450-mediated oxidation, two congeners of benzo[*a*]pyrene-7,8-dihydrodiol-9,10-epoxide are formed. Synthesis of radiolabeled analogs of this metabolite of benzo[*a*]pyrene has been performed, and the metabolite is able to form DNA and protein adducts with guanine and cysteine residues, respectively (Conney, 1982). Typically, if enough glutathione is present in the cell, it can "mop up" the electrophiles and prevent harm. Also, DNA repair can remove the damaged guanine. However, the presence of a promoter, such as 12-0-tetradecanoyl phorbol-13-acetate (TPA) may accelerate the cell cycle, resulting in replication before repair.

Excretion

Except for very low molecular weight substances or volatile gases, which are excreted from the lungs, the majority of endogenous and xenobiotic compounds are excreted either in the urine or feces. A small amount of compound may be present in sweat or breast milk as well. Basically, compounds or their conjugates that have a molecular weight of 300 Da or less are excreted in the urine. Compounds in excess of 450 Da are excreted in the feces. Compounds whose molecular weight falls in the range of 300–450 Da may be excreted in the urine or the feces.

For compounds that are excreted in the bile, there is the possibility that bacterial action in the intestines may affect the excreted compounds metabolically. In particular, *E. coli* and other facilitative anaerobes can cleave glucuronic acid and sulfate groups from excreted molecules through the action of β-glucuronidase and sulfatase enzymes. The resulting parent compound can then be reabsorbed by the cells lining the gut and reenter the systemic circulation. This is referred to as enterohepatic circulation.

Dose–Response Relationships

The concept of the dose–response relationship is the most fundamental concept in toxicology (see Dickerson et al., 1998). It is predicated on the idea that exposure to a given chemical results in a biochemical, physiological, or behavioral response, and that the magnitude of the response is proportional to the magnitude (a function of both duration and intensity) of the exposure. Dose–response relationships can be determined for both individuals and populations and involve a number of important assumptions. First, the response is due to the chemical that is administered. Second, the response is related to the dose; however, the response does not have to be monotonic (either increasing or decreasing but not both) and linear. A response does imply that there is a receptor molecule with which the compound interacts to cause a response, that the response and its degree is due to the presence of the agent at the target site, and that the concentration at the target site is related to the dose administered to the organism. Third, there is a quantitative method for determining and expressing the toxicity elicited by the substance.

Dose–response relationships must be defined for a specific response to a single compound or class of compounds. The selection of a toxic endpoint is crucial but not always straightforward. Among the possible endpoints are lethality, changes in biochemical endpoints such as SGOT (serum glutamic oxaloacetic transaminase) and SGPT (serum glutamic pyruvic transaminase), changes in physiological endpoints such as activity or respiration, and tissue damage as quantified by necropsy or histopathological examination. The time point at which the response is measured is also important. If the examination is made too early, damage may not have yet occurred. If the determination is made too late, the damage may have repaired or the organism may have expired.

Dose–response curves may be monotonic or multiphasic. A monotonic dose–response relationship is one in which there is only one dose that gives a particular response. A non-monotonic, or multiphasic, dose–response curve may have multiple doses that yield a given response. Many monotonic dose–response curves are sigmoidal in shape, with an area of no response in the beginning, followed by a region of rapid change with dose, and terminating in a plateau in which there is no further change with increasing dose. The area of no response is referred to as sub-threshold, and the inflection point is the threshold. Some multiphasic response curves show an initial increase with dose followed by a decrease in response as further increases in exposure occur. These dose–response curves are frequently referred to as inverted U-shaped. Others show an initial decrease followed by an increased response with applied dose and are called U-shaped dose–response curves. In both of these dose–response relationships, the same response may be seen with two differing doses.

The most common dose–response curve is sigmoidal (S-shaped) when plotted with dose as the abscissa and response as the ordinate. This is the kind of dose–response curve elicited by the majority of non-nutritive toxicants. Typically, there is a region of no response with increasing dose, followed by a region in which response increases rapidly with increasing response, followed by a plateau where response changes little with increasing dose. This may or may not be followed by a region of decreasing response with increasing dose. See figure 7-1 for a typical dose–response relationship of this type and a comparison to a U-shaped dose response.

A number of mathematical manipulations can be performed on a sigmoidal dose–response relationship to achieve a form more amenable to extrapolation. By plot-

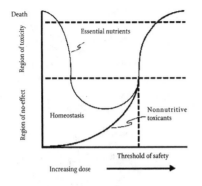

Figure 7-1. Typical dose–response curves for both essential nutrients and non-nutritive toxicants. Note the U-shaped dose response for the former and the sigmoidal shape for the latter. Reprinted with permission from Eaton Klaassen (1996).

ting the dose as the natural or base 10 logarithm (log) of dose, the curve becomes more linear in the middle yet remains nonlinear at very low doses and very high doses. A sigmoidal dose–response relationship approaches linearity when the dose is expressed as log dose and the effects are expressed as the probit or logit of response, as shown in figure 7-2. Note that the logit transformation predicts a higher risk at low exposures than does the probit transformation. At this point, a least-squares fit can be used to predict response at desired doses or the dose required to yield a specific response, such as the LD_{50}.

Linear dose–response relationships (in the absence of mathematical manipulations) are common when cancer is used as an endpoint. Generally, the number of individuals that develop cancer increases with dose, but the severity of the cancer is independent of dose. An example of a linear dose–response curve is depicted in figure 7-3. Note that at very low doses, it is common to find a nonlinear dose response. Here, a supralinear response indicates greater risk than implied by a linear model, whereas the sublinear and threshold responses are indicative of less risk than implied by the linear model.

Multiphasic dose–response curves show a break or inflection point in the response. This shape of curve is usually due to two or more competing mechanisms and is seen with vitamins, minerals, hormones, and a number of other toxicants.

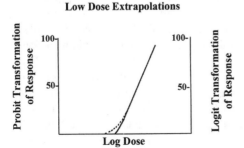

Figure 7-2. Transformation of a sigmoidal dose–response relationship. Reprinted with permission from Dickerson et al., 1998.

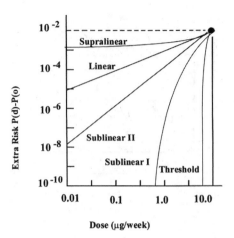

Figure 7-3. A dose–response curve for carcinogens may be linear at medium to high exposure. However, as this figure shows, the response at low exposures may show a threshold or be sublinear, linear, or supralinear. Reprinted with permission from Faustman and Omenn (1996).

These curves are sometimes characterized as U-shaped, inverted U-shaped, or hockey stick because of their appearance. For example, if one plots survival versus vitamin A exposure, one finds that the absence of vitamin A in humans is incompatible with life, but an excess causes toxicity, resulting in an inverted U-shaped dose response curve (figure 7-4). Other examples of non-monotonic dose–response curves can be seen with glucocorticoid treatment and immune function as well as E_2 concentration and uterine wet weight.

Xenobiotic Interactions

Often, mixtures of compounds are administered by design or accident. Almost all cases of environmental exposure are to mixtures of compounds. Moreover, these mixtures may be composed of chemicals from the same general class of compounds or composed of chemicals with widely differing structures. Whenever an organism is exposed to a mixture of compounds, these compounds may alter the way the organism responds to other compounds. The mechanism by which these interactions occur may be as simple as competition for the same binding site or as complex as altering the metabolism of the other compounds via a multistep pathway. Compounds may interact positively, where the overall beneficial or toxic response is increased, or negatively, where the beneficial effect or toxic response is blunted. Possible types of interaction include additive responses, antagonistic responses, synergisms, and potentiations.

For chemicals that interact additively, the effects of coadministering of compounds A and B can be predicted by considering the molar dose of A plus B to equal the molar dose of either compound. For example, giving a 5 mmol/kg dose of A plus a 5 mmol/kg dose of B is equivalent to a 10 mmol/kg dose of A or of B. Additive interactions are most common for compounds that interact with the same receptor or that have similar chemical structure. Examples include the organophos-

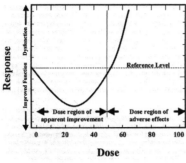

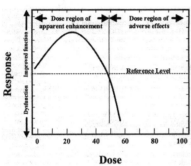

Figure 7-4. (Upper panel) Typical U-shaped dose–response curve. The curve has regions of apparent benefit as well as regions of adverse effect. (Lower panel) An inverted U-shaped dose–response curve. The left portion of this dose response shows apparent benefit, whereas the right portion indicates adverse effects. Reprinted with permission from Davis and Svendgaard (1982).

phorus insecticides and the halogenated aromatic hydrocarbons. With this kind of interaction, differences in potency can be accounted for by using linear scaling factors (toxic equivalency factors).

Antagonistic compounds oppose the action of each other. For example, if one administers 5 mmol/kg of compound A and 5 mmol/kg of compound C, it may be equivalent to only 1 mmol/kg of compound A. This kind of interaction is desirable in treating toxicity induced by one compound through the use of an antidote. For example, atropine is used to treat organophosphorus intoxication. There are four types or mechanisms of antagonism: functional, chemical, dispositional, and receptor antagonism. Functional antagonism occurs when the compounds cause opposite effects on the same physiological function. An example of functional antagonism is using norepinephrine to treat barbiturate-induced low blood pressure. Chemical antagonism relies on interaction between two compounds to detoxify one. An example is using dimercaprol to treat heavy metal poisoning. Dimercaprol chelates the heavy metal ions. Dispositional antagonism occurs when one compound directly effects the absorption, distribution, metabolism, or excretion of the other compound. Examples include syrup of ipecac causing emesis, charcoal-absorbing drugs in the stomach, and SKF-525A blocking the conversion of parathion to paraoxome. Finally, receptor antagonism occurs when both compounds bind competitively to the same receptor but one has much less potency but equal affinity. Examples include naloxone blocking morphine at the opioid receptor and 6-methyl-1,3,8-trichlodibenzofuran blocking the effects of TCDD at the aryl hydrocarbon (Ah) receptor (see chapters 10, 11).

Antagonism is different from tolerance in that the latter is acquired, whereas the former is a true interaction between two compounds. Tolerance may be either dispositional, where less compound reaches the target organ, or a reduced response from the target tissue.

Synergism between two toxic compounds occurs when the effect of the compounds when coadministered greatly exceeds that expected from additivity. Mathematically, if a 5 mmol/kg dose of A gives a 10% response and 5 mmol/kg dose of D gives a 20% response, the result of giving both together would be synergistic if a response higher than 30% was observed. Examples include the effects of ethanol on carbon tetrachloride or acetaminophen exposure.

Potentiation is similar to synergism except it involves exposure to a toxic compound plus a relatively nontoxic compound. Basically, the nontoxic compound potentiates the toxicity of the toxic compound. A classic example of this is the addition of piperonyl butoxide to pyrethrum-based insecticides. Piperonyl butoxide prevents the metabolism of the pyrethrums and greatly increases the toxicity by blocking P450 activity.

Toxicology and the Law

The Endocrine Disruptor Screening and Testing Advisory Committee (EDSTAC) was formed to develop strategies for evaluating the thousands of products and intermediates currently in use or in development that have the potential for human and/or environmental exposure. EDSTAC became necessary when the U.S. Congress mandated testing for endocrine-active substances in the Food Quality Protection Act (1996) and the Safe Drinking Water Reauthorization Act and Amendments (1996). These acts required that the EPA develop a screening program by August 1998, implement the program by August 1999, and report results back to Congress by August 2000. EDSTAC was chartered by the EPA administrator to provide advice and council to the EPA on these issues. This legislation increased the number of compounds likely to be tested from a few hundred to most chemicals in production or trials.

To accomplish the mandate of Congress, the major stakeholders (EPA, World Wildlife Fund, and the Chemical Manufacturers' Association) cosponsored three workshops to evaluate screening and testing methods for EDCs in mammals and other animals. These workshops assessed current methodologies for evaluating reproductive risk from chemicals and developed strategies to strengthen these methods as well as to develop rapid screening methods.

Reproductive and Developmental Toxicity Testing Strategies

Currently, the most widely used tests are the developmental toxicology test and multigenerational tests. The limitations of the developmental toxicology test are insufficient exposure during sexual differentiation and limited evaluation of reproductive and/or endocrine systems. Limitations of multigenerational tests include not enough diversity in the species tested, insufficient sensitivity of some endpoints, and failure to identify malformations elicited by known EDCs (e.g., eggshell thinning).

The two-tiered approach recommended by EDSTAC first tests potential EDCs in a battery of assays for effects on the gonadal and steroid axis. The second tier is designed to characterize the dose–response relationship of EDCs in wildlife and humans. Compounds are being selected (prioritized) for testing based on their production volume, potential for exposure, result of high throughput prescreening, structure, chemical class, and other relevant information. Once selected, the compounds will be evaluated by a series of in vitro and in vivo tests. The in vitro tests include ER binding/transcriptional activation, androgen receptor binding/transcriptional activation, and steroid hormone synthesis using minced testes. Proposed in vivo tests include uterotrophic assay in adult ovariectomized rat, pubertal female rat assay including thyroid tests, (anti)androgen assay in castrate/testosterone-treated male rat, frog metamorphosis assay for EDCs that may disrupt thyroid function, and a short-term fish gonadal recrudescence assay. There is some criticism of the proposed tests because of the lack of a developmental assay with a postnatal component (due to the potential for gestational exposure affecting continuing development of the reproductive tract). Other assays suggested by EDSTAC, but not included in a tier 1 testing strategy until sufficiently validated, are placental aromatase assays, uterotrophic assay using intraperitoneal rather than subcutaneous exposure, and a pubertal male assay.

Although the tests have been selected and screening is underway as this chapter was being written, some unresolved questions remain. These include the use of invertebrates (should they be included?), low dose considerations, the presence or lack of a threshold, interpretation of non-monotonic (U-shaped or inverted U-shaped) dose–response curves, and whether to include in utero (or in ovo) screening with postnatal (posthatch) evaluation. It is clear that as screening and testing proceed, reassessment and modifications must occur for the process to yield the information required by Congress yet not unduly affect the chemical and pharmaceutical industries.

References

Angel, P., Karim, M., 1991. The role of Jun, Fos, and the AP-1 complex in cell proliferation and transformation. Biochim. Biophys. Acta 1072, 129–157.

Arnold, S.F., Robinson, M.K., Notides, A.C., Guillette, L.J., McLachlan, J.A., 1996. A yeast estrogen screen for examining the relative exposure of cells to natural and xenoestrogens. Environ. Health Perspect. 104: 544–548.

Barrett, J.C., 1992. Mechanism of action of known human carcinogens. In: Vainio, H., Magee, P.N., McGregor, D.B., McMichael, A.J. (Eds.), Mechanism of Carcinogenesis in Risk Identification. International Agency for Research on Cancer, Lyon, France, Vol. 116, pp. 115–134.

Birnbaum, L.S., 1994. Endocrine effects of prenatal exposure to PCBs, dioxins and other xenobiotics: Implications for policy and future research. Environ. Health Perspect. 102, 676–679.

Bjerke, D.L., Brown, T.J., Maclusky, N.J., Hochberg, R.B., Peterson, R.E., 1994a. Partial demasculinization and feminization of sex behavior in male rats by in utero and lactational exposure to 2,3,7,8-tetraclorodibenzo-*p*-dioxin is not associated with alterations in estrogen receptor binding or volumes of sexually differentiated brain nuclei. Toxicol. Appl. Pharmacol. 127, 258–267.

Bjerke, D.L., Peterson, R.E., 1994. Reproductive toxicity of 2,3,7,8-tetrachlorodibenzo-*p*-dioxin in male rats: Different effects of in utero versus lactational exposure. Toxicol. Appl. Pharmacol. 127, 241–249.

Bjerke, D.L., Somner, R.J., Moore, R.V., Peterson, R.E., 1994b. Effects of in utero and lactational 2,3,7,8-tetrachlorodibenzo-*p*-dioxin exposure on responsiveness of the male rat reproductive system to testosterone stimulation in adulthood. Toxicol. Appl. Pharmacol. 127, 250–257.

Blanchard, B.A., Hannigan, J.H., 1994. Prenatal ethanol exposure: Effects on androgen and nonandrogen dependent behaviors and on gonadal development in male rats. Neurotoxicol. Teratol. 16, 31–39.

Bursch, W., Oberhammer, F., Schulte-Hermann, R., 1992. Cell death by apoptosis and its protective role against disease. Trends Pharmacol. Sci. 13, 245–251.

Calabrese, E.J., Baldwin, L.A., Mehendale, H.M., 1993. G2 subpopulation in rat liver induced into mitosis by low level exposure to carbon tetrachloride: An adaptive response. Toxicol. Appl. Pharmacol. 121, 1–7.

Childs, G.V., Marchetti, C., Brown, A.M., 1987. Involvement of sodium channels and two types of calcium channels in the regulation of adrenocorticotropin release. Endocrinology 120, 2059–2069.

Conney, A., 1982. Induction of microsomal enzymes by foreign chemicals and carcinogenesis by polycyclic aromatic hydrocarbons: G.H.A. Clowes Memorial Lecture. Cancer Res 42: 4875–4917.

Corcoran, G.B., Fix, L., Jones, D.P., 1994. Apoptosis: Molecular control point in toxicity. Toxicol. Appl. Pharmacol. 128, 169–181.

Davis, J.W., Svendgaard, D.J., 1982. U-shaped dose response curves: Their occurrence and implications for risk assessment. BELLE Newsletter 1, 1.

Dickerson, R.L., Brouwer, A., Gray, L.E., Grothe, D.R., Peterson, R.E., Sheehan, D.M., Sills-McMurry, C., Wiedow, M.A., 1998. Dose-response relationships. In: Kendall, R.J., Dickerson, R.L., Giesy, J.P., Suk, W.P. (Eds.), Principles and Processes for Evaluating Endocrine Disruption in Wildlife. SETAC Press, Pensacola, FL.

Eaton, D.L., Klaassen, C.D., 2001. Principles of Toxicology. In: Klaassen, C.D. (Ed.), Casarett and Doull's Toxicology: The Basic Science of Poisons, 6th edition. McGraw-Hill, New York, pp. 11–34.

Faustman, E.M., Omenn, G.S., 1996. Risk Assessment. In: Klaassen, C.D. (Ed.), Casarett and Doull's Toxicology: The Basic Science of Poisons, 6th edition. McGraw-Hill, New York, pp. 83–104.

Foster, P.M., Thomas, L.V., Cook, M.W., Walters, D.G., 1983. Effect of di-n-pentyl phthalate treatment on testicular steroidogenic enzymes and cytochrome P450 in the rat. Toxicol. Lett. 15, 265–271.

Gallo, M.A., 1995. History and scope of toxicology. In: Klaassen, C.D. (Ed.), Casarett and Doull's Toxicology: The Basic Science of Poisons, 6th edition. McGraw-Hill, New York, pp. 3–10.

Gomez, H.F., Dart, R.C., 1995. Clinical toxicology of snakebite in North America. In: Meier, J., White, J. (Eds.), Clinical Toxicology of Animal Venoms and Poisons. CRC Press, Boca Raton, FL, pp. 619–644.

Guillette, L.J., Gross, T.S., Masson, G.R., Matter, J.M., Percival, H.F., Woodward, A.R., 1994. Developmental abnormalities of the gonad and abnormal sex hormone concentrations in juvenile alligators from contaminated and control lakes in Florida. Environ. Health Perspect. 102, 680–688.

Guillette, L.J., Pickford, D.B., Crain, D.A., Rooney, A.A., Percival, H.F., 1996. Reduction in penis size and plasma testosterone concentrations in juvenile alligators living in a contaminated environment. Gen. Comp. Endocrinol. 101, 32–42.

Hakem, R., Hakem, A., Duncan, G.S., Henderson, J.T., Woo, M., Soengas, M.S., Elia, A., de la Pompa, J.L., Kagi, D., Khoo, W., Potter, J., Yoshida, R., Kaufman, S.A., Lowe, S.W., Penninger, J.M., Mak, T.W., 1998. Differential requirement for Caspase 9 in apoptotic pathways in vivo. Cell 94, 339–352.

Hamelink, J.L., 1977. Fish and chemicals: The process of accumulation. Annu. Rev. Pharmacol. Toxicol. 17, 167–177.

Herve, F., Gentin, M., Rajkowski, K.M., Wong, L.T., Hsia, C.J., Citanova, N., 1990. Estrogen binding properties of alpha 1-fetoprotein and its isoforms. J. Steroid Biochem. 36, 319–324.

Jobling, S., Reynolds, T., White, R., Parker, Mg., Sumpter, J.P., 1995. A variety of environmentally persistent chemicals, including some plasticizers, are weakly estrogenic. Environ. Health Perspect. 103, 582–587.

Jurgenmeier, J.M., Xie, Z., Devereaux, Q., Ellerby, L., Bredeson, D., Reed, J.C., 1998. Bax directly induces the release of cytochrome c from isolated mitochondria. Proc. Natl. Acad. Sci. USA 95, 4997–5002.

Kavlock, R.J., Allen, B.C., Faustman, E.M., Kimmel, C.A., 1995. Dose response assessments for developmental toxicity: IV. Benchmark doses for fetal weight changes. Fundam. Appl. Toxicol. 26, 211–222.

Kappus, H., 1986. Overview of enzyme systems involved in the bioreduction of drugs and in redox cycling. Biochem. Pharmacol. 35, 1–6.

Kelce, W.R., Stone, C., Laws, S., Gray, L.E., Kempainen, J., Wilson, E., 1995. Persistent DDT metabolite p,p'-DDE is a potent androgen receptor antagonist. Nature 375, 581–585.

Ketterer, B., 1988. Protective role of glutathione and glutathione transferases in mutagenesis and carcinogenesis. Mutat. Res. 202, 343–161.

Kimmel, C.A., Gaylor, D.W., 1988. Issues in qualitative and quantitative risk analysis for developmental toxicology. Risk Anal. 8, 15–20.

Kinzler, K.W., Vogelstein, B., 1996. Lessons from hereditary colorectal cancer. Cell 87, 159–170.

Lawley, P.D., 1994. Historical origins of current concepts of carcinogenesis. Adv. Cancer Res. 65: 17–111.

Lindroos, P.M., Zarnegar, R., Michalopoulos, G.K., 1991. Hepatocyte growth factor (hematopoietin A) rapidly increases in plasma before DNA synthesis and liver regeneration stimulated by partial hepatectomy and carbon tetrachloride administration. Hepatology 13, 743–750.

Mably, T.A., Bjerke, D.L., Moore, R.W., Gendron-Fitzpatrick, A., Peterson, R.E., 1992a. In utero and lactational exposure of male rats to 2,3,7,8-tetrachlorodibenzo-p-dioxin. 3. Effects on spermatogenesis and reproductive capacity. Toxicol. Appl. Pharmacol. 114, 118–126.

Mably, T.A., Moore, R.W., Goy, R.W., Peterson, R.E., 1992b. In utero and lactational exposure of male rats to 2,3,7,8-tetrachlorodibenzo-p-dioxin. 2. Effects on sexual behavior and the regulation of luteinizing hormone secretion in adulthood. Toxicol. Appl. Pharmacol. 114, 108–117.

Mably, T.A., Moore, R.W., Peterson, R.E., 1992c. In utero and lactational exposure of male rats to 2,3,7,8-tetrachlorodibenzo-p-dioxin. 1. Effects on androgen status. Toxicol. Appl. Pharmacol. 114, 97–107.

Makela, S., Davis, V.L., Tally, W.C., Korkman, J., Salo, L., Vihko, R., Santi, R., Korach, K.S., 1994. Dietary estrogens act through estrogen receptor-mediated processes and show no antiestrogenicity in cultured breast cancer cells. Environ. Health Perspect. 102, 572–578.

Makela, S., Santi, R., Salo, L., McLachlin, J.A., 1995. Phytoestrogens are partial estrogen agonists in the adult male mouse. Environ. Health Perspect. 103, 123–127.

Matsumura, F., Doherty, Y.G., Furukawa, K., Boush, G.M., 1975. Incorporation of 203 Hg in fish liver: Studies on biochemical mechanisms in vitro. Environ. Res. 10, 224–235.

McMaster, M.E., Munkittrick, K.R., Van Der Kraak, G.J., Flett, P.A., Servos, M.R., 1996. Detection of steroid hormone disruptors associated with pulp mill effluent using artificial exposures of goldfish. In: Servos, M.R., Munkittrick, K.R., Van Der Kraak, G.J. (Eds.), Environmental Effects of Pulp Mill Effluents. St. Lucie Press, Boca Raton, FL, pp. 425–438.

Miller, E.C., Miller, J.A., 1947. The presence and significance of bound aminoazo dyes in the livers of rats fed *p*-dimethylaminoazobenzene. Cancer Res. 7: 468–470.

Miller, J.A., 1970. Carcinogenesis by chemicals: An overview. G.H.A. Clowes Memorial Lecture. Cancer Res. 30, 559–576.

Nagata, S., Goldstein, P., 1995. The fas death factor. Science 267, 1449–1456.

Nicotera, P., Bellamo, G., Orrenius, S., 1992. Calcium-mediated mechanisms in chemically induced cell death. Annu. Rev. Pharmacol. Toxicol. 32, 449–470.

Noji, S., Tashiro, K., Koyama, E., 1990. Expression of hepatocyte growth factor gene in endothelial and Kupfer cells of damaged rat livers as revealed by in situ hybridization. Biochem. Biophys. Res. Comm. 173, 42–47.

Norton, S., 2001. Toxic effects of plants. In: Klaassen, C.D. (Ed.), Casarett and Doull's Toxicology. The Basic Science of Poisons, 6th edition. McGraw-Hill, New York, pp. 965–976.

Ojasoo, T., Vannier, B., Pasqualini, P.P. Jr., 1992. Effect of prenatally administered sex steroids and their antagonists on the hormonal responses of offspring in humans and animals. Marcel Dekker, New York.

Okajima, F., Akbar, M., Majid, M.A., Sho, K., Tomura, H., Kondo, Y., 1994. Genistein, an inhibitor of protein tyrosine kinase, is also a competitive antagonist for P-1-purinergic (adenosine) receptor in FRTL-5 thyroid cells. Biochem. Biophys. Res. Commun. 203, 1488–1495.

Olsen, S.C., Atluru, D., Erickson, H.H., Ames, T.R., 1994. Role of the tyrosine kinase inhibitor, genistein, in equine mononuclear cell proliferation and leukotriene B-4 synthesis. Biochem. Arch. 10, 11–16.

Oltavi, Z.N., Miliman, C.L., Korsmeyer, C.J., 1993. Bcl-2 heterodimerizes in vivo with a conserved homolog, bax, that accelerates programmed cell death. Cell 74, 609–619.

Parkinson, A., 2001. Biotransformation of xenobiotics. In: Klaassen, C.D. (Ed.), Casarett and Doull's Toxicology. The Basic Science of Poisons, 6th edition, McGraw-Hill, New York, pp. 133–224.

Piontek, M., Hengels, K.J., Porshen, R., Strohmyer, G., 1993. Antiproliferative effect of tyrosine kinase inhibitors in epidermal growth factor-stimulated growth of human gastric cancer cells. Anticancer Res. 13, 2119–2123.

U.S. EPA. 1995. Report: Workshop on Environmental Estrogens. U.S. Environmental Protection Agency, Research Triangle Park, NC.

Vonier, P.M., Crain, D.A., McLachlan, J.A., Guillette, L.J., Arnold, S.F., 1996. Interaction of environmental chemicals with estrogen and progesterone receptors from the oviduct of the American alligator. Environ. Health Perspect. 104, 1318–1322.

Whitten, P.L., Lewis, C., Russell, E., Naftolin, F., 1995. Potential adverse effects of phyto-estrogens. J. Nutrit. 125, S771–S776.

For Further Reading

Hodgson, E., Levi, P.E., 1994. Biochemical Toxicology, 2nd edition. Appleton & Lange, Norwalk, CT.

Klaassen, C.D. (Ed.), 2001. Casarett and Doull's Toxicology. The Basic Science of Poisons, 6th edition, McGraw-Hill, New York.

Kendall, R.J., Dickerson, R.L., Giesy, J.P., Suk, W. (Eds.), 1998. Principles and Processes for Evaluating Endocrine Disruption in Wildlife. SETAC Press, Pensacola, FL.

Thomas, J.A., Colby, H.D. (Eds.), 1997. Endocrine Toxicology. Taylor and Francis, Washington, DC.

8

Probabilistic Risk Assessment

Kenneth R. Dixon
Clyde F. Martin

An ecological risk assessment evaluates the potential adverse effects that human activities have on ecosystems. The risk assessment process provides a way to collect, analyze, and present scientific information so that it can be used for making environmental decisions. When conducted for an endocrine-disrupting chemical (EDC), ecological risk assessment can be used to identify sensitive species, analyze the effect of the EDC on those species, and analyze the resulting impact on populations and ecosystems. The ecological risk assessment process has been described by the U.S. EPA in a basic framework (U.S. EPA, 1992) and in a set of guidelines (U.S. EPA, 1998).

Risk assessment includes three primary phases: problem formulation, analysis, and risk characterization (figure 8-1). In the problem formulation phase, information on sources, stressors, effects, and ecosystems and receptor characteristics is integrated to produce assessment endpoints, conceptual models, and an analysis plan. In the analysis phase, data are evaluated in parallel and integrated processes to characterize exposure and ecological effects. Characterization of exposure involves developing measures of exposure and an analyzing exposure to produce an exposure profile. Characterization of ecological effects involves developing measures of effect and conducting an ecological response analysis to produce a stressor-response profile. In the risk characterization phase, risk estimation involves integrating exposure and stressor–response profiles and analyzing uncertainties. Risk description includes the evidence that supports causality between exposure and effects and a determination of how the observed or predicted effects cause adverse ecological effects.

Probabilistic risk assessments have been used for several years in other disciplines such as accident prediction, systems failure, and weather forecasting, but only

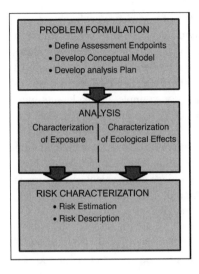

Figure 8-1. Ecological risk assessment framework. Modified from U.S. EPA (1998).

recently have they been used in ecological risk assessments (Vose, 1996). In the context of this book, risk can be defined as the probability of an effect occurring as a result of exposure to an EDC. By including random variables, with their associated probability distributions, in a risk assessment, probabilities are assigned to assessment endpoints so that the actual risk in the environment can be more closely evaluated.

Deterministic risk assessment involves the use of parameters that take on only a single value, thus determining the assessment's outcome. Probabilistic risk assessment, in contrast, involves the use of random variables and the associated probability distributions. Not only can risk parameters be functions of other variables, they can be functions of random variables and thus can be random variables themselves. A random variable can be thought of as a variable whose value depends on some probability distribution. Those methods that use random (stochastic) variables are called stochastic methods, and those that do not use random variables are called deterministic methods. Random variables are used to represent the random variation or unexplained variation in the endpoint or to state variables (those variables that describe the status of the system). Stochastic methods can include random variables expressed either as random inputs or as parameters with a random error term. For example, there may be unmodeled effects such as the deliberate removal of the effect of climate change in a population model. The error that is incurred in simulation when numerical calculations are performed can be included as a random variable. Before a random variable can be used effectively, however, the associated probability distribution must be determined. In the case of numerical error, the probability distribution is machine dependent, but most machines use a rounding procedure that produces a normal distribution. In the case of unmodeled effects, the problem is more difficult. The error, however, rarely is associated with a distribution that is well understood.

A discrete distribution describes the probability of occurrence of discrete events. For example, over a certain time period an individual may or may not give birth, be exposed to a toxicant, or die. The probability of such an event occurring is assigned a numerical value between zero and one, called the probability. One way of estimating the probability of such events is to observe the frequency of occurrence of the event in a large population of similar individuals. For example, if, in a large population of size N, m mortalities occur, then the probability of dying is approximately m/N. The "law of large numbers" tells us that as the number of observations becomes very large, this approximation becomes exact. One must be sure, however, that one is observing a representative segment of the population. For example, we could observe high school students in very large numbers and we could calculate the percentage that a given student would die in a given time period. This would give us little insight, however, into the true mortality rate of the general population.

An example of a discrete distribution is an empirical distribution used to describe reproduction in an individual-based population model (Dixon et al., 1999). For each age class, the probability of producing from zero to three offspring is defined (figure 8-2). The actual sampling is based on the cumulative distribution function (figure 8-3). For example, if the sample value of a random variate from a uniform probability distribution is between 0 and 0.15, the number of offspring would be 0; if the value is between 0.15 and 0.45, the number of offspring would be 1; and so on. The observed values that resulted from the sampling compared closely with the input probabilities (figure 8-4). The value of n represents the number of females in the given age class summed over time. In other words, it represents the number of decisions made concerning the number of offspring per female over the period of the simulation.

In some cases, state variables and parameters will be continuous variables (i.e., those that can assume values in an interval). For a continuous random variable X, we define the probability that X lies in the interval $[a, b]$ by

$$Prob(a < X \leq b) = \int_a^b f(X)dX. \tag{8-1}$$

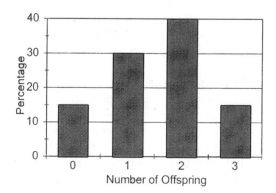

Figure 8-2. Discrete probability distribution of the probabilities of numbers of offspring occurring in females in age class 2.

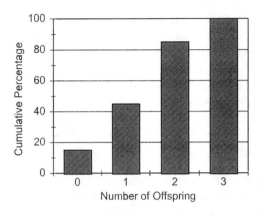

Figure 8-3. Discrete cumulative probability distribution of the probabilities of numbers of offspring occurring in females in age class 2.

The function $f(x)$ is called the probability density function. The cumulative distribution function $F(x)$ defines the range of possible values of the random variable x:

$$F(x) = P(X \leq x) = \int_{-\infty}^{x} f(u)du. \tag{8-2}$$

Monte Carlo methods are a class of numerical techniques used to estimate properties of a stochastic model using extensive simulation. For the random variables of the model, values are chosen using a probability distribution. Repeated simulations of the model will produce different outcomes, and, in effect, a mapping from the probability distributions of the random variables to the probability distribution of the outcome (assessment endpoint) is achieved. Parameter values may or may

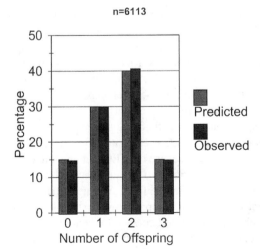

Figure 8-4. Observed and predicted percentage occurrence of the number of offspring, from zero to three, in a hypothetical species.

not come from a known probability distribution, such as uniform, normal, exponential, and so on. If the distribution is known and is approximately normal, with known mean, μ, and variance, σ^2, a parameter value P is determined by

$$P = \mu + R\sigma. \tag{8-3}$$

A probability distribution then can be calculated for a state variable in the model along with its mean and variance. Suppose the model has random variables for parameters $p_1, p_2, p_3, \ldots p_n$. The state variable will be a function of the n parameters:

$$X = f(p_1, p_2, p_3, \ldots, p_n). \tag{8-4}$$

Now we calculate a value for each parameter by sampling from its individual distribution function. We then obtain a value for the state variable X by running a simulation of the model. We repeat the process until we have N values of the state variable X. Finally, we determine μ and σ^2 for X.

In 1999, the Ecological Committee on FIFRA Risk Assessment Methods (ECOFRAM) described six different risk assessment methods (figure 8-5). Of the six methods, five could be considered probabilistic in that they use probability distributions in some way. The probabilistic methods ranged from the use of probability distributions in place of point estimates in the quotient method to overlapping distributions of exposure and toxicity to stochastic simulation models.

Quotient Method

The first method is the point estimate quotient method, which is the only completely deterministic method. The method is based on the ratio formed by the expected environmental concentration (EEC) divided by the toxicity of the contaminant of concern (e.g., LC_{50} or EC_{50}). If the quotient exceeds 1, then a significant risk may be indicated. The weakness of the this method is that none of the uncertainty of either the exposure or effect estimate is considered in the assignment of risk.

The fourth method, distribution-based quotients, replaces the point estimates of exposure and risk with probability distributions. Each distribution is sampled using Monte Carlo methods to form a distribution of quotients. Risk is then estimated as the probability that the quotient exceeds a given level. A major weakness of the method is that there is no cause-and-effect relationship between exposure and toxicity (i.e., distributions of exposure and effect are independent).

It must be remembered in applying the quotient method that there is uncertainty in both the numerator and the denominator. The EEC is estimated from field measurements and is likely to have a large variance. Because the EEC is in the numerator, the variance in the quotient is directly proportional to the variance of the EEC. The variance in toxicity should be small if it is obtained from laboratory measurements. Its total impact on the quotient variance will be small provided that the toxic impact of concern is not close to zero. It is not good science to declare that the quotient is greater than one. A valid claim would be that "with probability the quo-

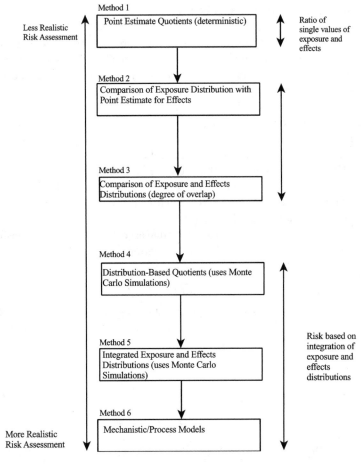

Figure 8-5. Overview of risk assessment methods. Adapted from ECOFRAM (1999).

tient is greater than one." This requires a careful estimate of the variance of the EEC. A simple calculation shows how the error enters the quotient when using the quotient method:

$$\frac{y+\alpha}{x-e}=(y+\alpha)\frac{1}{x}\left(1+\frac{e}{x}+...\right)=\frac{y}{x}+\frac{y}{xx}e+\frac{a}{x}+\frac{\alpha e}{xx}+...=\frac{y}{x}+\frac{y}{xx}e+\frac{\alpha}{x}. \quad (8\text{-}5)$$

Here y is the EEC and x is the expected toxic impact of concern. The variables α and e are errors or random variables. Thus, the expected value of the quotient is just the fraction y/x. However, the error term is more complicated, with error depending on the particular values of x and y. In the above formula the higher order terms have been repressed. It should be noted that if α is large with respect to e, then terms such as e may have to be retained.

Distributions of Exposure and Effects

The second method is the comparison of an exposure distribution with a point estimate for effects. In this method, a single distribution of exposure is compared with a single point estimate of toxic effects. The probability of the effect level occurring within the distribution of exposure determines the estimated risk. This method could be used if a complete probability distribution of toxicity effects is not available.

When probability distributions are available for both exposure and toxicity, risk can be based on the degree of overlap of the distributions (method 3). Overlapping probability distributions have been described in detail (Cardwell et al., 1993; SETAC, 1994; Parkhurst et al., 1995) and have been used in a number of ecological risk assessments (Solomon et al., 1996, 2001; Giesy et al., 1999). In this approach, cumulative frequencies of EECs and toxicity values (LC_{50}, LC_{10}, LC_5, etc.) are plotted on the same graph (figure 8-6).

In practice, toxicity values often are ranked in ascending order, then transformed to cumulative percentages using the transformation

$$\frac{100 \times i}{n+i},\tag{8-6}$$

where i is the ith observation of a total of n observations, starting with the lowest toxicity value. This method is often necessary because there are not enough data to

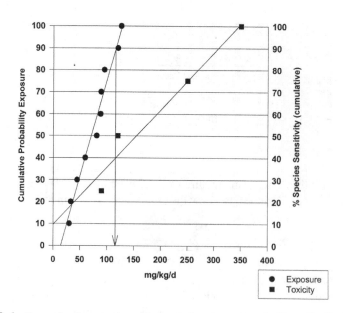

Figure 8-6. Example of comparison of cumulative exposure and effects distributions. The circles represent the cumulative exposure distribution and the squares represent squares the cumulative effects distribution. The vertical arrow indicates the concentration of the 90th percentile of the exposures.

determine the true probability distributions. The resulting plots should show an approximate linear relationship between cumulative frequencies and the exposure and toxicity data, and linear regression can be used to fit straight lines to the data. The area of overlap between the two lines (if any) then can indicate the level of risk to the organisms exposed to the EECs. In the example shown in figure 8-6, there is considerable overlap of the two distributions. The vertical arrow shows that at a value of 129 mg/kg/day, the 90th percentile of the exposure distribution, toxicity would be exceeded for 40% of the species. If the distribution of effects represents variability in the sensitivity of a single species, then there would be a 40% probability that the toxicity for that species would be exceeded in 90% of the exposure values.

Methods 2 and 3 do not provide for an estimate of the probability of the magnitude of effect. There also is no cause-and-effect relationship between exposure and effects.

The fifth method, integrated exposure and effects distributions, integrates the exposure and effect distributions. This method is similar to the distribution-based quotients method in that the exposure and effects distributions are sampled using Monte Carlo methods. Instead of forming a quotient, however, this method uses a dose–response function to obtain the probability of a certain magnitude of the response correlating with a certain exposure concentration value obtained from the exposure distribution.

The distributions of exposure concentrations, doses, or body burdens in a population are continuous probability distributions. It is not unusual to assume an exponential distribution (figure 8-7). The probability density function for the exponential distribution is defined by

$$f(x) = \begin{cases} \dfrac{1}{\beta} e^{-\frac{x}{\beta}} & \text{if } x \geq 0 \\ 0 & \text{otherwise} \end{cases}, \tag{8-7}$$

where β is the mean of the distribution and the only free parameter.

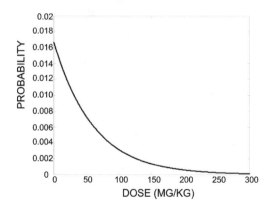

Figure 8-7. Probability density function for the exponential distribution. The mean of the distribution, β, is 60 mg/kg.

The actual dose to an individual is treated as a random variate that is generated from the cumulative distribution function for equation 8-7 (figure 8-8):

$$F(x) = \begin{cases} 1 - e^{-\frac{x}{\beta}} & \text{if } x \geq 0 \\ 0 & \text{otherwise} \end{cases}. \qquad (8\text{-}8)$$

The inverse-transform method can be used to obtain the random variate for dose, X (Law and Kelton 2000). To find F^{-1}, we set $u = F(x)$ and solve for x to obtain

$$F^{-1}(u) = -\beta \ln(1 - u). \qquad (8\text{-}9)$$

To generate the random variate X we first generate a random variate U from a uniform distribution $U(0,1)$. The second step is to return $X = -\ln(1 - U)$. We actually used U instead of $1 - U$ to save a subtraction. The relationship between the uniform random variate and the dose is shown in figure 8-9.

Once the predicted dose (or body burden) has been obtained, the predicted response is estimated from a dose–response function (figure 8-10). Suppose the dose obtained from equation 8-8 was 105 mg/kg. Then, using the dose–response function in figure 8-10, the probability of response is 60%. A uniform random variate is obtained and compared with the response probability. If the random variate is less than or equal to the value on the dose–response function (60%), a response is determined. If the random variate is greater than 0.60, no response is predicted.

Several iterations will yield a probability distribution of endpoint values, such as mortality percentage in the simulated population (figure 8-11). The cumulative distribution (figure 8-12) shows the probability of any level of response up to a certain level. Subtracting these probabilities from 1.0 gives the probability that a certain level of the endpoint will be exceeded. This curve can be compared with a graph of the threshold of acceptability defined by the risk manager to determine whether there is the potential for unacceptable risk (figure 8-13).

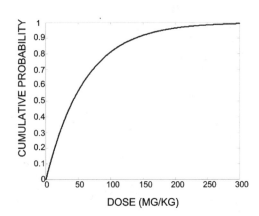

Figure 8-8. Cumulative probability distribution for the exponential probability density function.

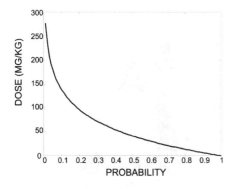

Figure 8-9. Inverse transform of the relationship between cumulative probability and dose. This form of the relationship is used in stochastic simulations to estimate dose.

Stochastic Simulation Models

The last method, mechanistic/process models, uses simulation models to generate estimates of exposure concentrations, dose levels, body burdens, and effects. Stochastic simulation models have been used in probabilistic ecological risk assessments for effects of the herbicide atrazine on aquatic ecosystems in the midwestern U.S. corn belt (Solomon et al., 1996) and for the effects of the insecticide chlorpyrifos, both on aquatic (Giesy et al., 1999) and terrestrial (Solomon et al., 2001) ecosystems in the midwestern United States. These models integrate exposure and effects, but in a way that provides cause-and-effect mechanisms or processes. Such

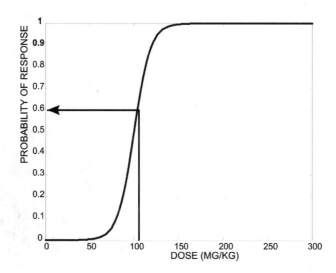

Figure 8-10. Dose–response function. The dose of 105 mg/kg gives a probability of response of 60%.

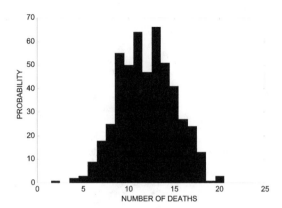

Figure 8-11. Frequency of the number of deaths in 500 simulated populations, or the probability that a given number of deaths will occur in a population.

models are likely to be individual-based models, and where the risk assessment concerns endocrine disruptors, they also should be physiologically-based pharmacokinetic (PBPK) models (Carson et al., 1983). This is because EDCs often affect specific organs or tissues and the PBPK models incorporate the partitioning of the EDC among the tissues and organs (figure 8-14).

One important mechanism that should be included in a model of endocrine systems is negative feedback, which regulates the secretion of most hormones (see chapter 1). This feedback regulatory mechanism can be illustrated using a block diagram (figure 8-15). A block diagram is a schematic representation of the cause-and-effect relationship between the input and output of a system. It offers a convenient and useful method of characterizing the functional relationships among the various components of the system. A negative feedback system is one in which the feedback signal is negative when summed with the reference set point (Goodman, 1994). Endocrine systems are negative feedback systems that maintain homeostasis. These systems have the tendency, if displaced from their normal values by an external stimulus or disturbance, to return to their normal values.

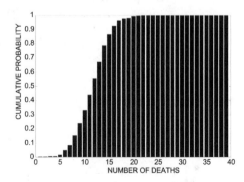

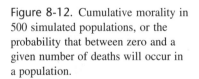

Figure 8-12. Cumulative morality in 500 simulated populations, or the probability that between zero and a given number of deaths will occur in a population.

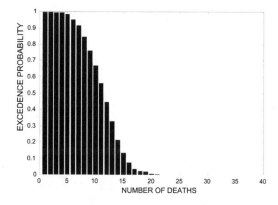

Figure 8-13. Probability of the number of deaths exceeding a given number.

In probabilistic risk assessments using simulation, as in method 5, probability distributions are measured (or estimated) for parameters to account for natural variation, lack of knowledge, or uncertainty. The actual parameter values used in a simulation are then obtained by sampling the distributions in a Monte Carlo process. The resulting model output then will contain endpoint values, one value for each set of parameter values in a given simulation. Several simulations will yield a probability distribution of endpoint values, such as mortality percentage in the simulated population. By altering the mean value of the model parameters and running additional sets of simulations, the percentage of outcomes that exceed a certain level of mortality can be estimated (figure 8-13). This curve can be compared with a graph of the threshold of acceptability defined by the risk manager to determine whether there is the potential for unacceptable risk.

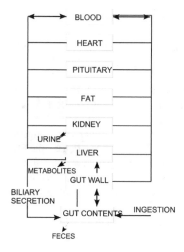

Figure 8-14. Flow diagram of distribution of an endocrine-disrupting compound in a physiologically-based pharmacokinetic model.

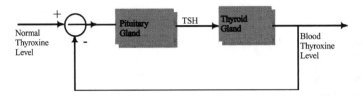

Figure 8-15. Block diagram of the feedback path between blood thyroxine level and secretion of thyroid-stimulating hormone (TSH) by the pituitary gland.

Risk assessment results provide a basis for comparing different management options, enabling decision makers and the public to make better informed decisions about the management of ecological resources.

References

Cardwell, R.D., Parkhurst, B.R., Warren-Hicks, W.,Volosin, J.S., 1993. Aquatic ecological risk. Water Environ. Technol. 5, 47–51.

Carson, E.R., Cobelli, C., Finkelstein, L., 1983. The Mathematical Modeling of Metabolic and Endocrine Systems/ Model Formulation, Identification, and Validation. John Wiley & Sons, New York.

Dixon, K.R., Huang, T-Y., Rummel, K.T., Sheeler-Gordon, L., Roberts, J.C., Fagan, J.F., Hogan, D.B., Anderson, S.R., 1999. An individual-based model for predicting population effects from exposure to agrochemicals. In: Thomas, M.B., Kedwards, T. (Eds.), Challenges in Applied Population Biology. Aspects of Applied Biology 53. Association of Applied Biologists, Horticulture Research International, Wellesbourne, Warwick, UK, pp. 241–251.

ECOFRAM, 1999. Terrestrial draft report. Ecological Committee on FIFRA Risk Assessment Methods.

Giesy, J.P., Solomon, K.R., Coates, J.R., Dixon, K.R., Giddings, J.M., Kenaga, E., 1999. Chlorpyrifos: Ecological risk assessment in North American aquatic environments. Rev. Environ. Contam. Toxicol. 160, 1–129.

Goodman, H.M. 1994. Basic Medical Endocrinology, 2nd edition. Raven Press, New York.

Law, A.M., Kelton, W.D., 2000. Simulation Modeling and Analysis, 3rd edition. McGraw-Hill, New York.

Parkhurst, B.R., Warren-Hicks, W., Etchison, T., Butcher, J.B., Cardwell, R.D., Volison, J.S., 1995. Methodology for aquatic ecological risk assessment. RP91-AER. Final report prepared for the Water Environment Research Foundation, Alexandria, VA.

SETAC (Society of Environmental Toxicology and Chemistry), 1994. Report of the aquatic risk assessment and mitigation dialogue group for pesticides. SETAC, Pensacola, FL.

Solomon, K.R., Baker, D.B., Richards, R.P., Dixon, K.R., Klaine, S.J., La Point, T.W., Kendall, R.J., Weisskopf, C.P., Giddings, J.M., Giesy, J.P., Hall, L.W., Jr., Williams, W.M., 1996. Ecological risk assessment of atrazine in North American surface waters. Environ. Toxicol. Chem. 15, 31–76.

Solomon, K.R., Giesy, J.P., Kendall, R.J., Best, L.B., Coats, J.R., Dixon, K.R., Hooper, M.J., Kenaga, E.E., McMurry, S.T., 2001. Chlorpyifos: Ecotoxicological risk assessment for birds and mammals in corn agroecosystems. Hum. Ecol. Risk Assess. 7, 497–632.

U.S. EPA, 1992. Framework for ecological risk assessment. EPA/630/R-92/001. U.S. Environmental Protection Agency, Washington, DC.

U.S. EPA, 1998. Guidelines for ecological risk assessment. EPA/630/R-95/002Fa. U.S. Environmental Protection Agency, Washington, DC.

Vose, D., 1996. Quantitative Risk Analysis: A Guide to Monte Carlo Simulation Modelling. John Wiley & Sons, New York.

9

Fish and Wildlife as Sentinels of Environmental Contamination

Lynn Frame
Richard L. Dickerson

F ish, wildlife, invertebrates, and microorganisms have been widely touted as sentinels for the health of the environment and thus as protectors of human health (Sheffield et al., 1998; Guillette, 2000; Fox, 2001). How sentinel systems can help us detect exposures that may result in adverse effects including endocrine disruption was the subject of a workshop sponsored by the U.S. Army Center for Environmental Health Research, the National Center for Environmental Assessment of the U.S. EPA, and the Agency for Toxic Substances and Disease Registry (van der Schalie et al., 1999). Sentinels could be used as broad indicators of ecosystem health (bioindicators) or as part of quantitative risk assessment (biomonitoring) (NRC, 1991; Lower and Kendall, 1992). What are sentinels? A common definition for a sentinel is any species (whether prokaryotic or eukaryotic, plant or animal, natural or transgenic), feral or domesticated, that can be used as an indicator of exposure to or toxicity from environmental contaminants and which can be used to assess impact on similar organisms or ecosystems (Lower and Kendall, 1992; Stahl, 1997).

The question is, however, How effective is wildlife as sentinels to protect other wildlife species and humans from excessive risk from environmental contaminants and/or disease? The simple answer is that environmental sentinels are signaling (LeBlanc, 1995); however, the degree of predictability relies on how well the toxicokinetics and toxicodynamics of the compound of interest are understood in the compared species. The problem of scientific prediction becomes exponentially more difficult when exposure to mixtures of dissimilar chemicals occurs. In this chapter, we explain the scientific basis of risk assessment using wildlife sentinels and biomarkers of effect, exposure, and susceptibility. We explain the difficulties in extrapolation using comparative biochemistry and physiology as factors in differential toxicokinetics and toxicodynamics. Additionally, we discuss how the new subdisciplines of toxicogenomics and proteomics can be expected to alter biomarker-based use of sentinel species.

Historical Perspective

The use of sentinel species to detect potential threats to human health is not recent. For example, canaries were used in coal mines for centuries to detect coal damp before this gas overcame coal miners (Burrell and Seibert, 1916; Schwabe, 1984). Domestic animals such as cattle, horses, and sheep developed adverse health effects to smog, fluoride, lead, and organophosphate compounds, thereby drawing attention to the hazardous effects of these compounds in the late 19th century and the first part of the 20th century both in the United States and England (Veterinarian, 1874a,b; Haring and Meyer, 1915; Holm et al., 1953; van Kampen et al., 1969). Abnormal behavior ("dancing") in domestic cats was the first sign of the danger from methylmercury contaminated fish in Minamata, Japan (Kurland et al., 1960). Carson (1962) reported the effects of organochlorine insecticides such as DDT on songbirds in the book *Silent Spring*. More recently, the effects of polybrominated biphenyls and polychlorinated dibenzo-*p*-dioxins were first noted in dairy cattle in Michigan and in horses in Missouri (Case and Coffman, 1973; Jackson and Halbert, 1974; Carter et al., 1975; Wellborn et al., 1975). Finally, abnormal male development in alligators drew attention to the reproductive hazards of endocrine-disrupting chemicals (EDCs) in the environment (Guillette et al., 1994; Matter et al., 1998). In the laboratory, toxicologists have been using laboratory rodents, a domesticated, often syngenic, wildlife species in medical and toxicological research for many years. European starlings and deer mice have been used as sentinels in biomonitoring programs with measurable success on a variety of contaminated sites (Akins et al., 1993; Dickerson et al., 1995). A recent, specific use for wildlife sentinels is to detect contaminant exposures that affect the endocrine system, and this use deserves research emphasis and priority funding (DeRosa et al., 1998). Currently, avian species are being used as sentinels to detect the spread of West Nile virus, thus allowing public health authorities to warn people of increased risk (Komar, 2001; McLean et al., 2002).

Basics

To understand fully the issues and limitations that define the prognostic use of wildlife as sentinels for ecosystem and human health, it is necessary to review some basic toxicology and risk assessment theory. A more detailed discussion on toxicology can be found in chapter 7. Much of the applicable toxicology was derived from knowledge gained from the field of pharmacology. In the following section, a brief review of toxicokinetics, toxicodynamics, and toxicogenomics is presented.

Toxicokinetics

Toxicokinetics is an important factor in risk assessment (Watanabe and Chen, 2001) and is the equivalent of pharmacokinetics, except that we are concerned with chemicals that elicit adverse effects rather than therapeutic ones. Pharmacokinetics can be defined as the study of the rate at which a drug or other compound reaches the target molecule that, in most cases, is a membrane bound cytosolic or nuclear receptor (Brody et al., 1998). Toxicants also may interact with lipid membranes and DNA in a nonspecific manner. Thus, toxicokinetics can be defined as the rate at which a toxicant reaches its target molecule. Much of the interspecies difference in sensitivity to a drug or contaminant can be explained by differences in kinetics. For both pharmacokinetics and toxicokinetics, the factors controlling the rate of toxicant interaction with intracellular targets are route of administration (exposure), dosing (exposure) schedule, adsorption, disposition, metabolism, and excretion. In wildlife, this may be complex depending on whether exposure occurs by ingestion, inhalation, dermal absorption, or a combination of these routes. If exposure is from ingestion of a long-lived contaminant in drinking water or from inhalation of a long-lived contaminant from air, the exposure is chronic and relatively constant. However, if the exposure is dermal or from a food source that is consumed sporadically, the exposure will be unpredictable.

For our purposes, routes of exposure to environmental contaminants may be simplified to inhalation, ingestion, and dermal exposure. Ingestion may be through drinking water or through the consumption of contaminated food. In general, water-soluble substances will end up in water supplies through runoff into streams and ponds or by leaching from soil into groundwater. In contrast, lipid-soluble compounds resistant to action of soil microorganisms will tend to bioconcentrate and biomagnify. Thus, the position of the sentinel species in the food web controls its exposure to a degree. Finally, it is quite conceivable that an organism may be exposed to the same contaminant by multiple routes of exposure. For example, wild rodents inhabiting a field containing elevated levels (above background) of DDT and DDE may be exposed to small amounts of DDE/DDT by inhalation, by dermal absorption (by living in burrows dug into contaminated soil), or by ingestion via drinking water containing silt with DDE/DDT adhered to it or by consuming insects with bioconcentrated DDE/DDT. Different foraging and burrowing behaviors among species of rodents may result in widely differing exposures in animals in the same habitat.

Another toxicokinetic factor is absorption, the process by which a toxicant crosses cell membranes and enters cells of the gut, lung, or skin. Absorption is affected by pH of the fluid in contact with stomach or intestinal epithelium, the thickness of the epidermis in the case of the skin, and the ventilation/perfusion rate in case of lung alveolar cells. Species differences in gut pH, dermal thickness, and cardiovascular/pulmonary system performance can affect bioavailability and toxicokinetics and thus sensitivity to a toxicant.

Distribution is the process by which a compound is transported from the site of original absorption to the target. In many cases, the target is within or a short distance away from the tissue that absorbs the toxicant. In these cases, diffusion dominates distribution. In other cases, the toxicant must be transported via the lymphatic or circulatory system to reach its target. For substances that are water soluble, this becomes solely a function of tissue perfusion rate. For substances that are lipophilic, another factor comes into play. This factor is the binding of the toxicant to plasma constituents, primarily proteins. This binding may be nonspecific or specific. Serum albumin is the most abundant protein in the plasma of most species. Other abundant proteins are α_1-acid glycoprotein and lipoproteins. Acidic toxicants bind primarily to serum albumin, basic toxicants to α_1-acid glycoprotein, and some heterocyclics bind to serum lipoproteins. In addition, some of the polychlorinated biphenyls (PCBs) can bind to specific carrier proteins such as those that transport steroid and thyroid hormones, resulting in the displacement of the native compounds. This can result in inappropriate absorption or metabolism of these endogenous hormones. There are marked species differences in the composition of plasma proteins, and this is a factor in toxicokinetic differences.

Once in the circulation, the toxicant can enter the cells containing the target molecule by diffusion, facilitated transport, or active transport. Potential toxicant actions within the cell are covered below in the section on toxicodynamics. The manner in which the toxicant in circulation is removed from the circulation is also a major focus of toxicokinetics.

Metabolism is the process by which an organism modifies an intrinsic compound or xenobiotic to enhance its water solubility and ease of excretion. A thorough discussion of metabolism is beyond the scope of this chapter (see Hodgson and Levi, 1994; Klaassen, 2001, for detailed information). There are two phases in metabolism: phase I, in which a functional group is added or exposed on the toxicant enzymatically, and phase II, in which a polar substrate such as glucose, glucuronic acid, glutathione, acetate, or sulfate is added to the toxicant. In most cases, metabolism reduces the toxicity of the parent compound, but it also may bioactivate the parent compound to a more toxic form. In addition, both sulfated and glucuronidated metabolites secreted into the bile may be cleaved by action of enzymes produced by enteric bacteria and the parent compound resorbed, a process known as enterohepatic recirculation. This increases the effective half-life of a toxicant. There are marked inter- and intraspecies differences in metabolic rates and pathways for many substrates. This is due in part to differences in expression and structure of phase I and II drug-metabolizing enzymes, predominantly the cytochromes P450. For example, species differences in rates of glucuronidation are well known and create uncertainties in risk assessment (Walton et al., 2001).

Excretion of a toxicant can be in the form of the unchanged parent compound or as a metabolite of the parent compound. In general, only very volatile or water-soluble compounds are excreted as the parent compound. Excretion of the unchanged parent molecule is referred to as elimination. All other molecules are metabolized to some extent. Often the metabolism and excretion terms of the toxicokinetic equation are lumped together as clearance. Therefore, the biological half-life of a compound is primarily a function of its absorption, distribution, metabolism, and excretion rates. Species differences in elimination and excretion have been reported.

Toxicokinetics is a means of predicting the rate at which a toxicant reaches a target molecule and the rapidity by which it is removed from the target molecule. Toxicity and resulting cellular damage are dependent on just these factors. In general, a species in which the delivery of a toxicant to a target is slow and the clearance is rapid will be less susceptible to that toxicant than a species in which delivery is rapid and clearance is slow. By using relative exposure and species toxicokinetic data, we can make a first step in predicting effects on one species through the use of a sentinel species. Moreover, as a database for species and a toxicant evolves, we can begin to assemble a biologically-based toxicokinetic (BBTK) model that greatly aids our understanding and ability to predict effects, which can help assess the risks of exposure to mixtures (Haddad and Krishnan, 1998).

Toxicodynamics

Toxicodynamics is related to its sister concept, pharmacodynamics, with the difference again that toxicodynamics deals with adverse effects of a compound and pharmacodynamics deals with therapeutic effects. Pharmacodynamics is broadly defined as the fundamental way in which a drug interacts with its biological effector (receptor) at a physiological, biochemical, or molecular level (Brody, 1998). Toxicodynamics is thus the way a toxicant interacts with a target molecule to elicit a response and is important in risk assessment. For many EDCs, the target molecule may be a hormone receptor or an enzyme involved in hormone biosynthesis. In contrast to pharmacodynamics, where the interaction is often binding to a specific cellular receptor normally occupied by an endogenous molecule, often a neurotransmitter or hormone, toxicant interactions run the gamut from gross tissue damage caused by interaction with caustic soda to binding of dioxin to the aryl hydrocarbon receptor eliciting CYP1A1 induction. The first example is nonspecific in that the toxicant interacts with the first molecule from which it can remove a proton. The second interaction is much more of a typical ligand–receptor interaction with clear structure–activity relationships (see chapter 2). In many cases, the molecular mechanism is not known, and effects are inferred from biochemical or physiological responses. Whereas there are similarities in many cases to the function and structure of the target molecule, there have been evolutionary changes as seen by alterations in amino acid sequence. These changes may partially be responsible for differences in response between species exposed to the same toxicant. For example, DDE is a CYP2B inducer in laboratory mice (*Mus musculus*), a CYP1A1 inducer in deer mice (*Peromyscus maniculatus*), and induces both CYP2B and CYP1A1 in the meadow vole (*Microtus ochrogaster*; Frame et al., 1999). CYP2B is a testosterone hydroxylase, and CYP1A1 is an estradiol hydroxylase. Thus DDE increases testosterone metabolism in the laboratory mouse, estradiol metabolism in the deer mouse, and both testosterone and estradiol in the vole.

Toxicogenetics

Toxicogenetics also is a term derived from is pharmacological equivalent, pharmacogenetics. Whereas pharmacogenetics is the field of pharmacology that examines the relationship between genetic factors and variation in response to drugs, toxicogenetics deals with how genetic variance and polymorphisms affect the response of individuals and populations to toxicants. A number of genetic differences have been described in receptors, metabolic enzymes, and transport proteins that affect responses to drugs and toxicants. Examples include the sequence differences in the nicotinic acetylcholine receptor between the human and mouse, which explain the differential sensitivity to curare. Other examples include human differences in CYP2D7, resulting in populations of poor and rapid metabolizers.

Toxicogenomics

Toxicogenomics is concerned with the use of genomic technologies to describe changes in gene expression resulting from exposure to xenobiotics, whether in humans or wildlife species and fishes (Lovett, 2000). This technology will allow for the identification of new toxicants as well as the discovery of the mechanism through which their biological activity arises (Nuwaysir et al., 1999). In the past, identification of genes and pathways was a formidable task; however, the advent of reasonably priced DNA microarrays and microarray readers have greatly accelerated the pace at which genes can be identified (Afshari et al., 1999; Burchiel et al., 2001). The application of toxicogenomics to toxicology in general is expected to yield high throughput and multiple endpoint capabilities, and its impact on research of sentinel species can be expected to provide significant benefits (Aardema and MacGregor, 2002). First, toxicogenomics will lead to improved ability to perform interspecies extrapolation due to the ability to monitor homologous gene activation across species. Second, the use of toxicogenomics will allow for the development of novel biomarkers by identifying genes or patterns of genes whose expression is altered by exposure to environmental toxicants. Third, toxicogenomics will allow researchers to identify subtle changes that may be masked by other biochemical responses. The major limitation of this technology is the relative scarcity of DNA microarrays for genes cloned from wildlife species (Olden et al., 2001).

Proteomics and Toxicant Exposure

Toxicant exposure alters the proteome of the exposed cells, tissues, or organism due to alterations in gene expression, such effects as induced by exposure to JP-8 vapor (Witzmann et al., 2000). The proteome is defined as the expressed genes of a cell, as differentiated from the genome. The study of these changes in the proteome can be described as toxicoproteomics. The genes encoding altered protein expression can be identified by a multistep procedure involving two-dimensional sodium dodecyl sulfate-pdyacrylamide gel (SDS-PAGE) electrophoresis. First, tissue extracts from exposed and control organisms are subjected to isoelectric focusing, which separates proteins by their isoelectric point (first dimension). The proteins thus separated are then run on a conventional SDS-PAGE gel and separated by molecular weight (second dimension). The gel is then typically transferred to a membrane for further analysis. This may include protein identification by MALDI-TOF (matrix assisted laser desorption ionization-time of flight) mass spectroscopy, which can yield sequence data. The ability to compare proteome changes after toxicant exposure across species aids in biomarker development and interspecies comparison of effects by identifying changes in protein expression.

The basic concepts of toxicokinetics, toxicodynamics, toxicogenetics, toxicogenomics, and toxicoproteomics give us the ability to understand inter- and intraspecies differences in response to the same concentration of toxicant. However, to extrapolate data between species, the fundamental concept underlying the use of sentinel species, we must build sufficient databases to develop predictive biologically-based toxicokinetic models and be able to modify them as genomic information becomes available. This is particularly true of toxicants that disrupt specific biochemical processes such as EDCs.

Conceptual Risk Assessment

Traditionally, risk assessment has consisted of four steps or phases. Phase I is described as hazard identification. For example, we know that compound X is a potent liver toxicant from laboratory studies. Therefore, we can look for compound X in the environment or in tissues from wildlife species or plants. Sentinel species may be used in hazard identification in this way by examining tissues from ill or dying individuals. Scientists noticed that there were reproductive difficulties in birds and reptiles; subsequent tissue analysis revealed elevated levels of DDT and DDE that were later shown to be causal. However, without exposure there is no risk. Therefore, care must be taken to ensure that measurement of a toxicant in environmental matrices is not an end in itself but a trigger for a more thorough investigation that determines whether the toxicant is bioavailable to organisms inhabiting the area.

Determining contaminant bioavailability is the second step in risk assessment and is referred to as exposure assessment. In this step investigators seek to determine the rate of exposure and the total body burden of each contaminant within an individual to determine the variability in contaminant concentration both intra- and interspecies. This is done by measuring biomarkers of exposure or effect in the species of interest.

A biomarker is a behavioral, biochemical, physiological, or toxic response elicited by exposure to a toxicant. A biomarker may also be the toxicant or a metabolite of the toxicant detected in the tissues, breath, feces, or urine. A biomarker may reflect exposure, an adverse effect, or susceptibility to another chemical or a pathogen. A number of biomarkers have been used in sentinel species work. These include blood and brain cholinesterase inhibition (from organophosphate and carbamate insecticides), induction/inhibition of cytochromes P450 (from a variety of phenolic, aromatic, alcoholic, and polyhalogenated aromatic hydrocarbon inducers and inhibition by metals and organometals), immunosuppression/immunostimulation (from metals and polynuclear/polychlorinated aromatic hydrocarbons), behavioral changes (from many xenobiotics), and alterations in endocrine function (from many xenobiotics). Each biomarker can be measured with a number of assays, and these must be optimized for each species.

A biomarker may be specific or quite nonspecific. For example, many animals respond to stress by increasing heart rate and respiratory rate. In contrast, cotinine in the urine or saliva is indicative of exposure to cigarette smoke. Similarly, a decrease in blood or brain acetylcholinesterase activity is indicative of exposure to an organophosphorus insecticide or nerve agent (e.g., sarin, used in the Tokyo subway bombings). Effective biomarkers are specific, easy to measure, and reproducible (accurate and precise). For work with endangered species, the biomarker method should be nonlethal. Examples include magnetic resonance imaging (MRI), computerized axial tomography (CAT), positron emission tomography (PET), and various assays on blood or saliva, fur or feathers, and urine or feces. Henshel et al.'s (1995) work with brain asymmetry using CAT in American bald eagles as a result of in-ovo polychlorinated dibenzo-*p*-dioxin exposure is one example of a nonlethal biomarker. Several laboratories have used skin/blubber biopsies to lethally measure contaminants and P_{450} enzyme induction in cetaceans (Newman et al., 1994; Fossi et al., 2000, 2001; Reddy et al., 2001). DNA damage in birds has been used as a biomarker for heavy metal exposure using the Comet assay on peripheral blood collected from storks and kites in an area contaminated by mining waste and in adjacent uncontaminated sites (Pastor et al., 2001).

In addition, biomarkers for acute exposure differ from biomarkers for assessing effects arising from chronic exposure. Biomarkers for acute exposure need to react quickly to the presence of a contaminant, remain in an altered state sufficiently long for detection, and then revert to pre-exposure values. Biomarkers for chronic exposure should react slowly to presence of a contaminant, and the degree of the change in the biomarker should reflect both the duration of exposure and magnitude of contaminant exposure and not return to pre-exposure levels. Such biomarkers may include development of DNA mutations and neoplasia, immune suppression, alterations in sexual development, and malformed offspring (LeBlanc and Bain, 1997). For example, Rodgers et al. (2001a,b) used a micronucleus assay and radionuclide uptake to determine whether radioresistance developed in rodents subchronically exposed to radionuclides in the Chernobyl exclusion zone.

Part of the challenge in using sentinel species is developing a sensitive, specific biomarker for the contaminant of interest in one or more species that inhabits the site of interest. Once this has been accomplished, sufficient numbers of individuals must be sampled in the contaminated area and an adjacent pristine area to determine the exposure range. Depending on the tendency for toxicant to bioconcentrate and biomagnify, it may be necessary to sample species from several trophic levels. For example, for toxicants like PCBs that are lipophilic and bind to sediment, this may require sampling a benthic organism, a fish species feeding upon that species, a more pelagic fish, and a fish-eating mammal or bird. This information can then be compared to dose–response relationships for the toxicant and a biochemical, physiological, or toxic response to determine a predicted effect level.

The third part of a conceptual risk assessment is dose–response assessment. Basically, data dose response provide a mathematical relationship between exposure (dose) and one or more adverse data effects. These effects may include reproductive success, immune suppression, or behavioral changes that directly impact survivability of the sentinel species or related species or that may provide a link to human problems such as an increase in cancer rate. Dose–response relationships are typically determined for sentinel species by using chemically naive, laboratory-reared animals that are dosed with the appropriate agent in a research facility. Conversely, native species may be trapped and released into a large enclosure or mesocosm followed by dosing with the agent of choice, or by enhancing populations of a desired sentinel and dosing the young. The latter approach works well for cavity-nesting avian species. After dosing, effects are determined using previously selected biomarkers.

The fourth part of the risk assessment involves analyzing field data in light of the dose–response assessment, predicting the risk, and formulating a risk management plan. This plan may involve deciding whether to remediate a contaminated area and determining what level of remediation is necessary to protect a specified number of species for a defined probability. For example, one might decide to decrease the amount of a chemical entering a receiving stream from a chemical plant until 95% of the species inhabiting that stream are protected 95% of the time. This concept is based on the inherent knowledge that protecting 100% of the species 100% of the time is not biologically or financially feasible. For a more detailed discussion of risk assessment, refer to DiGuilio and Monosson (1996) and Suter (1997).

Selection and Use of Sentinel Species

Selection of the most appropriate sentinel species is first based on the objective of the study. If the objective is to provide a qualitative assessment of ecosystem health or to provide warning of possible contamination of an environment, relatively few species need to be identified, and the population density of each species need not be great. If a quantitative risk assessment is desired for decision-making, more species are needed for a wider range of trophic levels, and there must be sufficient numbers of each species available for analysis to obtain the needed level of statistical significance.

Regardless of the objective, the selection of each species has several common themes. First, the species should be sensitive to the contaminant or contaminants of interest. It is not necessary that the sentinel be exquisitely sensitive, just more sensitive than the species one wishes to protect. Next, the species should have a fairly wide geographical distribution. This characteristic allows an investigator to compare data using the same species from multiple sites. Another consideration is the species' home range. With a migratory or wide-ranging species, it is difficult to determine where and when exposure occurred. Thus, it is advisable to select a species that has a restricted home range, is nonmigratory, and is possibly territorial. If the contaminant is bioconcentrated and/or biomagnified, the selection of a species that is on a higher trophic level is justified. Such species may be secondary or tertiary consumers. However, predators such as hawks, eagles, owls, mink, seals, and alligators are frequently protected or sparse at a typical site. Although protected and/or endangered species can still be used if nonlethal sampling occurs, too few individuals of a species will make it difficult to obtain statistically valid results.

A final consideration is the life span of the species. Short-lived species can be used for assessment of acute and subchronic effects of contaminant exposure, whereas long-lived species, such as turtles, are well-suited for assessing the effects of chronic exposure. Attempting to use an inappropriate species or inappropriate length of exposure can underestimate the effects of chronic exposure (LeBlanc and Bain, 1997).

Mammals have been used frequently as sentinels. The typical laboratory rodents *Mus musculus* and *Rattus norvegicus* have been the cornerstone for testing chemical substances for many years. These species can be used as environmental sentinels in several ways. First, samples of soil or other environmental matrices can be brought to the laboratory facility, and the rodents placed on the soil for variable lengths of time. If water is contaminated, it can be placed in the animals' drinking water to assess effects. Enclosures can be erected on contaminated sites and laboratory-reared rodents added. These approaches have the advantage that prior exposure is known and that acquired tolerance has not been established. However, it must not be automatically assumed that responses to toxicant exposure observed in laboratory rodents correlate to other mammals or even other rodents. Many species of indigenous rodents have been used in laboratory and field studies. These include the deer mouse (*Peromyscus maniculatus*), the white-footed mouse (*Peromyscus leucopus*), the voles (*Microtus townsendi* and *M. ochrogaster*), and the cotton rat (*Sigmodon hispidus*). Another mammal used in this manner is mink (*Mustela vision*) that have been fed contaminated Great Lakes fish (Fox , 2001). Mink and otters (*Lutra canadensis*) also have been used as sentinels for organochlorine compounds in trap-and-release studies. Other terrestrial mammals used include domestic animals such as dogs, goats, cattle, and sheep, as well as such diverse wild animals as shrews, deer, and moose.

Cetacean strandings also present an opportunity for collection of samples that can be used to determine body burdens of metals and organohalogen compounds as well as a complete necropsy to determine pathophysiology potentially associated with chemical exposure. Cetaceans are long-lived, feed on predatory fish species, and thus accumulate substantial body burdens of lipophilic compounds. For certain cetacean species, such as the beluga whales inhabiting the St. Lawrence seaway and some of the porpoises in Asia, body burdens of contaminants are so high and so much is excreted into milk that the first-born calf has exceedingly high mortality (Fox, 2001). Additionally, excessive incidence of tumors, immune suppression, and hermaphroditism have been reported for belugas and correlated with increased concentration of DDT and polychlorinated biphenyls, furans, and dioxins in blubber and other fat depots (Beland, 1993; DeGuise et al., 1994, 1995, 1996).

Avian species drew great attention in the 1960s and 1970s as sentinel species for the organochlorine pesticides, particularly DDT, when it was discovered that exposure to these pesticides resulted in wide-scale eggshell thinning (Ratcliffe, 1967; Hickey and Anderson 1968). Moreover, abnormal behavior, such as same-sex pairing, was observed in gulls and terns exposed to PCBs (Fry and Toone, 1981; Fry et al., 1987). As previously mentioned, brain asymmetry was observed in bald eagles inhabiting the Great Lakes region, presumably due to exposure to polychlorinated biphenyls (PCBs), polychlorinated dibenzo-*p*-dioxins (PCDDs), and polychlorinated dibenzofurans (PCDFs) (Henshel et al., 1995; see also chapter 11).

A wide variety of bird species, both altricial and galliform, have been used as sentinels. These include raptors such as the bald eagle, peregrine falcon, and sparrow hawk, and piscivirous species such as the brown pelican, great blue heron, double-crested comorant, gulls, and terns because of their high position in the food web. Owl species have also been suggested as sentinels. Species such as the northern bobwhite quail, eastern bluebird, European starling, and various warblers that use natural or man-made cavities for nesting also can be useful sentinels. It is possible to create enhanced populations of these species by providing appropriate nesting boxes on a contaminated site and nearby reference site (Akins et al., 1993; Hunt and Hooper, 1993; Trust et al., 1994). By providing sufficient number of nesting boxes on a reference site, a researcher can conduct dosing studies to establish dose–response relationships (Akins, 1995). A number of avian species are commercially available as either eggs or adults, including northern bobwhite quail, mallard ducks, and ringnecked pheasants. This allows researchers to expose either adults to contaminants for single or multigenerational studies or conduct egg injection–incubation studies to assess developmental effects of environmental agents (Dickerson et al., 1995; McMurry et al., 1999a,b; Peden-Adams, 1999). A number of these studies have focused on endocrine disruption.

Reptiles are not used as sentinels as frequently as are birds and mammals. However, reptilian species have characteristics that make them useful as sentinels. For instance, a number of lizard, turtle, and crocodilian species have environmental sex determination and are sensitive to estrogenic substances during embryonic development. Also, some of these species will readily breed in captivity. Much of the work on reptiles has been confined to measuring chemical residues in tissues and attempting to establish causal relationships with observed effects. Many studies have used the common snapping turtle, which has the advantage of being fairly long lived, high in the food-chain, and readily available. A number of researchers have performed dosing studies on reptilian eggs collected in the wild. Although these experiments are handicapped by the appreciable amount of lipophilic substances detectable in eggs gathered from apparently pristine surroundings, much valuable information has been obtained using turtle and alligator eggs (Bergeron and Crews, 1998; Matter et al., 1998). These studies have provided mechanistic and dose–response relationships for suspected EDCs. Turtles have been also used as sentinel species for ionizing radiation exposure at radionuclide-contaminated sites. Their long life span as compared to most wildlife species makes them ideal for ecological dosimetry using chromosome translocation as a biomarker (Ulsh et al., 2000).

Amphibians are receiving more attention as sentinels in light of the worldwide decline in amphibian numbers and reports of frogs with additional or deformed legs from several regions of the United States (Pechmann et al., 1991; Blaustein and Wake, 1995). Amphibians are susceptible to exposure by both dermal and ingestion routes due to their semiaquatic life history. Additionally, most species undergo a marked metamorphosis that can be influenced by chemicals that target thyroid hormone synthesis, transport, deiodination, or receptor activation. Studies to date indicate that anurans are sensitive to organochlorine pesticides and nonylphenol. Bantle (1995) has developed a predictive test using *Xenopus laevis* embryos brought to an early stage of development called FETAX (frog embryo teratogenesis assay–*Xenopus*). This assay has successfully identified a number of toxicants (Bantle, 1995). Several field and laboratory studies with environmentally relevant amphibian species have been reported (Hopkins et al., 1997; Jung and Walker, 1997; Clark et al., 1998; Mann and Bidwell, 1999; Diana et al., 2000; Johnson et al., 2000; Goleman et al., 2002; Hayes et al., 2002; Tavera-Mendoza et al., 2002a,b; Carr et al.,

2003). As funding for amphibian research continues to climb, increased information about amphibian biochemistry, endocrinology, and physiology will greatly increase the utility of amphibians as sentinel organisms.

Fish species have been used for years to assess aquatic health, particularly in freshwater ecosystems. In the Great Lakes, salmonids respond to PCBs and PCDDs by developing blue sac disease and goiter. Suckers downstream of bleached kraft papermills show elevated cytochrome P450 activity (McMaster et al., 1991, 1995). Other studies have shown that fish respond to general environmental stressors with elevated corticosteroid levels. Elevated levels of ethinylestradiol and natural estradiol from sewage treatment plants were first detected when feminization of male fish were observed on the basis of vitellogenin induction (Sumpter and Jobling, 1995; Jobling et al., 1996). Exposure to 4-nonylphenol and other alkylphenols also resulted in increased vitellogenin levels in the plasma of male fishes of diverse species (White et al., 1994; Lech et al., 1996; Pederson et al., 1999; Foran et al., 2000; Huang and Wang, 2001; Nichols et al., 2001; Schaiger et al., 2002). Freshwater and saltwater fishes have been used as sentinel species. Some of these, like sheepshead minnow, fathead minnow, and rainbow trout, are mandated by federal regulation in the testing of municipal and industrial plant outflows into receiving streams or in life-cycle tests for new compounds (U.S. EPA, 1991; ASTM, 1992; Adams, 1995). Species widely used in research and aquatic toxicity testing include trout, carp, suckers, mosquito fish, medaka, flounder, killifish, and zebrafish (Adams, 1995). Many of these species are easily maintained in the laboratory and can be used for dose–response assessment.

Invertebrates have received some attention as sentinel species. Programs such as Mussel Watch, administered by the U.S. National Oceanic and Atmospheric Administration, and the earthworm soil bioassay are examples of invertebrate monitoring (Goldberg and Bertine, 2000; Soto et al., 2000; Leland et al., 2001; O'Connor, 2002). Mussels can be used to monitor spatial and temporal changes in metal and organic contamination (Chase et al., 2001). Earthworms can be used to detect contamination and to monitor remediation efforts (Chang et al., 1997; Landrum et al., 2002). Invertebrates are used in laboratory dosing studies or in industrial effluent and sediment testing. Mysid shrimp, daphnia, and the common oyster have been used as sentinels for many years (Nimmo et al., 1978; Buikema et al., 1981; Passino-Reader et al., 1997; Naddy and Klaine, 2001; Roast et al., 2000; 2001; O'Connor, 2002). Freshwater hydra species have been used to detect developmental and other toxicants (Johnson et al., 1982, 1986; Wiger and Stottum, 1985; Karntanut and Pascoe, 2002). Hydra respond similarly to rat postimplantation embryos upon exposure to developmental toxicants (Mayura et al., 1991; Yang et al., 1993). Benthic organisms would also seem to have potential as sentinel species. Several studies have demonstrated that industrial and agricultural runoff impacts benthic population diversity and density (Sibley et al., 1997; Anderson et al., 2001; Ernst et al., 2001; Ivorra et al., 2002; Neumann and Dudgeon, 2002). Clearly, the use of invertebrates as sentinels is a field of immense potential that does not have the recognition it deserves.

The use of microorganisms as sentinels is a field that likewise has received little research emphasis. Many microorganisms respond to toxicants by altering gene transcription to form protective proteins and by increasing mutation rates. This principle is applied in the Ames assay for assessment of the mutagenic potential of chemicals, with and without S9 liver fraction bioactivation (Maron and Ames, 1983). Yeast also is used in this manner (Eckardt and von Borstel, 1985). The use of the S9 liver fraction, which contains phase I and II xenobiotic metabolizing enzymes, allows researchers to test compounds that may require bioactivation for genotoxicity in a bacterial test system. Researchers are using naturally occurring microorganisms as sentinels for environmental pollution and to accomplish remediation (Chan et al., 1999; Layton et al., 1999; Riaz-ul-Haq and Shakoori, 2000). It appears that microorganisms can be used as sentinels at a fairly low cost.

Limitations, Pitfalls, and Other Barriers

The major problem in using sentinel species is that less is known about their physiology, response to xenobiotic exposure, and interindividual variation than most common laboratory species. This is the critical factor that limits the ability of a researcher to extrapolate results using sentinel species to ecosystem or human health. For example, studies in laboratory rodents do not uniformly predict the results of exposure of the same chemical to feral rodents (Dickerson et al., 1999, Frame et al., 1999). A number of these factors have been compiled by NRC (1991), Sheffield et al. (1998), and van der Schalie et al. (1999) and include:

1. The techniques used to collect and analyze data from fish and wildlife environmental sentinel systems are not well standardized. Quality assurance ranges from extensive to nonexistent. Development of standardized methods, benchmark dose responses, and reporting formats would expedite cross-species extrapolation. Moreover, most of these existing studies are observational, and no attempt has been made to establish mechanism or causality. Exposure is often not well characterized regarding duration or intensity.
2. There is a lack of knowledge on the physiology, biochemistry, and endocrinology of potential sentinel species. Moreover, dose–response relationships for the majority of known environmental contaminants are lacking for most potentially useful species. There is not a concentrated effort to maintain a tumor registry in lower animals. The establishment of a centralized database in which wildlife and fish sentinel species data could be maintained along with data on livestock and companion animals would greatly facilitate the use of other species in protecting human health.
3. The data collected are, for the most part, fragmented and insufficient for use in risk assessment paradigms. Although efforts are underway to remedy this fragmentation, the amount of data needed is massive. The establishment of a centralized database also would aid in solving this problem.
4. The lack of toxicokinetic and toxicodynamic data for the contaminants of concern in most useful sentinel species prevents effective extrapolation to other species of concern. This lack of fundamental knowledge creates uncertainties in the prognostic use of data from sentinels that can range from a factor of 10 to 10,000. If the information obtained from sentinels is used to

design remediation processes, this may result in millions of dollars in unnecessary expense.

5. Whereas many environmental science departments offer training in basic risk assessment, few programs incorporate a thorough background in the appropriate design and use of sentinel species for biomonitoring and assessment strategies. Development of extramurally funded centers would provide training in biomarker-based sentinel species research.

6. Few scientific societies focus on sentinel species research. As a result, communication between scientists, engineers, and regulators is fragmented. In addition, this lack of communication results in disconnection among the needs of academia, federal and state regulatory agencies, and industry. Integrating data sets from studies would help identify mechanisms and causality.

7. The use of early developmental stages of an organism as opposed to adult animals is receiving much attention. Whereas exposure of an adult to a compound may result in transient effects, it is clear that many chemicals can produce irreversible damage if exposure occurs during gestation and prepubertal stages.

8. Development of alternative methods for assessing impacts on sentinel species using emerging technologies such as genomics and proteomics would expedite advances in the use of sentinel species to predict human health effects.

This list points out some crucial data gaps and opportunities for future research strategies. Although the topic of sentinel species has received some much needed attention by the National Institutes of Environmental Health Sciences, the U.S. EPA, and the American Chemistry Council, progress takes time, and money must be found for research. However, the future appears bright for such endeavors.

Summary

Fishes, invertebrates, and wildlife show great potential as sentinel species for the early detection of ecosystem contamination and dangers to human health. In many cases, it was adverse effects in sensitive species that drew attention to environmental problems. However, the full potential of sentinels will not be realized until data gaps in the biochemistry, endocrinology, and physiology of useful sentinels are filled and toxicokinetic, toxicodynamics, and toxicogenetic data are generated for these species for contaminants of present and future concern.

References

Aardema, M.J., MacGregor, J.T., 2002. Toxicology and genetic toxicology in the era of "toxicogenomics": impact of "-omics" technologies. Mutat. Res. 499, 13–25.

Adams, W.J., 1994. Aquatic toxicology test methods. In: Hoffman, D.J., Rattner, B.A., Burton, Jr. G.A., Cairns, J. Jr. (Eds.), Handbook of Ecotoxicology. Lewis Publishers, Boca Raton, FL, pp. 25–46.

Afshari, C.A., Nuwaysir, E.F., Barrett, J.C., 1999. Application of complementary DNA microarray technology to carcinogen identification, toxicology and drug safety evaluation. Cancer Res. 59, 4759–4760.

Akins, J.M., 1995. Porphyrin profiles in the nestling European starling (*Sturnus vulgarus*): A biomarker of field contaminant exposure. Ph.D. dissertation, Clemson University, Clemson, SC.

Akins, J.M., Hooper, M.J., Miller, H., Woods, J.S., 1993. Porphyrin profiles in the nestling European starling (*Sturnus vulgarus*): A potential biomarker of field contaminant exposure. J. Toxicol. Environ. Health 40, 47–59.

Anderson, B.S., Hunt, J.W., Phillips, B.M., Fairey, R., Roberts, C.A., Oakden, J.M., Puckett, H.M., Stephensons, M., Tjeerdema, R.S., Long, E.R., Wilson, C.J., Lyons, J.M., 2001. Sediment quality in Los Angeles Harbor, USA: A triad assessment. Environ. Toxicol. Chem. 20, 359–370.

ASTM (1992) In: Mayes, M.A. Barron, M.G., editors. Aquatic Toxicology and Risk Assessment: Volume 14, ASTM STP 1124, American Society for Testing and Materials, Philadelphia, PA.

Bantle, J.A., 1995. FETAX—a developmental assay using frog embryos. In: Rand, G.M. (Ed.), Fundamentals of Aquatic Toxicology, 2nd edition. Taylor and Francis, Washington, DC, pp. 207–230.

Beland, P., DeGuise, S., Girard, C., Lagace, A., Martineau, D., Michaud, R., Muir, D.C.G., Norstrom, R.J., Pelletier, E., Ray, S., Shugart, L.R., 1993. Toxic compounds and health and reproductive effects in St. Lawrence beluga whales. J. Great Lakes Res. 19, 766–775.

Bergeron, J.M., Crews, D., 1998. Effects of estrogenic compounds in reptiles: Turtles. In: Kendall, R.J., Dickerson, R.L., Giesy, J., Suk, W. (Eds.), Processes and Principles for Evaluating Endocrine Disruption in Wildlife. SETAC Press, Pensacola, FL.

Blaustein, A.R., Wake, D.B., 1995. The puzzle of declining amphibian populations. Sci. Am. 272, 52–63.

Brody, T.M., Larner, J., Minneman, K.P., 1998. Human Pharmacology: Molecular to Clinical, 3rd edition. Mosby, St. Louis, MO.

Buikema, A.L. Jr., Benfield, E.F., Niederlehner, B.R., 1981. Effects of pollution on freshwater invertebrates. J. Water Pollut. Contr. Fed. 53(6): 1007–1015.

Burchiel, S.W., Knall, C.M., Davis, J.W., Paules, R.S., Boggs, S.E., Afshari, C.A., 2001. Analysis of genetic and epigentic mechanisms of toxicity: Potentials of toxiogenomics and proteomics in toxicology. Toxicol. Sci. 59, 251–259.

Burrell, G.A., Seibert, F.M., 1916. Gases found in coal mines. Miners' Circular 14. Bureau of Mines, Department of the Interior, Washington, DC.

Carr, J.A., Gentles, A., Smith, E.E., Goleman, W.L., Urquidi, L.J., Thuett, K., Kendall, R.J., Giesy, J.P., Gross, T.S., Solomon, K.R., Van Der Kraak, G., 2003. Response of larval *Xenopus laevis* to atrazine: Assessment of growth, metamorphosis, and gonadal and laryngeal morphology. Environ. Toxicol. Chem. 22, 396–405.

Carson, R., 1962. Silent Spring. Houghton Mifflin, New York.

Carter, C.D., Kimbrough, R.D., Liddle, J.A., Cline, R.E., Zack, M.M., Jr., Barthel, W.F., Koehler, R.E., Philips, P.E., 1975. Teterachlorodibenzodioxin: An accidental poisoning episode in horse arenas. Science 188, 738–740.

Case, A.A., Coffman, J.R., 1973. Waste oil: Toxic for horses. Vet. Clin. N. Am. 3, 273–277.

Chan, C.M., Lo, W., Wong, K.Y., Chung, W.F., 1999. Monitoring the toxicity of phenolic chemicals to activated sludge using a novel optical scanning respirometer. Chemosphere 39, 1421–1432.

Chang, L.W., Meier, J.R., Smith, M.K., 1997. Application of plant and earthworm bioassays to evaluate remediation of a lead-contaminated soil. Arch. Environ. Contam. Toxicol. 32, 166–171.

Chase, M.E., Jones, S.H., Hennigar, P., Sowles, J., Harding, G.C., Freeman, K., Wells, P.G., Krahforst, C., Coombs, K., Crawford, R., Pederson, J., Taylor, D., 2001. Gulfwatch: Monitoring spatial and temporal patterns of trace metal and organic pollution in the Gulf of Mexico (1991–1997) with the blue mussel, *Mytilus edulis* L. Mar. Pollut. Bull. 42, 491–503.

Clark, E.J., Norris, D.O., Jones, R.E., 1998. Interactions of gonadal steroids and pesticides (DDT, DDE) on gonaduct growth in larval tiger salamanders, *Ambystoma tigrinum*. Gen. Comp. Endocrinol. 109, 94–105.

DeGuise, S., Bernier, J., Martienau, D., Beland, P., Fournier, M., 1996. Effects of in vitro exposure of beluga whale splenocytes and thymocytes to heavy metals. Environ. Toxicol. Chem. 15, 1357–1364.

DeGuise, S., Lagace, A., Beland, P., 1994. Tumors in St. Lawrence beluga whales (*Delphinaptera leucas*). Vet. Pathol. 3, 444–449.

DeGuise, S., Martineau, D., Beland, P., Fournier, M., 1995. Possible mechanisms of action of environmental contaminants on St. Lawrence beluga whales (*Delphinapter leucas*). Environ. Health Perspect. 103(suppl. 4), 73–77.

DeRosa, C., Richter, P., Pohl, H., Jones, D.F., 1998. Environmental exposures that affect the endocrine system: Public health implications. J. Toxicol. Environ. Health B Crit. Rev. 1, 3–26.

Diana, S.G., Resetarits, W.J., Schaeffer, D.J., Beckmen, K.B., Beasley, V.R., 2000. Effects of atrazine on amphibian growth and survival in artifical aquatic communities. Environ. Toxicol. Chem. 19, 2961–2967.

Dickerson, R.L., Hooper, M.J., Gard, N.W., Cobb, G.P., Kendall, R.J., 1995. Toxicological foundations of ecological risk assessment: Biomarker development and interpretation based on laboratory and wildlife species. Environ. Health Perspect. 102, 65–69.

Dickerson, R.L., McMurry, C.S., Smith, E.E., Cobb, G.P., Frame, L.T., 1999. Modulation of endocrine pathways by 4,4'-DDE in the deer mouse (*Peromyscus maniculatus*). Sci. Total Environ. 233, 97–108.

DiGuilio, R.T., Monosson, E. (Eds.), 1996. Interconnections between Human and Ecosystem Health. Chapman and Hall, London.

Eckardt, F., von Borstel, R.C., 1985. Mutagen testing of agricultural chemicals with yeast. Basic Life Sci. 34, 221–248.

Ernst, W., Jackman, P., Doe, K., Page, F., Julien, G., Mackay, K., Sutherland, T., 2001. Dispersion and toxicity to non-target organisms of pesticides used to treat sea lice on salmon in net pen enclosures. Mar. Pollut. Bull. 42, 433–444.

Foran, C.M., Bennett, E.R., Benson, W.H., 2000. Exposure to environmentally relevant concentration of different nonylphenol formulations in Japanese medaka. Mar. Environ. Res. 50, 135–139.

Fox, G.A., 2001. Wildlife as sentinels of human health in the Great Lakes-St. Lawrence basin. Environ. Health Perspect. 109(suppl. 6), 53–61.

Frame, L.T., Settachan, D.S., Dickerson, R.L., 1999. 4,4'-Dichlorodiphenylethylene (4,4'-DDE) is a 3-methylcholanthrene-type inducer in the deer mouse (*Peromyscus maniculatus*) but an Aroclor-like inducer in the vole (*Microtus ochrogaster*). Organohalogen Compounds, 44, 101–106.

Fry, D.M., Toone, C.K., 1981. DDT-induced feminization of gull embryos. Science 213, 922–924.

Fry, D.M., Toone, C.K., Speich, S.M., Peard, R.J., 1987. Sex ratio skew and breeding pattern of gulls: Demographic and toxicological considerations. Stud. Avian Biol. 10, 26–43.

Goldberg, E.D., Bertine, K.K., 2000. Beyond the Mussel Watch—new directions for monitoring marine pollution. Sci. Total Environ. 247, 165–174.

Goleman, W.L., Urquidi, L.J., Anderson, T.A., Smith, E.E., Kendall, R.J., Carr, J.A., 2002. Environmentally relevant concentrations of ammonium perchlorate inhibit development and metamorphosis in *Xenopus laevis*. Environ. Toxicol. Chem. 21, 424–430.

Guillette, L.J., Jr., 2000. Organochlorine pesticides as endocrine disruptors in wildlife. Cent. Eur. J. Public Health (suppl.) 8: 34–35.

Guillette, L.J., Jr., Gross, T.S., Masson, F.R., Matter, J.M., Percival, H.F., Woodward, A.R., 1994. Developmental abnormalities of the gonad and abnormal sex hormone concentrations in juvenile alligators from contaminated and control lakes in Florida. Environ. Health Perspect. 102, 681–688.

Haddad, S., Krishnan, K., 1998. Physiological modeling of toxicokinetic interactions: Implication for mixture risk assessment. Environ. Health Perspect. 106(suppl. 6), 1377–1384.

Haring, C.M., Meyer, K.F., 1915. Investigations of livestock conditions with horses in the Selby smoke zone. Calif. Hurrau Mines Bull. 98.

Hayes, T.B., Collins, A., Lee, M., Mendoza, M., Noriega, N., Stuart, A.A., Vonk, A., 2002. Hermaphroditic, demasculanized frogs after exposure to the herbicide atrazine at low ecologically relevant doses. Proc. Natl. Acad. Sci. USA 99, 5476–5480.

Henshel, D.S., Martin, J.W., Norstrom, R., Whitehead, P., Steeves, J.D., Cheng, K.M., 1995. Morphometric abnormalities in brains of Great blue heron hatchlings exposed in the wild to PCDDs. Environ. Health Perspect. 103(suppl. 4), 61–66.

Hickey, J.J., Anderson, D.W., 1968. Chlorinated hydrocarbons and eggshell changes in raptorial and fish-eating birds. Science 162, 271–273.

Hodgson, E., Levi, P. (Eds.), 1994. Biochemical Toxicology, 2nd edition. Appleton and Lange, Norwalk, CT.

Holm, L.W., Wheat, J.D., Rhodes, E.A., Firch, G., 1953. Treatment of chronic lead poisoning in horses with calcium disodium ethylenediamine tetraacetate. J. Am. Vet. Assoc. 123, 383–388.

Hopkins, W.A., Mendonca, M.T., Congdon, J.D., 1997. Increased circulating levels of testosterone and corticosterone in southern toads, *Bufo terrestris*, exposed to coal combustion waste. Gen. Comp. Endocrinol. 108, 237–246.

Huang, R.K., Wang, C.H., 2001. The effect of two alkylphenols on vitellogenin levels in male carp. Proc. Natl. Sci. Counc. Repub. China B 4, 248–252.

Hunt, K.A., Hooper, M.J., 1993. Development and optimization of reactivation techniques for carbamate-inhibited brain and plasma cholinesterase in birds and mammals. Anal. Biochem. 212, 335–343.

Ivorra, N., Hetrelaar, J., Kraak, M.H., Sabater, S., Admiraal, W., 2002. Responses of biofilms to combined nutrient and metal exposure. Environ. Toxicol. Chem. 21, 626–632.

Jackson, T.F., Halbert, F.L., 1974. A toxic syndrome associated with the feeding of polybrominated biphenyl-contaminated protein concentrate to dairy cattle. J. Am. Vet. Med. Assoc. 165, 437–439.

Jobling, S., Sheahan, D., Osborne, Ja., Matthiesen, P., Sumpter, J.P., 1996. Inhibition of testicular growth in rainbow trout (*Oncorhyncus mykiss*) exposed to estrogenic alkylphenolic chemicals. Environ. Toxicol. Chem. 15, 194–202.

Johnson, E.M., Gabriel, B.E., Christian, M.S., Sica, E., 1986. The developmental toxicity of xylene and xylene isomers in the hydra assay. Toxicol. Appl. Pharmacol. 82, 323–328.

Johnson, E.M., Gorman, R.M., Gavel, B.E., George, M.E., 1982. The hydra attenuate system for detection of teratogenic hazards. Teratog. Carcinog. Mutagen. 2, 263–276.

Johnson, M.S., Vodela, J.K., Reddy, G., Holladay, S.D., 2000. Fate and the biochemical effects of 2,4,6–trinitrotoluene exposure to tiger salamanders (*Ambystoma tigrinum*). Ecotoxicol. Environ. Safety 46, 186–191.

Jung, R.E.,Walker, M.K., 1997. Effects of 2,3,7,8-tetrachlorodibenzo-*p*-dioxin (TCDD) on development of anuran amphibians. Environ. Toxicol. Chem. 16, 230–240.

Karntanut, W., Pascoe, D., 2002. The toxicity of copper, cadmium and zinc to four different hydra (Cnidaria: Hydrozoa). Chemosphere 47, 1059–1064.

Klaasen, C.D. (Ed.), 2001. Casarett and Doull's Toxicology. The Basic Science of Poisons, 6th edition. McGraw-Hill, New York.

Komar, N., 2001. West Nile virus surveillance using sentinel birds. Ann. N.Y. Acad. Sci. 951, 58–73.

Kurland, L.T., Fara, S.N., Siedler, H., 1960. Minamata Disease. World Neurol. 1, 370–395.

Landrum, P.F., Gedeon, M.L., Barton, G.A., Greenberg, M.S., Rowland, C.D., 2002. Biological responses of *Lumbriculus variegates* exposed to fluoranthene-spiked settlement. Arch. Environ. Contam. Toxicol. 42, 292–302.

Layton, A.C., Gregory, B., Schultz, T.W., Sayler, G.S., 1999. Validation of genetically engineered bioluminescent surfactant resistant bacteria as toxicity assessment tools. Ecotoxicol. Environ. Safety 43, 222–228.

LeBlanc, G.A., 1995. Are environmental sentinels signaling? Environ. Health Perspect. 103, 888–890.

LeBlanc, G.A., Bain, L.J., 1997. Chronic toxicity of environmental contaminants: Sentinels and biomarkers. Environ. Health Perspec. 105(Suppl. 1), 65–80.

Lech, J.J., Lewis, S.K., Ren, L., 1996. In vivo estrogenic activity of nonylphenol in rainbow trout. Fundam. Appl. Toxicol. 30, 229–232.

Leland, J.E., Mullins, D.E., Berry, D.F., 2001. Evaluating environmental hazards of land applying composted diazinon using earthworm bioassays. Environ. Sci. Health B 36, 821–834.

Lovett, R.A., 2000. Toxicogenomics. Toxicologist brace for genomics revolution. Science 289, 536–537.

Lower, W.R., Kendall, R.J., 1992. Sentinel species and sentinel bioassays. In McCarthy, J.F., Shugart, L.R. (Eds.), Biomarkers of Environmental Contamination. Lewis Publishers, Chelsea, MI, pp. 309–331.

Mann, R., Bidwell, J., 1999. Toxicological issues for amphibians in Australia. In Campbell, A. (Ed.), Declines and Disappearances of Australian Frogs. Environment Australia, Canberra, pp. 185–201.

Maron, D.M., Ames, B.N., 1983. Revised methods for the Salmonella mutagenicity test. Mutat. Res. 113, 173–215.

Matter. J.M., Crain, A., McMurry, C.S., Pickford, D.B., Rainwater, T.R., Reynolds, K.D., Rooney, A.A., Dickerson, R.L., Guillette, L.J., Jr., 1998. Effects of endocrine disrupting contaminants in reptiles: Alligators. In: Kendall, R.J., Dickerson, R.L., Giesy, J., Suk, W. (Eds.), Processes and Principles Evaluating Endocrine Disruption in Wildlife. SETAC Press, Pensacola, FL, pp. 267–290.

Mayura, K., Smith, E.E., Clement, B.A., Phillips, T.D., 1991. Evaluation of the developmental toxicity of chlorinated phenols utilizing hydra attenuate and postimplantation rat embryos in culture. Toxicol. Appl. Pharmacol. 108, 253–266.

McLean, R.G., Ubico, S.R., Bourne, D., Komar, N., 2002. West Nile virus in livestock and wildlife. Curr. Top. Microbiol. Immunol. 267, 271–308.

McMaster, M.E., Van Der Kraak, G.J., Munkittrick, K.R., 1995. Exposure to bleached kraft pulp mill effluent reduces the steroid biosynthetic capacity of white sucker ovarian follicles. Endocrinology 112, 169–178.

McMaster, M.E., Van Der Kraak, G.J., Portt, C.B., Munkittrick, K.R., Sibley, P.K., Dixon, D.G., 1991. Changes in hepatic mixed function oxidase activity, plasma steroid levels and age at maturity of a white sucker (*Catostomus commersoni*) population exposed to bleached kraft pulp mill effluent. Aquat. Toxicol. 21, 199–218.

McMurry, C.S., Dickerson, R.L., 1999a. Effects of binary mixtures of six xenobiotics on hormone concentrations and morphometric endpoints in northern bobwhite quail (*Colinus virginianus*). Organohalogen Compounds 42, 93–96.

McMurry, C.S., Matter, J.M., Dickerson, R.L., 1999b. Effects of six xenobiotics on survival, hormone concentrations and morphometric endpoints in northern bobwhite quail (*Colinus virginianus*). Organohalogen Compounds 42, 87–92.

Naddy, R.B., Klaine, S.J., 2001. Effect of pulse frequency and interval on the toxicity of chlopyrifos to Daphnia magna. Chemosphere 45, 497–506.

Neumann, M., Dudgeon, D., 2002. The impact of agricultural runoff on stream benthos in Hong Kong China. Water Res. 36, 3103–3109.

Newman, J.W.Q., Vedder, Jm., Jarman, W.M., Chang, R.R., 1994. A method for the determination of environmental contaminants in living marine mammals using microscale samples of blubber and blood. Chemosphere 29, 671–681.

Nichols, K.M., Snyder, E.M., Snyder, S.A., Pierens, S.L., Miles-Richardson, S.R., Giesy, J.P., (2001). Effects of nonylphenol ethoxylate exposure on reproductive output and bioindicators of environmental estrogen exposure in fathead minnows Pimephales promelas. *Environ Toxicol Chem.* 20, 510–522.

Nimmo, D.R., Rigby, R.A., Bahner, L.H., Sheppard, J.M., 1978. The acute and chronic effects of cadmium on the estuarine mysid Mysidopsis bahia. Bull. Environ. Contam. 19(1), 80–85.

NRC (National Research Council), 1991. Animals as sentinels of environmental health hazards. National Academy Press, Washington, DC.

Nuwaysir, E.F., Bittner, M., Trent, J., Barrett, J.C., Afshari, C.A., 1999. Microarrays and toxicology: The advent of toxicogenomics. Mol. Carcinog. 24, 153–159.

O'Connor, T.P., 2002. Nation distribution of chemical concentrations in mussels and oysters in the USA. Mar. Environ. Res. 53, 117–143.

Olden, K., Guthrie, J., Newton, S., 2001. A bold new direction for environmental health research. Am. J. Public Health 110, A8–10.

Passino-Reader et al., 1997. Passino-Reader, D.R., Hickey, J.P., Olgivie, L.M., 1997. Bull. Environ. Contam. Toxicol. 59, 834–840.

Pastor, N., Lopez-Lazaro, M., Tella, J.L., Baos, R., Forrero, M.G., Hiraldo, F., Cortes, F., 2001. DNA damage in birds after the mining waste spill in southwestern Spain: A Comet assay evaluation. J. Environ. Pathol. Toxicol. Oncol. 20, 317–324.

Peden-Adams, M.M., 1999. Evaluation of xenobiotic-induced immunotoxicity and CYP450 activity in wildlife species. Ph.D. dissertation, Clemson University, Clemson, SC.

Pederson, S.N., Christainsen, L.B., Pedersen, K.L., Korsgaard, B., Bjerregard, P., 1999. In vivo estrogenic activity of branched and kinar alkylphenols in rainbow trout (*Oncorhyncus mykiss*). Sci. Total Environ. 15, 89–96.

Perchmann, J.H.K., Scott, D.E., Semlitsch, R.D., Caldwell, J.P., Vitt, L.J., Gibbon, J.W., 1991. Declining amphibian populations: The problem of separating human impacts from natural fluctuations. Science 253, 892–895.

Ratcliffe, D.A., 1967. Decrease in eggshell weight in certain birds of prey. Nature 215, 208–210.

Reddy, M.L., Dierauf, L.A., and Gulland, F.M.D., (2001). Marine mammals as sentinels of ocean health. In: Dierauf, L.S., Gulland, F.M.D., editors. *CRC handbook of marine mammal medicine.* 2nd Edition, CRC Press, Boca Raton, FL, pp. 3–9.

Riaz-ul-Haq, Shakoori, A.R., 2000. Microorganisms resistant to heavy metals and toxic chemicals as indicators of environmental pollution and their use in bioremediation. Folia Biol. (Krakow) 48, 143–147.

Roast, S.D., Widdows, J., Joes, M.B., 2000. Mysids and trace metals: disruption of swimming as a behavioural indicator of environmental contamination. Mar. Environ. Res. 50, 107–112.

Roast, S.D., Widdows, J., Jones, M.B., 2001. Environ. Toxicol. Chem. 20(5), 1078–1084.

Rodgers, B.E., Chesser, R.K., Wickliffe, J.K., Phillips, C.J., Baker, R.J., 2001a. Subchronic exposure of BALB/c and C57BL/6 strains of *Mus musculus* to the radioactive environment of the Chornobyl, Ukraine exclusion zone. Environ. Toxicol. Chem. 20, 2830–2835.

Rodgers, B.E., Wickliffe, J.K., Phillips, C.J., Chesser, R.K., Baker, R.J., 2001b. Experimental exposure of naïve bank voles (*Clethrionmys glareolus*) to the Chornobyl, Ukraine, environment: A test of radioresistance. Environ. Toxicol. Chem. 20, 1936–1941.

Schaiger, J., Mallow, U., Ferling, H., Knoerr, S., Braunbeck, T., Kalbfus, W., Negele, R.D., 2002. How estrogenic is nonylphenol? A transgenerational study using rainbow trout (*Oncorhynchus mykiss*) as a test organism. Aquat. Toxicol. 59, 177–189.

Schwabe, C.W., 1984. Animal monitors of the environment. In: Veterinary Medicine and Human Health, 3rd edition. Williams and Wilkins, Baltimore, MD, pp. 562–578.

Sheffield, S.R., Matter, J.M., Rattner, B.A., Guiney, P.D., 1998. Fish and wildife species as sentinels for environmental endocrine disrupters. In: Kendall, R.J., Dickerson, R.L., Giesy, J., Suk, W. (Eds.), Principles and Processes for Evaluating Endocrine Disruption in Wildlife, SETAC Press, Pensacola, FL, pp. 369–430.

Soto, M., Ireland, M.P., Marigomez, I., 2000. Changes in mussel biometry on exposure to metals: Implications in estimation of metal bioavailability in 'Mussel Watch' programmes. Sci. Total Environ. 247, 175–187.

Stahl, R.G., Jr., 1997. Can mammalian and non-mammalian "sentinel species" data be used to evaluate the human health application of environmental contaminants? Hum. Ecol. Risk Assess. 3, 329–335.

Sumpter, J.P., Jobling, S., 1995. Vitellogenesis as a biomarker for estrogenic contamination of the aquatic environment. Environ. Health Perspect. 103(suppl. 7), 173–178.

Suter, G.W., 1997. Integration of human health and ecological risk assessment. Environ. Health Perspect. 105, 1282–1283.

Tavera-Mendoza, L., Ruby, S., Brousseau, P., Fournier, M., Cyr, D., Marcogliese, D., 2002a. Response of the amphibian tadpole *Xenopus laevis* to atrazine during sexual differentiation of the ovary. Environ. Toxicol. Chem. 21, 1264–1267.

Tavera-Mendoza, L., Ruby, S., Brousseau, P., Fournier, M., Cyr, D., Marcogliese, D., 2002b. Response of the amphibian tadpole *Xenopus laevis* to atrazine during sexual differentiation of the testis. Environ. Toxicol. Chem. 21, 527–531.

Trust, K.A., Fowles, J.R., Hooper, M.J., Fairbrother, A., 1994. Cyclophosphamide effects on immune function of European starlings. J. Wildl. Dis. 30, 328–334.

US EPA (1991). Guidelines for developmental toxicity risk assessment. *Fed. Reg.* 56(234), 63798–63828.

van der Schalie, W.H., Gardner, H.S., Jr., Bantle, J.A., De Rosa, C.T., Finch, R.A., Reif, J.S., Reuter, R.H., Backer, L.C., Burger, J., Folmar, L.C., Stokes, W.S., 1999. Animals as sentinels of human health hazards of environmental chemicals. Environ. Health Perspect. 107, 309–315.

van Kampen, K.R., James, L.F., Rasmussen, J.L.F., Huffaker, R.H., Fawcett, M.O., 1969.

Organic phosphate poisoning of sheep in Skull Valley, Utah. J. Am. Vet. Med. Assoc. 154, 623–630.

Veterinarian, 1874a. The effects of fog on cattle in London. Veterinarian 47, 1–4.

Veterinarian, 1874b. The effects of the recent fog on the Smithfield Show and the London Dairies. Veterinarian 47, 32–33.

Walton, K., Dorne, J.L., Renwick, A.G., 2001. Uncertainty factors for chemical risk assessment: Interspecies differences in glucuronidation. Food Chem. Toxicol. 39, 1175–1190.

Watanabe, K.H., Chen, C., 2001. The role of physiologically based toxicokinetics models in biologically based risk assessment. Folia Histochem. Cytobiol. 39(suppl. 2), 50–51.

Wellborn, J.A., Allen, R., Byker, G., DeGrow, S., Hertel, J., Noordhoek, R., Koons, D., 1975. The contamination crisis in Michigan: Polybrominated biphenyls. Senate Special Investigating Committee, Lansing, MI.

White, R., Jobling, S., Hoare, S.A., Sumpter, J.P., Parker, M.G., 1994. Environmentally persistent alkylphenolic compounds are estrogenic. Endocrinology 135, 175–182.

Wiger, R., Stottum, A., 1985. In vitro testing for developmental toxicity using the hydra attenuate assay. NIPH Ann. 8, 43–47.

Witzmann, F.A., Carpenter, R.L., Ritchie, G.D., Wilson, C.L., Nordholm, A.F., Rossi, J., 2000. Toxicity of chemical mixtures: Proteomic analysis of persisting liver and kidney protein alterations induced by repeated exposure of rats to JP-8 jet fuel vapor. Electrophoresis 1, 2138–2147.

Yang, Y.G., Mayaura, K., Spainhour, C.B., Jr., Edwards, J.F., Phillips, T.D., 1993. Evaluation of the developmental toxicity of citrinin using Hydra attenuata and postimplantation rat whole embryo culture. Toxicology 31, 179–198.

Part III

Representative EDCs in Animals

Introduction to Part III

James A. Carr
David O. Norris

Endocrine-disrupting chemicals (EDCs) come from many different types of chemical classes, including pesticides, synthetic hormones, natural plant compounds, chemicals used in the plastics industry, and waste products of the aerospace industry such as perchlorate. In part III of this volume, several experts in the fields of toxicology and endocrinology discuss the ability of these various classes of contaminants to disrupt the endocrine system and affect the health of humans and wildlife. Some of the chemical classes examined here are well established as EDCs, whereas others (triazines, for example) are emerging as chemicals of concern. Each chapter presents information on the structure and physical characteristics of the chemical, the mechanism of action (if known), the environmental pathways through which exposure might occur, and an overview of what is known regarding the effects of the chemical on endocrine function in representative vertebrate groups. Most important, each chapter highlights ways in which the fields of endocrinology and toxicology can be applied in an integrated way to identify EDC exposure and outcome relationships.

The goal of this part is not to provide an all-inclusive list of known EDCs and their effects, but rather to focus on representative classes of EDCs for which a significant amount of peer-reviewed literature exists. Focusing on EDCs for which substantial data exist, several common themes emerge. First is the recurring link between wildlife and human health. Data collected on wildlife are not only important for gauging the health and reproductive success of wildlife populations, but for predicting potential effects on human health. A second theme is that EDCs have the potential to act at targets other than those directly involved in reproduction. Hontela and Lacroix (chapter 13) provide an elegant description of the impact of heavy metals on adrenal function. A third theme that develops from the chapters in this part is the challenge presented to scientists in identifying cause-and-effect relationships for EDCs. Both direct and indirect effects of EDCs are presented in every chapter in this part. The challenge is much broader than the difficulties in extrapolating from in vitro data to whole animal studies. One of the greatest hurdles in linking effects of EDCs to health status of humans and wildlife is the degree to which data from whole animal studies can be used to predict population-level effects. As pointed out by Damstra et al. (2002), a critical issue in understanding the health impacts of EDCs is whether adverse effects at the population level result from the proximity of the population to a point source or whether the effects might be occurring at a global level. This is one of the most important and hotly debated issues surrounding EDCs.

The chapters in this part illustrate two critical issues that must be considered when assessing EDC exposure. First is the need to consider the environmental persistence and degradation of EDCs when assessing cause-and-effect relationships. EDCs that degrade quickly may disappear in the environment before effects are exhibited in exposed animals. In other cases, environmental degradation may occur slowly, but degradation products may be biologically active and in some cases may be as potent or more potent at disrupting endocrine function as the parent compound. As discussed by Guillette et al. (chapter 12), DDT and its degradation product DDE have many of the same effects. Second is the need to consider the timing of exposure relative to key developmental events. The possibility of exposure during critical windows in development must be given serious consideration, as many EDCs have the potential to permanently alter reproductive development even after short exposures early in development. As described by Kloas (chapter 14), exposure of larval frogs to estrogenic chemicals early in development can permanently alter the direction of gonad differentiation, in some cases producing male frogs with female gonads (the "feminizing effect" observed with alkylphenols and other estrogenic contaminants; Qin et al., 2003; Levy et al., 2004).

References

Damstra, T., Barlow, S., Bergman, A., Kavlock, R., Van Der Kraak, G., 2002. Global assessment of the state-of-the-science of endocrine disruptors. World Health Organization, Geneva.

Levy, G., Lutz, I., Kruger, A., Kloas, W., 2004. Bisphenol A induces feminization in *Xenopus laevis* tadpoles. Environ. Res. 94, 102–111.

Qin, Z.F., Zhou, J.M., Chu, S.G., Xu, X.B., 2003. Effects of Chinese domestic polychlorinated biphenyls (PCBs) on gonadal differentiation in *Xenopus laevis*. Environ. Health Perspect. 111, 553–556.

10

Polychlorinated Dibenzo-*p*-Dioxins

Potential Mechanisms and Risk as Endocrine Disruptors

Mary K. Walker

Structure and Physical Characteristics

Polychlorinated dibenzo-*p*-dioxins (PCDDs) are environmental contaminants that belong to a large family of structurally related chemicals called halogenated aromatic hydrocarbons (HAHs). The polychlorinated dibenzofurans (PCDFs) and polychlorinated biphenyls (PCBs) represent other HAHs that are commonly found in the environment as mixtures with PCDDs. PCDDs are formed as trace contaminants in a variety of manufacturing processes that use chlorine, such as in the production of trichlorophenols, phenoxy herbicides, the wood preservative pentachlorophenol, and in the bleaching of pulp and paper (reviewed in Webster and Commoner, 1994). However, the majority of PCDDs entering the environment today are emitted from hospital and municipal waste incinerators (Fiedler, 1996; Thornton et al., 1996), where they are dispersed into the atmosphere and then deposited onto soil, water, and vegetation (figure 10-1).

Chemical and biological characteristics of PCDDs result in their persistence in the environment, their ability to bioaccumulate in the food chain, and exposure of both wildlife and humans. PCDDs exhibit low water solubility and vapor pressure, which allows them to partition into soil and sediment. They also are resistant to chemical and biological degradation and thus persist in the environment. Finally, PCDDs are lipophilic, and although they enter the environment in trace amounts, they partition into fatty tissues, causing them to bioaccumulate at higher concentrations as they are passed up the food chain. Thus, fish, wildlife, and humans are involuntarily exposed to HAHs through the diet. For humans, the consumption of meat, poultry, and dairy products accounts for more than 90% of total exposure.

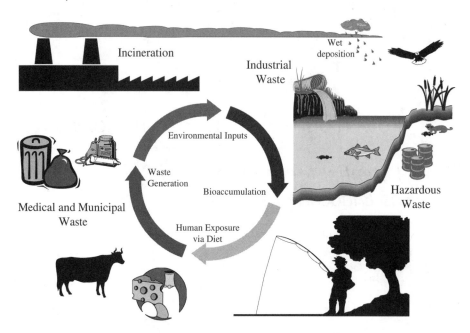

Figure 10.1. Polychlorinated dibenzo-*p*-dioxins enter the environment from industrial and hazardous waste sources as well as from hospital and municipal waster incinerators. These chemicals persist in the environment and bioaccumulate up the food chain, where humans are involuntarily exposed via the diet.

Mechanism of Action

Those PCDDs that are chlorinated on all the lateral positions (carbons 2, 3, 7, and 8) and lack chlorine substituents on three of the four internal positions (carbons 1, 4, 6, and 9) are the most potent in causing toxicity and are believed to act via a common mechanism. 2,3,7,8-Tetrachlorodibenzo-*p*-dioxin (TCDD) is the most potent of the chlorinated PCDDs and is used as a prototype for studying the toxicity of other PCDDs and HAHs with similar chlorination patterns.

TCDD binds to a cytosolic protein, the aryl hydrocarbon receptor (AhR), with high affinity (K_d = 1 nM; Swanson and Bradfield, 1993). The ligand-activated AhR then translocates to the nucleus, releases two molecules of heat shock protein 90, and interacts with a second protein, the AhR nuclear translocator (Arnt). Both the AhR and Arnt proteins belong to the basic helix-loop-helix, PAS (per-ARNT-Sim) family of DNA binding proteins. In the nucleus the AhR and Arnt proteins heterodimerize at their helix-loop-helix PAS domains, and then the basic amino acids of each protein bind a DNA consensus sequence (5' TNGCGTG 3'), called the dioxin or xenobiotic response element (DRE or XRE; Denison et al., 1988). DREs are found in the 5' enhancer region of genes whose expression is transcriptionally regulated by the AhR and Arnt proteins, including phase I monooxygenase enzymes such as cytochrome P4501A and 1B families and phase II enzymes such as glutathione-*S*-transferase.

Exposure of laboratory animals to TCDD and related compounds can cause a variety of toxic responses, including a wasting syndrome and body weight loss, chloracne, thymic atrophy, immunotoxicity, teratogenicity, reproductive toxicity, and cancer, and most effects are believed to be mediated by the AhR. The focus of this chapter is on the effects of PCDDs on endocrine-dependent reproduction and development in vivo, and known or proposed endocrine-disrupting mechanisms. This chapter is not meant as an exhaustive review, but emphasizes the fundamental principles of how PCDDs may function as endocrine disruptors in causing their toxic effects.

PCDDS as Antiestrogens

The antiestrogenic effects of PCDDs are well documented both from whole animal studies as well as from cell culture. The first evidence that suggested PCDDs may exhibit antiestrogenic activity came from carcinogenicity studies showing that female rats chronically fed a TCDD-containing diet exhibited a decreased incidence of spontaneous mammary and uterine tumors (Kociba et al., 1978). Because growth of these tumors can be estrogen-dependent, these results suggested that TCDD may have antiestrogenic properties. Subsequent studies of TCDD's effects on estrogen-dependent ovulation, uterine function, pregnancy, endometriosis, and embryonic development provided additional evidence that some of TCDD's toxic effects on reproduction and development may result from an antiestrogenic mechanism.

A number of mechanisms have been proposed for TCDD's antiestrogenic activity (reviewed in Zacharewski and Safe, 1998), including that (1) TCDD increases 17β-estradiol (E_2) metabolism, (2) TCDD decreases the affinity of E_2 for estrogen receptor (ER) binding and/or decreases ER protein expression, and (3) TCDD decreases E_2-induced gene expression by inhibiting ER transactivation at the estrogen response element. It should be noted that TCDD does not bind ERα, ERβ, or the progesterone receptor (PR; Romkes and Safe, 1988; Klinge et al., 1999) and thus does not cause antiestrogenic effects by functioning either as an ER antagonist or a PR agonist. Recent evidence from cell culture studies provides strong support for the ability of TCDD to disrupt ER transactivation (for additional detail, see Kharat and Saatcioglu, 1996; Safe et al., 1998; Klinge et al., 1999). Whether this mechanism accounts for the antiestrogenic effects of PCDDs observed in vivo remains to be determined.

The effects of PCDDs on reproduction and development are complex and likely involve physiological changes that represent endocrine-dependent and endocrine-independent mechanisms. Furthermore, the endocrine-disrupting effects of PCDD-mediated toxicity cannot be attributed completely to antiestrogenic mechanisms, but also may involve estrogenic, antiandrogenic, and antigonadotropic mechanisms. As a great deal of research has focused on PCDDs as antiestrogenic chemicals, I review the effects of PCDDs on reproduction and development, highlighting those effects where an antiestrogenic mechanism is known or proposed and noting those effects that are inconsistent with an antiestrogenic mechanism.

Effects of PCDDs on Ovulation

The anterior pituitary gonadotropin, follicle stimulating hormone (FSH), stimulates follicle growth, increases ovarian weight, and stimulates ovarian production of E_2. Elevated E_2 produced by the ovary stimulates a preovulatory surge of luteinizing hormone (LH) at mid-cycle, resulting in release of ova into the oviduct and the formation of corpora lutea in the ovary. The corpora lutea then produce progesterone. In laboratory animals TCDD and TCDD-like PCDDs inhibited gonadotropin-stimulated follicle growth, suppressed the E_2-stimulated preovulatory surge of LH, reduced the number of ovulated ova, reduced the number of corpora lutea, and decreased progesterone levels after ovulation.

Female Sprague-Dawley rats treated with graded doses of TCDD, or related AhR-active PCDDs, and then injected with gonadotropin to stimulate ovarian growth exhibited a dose-related failure of ovaries to increase in weight and a dose-related reduction in the number of ova in the oviducts after ovulation (Li et al., 1995; Gao et al., 1999). Histologically, the ovaries from TCDD-treated rats showed a large number of preovulatory follicles and the absence of corpora lutea. The TCDD-induced inhibition of ovulation was associated with changes in hormone levels throughout the reproductive cycle. Within 12 h of gonadotropin injection, TCDD stimulated a surge in both LH and FSH relative to controls. At the time of ovulation, TCDD suppressed the LH surge, and 16 h after ovulation the plasma levels of LH, FSH, and progesterone were reduced, while plasma E_2 levels were elevated relative to untreated control animals. The lowest TCDD dose that decreased ovarian weight and reduced the number of ovulated ova in the rat was 4 μg/kg, which did not cause overt toxicity typically reflected by a decrease in body weight (Gao et al., 1999).

These studies demonstrate that in the female rat TCDD and related PCDDs can suppress ovarian follicle growth stimulated by gonadotropins, inhibit the preovulatory LH surge, reduce ovulation, and significantly alter the pattern of hormone expression during the estrous cycle. The mechanism of TCDD-induced suppression of gonadotropin-stimulated follicle growth in the ovary has not been elucidated. The induction of LH and FSH within 12 h of gonadotropin injection in TCDD-treated animals suggests that the anterior pituitary could be a direct target of TCDD; however, the elevation of these gonadotropic hormones in the absence of ovarian weight gain also suggests that the ovary is a direct target and is consistent with an antigonadotropic mechanism. Although TCDD did not suppress serum E_2 levels before or after ovulation, the lack of a preovulatory LH surge in TCDD-treated animals is consistent with an antiestrogenic mechanism. TCDD has been shown to decrease ER mRNA and protein expression in the liver and uterus (Romkes and Safe, 1988). If TCDD also decreased ER expression in the pituitary, it could inhibit the E_2-stimulated preovulatory LH surge and prevent ovulation. It is interesting that the apparent antiestrogenic effects of TCDD on ovarian growth and ovulation are not observed in the mouse (Gallo et al., 1986; Cummings et al., 1996). Thus, species-specific differences in response to potential TCDD endocrine disruption are evident from these animal models.

Effects of PCDDs on Uterine Function

Before ovulation, E_2 produced by the growing follicles stimulates the proliferation of the endometrial uterine lining and the growth and coiling of endometrial blood vessels, resulting in an increase in uterine wet weight. E_2 also stimulates uterine peroxidase activity and epidermal growth factor receptor (EGFR) binding and increases the uterine expression of EGFR, PR, and ER. The antiestrogenic activity of TCDD and related PCDDs have been studied most extensively in the uterus, and laboratory data from both the rat and mouse have shown that TCDD decreases constitutive and E_2-stimulated hypertrophy in the uterus as well as other E_2-stimulated uterine responses (see review in Silvers and Rorke, 1998).

In both the rat and mouse, TCDD decreased constitutive uterine weight and antagonized the E_2-stimulated increase in uterine weight (Gallo et al., 1986; Romkes et al., 1987). These observations led investigators to assess whether TCDD also antagonized other E_2-stimulated responses in the uterus. For all the E_2-dependent uterine responses that were evaluated, TCDD decreased constitutive responses and antagonized the E_2-stimulated effects. In the rat and mouse, TCDD decreased uterine ER protein levels and suppressed E_2-dependent induction of ER protein (Romkes and Safe, 1988; DeVito et al., 1992), and in the rat TCDD also decreased constitutive and E_2-stimulated induction of uterine PR protein levels (Romkes and Safe, 1988), and decreased constitutive and E_2-stimulated uterine EGFR binding and EGFR mRNA levels (Astroff et al., 1990). Also in rats, TCDD suppressed constitutive and E_2-stimulated uterine peroxidase activity (Astroff and Safe, 1990). In mice, the lowest effective dose of TCDD that decreased uterine E_2 receptors and uterine weight (single dose of 10 µg TCDD/kg, or 12 multiple doses of 6 µg TCDD/kg over 1 month) did not cause overt toxicity or alter body weight (Gallo et al., 1986; DeVito et al., 1992). Although many of the studies in rats used overtly toxic doses of TCDD (20–80 µg/kg), which would reduce body weight and increase mortality 10–40 days after dosing, these same TCDD doses suppressed uterine responses to E_2 within 1–2 days of administration (Romkes and Safe, 1988; Astroff and Safe, 1990), suggesting that the effects of TCDD are not a result of overt toxicity. Furthermore, one study clearly demonstrated that 5 µg/kg TCDD inhibited E_2-stimulated effects in the rat uterus without causing overt toxicity and decreasing body weight (Astroff et al., 1990). Thus, taken together these data would suggest that TCDD exhibits antiestrogenic effects on the uterus at doses lower than those that produce overt toxicity.

The specific mechanisms responsible for TCDD suppression of E_2-stimulated effects in the uterus have not been fully elucidated; however, the induction of hepatic monooxygenase enzymes and increased metabolism of E_2 have been extensively studied as one potential explanation. The ED_{50} dose of TCDD in the female rat that suppressed uterine E_2 and progesterone receptors was 65–100-fold higher than the ED_{50} dose that induced hepatic aryl hydrocarbon hydroxylase (AHH) activity, demonstrating that doses causing antiestrogenic effects also maximally induced potential E_2 metabolizing enzymes in the liver (Astroff and Safe, 1988). However, a number of lines of evidence suggest that increased metabolism of E_2 is not likely responsible for TCDD's antiestrogenic effects. For example, the AhR partial ago-nist 6-methyl-1,3,8-trichlorodibenzofuran suppressed uterine ER and PR in the rat at doses that only minimally induced hepatic AHH activity (Astroff and Safe, 1988). Also, TCDD antagonized E_2-stimulated increases in ER and PR in isolated uterine strips in the absence of hepatic metabolic activity (Astroff and Safe, 1988). Finally, the doses of TCDD that suppressed uterine ER levels in the mouse failed to alter serum E_2 levels (DeVito et al., 1992). These studies suggest that TCDD's anti-estrogenic effects on the uterus likely do not result from the induction of hepatic monooxygenase enzymes and the subsequent increased metabolism of E_2.

Given that TCDD decreased the expression of ER in the uterus, the antiestrogenic effects of TCDD in the uterus may result from the downregulation of the ER and subsequent decreased ability of tissues to respond to E_2 stimulation. The promoter of the human ER gene contains numerous full or partial DNA consensus sequences corresponding to the DRE (White and Gasiewicz, 1993), suggesting that TCDD activation of the AhR could directly alter ER expression at the transcriptional level. However, in order for TCDD's antiestrogenic effects on the uterus to be mediated by a downregulation of ER levels, the reduction in uterine ER must precede the inhibition of E_2-stimulated uterine responses by TCDD. DeVito and co-workers (1992) showed that 30 μg TCDD/kg in mice reduced ER levels in the uterus as early as 24 h after treatment, but the same dose of TCDD did not exhibit an inhibition in E_2-stimulated uterine hypertrophy until 48 h after treatment, providing evidence that downregulation of ER may precede the antiestrogenic effect of TCDD on the uterus. Other studies suggest, however, that the time course of TCDD-induced ER downregulation is similar to that for TCDD-induced inhibition of E_2-stimulated effects. For example, Astroff and Safe (1990) demonstrated that 80 μg TCDD/kg in Sprague-Dawley rats inhibited E_2-stimulated uterine hypertrophy and peroxidase activity as early as 24 h after treatment, when TCDD-induced downregulation of ER was observed in the mouse; however, uterine ER levels were not determined in the rats in this study. In addition, Romkes and Safe (1988) showed that 80 μg TCDD/kg in Long Evans rats reduced uterine ER levels and inhibited E_2-stimulated increase in uterine PR levels as early as 24 h after treatment; however, they did not evaluate changes in either of these receptor levels at earlier time points to demonstrate con-clusively whether changes in ER levels could be detected before the inhibition of E_2-stimulated increase in uterine PR.

Effects of PCDDs on Maintenance of Pregnancy

Within approximately 3 days of implantation, the placenta begins to produce chori-onic gonadotropin and increase the secretion of relaxin. Chorionic gonadotropin maintains the corpus luteum and production of progesterone, increases E_2 produc-tion by the ovary, and ultimately maintains pregnancy. Given the antiestrogenic effects and disruption of endocrine hormone pathways induced by TCDD and re-lated PCDDs on ovarian and uterine function, it is not surprising that these chemi-cals also affect the ability of animals to maintain pregnancy. TCDD increased spontaneous abortions and altered reproductive hormone concentrations during pregnancy (Guo et al., 1999; McNulty, 1984). The studies described here focus on outcomes after exposure of pregnant monkeys. Multigenerational studies of TCDD on embryo development, subsequent reproductive function, and fertility in rats, mice, and hamsters are discussed in a subsequent section.

Pregnancies in 83% of rhesus monkeys (*Macaca mulatta*) exposed to 1 μg TCDD/kg during the first trimester (gestational days 25–40) terminated in spontaneous abortion (McNulty, 1984). Similarly, 50 and 100% of pregnant cynomolgus monkeys (*Macaca fascicularis*) treated with 1 or 2 μg TCDD/kg, respectively, on gestation day 12 experienced embryonic death within 12–20 days after TCDD exposure, followed by abortion within 12 days (Guo et al., 1999). In cynomolgus monkeys from both dosage groups in which abortion occurred, TCDD significantly decreased circulating bioactive chorionic gonadotropin and E_2 without significant changes in circulating progesterone and relaxin levels. Furthermore, the TCDD-induced changes in these reproductive hormones were evident 4–5 days after TCDD exposure and 7–16 days before embryonic death. It is noteworthy that TCDD did not alter the levels of immunoreactive chorionic gonadotropin as measured by enzyme-linked immunosorbent assay, but it significantly reduced the levels of chorionic gonadotropin that exhibited biological activity (i.e., bioactive) in an in vitro bioassay.

Although the mechanisms underlying the effects of TCDD on maintaining pregnancy have not been investigated in detail, the disruption of maternal endocrine hormones required for maintenance of pregnancy could account for embryonic mortality and subsequent abortions. Specifically, the TCDD-induced decrease in bioactive chorionic gonadotropin could result in a decrease in ovarian E_2 production and ultimately the loss of the fetus. When comparing the four animals treated with 1 μg TCDD/kg, two of which spontaneously experienced abortions, the levels of bioactive chorionic gonadotropin and E_2 levels were significantly reduced only in the those animals that aborted. In contrast, progesterone and relaxin levels did not differ between TCDD-treated animals that experienced abortions and those that did not. Although TCDD-induced spontaneous abortions were associated with a decrease in circulating E_2 levels, an antiestrogenic mechanism is not likely to be the primary cause. TCDD significantly reduced the level of bioactive chorionic gonadotropin, the hormone that regulates E_2 production by the ovary, suggesting that TCDD disrupts the production or structural function of bioactive chorionic gonadotropin, consistent with an antigonadotropic response. In addition, because TCDD exposure did not alter the chorionic gonadotropin-induced production of progesterone or relaxin from the ovary, the decrease in E_2 may result from multiple effects of TCDD on ovarian production of E_2.

Effects of PCDDs on Endometriosis

Endometriosis is a disease in which endometrial tissue grows outside the uterus. The etiology of this disease has been linked to retrograde menstruation carrying endometrial tissue from the fallopian tubes into the peritoneal cavity and to alterations in both humoral and cell-mediated immunity (Haney, 1990). In addition, the growth of endometrial lesions has been shown to be dependent on E_2 stimulation. Research has shown that TCDD increases the incidence and severity of endometriosis in rhesus monkeys in a dose-dependent manner (Rier et al., 1993). This observation

led to the development of both a rat and mouse model of endometriosis for studying the mechanism of TCDD's effects (Cummings et al., 1996).

Female rats and mice were treated with 3 or 10 µg TCDD/kg at five time points:
(1) 21 days before surgical induction of endometriosis, (2) at the time of surgery,
and (3) 3, 6, and 9 weeks after surgery (Cummings et al., 1996). When evaluated at
12 weeks after surgery, TCDD induced an increase in the diameter of the endometriotic lesions in both rats and mice; however, the effect in mice was significantly
greater and occurred at a lower dose than in the rat. Histologically, TCDD produced
necrotic and inflammatory changes in endometriotic sites in rats and primarily resulted in fibrotic changes in endometriotic sites in mice, more closely resembling
endometriotic lesions in humans (Cummings and Metcalf, 1995). A second study
revealed that mice exposed to TCDD perinatally and then in adulthood exhibited
increased size of endometriotic lesions compared to those mice only exposed to
TCDD as adults (Cummings et al., 1999). This perinatal effect was not observed in
rats. Finally, the observed effects of TCDD on the promotion of endometriosis
occurred at doses lower than those that resulted in overt toxicity, as evidenced by
the lack of an effect on body weight gain.

The mechanisms responsible for the ability of TCDD to promote the growth of
endometriotic lesions have not been elucidated but may be associated with its ability to disrupt both endocrine and immune function. The difference in TCDD's effects on endometriosis in rats and mice may be related to the differences in its effects
on ovarian and immune responses in these species. The dose of TCDD that significantly reduced ovarian weight in rats (an apparent antiestrogenic effect) failed to
increase endometriotic lesion size. In contrast, TCDD did not alter ovarian weight
in mice but did increase endometriotic lesion size (Cummings et al., 1996). Finally,
TCDD failed to suppress humoral immunity in rats, but it suppressed these responses
in mice (Smialowicz et al., 1994). It has been proposed that TCDD's antiestrogenic
effects on the ovary and its failure to suppress humoral immunity in the rat may be
related mechanistically to the failure of TCDD to promote proliferation of preex-
isting endometriotic lesions in the rat. Future studies that compare and contrast the
rat and mouse models will help elucidate the role of TCDD's endocrine disrupting
effects and immune suppression in its ability to promote the growth of endometriosis.

Effects of PCDDs on Adult Male Reproductive Function

PCDDs have been shown to alter reproductive function in adult male animals, resulting in decreased androgen concentrations, abnormal testicular development,
reduced spermatogenesis, and decreased fertility (Khera and Ruddick, 1973; Moore
et al., 1985). These studies demonstrated that TCDD disrupts steroid biosynthesis
in the adult testis and affects feedback mechanisms in the hypothalamus–pituitary
axis (Bookstaff et al., 1990a,b; Moore et al., 1991). Typically, however, these effects occurred at doses of TCDD that were overtly toxic, resulting in significant
decreases in body weight gain. In contrast, the effects of perinatal exposure of TCDD
on male reproductive development, function, and fertility have been shown to be
10–100-fold more sensitive than those effects observed in adult animals. Below I

review the effects of PCDDs on male reproductive development, function, and fertility after in utero and lactational exposure.

Effects of in Utero and Lactational Exposure to PCDDs

Research using many animal models has demonstrated that the developing embryo and fetus are more sensitive to the adverse effects of PCDDs than the adult (see review in Peterson et al., 1993). Thus, the recognition that (1) PCDDs have the potential to disrupt the endocrine system and alter reproductive function in adult animals and (2) the developing fetus is extremely sensitive to TCDD has resulted in extensive research on the effects of PCDDs on male reproductive development and function after in utero and lactational exposure. The results of those studies are summarized briefly here; however, given the large number of studies that have been conducted, refer to recent reviews on this topic for more detailed information (Birnbaum, 1998; Roman and Peterson, 1998a).

In the majority of studies, male rats, hamsters, and mice exposed to TCDD in utero and via lactation exhibited decreased number of ejaculated sperm and decreased weights of accessory sex organs, including prostate, seminal vesicle, and coagulating gland (Mably et al., 1992a,c; Bjerke et al., 1994b; Gray et al., 1995, 1997; Theobald and Peterson, 1997; reviewed in Roman and Peterson, 1998a); other responses have been shown to be both strain and species specific. Male Holtzman and Long Evans rats exposed in utero and lactationally to TCDD exhibited demasculinized sexual behavior in adulthood (Mably et al., 1992b; Gray et al., 1995), while only Holtzman rats exhibited slightly feminized sexual behavior, and male hamsters did not exhibit any alterations in sexual behavior (Gray et al., 1995). Although ejaculated sperm number was decreased in both Holtzman and Long Evans rats, only the Long Evans rats exhibited reduced fertility (Gray et al., 1995). Most of the effects described above occurred after perinatal exposure to TCDD doses of 1–2 μg/kg to the pregnant dam; however, doses as low as 0.064 μg TCDD/kg could significantly induce some of the adverse effects described above (Mably et al., 1992a,b).

Although the mechanisms responsible for mediating the developmental effects of in utero TCDD exposure on male reproductive development and function remain to be elucidated, some effects are consistent with an antiandrogenic mechanism and others with an antiestrogenic mechanism. Reduced sperm number, decreased accessory sex organ weights, and demasculinized sexual behavior all suggest androgenic deficiency, which could result from decreased androgen production, decreased androgen receptor (AR) expression, and/or decreased AR transactivation. Although one study demonstrated that circulating testosterone was significantly reduced in male rats both peri- and postnatally after in utero and lactational TCDD exposure (Mably et al., 1992c), the robustness of these results have not been established in subsequent studies (Gray et al., 1995; Roman et al., 1995). Thus, the effects of TCDD do not appear to result from an androgen deficiency.

Studies by Bjerke and co-workers (1994b) and Roman and Peterson (1998b) showed that the prostate from male Holtzman rats exposed to TCDD in utero and via lactation exhibited decreased responsiveness to androgen stimulation and a transient decrease in androgen-regulated mRNAs, respectively, in the absence of changes in circulating testosterone. These studies are consistent with a reduction in AR number or decreased AR signal transduction. Studies of the in utero and lactational TCDD-exposure of Long Evans rats demonstrated that AR number is not reduced in accessory sex organs, including the prostate (Gray et al., 1995). However, a systematic description of the effects of TCDD on prostate epithelial budding, proliferation, cell differentiation, and AR expression demonstrated that TCDD interferes with epithelial cell proliferation early in prostate development, delays cellular differentiation, and alters the spatial distribution of AR expression (Roman et al., 1998). It appears that TCDD disrupts epithelial differentiation and prostate morphogenesis during embryo development, which may result in the changes in spatial AR expression. These effects may be mediated independently of the endocrine system.

The decreases in sperm production and accessory sex organ weights and the demasculinization of sexual behavior by TCDD also are consistent with an antiestrogenic effect. Neonatal exposure to antiestrogens, such as tamoxifen, decreased sperm production and accessory sex organ weight, similar to responses induced by TCDD. However, tamoxifen also induced sterility and spermatogenic arrest and altered the development of the male-pattern central nervous system (Dohler et al., 1984; Taguchi, 1987)—effects that are not observed after in utero and lactational TCDD exposure (Bjerke et al., 1994a). Thus, these studies suggest that the effects of TCDD on the developing male reproductive system are not strictly antiestrogenic in origin. In summary, TCDD alters cellular differentiation and proliferation during embryonic development of the rat prostate, and these effects likely are mediated independently or downstream of endocrine receptor function; however, an endocrine-disrupting mechanism may contribute, in part, to the overall adverse effects of TCDD on male reproductive tract development and function.

Effects of in Utero and Lactational Exposure to PCDDs on Female Reproductive Development, Function, and Fertility

Although less research has been conducted on the effects of in utero and lactational exposure to PCDDs on the development of the female reproductive system and subsequent function, available data demonstrate that the female reproductive system is also highly sensitive to PCDD exposure during embryo development. Early studies demonstrated that prenatal exposure of Wistar rats to 0.5 µg TCDD/kg/day for 10 gestational days reduced embryo viability, neonatal growth, and fertility of the F_1 generation (Khera and Ruddick, 1973), and continuous exposure of Sprague-Dawley rats to dietary 0.01 µg TCDD/kg/day resulted in decreased embryo survival, neonatal survival, and fertility in the F_1 and F_2 generations (Murray et al., 1979).

The mechanism responsible for the effects of TCDD on female reproduction after in utero and lactational exposure have been investigated. Exposure of pregnant Holtzman rats on gestation day 15 to 1 µg TCDD/kg resulted in cleft clitoris, presence of a vaginal thread, decreased ovarian weight, and, in one study, delayed time to puberty (Gray and Ostby, 1995; Flaws et al., 1997). Similar results were observed after in utero and lactational exposure of Long Evans rats to 1 μg TCDD/kg (Gray and Ostby, 1995). The female reproductive system of hamsters has also been shown to be sensitive to in utero TCDD exposure. A dose of 2 μg TCDD/kg on gestation day 11.5 caused cleft clitoris, delayed puberty, abnormal vaginal estrous cycles, reduced F_1 fertility, and reduced F_2 survival (Wolf et al., 1999); however, these animals did not exhibit a vaginal thread or a decrease in ovarian weight, in contrast to the effects observed in rats. Additional species-specific effects were observed in mice. Female neonatal mice failed to exhibit cleft clitoris, a vaginal thread, a decrease in ovarian weight, or delayed time to puberty after in utero and lactational exposure to 15–60 μg TCDD/kg; however, females from the highest TCDD dose group exhibited a reduction in uterus weight without a change in body weight (Theobald and Peterson, 1997).

The mechanisms underlying TCDD's effects on the developing female reproductive system are not known, but disruption of endocrine signaling during reproductive development could contribute to these effects. It has been proposed that cleft clitoris may result from an estrogenic effect of TCDD, since the estrogen agonist diethylstilbestrol (DES) also produces cleft clitoris after in utero exposure (Vorherr et al., 1979). However, only in utero TCDD exposure, and not in utero DES exposure, produces a vaginal thread, suggesting that TCDD's effects on the female reproductive tract may or may not involve an estrogenic mechanism, and/or that they are mediated by independent mechanisms. The delayed time to puberty and presence of a vaginal thread after gestational TCDD exposure also has been proposed to result from an anti-estrogenic mechanism (Gray and Ostby, 1995). At the time of puberty, circulating E_2 levels increase and stimulate vaginal opening. A reduction or delay in the prepubertal increase in E_2 could delay both the time to vaginal opening and result in abnormal vaginal opening. TCDD, however, did not alter circulating E_2 levels nor alter the ability of the ovary to produce E_2 just before puberty (Gray et al., 1997). In addition, a more recent study has shown that the vaginal thread induced by TCDD was present at the time of birth and before puberty (Flaws et al., 1997). Thus, the formation of the vaginal thread does not result from abnormal vaginal opening at puberty, but rather results from a disruption of vaginal embryonic morphogenesis, which may or may not involve endocrine disruption.

The developing female reproductive system appears to be as equally or more sensitive than the developing male reproductive system to the adverse effect of in utero and lactational exposure to TCDD. In both males and females, adverse effects of in utero and lactational exposure occur at doses lower than those that cause reproductive toxicity in adults, and in females the adverse effects are often manifested as reproductive problems in the F_1 generation and reduced survival in the F_2 generation.

Endocrine Disrupting Effects of PCDDs on Vertebrate Wildlife

Typically, PCDDs occur in the environment as complex mixtures with other HAHs, such as PCDFs and PCBs. The mixtures consist of HAHs that mediate toxicity through the AhR (TCDD-like) and other HAHs that mediate toxicity through an AhR-independent mechanism (non–TCDD-like). The presence of these environmental mixtures makes it difficult to assess the risk that HAHs pose to vertebrate wildlife. For those TCDD-like HAHs in an environmental mixture, a procedure has been developed to estimate the TCDD equivalent concentration (TEQ), based on the amount of each congener in the mixture and its potency, relative to TCDD, to cause TCDD-like toxicity, called the toxic equivalency factor (TEF; Safe, 1990). This approach has been used to predict the toxicological risk that TCDD-like HAH mixtures pose to fish, birds, and other wildlife. In many locations, the contribution of TCDD-like PCBs to the total TEQ greatly exceeds the contribution of TCDD-like PCDDs and PCDFs. Thus, cause-and-effect relationships between HAH mixtures containing high levels of PCBs and toxicity in vertebrate wildlife have been studied in numerous environmental situations. Readers are referred to chapter 11 of this book for a review of the ability of PCBs and PCB mixtures to cause endocrine-disrupting effects on vertebrate wildlife.

One potential cause-and-effect relationship impacting vertebrate wildlife and primarily involving PCDDs will be discussed here: mortality of Lake Ontario lake trout during embryonic and posthatching development. In the 1940s lake trout populations were in decline in the Great Lakes, and by 1960 lake trout were deemed extinct in Lake Ontario and other Great Lakes. Stocking programs successfully reestablished sexually mature fish that produced fertile eggs; however, there was no apparent recruitment of young into the population. Many hypotheses, including the presence of chemical contaminants, were suggested for the failure of lake trout recruitment, but it was not until 1991 that the impact of PCDDs on Lake Ontario lake trout embryo survival was fully appreciated (Spitsbergen et al., 1991; Walker et al., 1991). Laboratory studies exposing uncontaminated lake trout eggs to TCDD and other PCDDs demonstrated that early life-stage development was extremely sensitive to PCDD-induced toxicity and mortality. Egg concentrations of 0.065 µg TCDD/kg increased embryo and posthatching mortality to 50% above control (LD_{50}), and the lowest-observable-adverse-effect level was 0.04 µg TCDD/kg (Walker and Peterson, 1991; Walker et al., 1991). Retrospective risk assessment studies conducted by the U.S. EPA in Lake Ontario predicted that the TEQ of TCDD and other PCDDs in lake trout eggs likely exceeded the LD_{50} for mortality throughout the 1960s and into the 1970s (Cook et al., 2003). It is not known if PCDD-induced early life-stage mortality in lake trout results from a disruption of the endocrine system during embryo development; however, it is clear from these studies that PCDDs have had a significant impact on lake trout populations in the Great Lakes.

Implications for Wildlife Populations and Human Health

Exposure of fish, wildlife, and humans to high levels of PCDDs and related HAHs has been associated with embryonic mortality, teratogenicity, reproductive dysfunction, chloracne, immunotoxicity, and cancer, with the reproductive and developmental effects representing some of the most sensitive adverse responses (Peterson et al., 1993; Giesy et al., 1994; Walker and Peterson, 1994). As reviewed in this chapter, some of the effects of PCDDs are consistent with an antiestrogenic mechanism; however, other effects suggest estrogenic, antiandrogenic, and antigonadotropic responses. Thus, it is unlikely that these reproductive and developmental effects are mediated exclusively through endocrine-disrupting mechanisms; rather, these toxic effects of PCDDs likely result from molecular and physiological changes representing both endocrine-dependent and -independent mechanisms. Regardless of the mechanism, however, PCDDs have the potential to impact vertebrate wildlife populations, as seen by the impact of PCDDs on lake trout reproduction and embryo survival in the Great Lakes, and they may impact human populations as well.

Adverse effects on human reproduction and development have been observed after accidental exposure of pregnant women to high levels of TCDD-like chemicals. The two most significant exposures occurred when rice oil was inadvertently contaminated with PCBs, PCDFs, and polychlorinated quaterphenyls in Japan in 1968 and again in Taiwan in 1979 (Hsu et al., 1994; Masuda, 1994). In Taiwan, 20% of babies born to women who had ingested the contaminated rice oil during their pregnancy died of pneumonia, bronchitis, sepsis, or prematurity. Those babies that survived typically exhibited low birthweights, hyperpigmentation of the skin, hypersecretion of meibomian gland, swelling around the eyes, respiratory distress, and pneumonia, whereas children exposed after birth were also more likely to develop bronchitis and had an increased incidence of abnormal pulmonary auscultation. At 7 years of age these children also exhibited neurological deficits, including reduced scores on IQ tests, reduced scores on cognitive tests, hyperactivity, and behavioral problems (Masuda, 1994). In addition, preliminary data suggest that perinatally exposed boys exhibited elevated circulating E_2 levels and significantly smaller penises at puberty (Guo et al., 1995).

These children were exposed to both TCDD-like and non–TCDD-like chemicals. Although it is difficult to distinguish the specific contribution of TCDD-like and non–TCDD-like chemicals to the developmental effects described above, many of the effects observed in the children from Japan and Taiwan resemble TCDD toxicity seen in fetal and adult laboratory animals. In addition, although it is clear from these accidental exposures that high doses of TCDD-like chemicals can have adverse effects on human reproductive outcomes, the question remains whether low exposure levels pose a risk to human reproduction and development. Adult humans gradually accumulate HAHs over their lifetime, and in Western industrialized countries the average background body burden of PCDDs and related HAHs is 8–13 ng TEQ/kg body weight, while more highly exposed populations have body burdens upward of 100–7000 ng TEQ/kg (DeVito et al., 1995). The most sensitive adverse effects of PCDDs on reproduction and development in laboratory animals occur at doses ranging from 64 to 1,000 ng TCDD/kg, fivefold higher than the average background but within the range of human body burdens of more highly exposed populations. Thus, the average human body burden of TCDD-like chemicals is within one order of magnitude of the most sensitive adverse effects of TCDD on development in laboratory animals, while the body burden of more highly exposed human populations is within the range of those TCDD doses shown to disrupt reproduction and development in multiple laboratory species.

Whether human reproduction and development are at significant risk to the adverse effects of PCDDs shown to occur in laboratory animals is uncertain. The TEFs used to calculate human body burden of AhR-active chemicals are rarely based on reproductive and developmental endpoints. Future studies need to determine the TEFs of those PCDDs, PCDFs, and PCBs that constitute the largest percentage of the human TEQ body burden, based on reproductive and developmental toxicity as endpoints. Finally, understanding the mechanistic basis for the endocrine-disrupting effects of PCDDs will help identify those toxic effects that are most sensitive to PCDD exposure, aid in the development of biomarkers to assess whether toxic effects observed in laboratory studies also occur in humans exposed to low doses of PCDDs and related chemicals, and ultimately decrease the uncertainty in predicting the risk that exposure to PCDDs and related chemicals poses to human reproduction and development.

References

Astroff, B., Rowlands, C., Dickerson, R., Safe, S., 1990. 2,3,7,8-Tetrachlorodibenzo-*p*-dioxin inhibition of 17 beta-estradiol-induced increases in rat uterine epidermal growth factor receptor binding activity and gene expression. Mol. Cell. Endocrinol. 72, 247–252.

Astroff, B., Safe, S., 1988. Comparative antiestrogenic activities of 2,3,7,8-tetrachlorodibenzo-*p*-dioxin and 6-methyl-1,3,8-trichlorodibenzofuran in the female rat. Toxicol. Appl. Pharmacol. 95, 435–443.

Astroff, B., Safe, S., 1990. 2,3,7,8-Tetrachlorodibenzo-*p*-dioxin as an antiestrogen: effect on rat uterine peroxidase activity. Biochem. Pharmacol. 39, 485–488.

Birnbaum, L.S., 1998. Developmental effects of dioxin. In: Horach, K.S. (Ed.), Developmental and Reproductive Toxicology. Marcel Dekker, New York, pp. 87–112.

Bjerke, D.L., Brown, T.J., MacLusky, N.J., Hochberg, R.B., Peterson, R.E., 1994a. Partial demasculinization and feminization of sex behavior in male rats by in utero and lactational exposure to 2,3,7,8-tetrachlorodibenzo-*p*-dioxin is not associated with alterations in estrogen receptor binding or volumes of sexually differentiated brain nuclei. Toxicol. Appl. Pharmacol. 127, 258–267.

Bjerke, D.L., Sommer, R.J., Moore, R.W., Peterson, R.E., 1994b. Effects of in utero and lactational 2,3,7,8-tetrachlorodibenzo-*p*-dioxin exposure on responsiveness of the male rat reproductive system to testosterone stimulation in adulthood. Toxicol. Appl. Pharmacol. 127, 250–257.

Bookstaff, R.C., Kamel, F., Moore, R.W., Bjerke, D.L., Peterson, R.E., 1990a. Altered regulation of pituitary gonadotropin-releasing hormone (GnRH) receptor number and pituitary responsiveness to GnRH in 2,3,7,8-tetrachlorodibenzo-*p*-dioxin-treated male rats. Toxicol. Appl. Pharmacol. 105, 78–92.

Bookstaff, R.C., Moore, R.W., Peterson, R.E., 1990b. 2,3,7,8-tetrachlorodibenzo-*p*-dioxin increases the potency of androgens and estrogens as feedback inhibitors of luteinizing hormone secretion in male rats Toxicol. Appl. Pharmacol. 104, 212–224.

Cook, P.M., Robbins, J., Endicott, D.D., Lodge, K., Guiney, P.D., Walker, M.K., Zabel, E.W., Peterson, R.E., 2003. Effects of aryl hydrocarbon receptor medicated early life stage toxicity on lake trout populations in Lake Ontario during the 20th century. Environ. Sci. Tech., 37, 3864–3877.

Cummings, A.M., Hedge, J.M., Birnbaum, L.S., 1999. Effect of prenatal exposure to TCDD on the promotion of endometriotic lesion growth by TCDD in adult female rats and mice. Toxicol. Sci. 52, 45–49.

Cummings, A.M., Metcalf, J.L., 1995. Induction of endometriosis in mice: A new model sensitive to estrogen. Repro. Toxicol. 9, 233–238.

Cummings, A.M., Metcalf, J.L., Birnbaum, L., 1996. Promotion of endometriosis by 2,3,7,8-tetrachlorodibenzo-*p*-dioxin in rats and mice: time-dose dependence and species comparison Toxicol. Appl. Pharmacol. 138, 131–139.

Denison, M.S., Fisher, J.M., Whitlock, J.P., Jr., 1988. The DNA recognition site for the dioxin-Ah receptor complex: nucleotide sequence and functional analysis. J. Biol. Chem. 263, 17221–17224.

DeVito, M.J., Birnbaum, L.S., Farland, W.H., Gasiewicz, T.A., 1995. Comparisons of estimated human body burdens of dioxinlike chemicals and TCDD body burdens in experimentally exposed animals. Environ. Health Perspect. 103, 820–831.

DeVito, M.J., Thomas, T., Martin, E., Umbreit, T.H., Gallo, M.A., 1992. Antiestrogenic action of 2,3,7,8-tetrachlorodibenzo-*p*-dioxin: tissue-specific regulation of estrogen receptor in CD1 mice. Toxicol. Appl. Pharmacol. 113, 284–292.

Dohler, K.D., Srivastava, S.S., Shryne, J.E., Jarzab, B., Sipos, A., Gorski, R.A., 1984. Differentiation of the sexually dimorphic nucleus in the preoptic area of the rat brain is inhibited by postnatal treatment with an estrogen antagonist. Neuroendocrinology 38, 297–301.

Fiedler, H., 1996. Sources of PCDD/PCDF and impact on the environment. Chemosphere 32, 55–64.

Flaws, J.A., Sommer, R.J., Silbergeld, E.K., Peterson, R.E., Hirshfield, A.N., 1997. *In utero* and lactational exposure to 2,3,7,8-tetrachlorodibenzo-*p*-dioxin (TCDD) induces genital dysmorphogenesis in the female rat. Toxicol. Appl. Pharmacol. 147, 351–362.

Gallo, M.A., Hesse, E.J., Macdonald, G.J., Umbreit, T.H., 1986. Interactive effects of estradiol and 2,3,7,8-tetrachlorodibenzo-*p*-dioxin on hepatic cytochrome P-450 and mouse uterus. Toxicol. Lett. 32, 123–132.

Gao, X., Son, D.S., Terranova, P.F., Rozman, K.K., 1999. Toxic equivalency factors of polychlorinated dibenzo-*p*-dioxins in an ovulation model: validation of the toxic equivalency concept for one aspect of endocrine disruption. Toxicol. Appl. Pharmacol. 157, 107–116.

Giesy, J.P., Ludwig, J.P., Tillitt, D., 1994. Dioxins, dibenzofurans, and PCBs in wildlife. In: Schecter, A. (Ed.), Dioxins and Health. Plenum Press, New York, pp. 249–307.

Gray, L.E.J., Kelce, W.R., Monosson, E., Ostby, J.S., Birnbaum, L.S., 1995. Exposure to TCDD during development permanently alters reproductive function in male Long Evans rats and hamsters: Reduced ejaculated and epididymal sperm numbers and sex accessory gland weights in offspring with normal androgenic status. Toxicol. Appl. Pharmacol. 131, 108–118.

Gray, L.E., Jr., Ostby, J.S., 1995. In utero 2,3,7,8-tetrachlorodibenzo-*p*-dioxin (TCDD) alters reproductive morphology and function in female rat offspring. Toxicol. Appl. Pharmacol. 133, 285–294.

Gray, L.E., Ostby, J.S., Kelce, W.R., 1997. A dose-response analysis of the reproductive effects of a single gestational dose of 2,3,7,8-tetrachlorodibenzo-*p*-dioxin in male Long Evans Hooded rat offspring. Toxicol. Appl. Pharmacol. 146, 11–20.

Gray, L.E., Jr., Wolf, C., Mann, P., Ostby, J.S., 1997. In utero exposure to low doses of 2,3,7,8-tetrachlorodibenzo-*p*-dioxin alters reproductive development of female long Evans hooded rat offspring. Toxicol. Appl. Pharmacol. 146, 237–244.

Guo, Y., Hendricks, A.G., Overstreet, J.W., Dieter, J., Stewart, D., Tarantal, A.F., Laughlin, L., Lasley, B.L., 1999. Endocrine biomarkers of early fetal loss in cynomolgus macaques (*Macaca fascicularis*) following exposure to dioxin. Biol. Reprod. 60, 707–713.

Guo, Y.L., Lambert, G.H., Hsu, C.C., 1995. Growth abnormalities in the population exposed in-utero and early postnatally to polychlorinated-biphenyls and dibenzofurans. Environ. Health Perspec. 103, S6, 117–122.

Haney, A.F., 1990. Etiology and histogenesis of endometriosis. Prog. Clin. Biol. Res. 323, 1–14.

Hsu, C-C., Yu, M-L.M., Chen, Y-C.J., Guo, Y-L.L., Rogan, W.J., 1994. The Yu-cheng Rice Oil Poisoning Incident. In: Schecter, A. (Ed.), Dioxins and Health, Plenum Press, New York, pp. 661–684.

Kharat, I., Saatcioglu, F., 1996. Antiestrogenic effects of 2,3,7,8-tetrachlorodibenzo-p-dioxin are mediated by direct transcriptional interference with the liganded estrogen receptor. Cross-talk between aryl hydrocarbon- and estrogen-mediated signaling. J. Biol. Chem. 271, 10533–10537.

Khera, K.S., Ruddick, J.A., 1973. Polychlorodibenzo-p-dioxins: perinatal effects and the dominant lethal test in Wistar rats. In: Blair, E.H. (Ed.), Chlorodioxins—Origin and Fate. American Chemical Society, Washington, DC, pp. 70–84.

Klinge, C.M., Bowers, J.L., Kulakosky, P.C., Kamboj, K.K., Swanson, H.I., 1999. The aryl hydrocarbon receptor (AHR)/AHR nuclear translocator (ARNT) heterodimer interacts with naturally occurring estrogen response elements. Mol. Cell. Endocrinol. 157, 105–119.

Kociba, R.J., Keyes, D.G., Beyer, J.E., Carreon, R.M., Wade, C.E., Dittenber, D.A., Kalnins, R.P., Frauson, L.E., Park, C.N., Barnard, S.D., Hummel, R.A., Humiston, C.G., 1978. Results of a two-year chronic toxicity and oncogenicity study of 2,3,7,8-tetrachlorodibenzo-p-dioxin in rats. Toxicol. Appl. Pharmacol. 46, 279–303.

Li, X., Johnson, D.C., Rozman, K.K., 1995. Reproductive effects of 2,3,7,8-tetrachlorodibenzo-p-dioxin (TCDD) in female rats: Ovulation, hormonal regulation, and possible mechanism(s). Toxicol. Appl. Pharmacol. 133, 321–327.

Mably, T.A., Bjerke, D.L., Moore, R.W., Gendron-Fitzpatrick, A., Peterson, R.E., 1992a. *In utero* and lactational exposure of male rats to 2,3,7,8-tetrachlorodibenzo-p-dioxin. 3. Effects on spermatogenesis and reproductive capability. Toxicol. Appl. Pharmacol. 114, 118–126.

Mably, T.A., Moore, R.W., Goy, R.W., Peterson, R.E., 1992b. In utero and lactational exposure of male rats to 2,3,7,8-tetrachlorodibenzo-p-dioxin. 2. Effects on sexual behavior and the regulation of luteinizing hormone secretion in adulthood. Toxicol. Appl. Pharmacol. 114, 108–117.

Mably, T.A., Moore, R.W., Peterson, R.E., 1992c. In utero and lactational exposure of male rats to 2,3,7,8-tetrachlorodibenzo-p-dioxin. 1. Effects on androgenic status. Toxicol. Appl. Pharmacol. 114, 97–107.

Masuda, Y., 1994. The Yusho Rice Oil Poisoning Incident. In: Schecter, A. (Ed.), Dioxins and Health. Plenum Press, New York, pp. 633–684.

McNulty, W.P., 1984. Fetotoxicity of 2,3,7,8-tetrachlorodibenzo-p-dioxin (TCDD) for rhesus macaques (*Macaca mulatta*). Am. J. Primatol. 6, 41–47.

Moore, R.W., Jefcoate, C.R., Peterson, R.E., 1991. 2,3,7,8-Tetrachlorodibenzo-p-dioxin inhibits steroidogenesis in the rat testis by inhibiting the mobilization of cholesterol to cytochrome P450scc Toxicol. Appl. Pharmacol. 109, 85–97.

Moore, R.W., Potter, C.L., Theobald, H.M., Robinson, J.A., Peterson, R.E., 1985. Androgenic deficiency in male rats treated with 2,3,7,8-tetrachlorodibenzo-p-dioxin. Toxicol. Appl. Pharmacol. 79, 99–111.

Murray, F.J., Smith, F.A., Nitschke, K.D., Humiston, C.G., Kociba, R.J., Schwetz, B.A., 1979. Three generation reproduction study of rats given 2,3,7,8-tetrachlorodibenzo-p-dioxin (TCDD) in the diet. Toxicol. Appl. Pharmacol. 50, 241–252.

Peterson, R.E., Theobald, H.M., Kimmel, G.L., 1993. Developmental and reproductive toxicity of dioxins and related compounds: cross-species comparisons. Crit. Rev. Toxicol. 23, 283–335.

Rier, S.E., Martin, D.C., Bowman, R.E., Dmowski, W.P., Becker, J.L., 1993. Endometriosis in rhesus monkeys (*Macaca mulatta*) following chronic exposure to 2,3,7,8-tetrachlorodibenzo-*p*-dioxin. Fundam. Appl. Toxicol. 21, 433–441.

Roman, B.L., Peterson, R.E., 1998a. Developmental male reproductive toxicology of 2,3,7,8-tetrachlorodibenzo-*p*-dioxin (TCDD) and PCBs. In: Horach, K.S. (Ed.), Developmental and Reproductive Toxicology. Marcel Dekker, New York, pp. 593–624.

Roman, B.L., Peterson, R.E., 1998b. In utero and lactational exposure of the male rat to 2,3,7,8-tetrachlorodibenzo-*p*-dioxin impairs prostate development 1. Effects on gene expression. Toxicol. Appl. Pharmacol. 150, 240–253.

Roman, B.L., Sommer, R.J., Shinomiya, K., Peterson, R.E., 1995. *In utero* and lactational exposure of the male rat to 2,3,7,8-tetrachlorodibenzo-*p*-dioxin: impaired prostate growth and development without inhibited androgen production. Toxicol. Appl. Pharmacol. 134, 241–250.

Roman, B.L., Timms, B.G., Prins, G.S., Peterson, R.E., 1998. In utero and lactational exposure of the male rat to 2,3,7,8-tetrachlorodibenzo-*p*-dioxin impairs prostate development. 2. Effects on growth and cytodifferentiation. Toxicol. Appl. Pharmacol. 150, 254–270.

Romkes, M., Piskorska-Pliszczynska, J., Safe, S., 1987. Effects of 2,3,7,8-tetrachlorodibenzo-*p*-dioxin on hepatic and uterine estrogen receptor levels in rats. Toxicol. Appl. Pharmacol. 87, 306–314.

Romkes, M., Safe, S., 1988. Comparative activities of 2,3,7,8-tetrachlorodibenzo-*p*-dioxin and progesterone as antiestrogens in the female rat uterus. Toxicol. Appl. Pharmacol. 92, 368–380.

Safe, S., 1990. Polychlorinated biphenyls (PCBs), dibenzo-p-dioxins (PCDDs), dibenzofurans (PCDFs), and related compounds: Environmental and mechanistic considerations which support the development of toxic equivalency factors (TEFs). Crit. Rev. Toxicol. 21, 51–88.

Safe, S., Wang, F., Porter, W., Duan, R., McDougal, A., 1998. Ah receptor agonists as endocrine disruptors: antiestrogenic activity and mechanisms. Toxicol. Lett. 102–103, 343–347.

Silvers, K., Rorke, E.A., 1998. TCDD and uterine function. In: Horach, K.S. (Ed.), Developmental and Reproductive Toxicology. Marcel Dekker, New York, pp. 413–430.

Smialowicz, R.J., Riddle, M.M., Williams, W.C., Diliberto, J.J., 1994. Effects of 2,3,7,8-tetrachlorodibenzo-*p*-dioxin (TCDD) on humoral immunity and lymphocyte subpopulations: differences between mice and rats. Toxicol. Appl. Pharmacol. 124, 248–256.

Spitsbergen, J.M., Walker, M.K., Olson, J.R., Peterson, R.E., 1991. Pathologic alterations in early life stages of lake trout, *Salvelinus namaycush*, exposed to 2,3,7,8-tetrachlorodibenzo-*p*-dioxin as fertilized eggs. Aquat. Toxicol. 19, 41–72.

Swanson, H.I, Bradfield, C.A., 1993. The AH-receptor: Genetics, structure and function. Pharmacogenetics 3, 213–230.

Taguchi, O., 1987. Reproductive tract lesions in male mice treated neonatally with tamoxifen. Biol. Reprod. 37, 113–116.

Theobald, H.M., Peterson, R.E., 1997. In utero and lactational exposure to 2,3,7,8-tetrachlorodibenzo-*p*-dioxin: Effects on development of the male and female reproductive system of the mouse. Toxicol. Appl. Pharmacol. 145, 124–135.

Thornton, J., McCally, M., Orris, P., Weinberg, J., 1996. Hospitals and plastics. Dioxin prevention and medical waste incinerators. Public Health Rept. 111, 299–313.

Vorherr, H., Messer, R.H., Vorherr, U.F., Jordan, S.W., Kornfeld, M., 1979. Teratogenesis and carcinogenesis in rat offspring after transplacental and transmammary exposure to diethylstilbestrol. Biochem. Pharmacol. 28, 1865–1877.

Walker, M.K., Peterson, R.E., 1991. Potencies of polychlorinated dibenzo-*p*-dioxins, dibenzofurans, and biphenyls, relative to 2,3,7,8-tetrachlorodibenzo-*p*-dioxin, for producing early life stage mortality in rainbow trout (*Oncorhynchus mykiss*). Aquat. Toxicol. 21, 219–238.

Walker, M.K., Peterson, R.E., 1994. Aquatic toxicity of dioxins and related chemicals. In: Schecter, A. (Ed.), Dioxins and Health. Plenum Press, New York, pp. 347–387.

Walker, M.K., Spitsbergen, J.M., Olson, J.R., Peterson, R.E., 1991. 2,3,7,8-Tetrachlorodibenzo-*p*-dioxin (TCDD) toxicity during early life stage development of lake trout (*Salvelinus namaycush*). Can. J. Fish. Aquat. Sci. 48, 875–883.

Webster, T., Commoner, B., 1994. Overview: the dioxin debate. In: Schecter, A. (Ed.), Dioxins and Health. Plenum Press, New York, pp. 1–50.

White, T.E., Gasiewicz, T.A., 1993. The human estrogen receptor structural gene contains a DNA sequence that binds activated mouse and human Ah receptors: A possible mechanism of estrogen receptor regulation by 2,3,7,8-tetrachlorodibenzo-*p*-dioxin Biochem. Biophys. Res. Commun. 193, 956–962.

Wolf, C.J., Ostby, J.S., Gray, L.E.J., 1999. Gestational exposure to 2,3,7,8-tetrachlorodibenzo-*p*-dioxin (TCDD) severely alters reproductive function of female hamster offspring. Toxicol. Sci. 51, 259–264.

Zacharewski, T., Safe, S.H., 1998. Antiestrogenic activity of TCDD and related compounds. In: Korach, K.S. (Ed.), Reproductive and Developmental Toxicology. Marcel Dekker, New York, pp. 431–448.

11

Toxicology of PCBs
and Related Compounds

John P. Giesy
K. Kannan
Alan L. Blankenship
Paul D. Jones
J. L. Newsted

Polychlorinated Biphenyls

Polychlorinated biphenyls (PCBs) are among the most studied environmental contaminants. PCBs have the potential to modulate the endocrine system through a number of mechanisms and have been designated as endocrine-disrupting compounds (EDCs) by a number of researchers and government agencies (Kendall et al., 1998; Knobil et al., 1999). Effects can be either direct or indirect. Here we present evidence that PCBs are EDCs. PCBs and their metabolites are direct-acting estrogen agonists. PCBs also can bind to plasma hormone-binding globulins, such as transthyretin (TTR), thus displacing both retinol (vitamin A) and thyroxine (T_4). PCBs induce enzymes that can, in turn, change concentrations of hormones or substrates involved in steroid hormone synthesis or signal transduction in cells. These enzymes include uridine diphosphate glucuronyl transferase (UDPGT) and testosterone hydroxylases that are important in modulating hormone concentrations. PCBs also cause neurobehavioral deficits and developmental deformities. The mechanism by which such effects are caused is less well known, but it has been. speculated that interactions with the endocrine system might be involved during early development (Gillman et al., 1991; Colborn and Clement, 1992; Dohlen and Jarzab, 1992; Colborn et al., 1993, 1996; Walker and Peterson, 1992). Major issues are determining the threshold for these effects in wildlife and determining if they can cause population-level effects now or in the future.

The purpose of this chapter is to discuss the mechanisms by which PCBs can disrupt the endocrine systems of wildlife, provide examples of these effects, and summarize available toxicological data for PCBs that are potentially useful for risk assessments of wildlife. It is beyond the scope of this chapter to provide a comprehensive review of PCB toxicity data or to discuss effects in humans or biomedical studies based on rodent models. In this chapter, data relating to wildlife organisms, such as freshwater and marine invertebrates, fish, birds, and mammals that could be exposed to PCBs are evaluated. However, when no wildlife-specific information is available, effects on animal models are evaluated, especially for elucidating mechanisms of action relative to endocrine disruption. Specifically, the availability of both dietary exposure and tissue residue-based toxicological data were evaluated for unweathered technical Aroclor mixtures, total PCBs, and individual PCB congeners. In addition, because of its structural and toxicological similarities to some PCB congeners, 2,3,7,8-tetrachlorodibenzo-*p*-dioxin (TCDD) will be considered. Effects of polychlorinated-dibenzodioxins (PCDDs) and -dibenzofurans (PCDF) on wildlife are discussed in chapter 10.

The common mechanism of toxic action between some PCB congeners and TCDD will be discussed, relative to how it can be considered in ecological risk assessments of complex mixtures of weathered PCBs. Weathering is the process by which the absolute and relative concentrations of individual congeners released in technical mixtures change due to differential solubility, volatility, and biotic and abiotic transformation processes. In addition to TCDD, dioxin equivalents also will be discussed. Dioxin equivalents are defined as two types, those that are calculated from concentrations of individual congeners by use of relative potency factors (RePs) or toxic equivalency factors (TEFs) and those that are determined as the response of either in vitro or in vivo systems to complex mixtures containing TCDD-like activity. TEFs are the relative potencies of the congeners to cause specific effects relative to the prototypical congener, TCDD. The total activity in a sample, calculated as the sum of the products of concentrations of individual congeners multiplied by their TEFs, is referred to as TEQs, and those determined by the use of integrating bioassays are designated as TCDD-EQ. The limitations of these toxicity data are discussed. Specific recommendations on the use of this information for the purposes of conducting ecological risk assessments, feasibility studies, and related activities are also addressed.

PCBs are members of the group of polyhalogenated diaromatic hydrocarbons (PHDHs). This group also includes a number of structurally similar compounds, including PCDDs, PCDFs, and polychlorinated napthalenes (PCNs) (Giesy et al., 1994a,b,c). There are theoretically 209 possible different combinations of chlorine substitutions based on the number and position of the substitutions (figure 11-1). Each combination is referred to as a congener. PCBs with the same number of chlorine atoms but with different substitution patterns are called isomers. The sum of all isomers with the same number of chlorine atoms is referred to as a homolog group. A systematic numbering scheme has been applied to the individual chlorobiphenyl congeners with numbers ranging from 1 to 209 referred to as CB (chlorobiphenyl) or IUPAC (International Union of Pure and Applied Chemists) numbers (Ballschmiter and Zell, 1980). However, it should be noted that 11 congeners (nos. 33, 34, 76, 98, 122, 123, 124, 125, 177, 196, and 201) have different designations in the Ballschmiter and Zell and IUPAC numbering systems (Erickson, 1997). For instance, IUPAC congener no. 77 is a tetrachlorobiphenyl with chlorines in the 4, 4', 5, 5' positions.

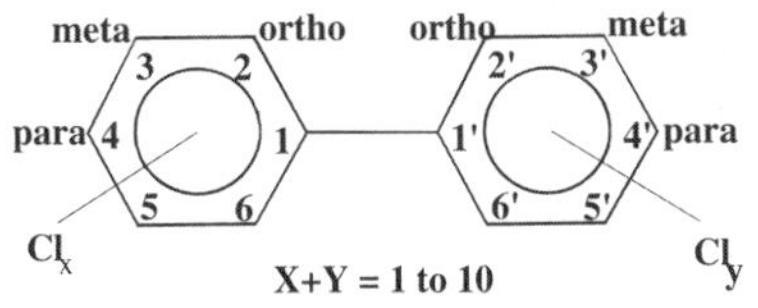

Figure 11-1. General structure of polychlorinated biphenyls (PCBs)

This molecule thus has chlorines in *meta* and *para* positions, but not in the *ortho* positions (figure 11-1). Congeners of this type are referred to as non-*ortho*-substituted congeners. Because there is no structural hindrance for the biphenyl rings to rotate, they can attain a planar configuration. These congeners are referred to as coplanar congeners. Mono-*ortho*-substituted congeners are less capable of attaining a planar configuration but are not as hindered as di-*ortho*-substituted congeners. In addition to whether congeners are *ortho*-substituted, lateral substitution is also important. Congeners that are coplanar and substituted in the lateral positions 3, 4, or 5 have toxicities similar to the PCDDs and PCDFs. The PCB congeners that are most similar to PCDDs in their toxicities are 3, 3', 4, 4', 5, 5'-hexachlorobiphenyl (IUPAC 169) and 3, 3', 4, 4', 5-pentachlorobiphenyl (IUPAC 126) because they are laterally substituted in four positions and non-*ortho*-substituted (Giesy and Kannan, 1998). Of the 209 theoretically possible congeners, only about 130 individual congeners have been identified in commercial PCB mixtures at concentrations ≥0.05% of the total mixture.

PCBs were produced commercially by the chlorination of biphenyl, which resulted in technical mixtures containing a given chlorine content, depending on the selected degree of chlorination (Hutzinger et al., 1974). Although all 209 PCB congeners can be synthesized in the laboratory, the reaction conditions in commercial processes favored specific substitution reactions leading to particular compositions in the technical mixtures, which were marketed according to their chlorine percentage by weight. For example, Aroclors 1221, 1242, 1248, 1254, 1260, and 1268 were commercial PCB preparations produced by the Monsanto Chemical Company in the United States (St. Louis, Missouri), and contain 21, 42, 48, 54, 60, and 68% chlorine by weight, respectively, as indicated by the last two digits in the numerical designation. Aroclor 1016, introduced in 1970, was an Aroclor 1242 substitute containing 42% chlorine by weight but processed in such a way that PCDD and PCDF impurities and coplanar PCB congeners had been removed. Use of the Aroclor tradename was not restricted to PCBs but also was applied to other polyhalogenated aromatic mixtures. For example, Aroclor 5460 was a polychlorinated terphenyl mixture. In this chapter, however, "Aroclor" is used to refer only to commercial, technical PCB mixtures. Technical PCB mixtures were also produced by manufacturers in other countries, and these include Clophens (Bayer, Germany), Phenoclors and Pyralenes (Prodelec, France), Fenclors (Caffaro, Italy), Fenochlors (Cross, S.A., Spain), Kanechlors (Kanegafuchi, Japan), Sovol (Sovol, former USSR), Delor (Chemko, former Czechoslovakia) and Chlorofen (Poland). The production of PCBs in OECD (Organization for Economic Cooperation and Development) member countries is estimated to have been approximately 1.2 million metric tons, which is believed to contribute to the total world production substantially (Bletchly, 1984; Tanabe, 1988). In addition, the former USSR produced 100,000 metric tons of Sovol, a technical PCB mixture resembling Aroclors 1242 and 1254, from the 1940s to the 1990s (Ivanov and Sandell, 1992). The extent to which PCBs were produced in developing countries in Asia, Africa, and South America is not known.

The beneficial chemical properties primarily responsible for the many industrial applications of PCBs, such as inflammability, chemical and thermal stability, dielectric properties and miscibility with organic compounds, are also the properties that have contributed to them being of environmental concern. PCBs were manufactured beginning in 1930 and used widely in industry as heat transfer fluids, hydraulic lubricants, dielectric fluids for transformers and capacitors, organic diluents, plasticizers, pesticide extenders, adhesives, dust-reducing agents, cutting oils, flame retardants, sealants, and as ink carriers in carbonless copy paper (Hutzinger et al., 1974). PCBs have been released into the environment by electrical industries, pulp and paper mills, waste incinerators, chlor-alkali plants, and mining industries. In addition, the use of PCBs in mining and metal industries has contributed to their occurrence in the environment.

Individual PCB congeners exhibit different physicochemical properties, which result in different environmental distributions and toxicity profiles. PCBs have low water solubility, which decreases with increasing degree of chlorination. For example, the water solubilities of monochlorobiphenyl congeners range from 1000 to 5000 mg/l but that of decachlorobiphenyl is only 0.015 mg/l (Mackay et al., 1992). Vapor pressure and degradability also decrease with increasing chlorine content (Loganathan and Kannan, 1994). Susceptibility to degradation and bioaccumulation depends on the structural arrangement of chlorine atoms among isomers. The various commercial PCB preparations vary in congener composition and thus in physicochemical and biochemical properties, such as metabolism and biodegradation of individual PCB congeners. Thus, the relative composition of environmental extracts vary among location, time, and trophic levels and vary from the original technical mixtures (Ballschmiter and Zell, 1980; Waid, 1987; McFarland and Clarke, 1989; Schulz et al, 1989; Kannan et al, 1997a; Giesy and Kannan, 1998).

PCBs were first detected in environmental samples in 1966 (Jensen, 1966). PCBs are persistent in the environment and are readily transported from localized or regional sites of contamination to remote areas via the atmosphere (Risebrough et al., 1968; Atlas and Giam, 1981). This long-range atmospheric transport has led to their presence in almost every compartment in the environment (Tanabe, 1988). Moreover, due to their lipophilicity, these compounds bioaccumulate and biomagnify through food chains.

Weathering of PCBs

Although PCBs were originally thought to be refractory organic compounds, certain PCB congeners can be degraded under aerobic conditions or microbially dechlorinated by reductive dechlorination under anaerobic conditions in aquatic environments (Abramowicz, 1990). Toxic organic chemicals can also be transformed by hydrolysis, photolysis, and biodegradation. Thus, the compositions of PCB congener mixtures that occur in the environment differ substantially from those of the original technical Aroclor mixtures released (Giesy and Kannan 1998; Zell and Ballschmiter 1980; Newman et al., 1998). As discussed previously, this is because the composition of PCB mixtures changes over time after release into the environment due to several processes collectively referred to as environmental weathering. The weathered multicomponent mixtures may have significant differences in peak patterns compared to Aroclor standards. Aerobic degradation of PCBs is therefore limited to the overlying water and possibly the top few millimeters of oxic sediments, whereas anaerobic dechlorination of PCBs may occur in sediments.

Weathering is a result of the combined effects of processes such as differential volatilization, solubility, sorption, anaerobic dechlorination, and metabolism resulting in changes in the composition of the PCB mixture over time and between trophic levels (Froese et al., 1998). Less chlorinated PCBs often are lost most rapidly due to volatilization and metabolism, whereas more highly chlorinated PCBs often are more resistant to degradation and volatilization and sorb more strongly to particulate matter. More highly chlorinated PCBs tend to bioaccumulate to higher concentrations than low molecular weight PCBs in tissues of animals and may biomagnify in food webs.

Under anaerobic conditions, catabolism leaves the biphenyl ring intact while removing chlorines from the ring, thereby producing less chlorinated congeners. Although details of the dechlorination process are not fully understood, reductive dechlorination proceeds primarily through selective removal of *meta* (3,5) and *para* (4) chlorines (Abramowicz 1990, 1994; Rhee et al., 1993a,b; Quensen et al., 1998). Anaerobic dechlorination has been observed in a number of sites, including the upper Hudson River, but is not thought to be important at sites with lower PCB contamination levels, possibly due to PCB concentrations being below a threshold value. Microbial reductive dechlorination of PCBs occurs in variety of anaerobic environments (Bedard and Quensen, 1995). This process does not remove all of the chlorines and does not alter the basic structure of the biphenyl. Because reduction dechlorination results in a decrease of some congeners and an increase in concentrations of others, there is generally not a great change in the total molar concentration of PCBs in sediments. The major conclusion from aerobic degradation studies (Brown et al., 1987; Bedard and Haber, 1990; Benard and Quensen, 1995; Bedard and May, 1996; Bedard et al, 1996, 1997) is that biodegradation of PCBs can occur by the attack of a dioxygenase enzyme at an unchlorinated 2,3 (or 5,6) site or at an unchlorinated 3,4 (or 4,5) site. These attacks result in cleavage of the biphenyl ring and can be carried out by a variety of naturally occurring bacteria. Congeners with chlorines at both *ortho* (2,6) positions on either ring are generally not degraded as readily as congeners lacking this characteristic.

The degree and position of chlorine substitution (figure 11-1) not only influence physicochemical properties but also the toxic effects. Thus, it is important to consider not only the total PCB concentration in a sample, but also to characterize the distribution of individual PCB isomers present in a sample. Because reductive dechlorination occurs preferentially for chlorines in the *meta* and *para* positions, the process selectively reduces the relative proportions of the PCBs that are laterally substituted. These congeners are also those that tend to have the greatest potency to cause aromatic hydrocarbon receptor (AhR)-mediated effects. One of the most potent of the laterally substituted, non-*ortho*-substituted congeners is congener 126 (3,3',4,4',5-pentachlorobiphenyl). The absolute and relative concentrations of this congener decrease by as much as 10- to 100-fold due to reductive dechlorination (Mousa et al., 1998; Quensen et al., 1998). The total concentration of TEQ in sediments, either determined by the bioassay (TCDD-EQ) or by application of TEFs (TEQ) to concentrations of individual congeners were reduced by 100-fold by reductive dechlorination (Quensen et al., 1998).

Metabolic Products of PCBs

Metabolic products of PCBs, such as hydroxylated PCBs (OH-PCBs), and their methyl-sulfonyl, glucuronide, and sulfate conjugates (Bergman et al., 1994) are produced by phase I and phase II xenobiotic transformation enzymes. Methyl sulfonyl metabolites are retained in the liver, lung, kidney, uterine fluid, and fetal tissues (Darnerud et al., 1986; Bergman et al., 1992; Morse et al., 1995) and exhibit toxic effects. OH-PCBs are formed via an arene oxide intermediate catalyzed by cytochrome P450s 1A and 2B (Sundström et al., 1976; Bergman et al., 1994). Typically, OH-PCBs are readily excreted in feces and urine as the sulfate and glucuronide conjugates (Bergman et al., 1994) or unconjugated (Koga et al., 1990). However, OH-PCBs such as 4-OH-3,3',4',5-C_{14}-biphenyl that contain a 4-OH-3,5-C_{12} moiety are retained in plasma due to binding to plasma proteins such as TTR (Brouwer et al., 1990) or partitioning into lipids. Hydroxylated metabolites of PCBs are retained in the blood of wildlife (Klasson-Wehler et al., 1998), but little information is available on the in vivo effects of OH-PCBs on wildlife. The affect of OH-PCB on the observed feminization of birds is unknown. This effect has been attributed to the estrogenic effects of 1,1-dichloro-2,2-*bis*(*p*-chlorophyenyl)-ethylene (*o,p*-DDE; Feyk and Giesy, 1998). However, because the birds in which these effects were observed were exposed simultaneously to both DDE and PCBs and their metabolites, it is impossible to know what role OH-PCBs might have played. There have been no controlled laboratory investigations of the effects of OH-PCBs in any wildlife species. Some OH-PCBs exhibit an estrogenic uterotrophic response in mice (Korach et al., 1988), whereas others have exhibited estrogenic, antiestrogenic or no activity in vitro (Connor et al., 1997; Fielden et al., 1997; Kramer et al., 1997; Kramer and Giesy, 1999; Moore et al., 1997).

Extrapolation of Dose Ranges and Routes of Exposure

The vector of exposure is important in determining the absolute and relative magnitudes of TEFs. We have relied when possible, primarily on concentrations in tissues of experimental organisms or their diets. Studies that used waterborne exposures or intraperitoneal injections or that did not report tissue concentrations were not used in deriving toxicity reference values (TRVs). To interpret the results of these studies would require understanding of the biological availability and potential for changes in relative concentrations of PCB congeners in tissues. Much of the information used for establishing TEFs has come from in vitro studies of the induction of monooxygenases and, more recently, from subchronic toxicity studies. Most of the in vivo studies have produced information on acute effects induced at the greater dose ranges such as lethality. Because, in real-world scenarios, biological effects at chronic, low-level exposures are more relevant, TEFs derived from large doses may be questionable. The dose regimes used in exposure studies are different, which may improve the derived TEF value and eventually the risk assessment process.

Toxicology of PCBs

Initially, all PCB congeners were regarded as toxic. Early textbooks suggested that the toxicity of PCB congeners was proportional to the degree of chlorination. However, in vivo studies conducted with rodents showed that the toxicity of PCB congeners varied greatly, and only small group of congeners had great toxic potential (Safe, 1990, 1994; Silberhorn et al., 1990).

Studies with wildlife have demonstrated a causal link between adverse health effects and PCB exposure (Giesy et al., 1994a,c; Bowerman et al., 1995). However, the observed toxicity to birds and mammals correlates more strongly with TEQs than with total PCBs (Giesy et al., 1994c; Leònards et al., 1995). PCBs often are among the predominant chemicals of concern at contaminated sites due to their relative abundance compared to other PCDHs, persistence, bioaccumulation and biomagnification potential, and their potential to adversely affect the health of humans and wildlife (Giesy et al., 1994b,c). Due to their bioaccumulation potential, a site-specific exposure model that uses a food web analysis is often developed for ecological risk assessments for PCBs. At most sites, PCBs are predominantly bound to particles or strongly associated with an organic carbon fraction. Aquatic organisms are exposed to a combination of dissolved, sediment-associated, and food-associated PCBs, whereas, in terrestrial ecosystems, lower trophic-level organisms are exposed to PCBs primarily through ingestion of soil and prey, although dermal absorption and inhalation may be important routes of exposure for certain species. At each higher trophic level, certain PCB congeners are selectively enriched or depleted due to the mechanisms discussed previously. As a result, organisms at the top of the food chain are generally at greatest risk of adverse effects due to exposure to PCBs. However, foraging preferences, species sensitivity, and other site-specific factors can modify the magnitude of the effect at any trophic level. Differences in the composition of PCB residues in environmental matrices have implications for quantification and hazard evaluation, particularly when considering differences in the biological activity, both qualitatively and quantitatively, among isomers as well as congeners. Several studies have demonstrated differences in both mechanisms and toxic potentials of individual PCB congeners (Safe, 1984, 1994; Strang et al., 1984; Seegal, 1996; Giesy and Kannan, 1998). Thus, the impacts of PCBs on the environment and biota are due to the individual components of these mixtures and their additive and/or nonadditive (synergistic or antagonistic) interactions among themselves and other chemical classes of pollutants (Sanderson and Giesy, 1998). Therefore, the development of scientifically based regulations for the risk assessment of PCBs requires analytical and toxicological data on individual PCB congeners present in any technical mixture and information regarding interactive effects.

Commercial PCB mixtures elicit a broad spectrum of toxic responses that depend on several factors including chlorine content, purity, dose, species and strain, age and sex of animal, and route and duration of exposure. Immunotoxicity, carcinogenicity, and developmental toxicity as well as biochemical effects of commercial PCB mixtures have been investigated extensively in various laboratory animals, fishes, and wildlife species. The mechanisms of PCB toxicity and their dioxinlike effects and carcinogenicity have been reviewed (Poland et al., 1982; Safe, 1984; Silberhorn et al., 1990; Barrett, 1995). However, information on the toxicological risk assessment of PCBs in wildlife is limited (Delzell et al., 1994). Particularly, with the recent developments in the understanding of non-dioxinlike effects of PCBs, it is pertinent to examine the effective doses at which *ortho*-substituted PCBs could elicit non-dioxinlike effects in animals (see below).

We have taken a rather broad view of the endocrine-disrupting potential of PCBs. Initially EDCs were compounds that could function as hormone receptor agonists, particularly with the estrogen receptor (ER). Subsequently, research into the effects of EDCs has been expanded to include both agonists and antagonists and included additional receptor systems, such as the androgen receptor (AR) and thyroid hormone receptor (TR). The primary issue is whether low-level exposures to PCBs can alter signal transduction pathways in a way that results in adverse outcomes during long-term or chronic exposures. The potential for PCBs to act as hormone mimics in wildlife has been reviewed (see Kendall et al., 1998). Thus, we have included information on the effects of PCBs on signal transduction pathways and on the hormone status of individuals. These effects can be at the hormone receptor or on other factors that affect hormonal homeostasis, such as steroidogenesis. Furthermore, because of the close linkage of the neural and endocrine systems into an integrated neuroendocrine system, we have discussed the neurobehavioral effects of PCBs.

Mechanisms of Toxic Action

Aromatic Hydrocarbon Receptor-Mediated Effects of PCBs

Some PCB congeners are structurally related to TCDD and can invoke various common toxic responses, such as induction of aryl hydrocarbon hydroxylase (AHH) and ethoxyresorufin-*O*-deethylase (EROD), a mixed function oxidase, weight loss, hypothyroidism, decreased hepatic or plasma vitamin A levels, porphyria, thymic atrophy, immunotoxicity, and teratogenicity (Poland and Glover, 1977; Parkinson et al., 1980, 1983; Bandiera et al., 1982; Safe, 1984; Leece et al., 1985; Yoshimura et al., 1985; Goldstein and Safe, 1989). The non-*ortho*-substituted coplanar PCBs,

3,4,4',5-tetraCB (PCB 81), 3,3',4,4'-teraCB (PCB 77), 3,3',4,4',5-pentaCB (PCB 126) and 3,3',4,4',5,5'-hexaCB (PCB 169), which are substituted in both *para*, at least two *meta*, and no *ortho* positions, are the most toxic PCB congeners. It is hypothesized that the lack of chlorine substitution at opposing *ortho* positions allows the phenyl rings to rotate into the same plane, and so these congeners are commonly referred to as coplanar PCBs (Safe, 1994).

The responses listed above are mediated through the AhR, which is a high-affinity cytosolic protein, Poland et al., 1976, 1979; Poland and Glover, 1977; Poland and Knutson, 1982; Safe, 1984; Goldstein and Safe, 1989). A schematic representation of the proposed mechanism of action is given in figure 11-2. Although most of the pleiotropic effects of the AhR-active compounds are thought to be due to genomic effects, other short-term effects are observed. Because AhR-active compounds can modulate certain signal transduction pathways, this would qualify as an endocrine-like effect. The relatively great range in sensitivities among organisms to exposures to PCBs may be due in part to differences in expression of the AhR (Karchner et al., 2000).

Represented in figure 11-2 are two additional important factors. First is the involvement of the aromatic receptor nuclear transport (ARNT) protein. Interference in pathways involving this protein are discussed below. Genetically modified cell lines that use the firefly tail enzyme luciferase gene as a reporter gene have been developed to be sensitive, selective indicators of genomic responses to individual ligands. A key concept is that reporter systems have been developed for several classes of animals so that comparisons of among-species toxicity can be rapidly made. Furthermore, cell lines have been developed to screen for receptor agonists and antagomsts at the same time. Below, we discuss the effects of PCBs on the AhR-mediated system using the H4IIE-luc rat hepatoma cell line, which responds to AhR-active and estrogenic or antiestrogenic compounds.

Effects of PCBs on Cross-Talk among Signal Transduction Pathways Requiring ARNT

Although the coplanar TCDD-like PCB congeners can interact with the AhR, it is not known how interaction with this one receptor can cause so many pleiotropic effects in organisms. Induction of CYP1 monooxygenase enzymes is one response, but it is unlikely that this is the cause of most of the effects observed. Induction of these enzymes can cause production of free radicals that in turn can cause oxidative stress and ultimately result in damage to developing tissues, leading to cell death and deformities (Wright and Tillitt, 1999).

The nuclear transduction factor ARNT is part of the transduction process initiated when a PCB congener binds to the AhR (figure 11-2). The ARNT protein, however, does not function uniquely in the AhR signaling pathway, but also pairs with hypoxia inducible factor 1α (HIF-1α) to regulate genes active in response to low oxygen stress (Semenza, 1994; Guillemin and Krasnow, 1997; Wenger and Gassmann, 1997). HIF-1α is continuously synthesized and degraded under normal oxygen tension. Hypoxic conditions inhibit the degradation of HIF-1α by the ubiquitin proteasome system and triggers the nuclear localization of HIF-1α (Kallio

Mechanism of Action for ER-Activation

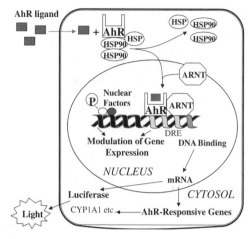

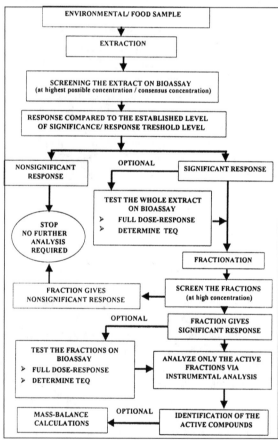

Figure 11-2. Mechanism of action for estrogen receptor (ER) binding. ERE, estrogen response element.

et al., 1997; Salceda and Caro, 1997; Huang et al., 1998). Transition metals, such as cobalt, and iron chelators, such as desferrioxamine (Dfx), elicit the same response in cells, which suggests that these stimuli work at different stages of a common oxygen sensing pathway (Guillemin and Krasnow, 1997). Inside the nucleus, HIF-1α heterodimerizes with ARNT to form a transcription factor complex HIF-1 (Jiang et al., 1996). The binding of HIF-1 to the core DNA sequence, TACGTG, of the hypoxia response enhancer (Semenza et al., 1991; Wang and Semenza, 1993; Jiang et al., 1996) and subsequent transactivation enhance expression of genes such as *Epo* for erythropoiesis (Semenza, 1994), vasculature epidermal growth factor (VEGF) for angiogenesis (Shweiki et al., 1992; Goldberg and Schneider, 1994; Forsythe et al., 1996; Maxwell et al., 1997), and GLUT-1 for glucose transport (Semenza, 1994; Wenger and Gassmann, 1997).

In vitro, ARNT can also form a homodimer or heterodimers with a number of PAS (Per-ARNT-Sim) proteins, which include AhR, ARNT, HIF-1α, and transcription factors involved in various gene regulation pathways (Hogenesch et al., 1997; Hahn, 1998). These results suggest ARNT might be a central regulator of many cellular pathways through its formation of functional transcription activators or repressors by dimerizing with other members of the PAS family. In adult animals, ARNT is ubiquitously expressed in all tissues. The essential role of ARNT is supported by studies on ARNT null mice, which are not viable beyond day 10.5 of gestation (Kozak et al., 1997; Maltepe et al., 1997).

The importance of ARNT during development and the observation that ARNT dimerizes with various PAS proteins suggest that availability of ARNT could be critical for various cellular activities, and competitive recruitment of ARNT may repress the activation of certain genes. AhR-ligands are potent inducers of AhR-mediated gene activation and can cause prolonged activation of AhR. It is possible that exposure to and subsequent recruitment of ARNT through AhR may inhibit other signal transduction pathways depending on ARNT. Such effects have been described recently (Gradin et al., 1996; Gassmann et al., 1997; Kallio et al., 1997; Chan et al., 1999). We hypothesize that the pleiotropic effects of AhR-ligands could be due to induced ARNT deficiencies in multiple ARNT-requiring pathways. Because the AhR and HIF-1 pathways are two of the most well understood pathways involving ARNT, they were chosen as a model to study the potential effect of AhR-ligands on linked pathways. The interaction between the AhR and hypoxia pathways was investigated at the level of transcription, nuclear protein complex formation, and DNA-binding activities.

Induction of the AhR signaling pathway is represented by the induction of the *CYP1A1* monooxygenase gene, expression of which is controlled by the enhancer sequence containing dioxin-responsive enhancers (DREs). One explanation of the pleiotropic responses of animals to exposure to AhR ligands is that multiple DRE-controlled genes are regulated by the AhR-ligand complex. This mechanism of action confines the effect of the AhR ligands to the regulation of genes specified by the DRE enhancer sequences. This limited scope of action, however, cannot explain the wide spectrum of toxic responses, ranging from cell proliferation to apoptosis, caused by the AhR ligands such as PCDHs. The correlation between *CYP1A1* gene induction and species sensitivity to AhR ligands, such as coplanar

PCB congeners, is poor. For example, guinea pigs, which are the most sensitive species to the lethal effects of AhR ligands, do not show induction of liver metabolizing enzymes (Poland and Knutson, 1982). It is, however, unlikely that all the effects caused by TCDD-like PCB congeners are due to the direct response of DREs. The fact that many effects of AhR ligands can be described as alteration of cell growth and differentiation (Okey et al., 1994) suggests an interaction among multiple signaling pathways. Therefore, an alternative hypothesis for the pleiotropic responses of animals to coplanar PCB congeners is cross-talk among pathways with common mechanisms of signaling transduction. Okey et al. (1994) demonstrated that the AhR and hypoxia pathways interact at the level of gene expression, and such interaction is correlated with interference of DNA binding activity and nuclear complex formation. Reciprocal repression of gene expression of the two pathways was observed in B-1 cells. Inhibitory effects of the inducers of the hypoxia response pathway on AhR-mediated gene expression also were observed in mouse liver cells (Hepa 1 and HIL1.1c2 cells). This suggests that such an interaction is not unique to B-1 cells. Although cytochrome P450s, such as CYP1A1, are heme proteins whose production requires iron, repression of EROD activity in B-l and Hepa 1 cells by Dfx was not caused by direct inhibition of protein synthesis because the luciferase reporter enzyme whose production is independent of iron was also reduced by Dfx in H1L1.1c2 cells. Furthermore, H4IIE-luc cells did not exhibit effects of Dfx on either EROD or luciferase activity when cells were treated with same concentrations of Dfx. These results suggest that the concentrations of Dfx used were sufficient to activate the hypoxia response pathway, yet they did not affect CYP1A1 protein synthesis.

AhR, ARNT, and HIF-1α belong to bHLH (basic-helix-loop-helix)/PAS protein family. The bHLH motif is characteristic of a family of proteins that function as modulators of cell proliferation and differentiation. PAS proteins are found in representative organisms of all five kingdoms and may play a role in determining target gene specificity (Hahn, 1998; Hahn et al., 1992). PAS proteins are involved in development and differentiation (Sim group protein, tracheous) (Nambu et al., 1991; Isaac and Andrew, 1996), regulation of circadian clocks (Per, CLOCK) (Huang et al., 1995; King et al., 1997), sensing and responding to oxygen tension (HIF-1α, endothelial per-ARNT-Sim (PAS) domain protein EPAS-l/HIF) (Tian et al., 1997; Wenger and Gassmann, 1997), and steroid receptor signaling (SRC-1) (Yao et al., 1996). The myogenic bHLH proteins autoregulate their expression and cross-regulate the expression of other family members (Olson, 1990). The PAS domain provides the specificity for dimerization among PAS/bHLH proteins (Pongratz et al., 1998). PAS proteins, therefore, may behave in a way similar to the myogenic bHLH proteins and interact with each other through the PAS domain. As ARNT dimerizes with many PAS proteins such as AhR, Sim1, Sim2, HIF-1α, and EPAS-1 (Sogawa et al., 1995; Jiang et al., 1996; Hogenesch et al., 1997; Moffett et al., 1997; Tian et al., 1997), ARNT may act as a central regulator of PAS protein-dependent pathways.

Eukaryotic gene regulation can be viewed as an interplay between activating and repressing influences. Transactivation of genes is controlled at many levels but typically involves binding by transcription factors to specific regulatory cis elements. Although the cis elements can be specific to a particular set of genes, the transcription factors may be less diverse and shared by several pathways. Sharing a few common transcription factors by many transcription regulation pathways also provides the advantage of network regulation and more sophisticated, fine-tuned controls. This sharing also leads to the observed phenomenon known as squelching, which refers to alteration of transcription by sequestering limiting components required for transcriptional activation or repression away from the promoter in the affected gene (Cahill et al., 1994). For instance, the estrogen hormone receptor inhibits the transcriptional activation mediated by the progesterone and glucocorticoid receptors (Meyer et al., 1989). Squelching could be a common mechanism used by transcription factors to downregulate promoters whose activity is governed by coactivators of low abundance.

Reciprocal repression of the AhR- and HIF-1-mediated pathways was observed in B-1 cells, but the degrees of interaction were different. Inhibition of AhR-mediated gene expression by induction of the hypoxia response pathway was greater than the inhibition of the hypoxia response pathway by TCDD. HIF-lα binds to ARNT with greater affinity than does AhR (Gradin et al., 1996). This differential binding affinity to ARNT may explain the stronger inhibition by hypoxia response inducers on the AhR pathway and may serve as a regulatory mechanism in addition to dimerization specificity.

Results of electrophoresis mobility shift assays further elucidate the mechanism of interaction between the two pathways. The reduction of hypoxia-induced DNA-binding activity by AhR ligands and TCDD-induced DNA-binding activity by Dfx supports the hypothesis that interaction between the AhR- and HIF-1-mediated pathways is due to recruitment of ARNT (Hogenesch et al., 1997; Chan et al., 1999). Therefore, ARNT may serve as a nuclear integrator and modulator of various signal transduction pathways. Alternatively, these data do not exclude the possibility that other commonly involved transcription activators and coactivators are also recruited. For instance, the coactivator CBP/p300 is associated with both the AhR- and HIF-1-mediated pathways, acting synergistically with the transcription factors and basal transcription machinery (Kobayashi et al. 1997; Bunn et al., 1998; Ebert and Bunn, 1998; Kallio et al., 1998). Therefore, besides potential competition for ARNT, the AhR- and HIF-1-mediated pathways also may interact through squelching of CBP or other shared coactivators.

Unlike in B-1, Hepa 1, and H1L1.1c2 cells, AhR-mediated gene expression in H4IIE-luc cells was not affected by the induction of the hypoxia response pathway. This suggests that the interaction between the AhR and hypoxia pathways might be different among cell lines, tissues, or species (Gassman et al., 1997; Chan et al., 1999). The degree of interaction among ARNT-dependent pathways may depend on the abundance of ARNT in the cells. In H4IIE cells, the ratio of AhR to ARNT is 0.3, compared to 10 in Hepa 1 cells (Holmes and Pollenz, 1997), which indicates that ARNT exists in excess in H4IIE cells. However, to elucidate this hypothesis, more needs to be known about the abundance of HIF-lα in cells. Another explanation for the lack of effect of hypoxia inducers on the induction by TCDD is that H4IIE-luc cells are defective in the hypoxia pathway or insensitive to hypoxia inducers.

In conclusion, there is evidence for a potential mechanism of action for AhR ligand toxicity. In this model, the AhR ligand acts as a disruptor of multiple gene regulation pathways through its recruitment and sequestering of ARNT by activated AhR. TCDD-like PCBs, by altering the transcription regulation network, can cause diverse toxicological effects by changing the relative and absolute rates of transcription of a variety of genes expressed under normal physiological conditions. This model offers a plausible explanation for the wide range of toxic effects of PCB. For example, VEGF, a growth factor that regulates angiogenesis, is regulated by HIF-1α and ARNT in response to oxygen availability. ARNT- and HIF-1α-null mice cannot survive gestation due to defects in vasculature development (Maltepe et al., 1997; Iyer et al., 1998). TCDD causes reduced vasculature in developing chicks and medaka (*Oryzias latipes*; teleost fish) (Cantrell et al., 1996; Walker et al., 1997). TCDD is also known to elicit a wasting syndrome, which is partially attributed to downregulation of glucose transporters (Matsumura, 1995). At least one of the glucose transporters, GLUT-1, is regulated through HIF-α and ARNT (Ebert and Bunn, 1998). Furthermore, this hypothesis also could explain the wide range of sensitivities among tissues and species. As more genes regulated by ARNT are discovered and more ARNT-interacting transcription factors are characterized, it should be possible to offer better and more complete explanations of TCDD toxicity.

Nongenomic Effects of PCBs in Signal Transduction Pathways

As has been discussed, the major mechanism of action for the toxicity of PCBs is related to their ability to bind to and activate the AhR, which is a cytosolic, ligand-activated transcription factor (figure 11-2; Poland and Knutson, 1982). The most characterized pathway involves translocation of the activated, cytosolic AhR to the nucleus, where it binds with ARNT to form a heterodimer, which subsequently binds to DREs. Binding of the heterodimer results in modulation of transcription of genes that contain one or more DRE(s). However, other pathways have been suggested that involve AhR-regulated gene expression but differ in the biochemical events after activation of the AhR (Blankenship and Matsumura, 1994; Delescluse et al., 2000). For example, an AhR-dependent pathway has been characterized involving protein phosphorylation that can occur in the absence of DNA binding of the AhR–ARNT complex, and thus this pathway can be considered a DRE-independent pathway. This DRE-independent pathway has the potential to modulate gene transcription through a variety of mechanisms, such as altering DNA-binding activities of transcriptional factors (Enan and Matsumura, 1995), modulating the responsiveness of growth factor receptor signaling pathways (Madhukar et al., 1984; Abbott and Birnbaum, 1990), and inducing membrane translocation and activity of tyrosine kinase pp60src (Blankenship and Matsumura, 1997a,b; Kohle et al., 1999), without necessarily requiring DREs in the promoter region of the affected genes. Other mechanisms such as sequestration of common accessory factors or coactivator proteins used by other transcription factors have been postulated to represent potentially significant pathways leading to changes in gene transcription. An example of this, recruitment of ARNT away from other ARNT-requiring pathways, is discussed later in this chapter.

The transcriptional response of cells to TCDD has been characterized using microarray hybridization approaches, and TCDD caused at least a twofold change in the expression of 310 genes (Puga et al., 2000). Of these 310 genes affected by TCDD, 108 were considered primary effects of TCDD because they were affected in the presence of cycloheximide, and thus protein synthesis was blocked. It remains to be determined how many of these genes are modulated by the mechanism relating to the central dogma of AhR activation involving binding of activated AhR complex to DREs. Most of the genes regulated through DREs are related to metabolic pathways and include some genes responsible for synthesis of the cytochrome P450 enzymes (i.e., CYP1AI, CYP1A2, etc.), glutathione S-transferase, UDP-glucuronosyltransferase, and aldehyde dehydrogenase. However, it does not appear that metabolic enzyme induction is sufficient to explain the pleiotropic toxic effects of AhR ligands such as changes in cell growth and/or differentiation, which represent cell-, tissue-, sex-, species-, and developmental stage-specific responses. It is much more likely that multiple mechanisms are responsible for AhR-mediated toxicity. Thus, it remains plausible that additional pathways involving AhR-regulated gene expression, such as changes in protein phosphorylation, may represent biologically significant mechanisms of AhR-mediated toxicity. These additional pathways deserve further attention and investigation not only for TCDD, but for structural analogs such as some of the PCB congeners.

Non–AhR-Mediated Effects of PCBs in Signal Transduction Pathways

The TEF approach, when based solely on AhR-mediated effects, does not address potential non-TCDD-like effects of PCBs. Because only a small portion of the total mass of PCB mixtures are coplanar non-*ortho* congeners that elicit dioxinlike activities (Safe, 1990; Neubert et al., 1992; Birnbaum and DeVito, 1995; Neumann, 1996), the TEF approach based solely on AhR-mediated responses cannot be applied to the risk assessment of non–AhR–mediated toxic effects. The non- and mono-*ortho*-substituted PCB congeners make up a large portion of the total mass of both technical mixtures and weathered mixtures of PCBs (Giesy and Kannan, 1998). Thus, ignoring the non-dioxinlike effects of PCBs could result in underestimating the potential adverse effects of environmental mixtures. If the dioxinlike PCBs are the critical contaminant, then variation among mixtures can be reduced by the TEQ approach. However, if the critical mechanism of action, that occurring at lesser concentrations relative to environmental exposures, is caused by non-TCDD-like compounds, the use of the TEQ approach would not be accurate either.

Nonplanar *ortho*-substituted PCBs elicit a diverse spectrum of non–AhR-receptor mediated toxic responses in experimental animals, including neurobehavioral (Schantz et al., 1992), neurotoxic (Seegal et al., 1990; Kodavanti and Tilson, 1997), carcinogenic (Barrett, 1995), and endocrine changes (Brouwer, 1989; van Birgelen et al., 1992). Recent studies have provided data on the potency and possible mechanisms by which *ortho*-substituted congeners exhibit toxic effects (Kodavanti and Tilson, 1997). In addition, certain metabolites of PCBs have antiestrogenic properties

(Kramer et al., 1997) and cause hypothyroidism and decreased plasma vitamin A levels (Brouwer et al., 1989). These alterations in vitamin A and thyroid hormone (TH) concentrations may significantly modulate tumor promotion and developmental and adult neurobehavioral changes (Ahlborg et al., 1992).

Hormone Status

The most subtle and important biological effects of TCDD and the TCDD-like PCB congeners on wildlife are their effects on endocrine hormones and vitamin homeostasis (Colborn and Clement, 1992). There are some studies on the effects of PCBs on wildlife, but most of the studies of mechanisms have been conducted in animal models, particularly rodents. We present the evidence of effects from these studies, not to demonstrate that they occur in wildlife, but rather to establish plausible hypotheses of mechanisms that might apply to wildlife so that further studies can be conducted.

Effects on Androgens and Estrogens

TCDD and structurally analogous PCB congeners are known to have effects on the actions of both male and female steroid hormones (Mably et al., 1992a,b,c). For instance, TCDD has both estrogenic and antiestrogenic effects in different tissues, depending on timing of exposures during development (Peterson et al., 1993). The induction of the mixed-function oxidase system can also reduce the concentrations of circulating steroid hormones, which can have adverse effects on the reproduction and development of wildlife (Hodson et al., 1992). Furthermore, TCDD can affect hepatic microsomal testosterone hydroxylase activity in birds (Sanderson et al., 1997). This in turn can affect the status of plasma steroid hormones, The TCDD-like PCB congeners, because they can bind to the AhR, are expected to cause the same types of effects on hormone status as does TCDD.

Coplanar PCBs can inhibit estradiol-induced cell clumping in MCF-7 breast cancer cells (Gierthy and Crane, 1984). This suggests that coplanar PCBs are antiestrogenic (based on in vitro bioassays). This inhibition of clumping was due to the metabolism of estradiol to 2- and 4-hydroxy estradiol, which alters the synthesis of biogenic amines (Lloyd and Weisz, 1978; Foreman and Porter, 1980). Similarly, the hydroxy metabolites of PCBs are antiestrogenic (Kramer et al., 1997), which could alter steroid hormone homeostasis and consequently affect neurochemical behavior. Hypothalamic brain tissues convert endogenous estrogens to catechol estrogens (Foreman and Porter, 1980), which have been shown to inhibit tyrosine hydroxylase, a rate-limiting enzyme for dopamine synthesis. Therefore, alterations in endogenous estrogen concentrations and metabolism in animals could modulate biogenic amines in brain, which would subsequently lead to neurotoxic effects.

Thyroid Hormone

Thyroid hormones (T_4; triiodothyronine, T_3) are important both in adult and developing animals (Morse, 1995; see chapters 3 and 4 for discussion of the hypothalamus–pituitary–thyroid axis). The thyroid gland as well as plasma concentrations of TH can be influenced by exposure to coplanar PCBs (McKinney et al., 1985; Brouwer, 1987; Brouwer et al., 1998; Hauser et al., 1998), but ultrastructural examination of the thyroid gland of rats exposed to Aroclor 1254 revealed that the effects are not similar to those caused by stimulation by thyroid-stimulating hormone (TSH) or by iodine deficiency (Collins and Capen, 1980). These observations indicate that PCBs can have direct effects on the thyroid gland.

TCDD and TCDD-like PCB congeners mimic the effects of T_4 as a key metamorphosis signal (McKinney et al., 1985). TCDD has also been shown to downregulate the epidermal growth factor receptor (Newsted and Giesy, 1991, 2000), which may result in disruption of embryonic development at critical stages. Altered concentrations of thyroid and steroid hormones and vitamin A are frequently reported to accompany embryonic abnormalities in wildlife populations exposed to planar PCDHs (Gilbertson et al., 1991). Individuals from these populations have altered sexual development, including demasculinazation (Fry and Toone, 1981; Colborn and Clement, 1992), sexual dysfunction as adults (Colborn et al., 1993), and immune system suppression (Brouwer, 1989; Nebert, 1990). The observations on adult sexual dysfunction are especially significant because young that appear to be normal while raised by exposed parents may become reproductively dysfunctional when they mature (Brunström, 1988). Poor reproductive efficiencies and adventive, opportunistic diseases are characteristic of the wild animals in exposed populations of the Great Lakes region (Beland et al., 1993). Both TH and vitamin A were decreased in the blood of harbor seals (*Phoca vitulina*) when they were fed PCDH-containing fish from the Baltic Sea (Brouwer et al., 1989). Although PCB concentrations were relatively high in these fish, because of the co-occurrence of other residues such as DDTs, PCDDs, and PCDFs, it is impossible to determine the role of PCBs per se. PCBs also decreased plasma concentrations of T_4 in mink (*Mustela vison*) and otters (*Lutra lutra*) fed PCB-containing fish (Heaton et al., 1995a,b; Murk et al., 1998).

PCBs competitively bind to TH-binding proteins (e.g., transthyretin; TTR), such that THs are not bound and are cleared more rapidly from the body. As a result, they can cause a functional TH deficiency resulting in goiter. Thus, the occurrence of goiter might be a good indicator of effects of PCBs on wildlife under field conditions. Goiter, an enlargement of the thyroid gland without hyperthyroidism, is endemic in Great Lakes wildlife (Moccia et al., 1986). Goiter is commonly caused by an iodine deficiency. However, chemicals or defects that interfere with TH synthesis, metabolism, or regulation can also cause the disease. Goiter can have a histological appearance that varies depending on its cause. Goiters caused by iodide deficiency usually exhibit epithelial follicle cells that are cuboidal to columnar, and the colloid is reduced and pale staining. When there is a relative lack of THs in the body, thyrotropin (TSH) can stimulate thyroid hypertrophy and hyperplasia. In contrast, goiters caused by thyrotropin-releasing hormone (TRH) or TSH deficiency are characterized by squamous follicle cells and increased colloid accumulation. Whereas mammalian goiter caused by PCBs is a hyperplastic goiter resembling that caused by iodide deficiency, the histological picture in birds is quite different. Thyroid weight, follicle size, and colloid are all increased in PCB-exposed birds, and epithelial cells are flattened. At high PCB doses, thyroid atrophy is observed (Jefferies and Parslow, 1976).

Effects of PCBs on Vitamin A

Vitamin A (retinal, a precursor for retinoic acid) is important in many functions in animals, such as embryonic development, vision, maintenance of dermally derived tissues, immune competence, hemopoiesis, and reproductive functions (Brouwer, 1987). One of the retinoic acid receptors (RXR) forms a heterodimer with TH receptors as a normal part of the mechanism of action for THs (see chapter 4). Changes in the status of vitamin A in the plasma or liver may be responsible for the birth defects observed in birds that have been exposed to planar PCDH. TCDD-like coplanar PCDHs, including PCBs, affect concentrations of vitamin A in both the blood and liver of exposed organisms. These effects are thought to be due to at least two processes. In blood, some of the OH-PCBs bind to TTR (Brouwer et al., 1989). In the liver, induction of hepatic enzymes such as acyl-CoA–retinol acyltransferase and UDPGT is thought to alter the metabolic pathways involved in the storage and mobilization of vitamin A and result in the observed depletion of retinols in the liver. Laboratory studies have determined that both vitamin A and its storage form in the liver (retinal palmitate) were depleted in birds exposed to sublethal doses of the TCDD-like coplanar PCB congener no. 77 (Spear and Moon, 1985).

Neurobehavioral Effects of PCBs

There is no information about neurobehavioral effects of PCBs in wildlife, but developmental and cognitive dysfunctions have been observed in children exposed to PCBs either *in utero* or via breastfeeding. Details of the evidence linking prenatal exposure to PCBs with neurobehavioral developmental deficits epidemiological studies of children are given elsewhere (Kodavanti et al., 1994; Maier et al., 1994; Eriksson and Fredriksson, 1996a,b; Chishti et al., 1996; Morse et al., 1996a,b; Gasiewicz, 1997; Jacobson and Jacobson, 1997; Wong et al., 1997). The neurotoxic effects observed in epidemiological studies of children born to mothers who had been accidentally exposed to PCBs in rice oil such as that consumed at Yusho and Yu-Cheng or where mothers consumed PCBs from Great Lakes fish or from a nonfish diet indicated that the developing nervous system is sensitive to PCBs (Kuratsune et al., 1972; Rogan and Gladen, 1992; Seegal and Schantz, 1994; Huisman et al., 1995; Lonky et al., 1996; Seegal, 1996). The correlation between the presence of AhR-mediated effects, such as chloracne or hyperpigmentation, and observed cognitive dysfunctions was poor or nonexistent. This observation suggested the possibility that the alterations in neurological function in the Yusho and Yu-Cheng children may be due to exposure to nonplanar *ortho*-substituted PCB congeners present in many commercial mixtures of PCBs, rather than the coplanar contaminants that interact with the AhR (Yu et al., 1998; Rogan and Gladen, 1992). Concentrations of PDCFs were especially great in the contaminated rice oil, thus, it is impossible to determine if the effects were primarily due to AhR-mediated effects or to other possible mechanisms of non-AhR-active PCB congeners or both.

Modulation of TH status in the developing organism or, in the case of viviparous organisms, the adult may result in neurobehavioral deficits in the developing organism (Morse et al., 1995).

The only controlled laboratory study of the neurobehavioral effects of contaminants in fishes found that rats that ate fish from Lake Ontario exhibited abnormal responses in certain behavioral tests (Daly, 1993). However, in these cases, individuals were exposed to co-occurring contaminants, including organochlorine pesticides and mercury as well as PCBs, PCDDs, and especially PCDFs.

Several studies have provided data on the potency and possible mechanisms by which *ortho*-substituted PCB congeners exhibit toxic effects (Seegal, 1996; Kodavanti and Tilson, 1997; Tilson et al., 1990). Aroclor 1254 reduces cellular dopamine, conent of pheochromocytoma (PC12) cells (Greene, and Rein, 1977; Kittner et al., 1987; Seegal et al., 1989). This is a continuous cell line derived from a rat adrenal-gland tumor that can synthesize, store, release, and metabolize biogenic amine neurotransmitters, including dopamine, in a manner similar to that of the mammalian central nervous system. Subsequent studies using a continuous mouse neuroblastoma cell line (NIE-N115), deficient in L-aromatic amino acid decarboxylase, which converts L-DOPA (L-dihydroxyphenylalanine) to dopamine, demonstrated that exposure to 2,2'-diCB (PCB 4) resulted in a significant decrease in media concentrations of L-DOPA, suggesting decreased activity of the rate-limiting enzyme for dopamine synthesis, tyrosine hydroxylase. Additional studies examined the relationship between the structure of individual PCB congeners and their ability to alter PC12 cellular dopamine content.

About 50 individual PCB congeners have been tested for their ability to reduce cellular dopamine content in PC12 cells (table11-1). Di-*ortho*- through tetra-*ortho*-substituted congeners were the most potent at reducing cellular dopamine, whereas coplanar PCB congeners were ineffective (Shain et al., 1991). In addition, chlorine substitution in a *meta* position decreased the potency of *ortho*-substituted congeners, but *meta* substitution had little effect on congeners with both *ortho* and *para* substitutions. Further experiments with PCB 4, a di-*ortho*-substituted congener that was potent in decreasing dopamine concentrations in vitro, indicated that the active agent was not a metabolite (Shain et al., 1991). Investigations of the effects of various PCB congeners on Ca^{2+} homeostasis and protein kinase C (PKC) translocation in cerebellar granule cells found a similar structure–activity relationship, which showed that *ortho*-substituted PCBs alter Ca^{2+} homeostasis in brain, but AhR-active congeners were inactive (Kodavanti et al., 1995). Based on these studies, two different PCB binding sites in the brain were suggested. Further, the effect of *ortho*-substituted PCBs in reducing dopamine levels in brain were additive (Seegal et al., 1990). As an example, a mixture of PCB congeners 2,4,4'- (PCB 28), 2,2',4,4'- (PCB 47) and 2,2',5,5'- (PCB 52) was more potent in reducing brain dopamine content than equal amounts of each congener in in vitro systems (Seegal et al., 1990). These results suggest that PCB congeners predicted to have little dioxinlike activity, based on structural configuration, decrease dopamine levels in the nervous system and that neurotoxicity might be due to a mechanism independent of AhR activation.

Neurotoxicological effects of individual PCB congeners and technical PCB mixtures in laboratory animals and in in vitro studies and their effects and effective

Table 11-1. EC_{50} values for PCB-congener mediated decreases in dopamine content in PC-12 cells in vitro and [^{3}H]phorbol ester binding in rat cerebellar granule cells and IC_{50} values for microsomal labeled Ca^{2+} uptake in rat cerebellar granule cells.

Congener/PCB Mixture	IUPAC No.	EC_{50} (Dopamine Content; μM)	EC_{50} ([^{3}H]phorbol Binding; μM)	IC_{50} (Ca2+ Uptake; μM)
2,2'-	4	64	43	80 (62)[a]
2,2',4,6-	50	71	41	7.3
2,3',4,6-	69	78	na	na
2,2',4,6,6'-	104	93	38	5.5
2,2',5-	18	82	na	na
2,2',5,5'-	52	86	28	4.9
2,2',4,5'-	49	97	na	na
2,2',4,4'-	47	115	89	5.8
2,6-	10	106	na	na
2,4,4',6-	75	118	na	na
2,2',3,3',4,4',6-	171	134	na	na
2,4,6-	30	150	na	na
2,2',3,5'-	44	114	na	na
2-	1	182	na	na
2,4,4'-	28	196	>100	6.9
2,4-	7	200	na	na
2,4'-	8	200	na	na
2,2',4,4',6,6'-	155	156	na	na
2,2',4,4',6-	100	158	na	na
2,2',4,5',6-	103	157	na	na
2,3-	5	173	na	na
2,4',5-	31	176	na	na
2,3'-	6	173	na	na
2,3',5-	26	161	na	na
2,3,6-	24	160	na	na
3,4-	12	169	na	na
3,3'-	11	195	60	13
2',3,4-	33	185	na	na
4-	3	335	na	na
2,3',4,4'-	66	>200	na	na
2,3',4-	25	>200	na	na
2,2',3,3'-	40	>200	na	na
2,3,4-	21	>200	na	na
2,2',3,4,4',5,6-	181	370	na	na
2,5-	9	>200	na	na
3,5-	14	>200	74	17
3-	2	300	na	na
3,4'-	13	410	na	na
3,4',5-	39	310	na	na
4,4'-	15	>1000	>100 (NEO)	>100 (NEO)
2,2',6,6'-	54	>1000	>100 (NEO)	>100 (NEO)
3,3',4,4'-	77	>1000	>100(NEO)	>100(NEO)
3,3',4,4',5-	126	>1000	>100 (NEO)	>100 (7.6)
2,2',6-	19		58	7
2,2',4,6'-	51		50	2.4
3,3',5,5'-	80		72	>100

(*continued*)

Table 11-1. Continued

Congener/PCB Mixture	IUPAC No.	EC_{50} (Dopamine Content; μM)	EC_{50} ([^3H]phorbol Binding; μM)	$1C_{50}$ (Ca2+ Uptake; μM)
2,3,3',4,4'-	105		95	5.3
2,3',4,4',5-	118		>100	6.6
2,2',3,3',4,4'-	128		>100	4.9
2,2',3,3',5,5'-	133		>100	5.1
2,2',3,3',6,6'-	136		58	6.3
2,2',4,4',5,5'-	153		>100	6.6
2,3,3',4,4',5-	156		>100	5.4
3,3',4,4',5,5'-	169		>100 (NEO)	>100 (NEO)
2,2',3,4,4',5,5'-	180		>100 (NEO)	4.8
Aroclor 1016			71	6.8
Aroclor 1254			56	6.3
Aroclor 1260			>100	7.6

From Shain et al. (1991); Kodavanti et al. (1995, 1996). Only mean values are presented. NEO, no effect observed up to 100 μM; na: Data not available.

[a]Values in parentheses are from Kodavanti et al.*(1993a).

doses have been compiled (Giesy and Kannan, 1998). These data suggest that *ortho*-substituted PCB congeners are potential neurotoxicants when exposure is prenatal. Neurotoxicological effects may be of considerable significance after neonatal exposure and acute accidental exposures. In general, developmental exposure to PCB mixtures or congeners could alter motor activity (Schantz et al., 1992), neurological development, and cognitive function (Holene et al., 1995) in offspring. After acute exposure to PCBs in mice, altered content of neurotransmitters such as dopamine is associated with neurobehavioral changes (Seegal et al., 1985, 1989). Studies conducted in adult nonhuman primates also suggested that *ortho*-substituted congeners are capable of reducing brain dopamine concentrations. Adult pig-tailed macaques (*Macaca nemestrina*) were exposed to Aroclor 1016 or Aroclor 1260 at doses of 0.8, 1.6, or 3.2 mg/kg body weight per day for 20 weeks. Reductions in dopamine content were observed in brain areas that synthesize dopamine. The congeners that accumulated in the brain were 2,4,4'- (PCB 28), 2,2',4,4'. (PCB 47), and 2,2',5,5'- (PCB 52). The dopamine-reducing effects of PCBs persisted even after exposure was terminated, suggesting that the neurobehavioral alterations after exposure to *ortho*-substituted PCBs may be long-term and irreversible (Seegal et al., 1994). These results also imply that the potency of *ortho*-substituted PCBs in reducing brain dopamine may be species specific. Similarly, the effects of PCBs on learning in rats are sex specific, with females being more sensitive than males (Schantz et al., 1992). Neurotoxic effects were prominent after prenatal exposure, whereas adults were relatively less susceptible to neurotoxic effects following exposure to PCBs (Seegal, 1996).

In addition to reducing dopamine, *ortho*-substituted PCBs alter the translocation/activation of PKC and intraneuronal sequestration of Ca^{2+} in brain cerebellar granule cells (Kodavanti et al., 1993a,b, 1994, 1995, 1996). *Ortho*-substituted congener 2,2',3,5',6- (PCB 95) altered microsomal Ca^{2+} transport by interfering with the ryanodine receptor in rat brain (Wong et al., 1997). Similarly, 2,2'-diCB interfered with oxidative phosphorylation by inhibiting mitochondrial Mg^{2+}-ATPase activity in mitochondrial and synaptosomal preparations of rat brain (Maier et al., 1994). Alterations in hormone levels, including THs, may be responsible for PCB-induced neurotoxicity (Seegal, 1996). Particularly, alterations in hormone levels during early development of animals or humans may have long-term consequences on the behavior and neurochemistry of the animal in adulthood (Seegal, 1996). Because of these multiple mechanisms of action, a simple quantitative structure–activity relationship for the neurotoxic effects of PCBs may not be possible with the limited data available. Because no ReP values for PCB congeners to cause neurotoxic effects were available, Giesy and Kannan (1998) did an analysis to determine if PCBs causing AhR-mediated effects or those causing non-AhR-mediated effects were likely to be the critical contaminants in ecological risk assessments. Giesy and Kannan sought to determine the mechanism that, based on the RePs to cause various effects and the relative concentrations of congeners in weathered mixtures, would exceed the threshold TRV for that effect by the greatest degree and thus result in the greatest hazard quotient (HQ). That analysis found that ecological risk assessments based upon AhR-mediated effects would result in the least allowable exposure to the total mixture and would be more protective of wildlife than assessments based upon non-AhR-mediated effects.

In addition to *ortho*-substituted PCBs, the non-*ortho* coplanar congener 3,3',4,4'-PCB 77 alters dopamine concentrations, depending on the species, the developmental status of the animal at the time of exposure, and the dose (Agarwal et al., 1981; Eriksson et al., 1991; Chishti and Seegal, 1992). Early postnatal exposure of mice to non-*ortho* coplanar congeners has been observed to alter cholinergic function (Eriksson et al., 1991; Eriksson and Fredriksson, 1996a,b, 1998). A decrease in cellular dopamine concentrations in PC12 cells after exposure to 3,3',4,4',5- PCB 126 has been reported (Angus and Contreras, 1996). Although the decrease in dopamine by PCB 126 in this study was attributed to cytotoxicity, this implies that the non-*ortho* PCBs could be lethal to cells at concentrations that are neurotoxic for certain *ortho*-substituted PCBs.

A few studies have examined behavioral alterations in rats after eating contaminated fish from the Great Lakes (Hertzler, 1990; Daly, 1993). Rats fed different rations of Great Lakes fish (8, 15, and 30% of the diet) for 20 days exhibited behavioral alterations. The effects included reduced exploratory activity and decreased rearing and nose-poke behavior in comparison with controls. PCB concentrations in the fish were eaten in the range of 4–19 µg/g wet weight, and total PCB concentrations in rat brain after the exposure period were 50–78 ng/g wet weight. In contrast, there was no significant effect in behavioral measures after a 90-day subchronic exposure to PCB-contaminated Great Lakes fish, although the accumulation of *ortho*-substituted congeners such as 2,2',4,4'- (PCB 47), 2,2',5,5'- (PCB 48), 2,2',4,4',5,5'- (PCB 153), 2,2',5,5'- (PCB 52), 2,4,4',5- (PCB 74), and 2,2',4,5'- (PCB 49) was in the range of 2.5–18 ng/g wet weight in the brains of the rats (Beattie et al., 1996). Confounding factors in these studies could be the presence of other contaminants such as methyl mercury in the diet. Synthetic pyrethroids, organophosphorus pesticides, organometallics such as tributyltin and methyl mercury, and aluminium have been shown to alter Ca^{2+} homeostasis and neurobehavioral responses in exposed laboratory animals (Evangelista de Duffard 1996; Kodavanti et al., 1993a,b,). Thus, the results of these studies are considered equivocal.

Immunological Effects of PCBs

The immune system is a target for toxicity of compounds that act through the AhR, including non-*ortho* PCBs, and the immune system is closely linked with the endocrine system (see chapters 4–6). This evidence has been derived from numerous studies in various animal species, primarily rodents, but also guinea pigs, rabbits, monkeys, fishes, and birds. However, because of the widely differing experimental designs, exposure protocols, immunologic assay, and endpoints, it has been difficult to define a PCB-induced immunotoxic syndrome in a single species, let alone across species. Several studies have established immunotoxic effects of PCBs (see reviews by Kerkvliet, 1994; Fox and Grasman, 1999), but other studies suggest lack of effects of PCBs on the immune system (Fowles et al., 1997; Omara et al., 1997; Rice et al., 1998; Hutchinson et al., 1999). The suppressive effects of PCBs on the human immune system remain controversial because of the conflicting data reported from different laboratories, particularly in in vitro studies (Nagayama et al., 1998; Yu et al., 1998). Furthermore, it is not clear that suppression observed in the laboratory of one or more immune system parameters has a consequence in nature.

Thymic involution is one of the hallmarks of exposure to halogenated aromatic hydrocarbons (HAHs) in all species examined (Kerkvliet, 1994). Because the thymus plays a critical role in the prenatal development of T lymphocytes, HAH-induced thymic atrophy is referred to an immunotoxic effect. However, although an intact thymus is crucial to the development of the T-cell receptor repertoire during the prenatal and early postnatal periods in rodents as well as in humans, the physiological role played by the thymus in adult life has not been established (Kerkvliet, 1994). Exposure to PCBs during the pre-/neonatal stages in rodents resulted in reduction in the number and function of T lymphocytes, but little effect was noticed in adult animals. The mechanism for PCB-induced thymic atrophy has not been elucidated and may involve multiple mechanisms (Kerkvliet, 1994). Immunotoxicity was related to AhR-dependent activity in mice, although such relationships are not available for low-level, chronic exposures (Kerkvliet, 1994). Studies have shown that TCDD was the most immunotoxic of HAHs, and coplanar PCBs have similar, albeit less severe, effects. Studies also have shown that *ortho*-substituted PCBs can affect the immune system through various mechanisms. Rat neutrophils exposed to a di-*ortho*-substituted PCB congener (2,2',4,4'-) altered neutrophil function by both stimulating degranulation and the production of superoxide anion (Brown and Ganey, 1995). Perturbation of Ca^{2+} homeostasis in human granulocytes exposed to *ortho* PCBs resulted in the activation of phospholipase C, which is suggested as a mechanism for immune suppression (Voie and Fonnum, 1998).

In addition to several studies that documented immunotoxic effects of PCBs in laboratory mammals, PCBs have been shown to cause lymphoid depletion in chicks (Andersson et al., 1991) and impairment of resistance to hepatitis virus in PCB-treated ducks (Friend and Trainer, 1970). The nonspecific immune responses by phagocytes in earthworms, including their role in wound healing, decrease dramatically after exposure to Aroclor 1254 (Ville et al., 1995).

The difficulty in demonstrating consistent, direct effects of PCBs *in vitro* on lymphocytes, the dependence of those effects on serum components, and the requirement for high concentrations of PCBs, are all consistent with an indirect mechanism of PCBs on the immune system. One potentially important indirect mechanism is via PCB effects on the endocrine system, especially the hypothalamus–pituitary–adrenal axis (see chapter 5).

Methods for Evaluating PCB Exposure

Toxic Equivalency Factor Approach

One approach to congener-specific hazard assessment of complex mixtures of PCBs that are different from the original technical mixtures is to develop RePs for individual congeners. In this way the concentration of each congener can be corrected for its toxicity and the products of the corrected concentrations summed to give an index of the toxicity of the entire altered mixture. This method requires knowledge of and depends on the mechanism of action of the various PCB congeners. If each congener causes different toxic responses and acts via independent mechanisms, then the relative toxicities of every congener must be determined separately. Due to the highly complex nature of PCB mixtures in environmental and biological samples, this would be a daunting, if not impossible, task (Sanderson and Giesy, 1998).

The structure–receptor-binding relationships for different classes of PCDHs, including certain PCB congeners, have been developed using [^{3}H]2,3,7,8-TCDD as the radioligand and rat hepatic cytosol as a source of the AhR (Poland and Knutson, 1982). TCDD is the ligand that has the greatest affinity for the AhR, and it is also the most toxic member of the PCDHs. Congeners of PCDDs, PCDFs, and PCBs that are structurally similar to TCDD cause similar effects but with varying potencies. Based on studies that indicated the pivotal role of the AhR in mediating most, if not all, of the toxic and biochemical effects induced by PCBs, PCDDs, and PCDFs and the structure–activity relationships, a TCDD TEF approach has been developed (Safe 1990; van den Berg et al., 1998). This approach allows the toxic potential of a complex mixture of individual congeners to be expressed as one integrated parameter, TEQ. In a TEQ, the toxic potency of the mixture corresponds to the potency of the most toxic congener, TCDD. In this way, definitive studies of TCDD for several species and endpoints can be used to derive a maximum allowable toxicant concentration (MATC). If relative potencies can be derived for PCB congeners for a few endpoints and species that are intercorrelated, and if congeners can be established to have the same rank-order among endpoints and species, the relative potencies can be used to develop a TEF for each congener.

As an example of the technique, if the ED_{50} values for iminunosupressive activity of TCDD and 1,2,3,7,8-penta-CDD were 1 and 2 mg/kg, respectively, then the TEF for the latter compound would be the ratio ED_{50} (TCDD) to ED_{50} (1,2,3,7,8-

pentaCDD), or 0.5. TEF values have been determined for several different AhR-mediated responses. However, for every PCB congener tested, the TEF values are response- and species-dependent (Safe, 1990). As an example, TEFs for 2,3,7,8-TCDF obtained from in vivo and in vitro studies varied from 0.17 to 0.016 and 0.43 to 0.006, respectively (Safe, 1994).

There are some limitations to the determination of relative potency factors due to violations of the assumptions of dose–response relationships (Villeneuve et al., 2000). Regulatory agencies have chosen consensus TEF values for individual congeners (van den Berg et al., 1998). TEFs are different from RePs, in that they are consensus values used in risk assessments. Thus, TEFs are based on RePs, a number of species and endpoints, and are meant to overestimate the actual potency of each congener. For instance, the greatest ReP was selected from all of the species- and endpoint-specific RePs and often rounded up to the nearest tenth. Selection criteria have been based on the importance of data obtained for specific responses (e.g., carcinogenicity, reproductive and developmental toxicity).

Although there is an inherent uncertainty in the TEQ approach due to potential species-specific differences in the relative toxicity of PCBs and TCDD and deviations from a simple additive model (Villeneuve et al., 2000), it is still useful for ecological risk assessments (Sanderson and Giesy, 1998). The currently used World Health Organization (WHO) TEF values are intended only for use as order-of-magnitude estimates of risk, and not as predictors of actual species-specific toxic responses (table 11-2; van den Berg et al., 1998). These values are tentative and will be updated when more data are available. Some of the TEFs for fish and birds were derived mainly from in vitro studies, and more in vivo studies are needed to validate avian and teleost TEFs. Despite these acknowledged limitations, the TEQ approach is an effective way to predict the overall toxicity of complex mixtures of TCDD-like chemicals. TEQs can be estimated for any sample for which

Table 11-2. TCDD equivalency factors (TEFs) for several dioxinlike PCB congeners for fish, birds, and mammals (from van den Berg et al., 1998).

PCB Congener (IUPAC No.)	Fish TEF	Bird TEF	Mammal TEF
2,3,7,8-TetraCDD	1	1	1
3,3',4,4'-TetraCB (77)	0.0001	0.05	0.0001
3,4,4',5-TetraCB (81)	0.0005	0.1	0.0001
3,3',4,4',5-PentaCB(126)	0.005	0.1	0.1
3,3',4,4',5,5'-HexaCB (169)	0.00005	0.001	0.01
2,3,3',4,4'-PentaCB (105)	<0.000005	0.0001	0.0001
2,3,4,4',5-PentaCB (114)	<0.000005	0.0001	0.0005
2,3',4,4',5-PentaCB (118)	<0.000005	0.00001	0.0001
2',3,4,4',5-PentaCB (123)	<0.000005	0.00001	0.0001
2,3,3',4,4'5-HexaCB (156)	<0.000005	0.0001	0.0005
2,3,3',4,4',5'-HexaCB (157)	<0.000005	0.0001	0.0005
2,3,4,4',5,5'-HexaCB (167)	<0.000005	0.00001	0.00001
2,3,3',4,4',5,5'-HeptaCB (189)	<0.000005	0.00001	0.0001

congener-specific PCDH concentrations and either endpoint- and species-specific RePs or a TCDD equivalency factors (TEFs) are available. TEQ concentrations for a sample containing PCDH congeners can be calculated using the equation

$$\text{ng TEQ/kg sample (lipid)} = \sum_{i=1}^{n} \left[\text{PCDH}_i \times \text{TEF}_i \right] \qquad (11\text{-}1)$$

where TEQ = equivalent TCDD concentration; PCDH = AhR-active PCDH; n = any PCDH congener, and TEF = TCDD equivalency factor (based on RePs).

The TEF approach was first used to assess the risks associated with air emissions of PCDDs and PCDFs formed during high-temperature incineration of industrial and municipal waste (Giesy and Kannan, 1998). Subsequently, the U.S. EPA proposed interim guidelines for estimating risks associated with mixtures of PCDDs and PCDFs for other media as well. Several international agencies have also adopted the TEF approach for assessing risks of PCDDs and PCDFs (Olie et al., 1983; Ahlborg, 1989; Kutz et al., 1990; Barnes, 1991). The mechanistic considerations for development of TEFs for risk assessment of PCBs have been described elsewhere (Safe, 1994, 1990). A brief description of the development of TEFs using mammalian models (Safe, 1994) and the recent progress in studies relating to fish- and bird-specific TEFs for PCBs are presented below.

The toxic potencies derived from in vivo and in vitro assays for coplanar PCBs are variable and depend on both the species (rat, mouse, monkey) and endpoint (Safe, 1994). For example, the toxic potency ratios of PCB 126: TCDD for different responses were 66 (body weight loss, rat); 8.1 (thymic atrophy, rat); 10 (mouse fetal thymic lymphoid development); 125 (AHH induction, rat) and 3.3 (AHH induction in H4IIE cells). Based on these toxicity data, TEFs in the range of 0.008–0.3 could be derived, depending on the species and endpoint selected. A consensus TEF for mammals of 0.1 was assigned to this congener (van den Berg et al., 1998).

Similar to non-*ortho* coplanar PCBs, chlorobiphenyl congeners with chlorine substitution at only one *ortho* position (mono-*ortho* PCBs) can achieve partial coplanarity and also exhibit AhR agonist activity. Based on the potency of PCB congeners relative to TCDD for several Ah-mediated responses in in vivo and in vitro mammalian models, TEFs have been proposed for non-*ortho* and mono-*ortho* PCBs (Safe, 1994; van den Berg et al., 1998). As mentioned earlier, the potency ranges of these congeners varied by two to three orders of magnitude depending on the species and the endpoint used to derive values. Data that were considered in determining TEF values were prioritized based on in vivo studies being given greater weight than in vitro studies, and effects that are clearly adverse were given more strength than biochemical changes. In mammalian models, long-term in vivo exposures were given more weight than acute exposure studies. Currently, TEF values assigned to PCB congeners are tentative and subject to modification as new data become available. Recognizing the need for a more consistent approach for setting internationally accepted TEFs, the WHO European Centre for Environment and Health (WHO-ECEH) and the International Program of Chemical Safety (IPCS) initiated a project in the early 1990s to create a database relevant to setting TEFs and to assess the relative potencies and derive consensus TEFs for halogenated aromatics (Ahlborg et al., 1994). The first international TEFs for effects of dioxinlike

PCBs on mammals were proposed in 1994 and have been revised and updated (table 11-2; van den Berg et al., 1998).

Applications of the TEF Approach

TEFs have been used to assess the risk associated with mixtures of PCB congeners measured in biota or environmental matrices. This is done by multiplying the concentration of each non- or mono-*ortho* congener detected in the biota by the corresponding TEF to yield a TCDD equivalent concentration, or TEQ (table 11-3; Tanabe et al., 1989). A total TEQ for all toxic congeners in the sample can be calculated by summing all of the individual TEQs. TEFs were first used to determine non-*ortho* coplanar PCB-derived TEQs (Tanabe et al., 1987a,b, 1989; Kannan et al., 1989) to compare dioxinlike activities in environmental samples with those obtained for PCDDs/PCDFs, using TEFs derived from the relative potencies of PCB congener-induced AHH and EROD activities in H4IIE cells (Sawyer and Safe, 1982). Concentrations of TEQs contributed by PCBs in most extracts from environmental samples or human tissues exceeded the TEQs contributed by PCDDs/PCDFs in these same extracts (Kannan et al., 1989; Tanabe et al., 1989). Comparable results have been obtained in other studies (Kubiak et al., 1989; Tarhanen et al., 1989; Asplund et al., 1990; Dewailly et al., 1991; Harris et al., 1993, 1994).

The utility of the TEF approach to environmental risk assessment is shown by the correlation between total TEQs and adverse effects in populations of birds ((Tillitt et al., 1992, 1993; Giesy et al., 1994a,b). A negative correlation was reported between the incidence of deformities in double-crested cormorant (*Phalacrocorax auritus*) populations from the North American Great Lakes and the total TEQ in their eggs (Yamashita et al., 1993). Poor hatching success in a population of Forster's tern (*Sterna forsteri*) was directly correlated with TEQs in the eggs (Kubiak et al., 1989). In a laboratory and field study on populations of common terns (*Hydroprogne caspia*) from the Netherlands, egg volume was negatively correlated with total TEQ in the egg yolk (Boseveld and van den Berg, 1994). A weak negative correlation

Table 11-3. An example for deriving 2,3,7,8-TCDD equivalents (TEQs) by the TEF approach.

Congener	TEF	Concentration (pg/g, Wet Weight)	TEQ (pg/g, Wet Weight)
Dioxins			
2,3,7,8-tetraCDD	1	3.7	3.7
1,2,3,7,8-pentaCDD	1	6.4	6.4
1,2,3,4,7,8-hexaCDD	0.1	3.9	0.39
1,2,3,6,7,8-hexaCDD	0.1	34	3.4
1,2,3,7,8,9-hexaCDD	0.1	5.7	0.57
1,2,3,4,6,7,8-heptaCDD	0.01	33	0.33
OCDD	0.0001	510	0.051
Furans			
2,3,7,8-tetraCDF	0.1	3.1	0.31

was observed between total TEQs and survival of early life stages in populations of lake trout (*Salvelinus namaycush*) from the North American Great Lakes (Wright and Tillitt, 1999).

Limitations of TEF Approaches

Interactive Effects Despite the ability of the TEF approach to predict the potency of some mixtures of planar HAHs, there are limitations to its application (van den Berg et al, 1998). The assumption that toxic responses to planar HAHs are additive and that other classes of contaminants do not modify or add to the toxicity may or may not be valid (Giesy et al., 1994a; Giesy and Kannan, 1998). Both additive and nonadditive interactions among planar HAHs have been observed (Safe, 1994). Data from rodent studies indicate that toxic responses to mixtures of planar PCDHs are additive (Sawyer and Safe, 1982; Pluess et al., 1988). However, there are other rodent data showing either less-than-additive (antagonistic) responses (Haake et al., 1987; Biegel et al., 1989) or greater-than-additive (synergistic) responses (Birnbaum et al., 1985; Bannister and Safe, 1987). Studies have also reported both additive (Walker et al., 1996) and other interactive effects (van Birgelen et al., 1992, 1996; Harper et al., 1995; Pohjanvirta et al., 1995; Li, 1996a,b) of planar congeners in experimental animals or in cell lines derived from various animals, including mammals and fishes. Although estimated TEQs based on instrumental analyses do not account for these interactions, bioassay-derived TCDD-EQs integrate potential additive and nonadditive interactions among AhR agonists and between AhR agonists and other compounds by measuring a final receptor-mediated response (Tillitt et al., 1991; Sanderson et al., 1998). Comparison of bioassay-derived TCDD-EQs with those 'of instrumental TEQs estimated for the same samples also suggests the existence of both nonadditive and additive interactions in biota (Tillitt et al., 1992; William and Giesy, 1992). Details about bioassays and their applications in risk assessment are presented below. The exclusion of nonadditive mixture interactions in the present TEF approach has been justified because (1) the antagonistic or synergistic effects are observed only at high dose levels, and the magnitude of these interactions is smaller than the uncertainties already present in the TEF values; (2) the observed nonadditive effects are highly species, response, and dose dependent, and their relevance might be of minimal importance; and (3) the mechanisms responsible for these nonadditive effects are unknown (Ahlborg et al., 1992). In general, complex mixtures of PCBs are slightly less than additive, so the TEQ of an additive model is conservative (protective).

Species- and Endpoint-Specific Variations The TEF approach assumes that the rank order of relative potencies of congeners are the same among species. However, there are quantitative differences in the relative potencies of PCB congeners among species and endpoints. There are considerable variations in the potency of mono- and non-*ortho* PCBs among, as well as within, mammalian, teleostean, and avian models (Jenz and Metcalfe, 1991). The application of mammalian-derived TEFs from rodent bioassays for the assessment of risks in other mammals (e.g., dolphins, whales) may not be appropriate due to differences in the responsiveness of these animals to

PCB congeners. Similarly, the differences in the potencies of PCB congeners for various endpoints lead to a range of relative potency values from which a congener-specific TEF is derived. Therefore, the predictive ability of the TEF approach is species and endpoint dependent (Safe, 1994; Metcalfe and Haffner, 1995; Seed et al., 1995). Uncertainties of a few orders of magnitude between species and for specific endpoints are a major drawback in using the TEF approach in risk assessment. Once TEFs are established, total class-specific concentrations of TEQ can be calculated in a tissue of interest. However, to conduct a risk assessment, this exposure concentration needs to be compared to a species-, tissue-, and endpoint-specific toxicant reference value (TRV). It is important that the measure of exposure be compared to the appropriate TRV. For instance, because of changes in the absolute and relative concentrations of individual congeners between trophic levels, a dietary TRV cannot be compared to the concentration of TEQ in liver of adults or egg.

Age- and sex-specific differences in sensitivities could also influence the toxic effects of PCB exposure. EROD induction potencies of planar HAHs in primary chicken hepatocyte cultures were age dependent (Bosveld et al., 1997). EROD activities were lower in hepatocyte cultures prepared from 14-day-old embryos than those from 19-day-old embryos or 1-day-old hatchlings. In the white leghorn chicken, TCDF was 1.2- to 3.4-fold more potent than TCDD, which was different from the potency observed in mammalian and teleostean cell lines (Bosveld et al., 1997). It is also imperative to note that the induction of P450 enzymes may not necessarily indicate a toxic effect but may be an adaptive mechanism. Moreover, the induction of P450 enzymes is sometimes nonspecific.

Toxicokinetics   The TEF values for dioxinlike PCBs have been derived mainly from short-term tests and in vitro assays (Safe, 1994). Such studies may not reflect delivery of a toxicant to a target organ due to pharmacokinetics, metabolism, and excretion (De Vito et al., 1995; Lawrence and Gobas, 1997). Also, for extrapolations among species, the toxicokinetics must be identical, or differences have to be taken into account. Some of the factors that would affect interspecies differences have been reviewed (Barrett, 1995). In addition, species- and tissue-specific differences in the binding properties, specificity, and physicochemical properties of the AhR and the contribution of other P-450 genes to HAH-induced activities challenge the generalities of assumptions of the TEF approach.

Dose–Response Relationships   The TEF approach assumes that the relative potencies of individual congeners can be derived. To develop relative potency values, several assumptions must be made. Regardless of the methods applied, the maximum achievable response for the endpoint of interest is identical for the chemicals evaluated and TCDD. That is, the congener of interest must have the same efficacy as TCDD. A second assumption of parallel lines and slope-ratio methods is that the dose–response relationships are parallel or that they have the same origin. Based on both theoretical analyses and empirical examples from certain studies that developed TEFs, these assumptions for dose–response relationships are seldom met (Putzrath, 1997). Furthermore, the slopes of the dose–response curves for many endpoints were different (De Vito et al., 1994). It has been suggested that the relative potencies among chemicals would be more accurately represented by a function rather than by a point estimate such as the EC_{50} or LD_{50}, which are generally used to estimate relative potency (Neubert et al., 1992; Putzrath, 1997). This can be accomplished by using probability functions. The assumptions for deriving RePs have been reviewed (Villeneuve et al., 2000).

Sources of Uncertainties in Deriving RfDs for Non-dioxinlike Effects of PCBs Laboratory studies describing neurotoxicological effects of PCBs were based on dietary exposure of rats, mice, or nonhuman primates with technical mixtures of PCBs, which may not represent the PCB mixtures in environmental matrices. Doses of PCBs to laboratory animals in these studies were greater than those observed in real-world situations. Similarly, the in vitro assays with rat cerebellar granules or PC12 cells have used greater doses, and the EC_{50} values for various endpoints were generally large (>50 µM). The EC_{50} values based on in-vitro studies of neurotoxicological effects of PCBs are presented in (table 11-1). In in vitro systems, less chlorinated *ortho*-substituted congeners were more potent in reducing dopamine than were more chlorinated congeners. Aroclor 1016 (42% chlorine by weight) was more potent than the more chlorinated Aroclor 1260 (60% chlorine) in producing neurobehavioral effects and reducing brain dopamine in pig-tailed macaques (Seegal et al., 1990). This was attributed to the greater abundance of lesser chlorinated *ortho*-substituted PCBs in Aroclor 1016 than in Aroclor 1260 (table 11-4). Accumulation of lesser chlorinated *ortho*-substituted PCB congeners, IUPAC nos. 28, 47, and 52, in brains of pig-tailed macaques after exposure to 3.2 mg/kg body weight per day for 20 weeks also has been observed (Seegal et al., 1990).

Few studies have examined the presence of PCB congeners in brains of humans and wildlife. PCBs were not detected in brain tissues obtained from two men with Parkinson's disease (Corrigan et al., 1996). The concentration of PCBs in the brain of a Yu-Cheng victim was 80 ng/g, whereas concentrations in fat tissues ranged up to 11 µg/g (Chen and Hsu, 1986). PCB 153 (2,2',3,4,4',5,5'-HxCB 1.6 ng/g, wet weight) and PCB 138 (2,2',3,4,4',5'-HxCB 0.96 ng/g, wet weight) were the only two congeners detected in the brain of gray seals (*Halichoerus grypus*), at a concentration of 1% of that measured in the blubber (Jenssen et al., 1996). Similarly, the PCB profile in brain tissue resembled those in other body tissues, with PCB 153 > PCB 138 > PCB 187 in harbor porpoises (Tilbury et al., 1997), suggesting there was no preferential enrichment of lesser chlorinated *ortho*-substituted PCBs in wildlife. PCB concentrations in brain were 1.5% of that found in the blubber. Similarly, concentrations of total PCBs in the brain of mammals from Greek waters were 1–2% of those found in the blubber (Georgakopoulou-Gregoriadou et al., 1995). Less chlorinated *ortho*-substituted PCB congeners are metabolized in humans (Tanabe et al., 1988), birds (Walker, 1990), and dolphins (Tanabe et al., 1988; Kannan et al., 1993, 1994; Boon et al., 1997; Leonards et al., 1997). These results suggest that the accumulation of the lesser chlorinated PCBs is small after chronic exposure. Laboratory studies have shown the presence of 1 µg/g, wet weight, of PCBs in brains of exposed rats and mice, which could be due to the exposure at greater concentrations. Therefore, neurotoxicological effects in laboratory animals may be expected to occur only with relatively great exposure.

Table 11-4. Abundance of ortho-substituted PCB congeners in various Aroclor mixtures.[a]

Chlorobiphenyl (CB) Congener	o,o'-Cl	Composition (Weight %)				
		Aroclor 1016	Aroclor 1242	Aroclor 1254	Aroclor 1260	Aroclor 1268
Di-CB	1	14.3	10.2	—	—	—
	2	4.26	3.21	—	—	—
Tri-CB	1	30.6	21.9	0.61	0.1	—
	2	20.7	14.1	0.6	—	—
	3	0.96	0.53	—	—	—
Tetra-CB	1	4.28	11.3	5.84	0.09	—
	2	20.3	18.4	10.7	0.9	—
	3	3.27	2.52	0.09	—	—
Penta-CB	1	—	3.18	12.3	0.94	—
	2	0.15	6.85	29.7	9.28	—
	3	0.84	3.76	8.9	3.29	—
	4	—	—	0.08	—	—
Hexa-CB	1	—	0.09	1.83	1.28	—
	2	0.19	1.22	13.3	24	—
	3	—	1.01	7.63	18.5	—
	4	—	0.07	1.12	2.23	4
Hepta-CB	1	—	—	—	0.11	—
	2	—	0.17	0.82	13.5	—
	3	—	—	3.03	17.5	8
	4	—	—	0.53	2.74	—
Octa-CB	2	—	—	—	1.45	3.5
	3	—	—	—	3.76	31
	4	—	—	0.68	2.06	11
Nona-CB	3	—	—	—	0.45	21
	4	—	—	—	0.22	14
Deca-CB	4	—	—	—	0.05	4.8

[a]Congeners that contributed to <0.05% of the total composition were not included. Data from Schulz et al. (1989) for Aroclors 1016, 1242, 1254 and 1260 and from Kaninan et al. (1997) for Aroclor 1268.

The results reviewed here indicate that it is unlikely that the effects of *ortho*-substituted PCBs will be the critical toxic effects of PCBs. This is due to the following factors: First, relatively great concentrations of lesser chlorinated di-*ortho*-substituted congeners need to accumulate in brain to cause the observed effects. Second, the congeners that are active do not tend to be accumulated in brains of animals exposed to complex mixtures of PCBs in the environment (field studies). Finally, the most neurotoxic congeners are the less chlorinated PCBs, which are more easily degraded in the environment and less bioconcentrated and more readily metabolized and excreted. The relative potential for adverse effects through AhR-mediated and non–AhR-mediated effects of environmentally weathered mixtures is discussed in subsequent sections. Exposure of experimental animals to weathered PCBs may provide more realistic estimates for the risk assessment of *ortho*-substituted PCB congeners.

Toxicity Reference Values

The TRV is the concentration of a chemical in water, food, or the tissues of the organism that will not cause toxicological effects in organisms of concern. Ideally, TRVs are derived from chronic toxicity studies in which an ecologically relevant endpoint was assessed in the species of concern or in a closely related species. Although TRVs can be expressed or defined based on no-observable adverse effect levels (NOAELs), the use of lowest-observable-adverse-effect levels (LOAELs) is generally preferred because NOAELs by definition incorporate greater uncertainty than LOAELs. Alternatively, TRVs can be expressed as the geometric mean of the NOAEL or LOAEL to provide a conservative estimate of a threshold of effect (Tillitt et al., 1996).

There are three potential limitations of extrapolation of laboratory toxicity data to species exposed in the environment. The first is the wide range of sensitivities that even closely related species can exhibit when exposed to AhR-active chemicals. The second limitation is that most laboratory studies of toxicity are based on exposure to a parent Aroclor mixture that may be substantially different from the congener mixture to which animals in the wild are exposed. The third limitation applies to the TEF/TEQ approach. When using a TEF/TEQ approach, care should be taken when reviewing the literature for TRVs based on TEQs, which are based on the most appropriate set of TEFs. For example, the TEQ-based TRVs for bald eagles (*Haliaeetus leucocephalus*) derived by Elliott et al. (1996) were calculated from a mammalian-based set of TEFs because avian-specific TEFs were not available. In such situations, an appropriate TEQ-based TRV can be recalculated if the congener-specific data are available.

It is therefore essential to critically evaluate the applicability of the toxicological data to the site-specific organisms of concern and to exposure pathways. TRVs derived in the same species are not available for the majority of wildlife, and therefore TRVs must be derived using toxicological data for surrogate species in combination with uncertainty factors. Uncertainty concerning interpretation of the toxicity test information among different species, different laboratory endpoints, and different experimental designs, age of test animals, duration of test, and so on, are addressed by applying uncertainty factors to the toxicology data to derive the final TRV.

In addition to dietary- and media-specific TRVs for PCBs, TRVs based on tissue residues are being used increasingly to evaluate the potential for adverse effects due to PCBs. For the purposes of this chapter, the term "tissue residue-based TRV" is synonymous with the MATC. In this chapter, we have compiled tissue residue-based TRVs for fishes, birds, and mammals. To derive tissue residue-based TRVs, site-specific parameters for exposure to upper-trophic-level organisms, including concentrations of contaminants in prey, we used to estimate relevant tissue concentrations in the organisms of concern. It is important to note that the accuracy of this approach depends on the availability of sufficient data to develop food-chain exposure models. Tissue residue-based effect level data are gaining increasing regulatory acceptance as evidenced in the Canadian Tissue Residue Guidelines (TRG) for Polychlorinated Biphenyls for the Protection of Wildlife Consumers of Aquatic Biota (CCME, 1998).

There are two major problems with extrapolation laboratory toxicity data to wild species. The first is the wide range of species sensitivity to AhR-active chemicals, such as PCBs (Giesy and Kannan, 1998). The second is somewhat unique to PCBs, in that most laboratory toxicity studies are primarily based on exposure to parent Aroclors, whereas wild species are likely exposed to congener patterns that are very different from the parent Aroclor due to environmental weathering (discussed earlier in chapter). When deriving TRVs for a particular species, there are a number of uncertainties that need to be taken into account. One way to do this is by assigning uncertainty factors to assure that the derived TRV protects the wildlife species of interest. It is important to remember that the derivation and application of TRVs is part of a conservative process, meant to be protective rather than to accurately predict the level of effects. For this reason, application of a tiered risk assessment approach that allows for more and more refined estimates of TRVs is suggested. Each level of assessment requires more sophisticated estimates of both exposure and response. The TRV provides information only on the response parameters. Refining TRV estimates may require additional toxicity testing, either for a particular mechanism of action, species, or vector of exposure.

As an example of the application of uncertainty factors, TRV values were derived for bald eagles based on reproductive productivity. For bird species, the most predictive dose metric for dioxinlike chemicals is egg concentrations rather than adult tissue concentrations (Giesy et al., 1994a). This is due in part to stage-specific sensitivity of many species (birds, fishes, and mammals) during development and relative tolerance of adults to similar levels of dioxinlike chemicals. In other words, effects can be seen at lower doses for a young, developing animal than for an adult. For example, the LD_{50} (the dose that is lethal to 50% of a test population) for chicken eggs (Henshel et al., 1993) is 200-fold less than the LD_{50} for an adult chicken (on a wet weight basis; Greig et al., 1973). Furthermore, LOAEL values for developmental toxicity occur at doses that are approximately 10-fold lower than LD_{50} endpoints. Thus, when trying to characterize risk to avian species for dioxinlike compounds, the most sensitive endpoint is developmental toxicity.

The TRV derived for bald eagles was based on developmental toxicity in populations of bald eagles exposed to known concentrations of TEQs (Elliott et al., 1996). However, Elliott et al. (1996) calculated bald eagle TRVs based on hepatic cytochrome P450 1A enzyme induction of 100 and 210 ng TEQ/kg egg (wet weight basis) for the no-observable-adverse-effect concentration (NOAEC) and lowest-observable-adverse-effect concentration (LOAEC), respectively. No significant concentration-related effects for morphological, physiological, or histological parameters, such as chick growth, edema, or density of thymic lymphocytes, were observed at these concentrations (Elliott et al., 1996). Thus, hepatic CYP1A induction appears to be more sensitive than these developmental endpoints. Therefore, these TRVs would be expected to be conservative and would be expected to protect eagle chicks from the effects of Ah receptor-active PCB congeners, which cause embryo lethality or deformities. The TEQ-based TRVs from Elliott et al. (1996) were calculated from a mammalian-based set of TEFs. Because exposure calculations in this study are based on 1998 WHO TEFs for birds, the TRV should be based on the same set of TEFs. After modifying TRVs from Elliott et al. (1996) by use of the most recent and bird-specific TEFs, TRVs were 134 and 400 ng TEQ/kg on a whole egg (wet weight basis) for the NOAEC and LOAEC, respectively. Based on a mean lipid concentration in bald eagle eggs of 7.5% (standard deviation of 2.9%), the lipid-normalized TRVs were 1786 and 5329 ng TEQ/kg for the NOAEC and LOAEC, respectively.

Because these values are suitable for comparison in the current analyses (i.e., same species, same exposure route, similar mixture, etc.), no uncertainty factors have been applied to the TRV in the risk characterization. When TRVs are not available for the species of concern, it may be necessary to modify the TRV (U.S. EPA, 1996). Uncertainty concerning interpretation of the toxicity test information among different species, different laboratory endpoints, and differences in experimental design (age of test animals, duration of test, etc.) are typically addressed by applying uncertainty factors to literature-based toxicity data to calculate the final TRV. Methods for applying uncertainty factors have been published (Opresko et al., 1994; U.S. EPA, 1995). Here we have used the method recently published by U.S. EPA Region VIII for the Rocky Mountain Arsenal (RMA) (U.S. EPA, 1998). The RMA procedure uses three uncertainty factors:

- Intertaxon variability extrapolation, where values range from 1 to 5 (category A);
- Exposure duration extrapolation, where values range from 0.75 to 15 (category B);
- Toxicologic endpoint extrapolation, where values range from 1 to 15 (category C); and
- Modifying factors (category D), which incorporates other sources of uncertainty, including:
  1. Threatened, or listed, and endangered species, where values range from 0 to 2 ($d1$);
  2. Relevance of endpoint to ecological health, where values range from 0 to 2 ($d2$);
  3. Extrapolation from laboratory to field, where values range from −1 to 2 ($d3$);
  4. Study conducted with relevant co-contaminants, where values range from −1 to 2 ($d4$);
  5. Endpoint is mechanistically unclear (vs. clear), where values range from 0 to 2 ($d5$);
  6. Study species is either highly sensitive or highly resistant, where values range from −1 to 2 ($d6$);
  7. Ratios used to estimate whole body burden from tissue or egg, where values range from 0 to 2 ($d7$);
  8. Intraspecific variability, where values range from 0 to 2 ($d8$); and
  9. Other applicable modifiers, where values range from −1 to 2 ($d9 \ldots n$).

The TRV is calculated using equation 11–2:

$$TRV = \frac{Study\ dose}{A*B*C*D},\qquad (11\text{-}2)$$

where

$$D = (d1 + d2 + d3 + \ldots dn).\qquad (11\text{-}3)$$

The assignment of uncertainty factors to the NOAEL for the calculation of a TRV for bald eagles is as described below.

Intertaxon variability (A), is used to estimate a concentration or dose for a wildlife species from laboratory test data from surrogate species. The field study that provided the NOEL used the bald eagle as the test organism (Elliott et al., 1996), so there is no extrapolation among species, and thus $A = 1$.

Exposure duration (B), is used to estimate the LOAEC or LOAED of a chemical when only acute (short-term) toxicity test data are available. The field study that provided the NOEL (Elliott et al., 1996) used eagle eggs collected from the wild, so the doses reflect actual chronic exposures, and thus $B = 1$.

Toxicologic endpoint (C), is used to estimate NOAEL and/or LOAED values from studies that report other endpoints. The eagle egg study calculated both NOEL and LOAEL doses (Elliott et al., 1996) for induction of hepatic cytochrome P4501A, as expressed by EROD, which is more sensitive than the developmental toxicity endpoints of chick growth, edema, or density of thymic lymphocytes. The importance of enzyme induction at the population level is unknown, but there was no extrapolation from other endpoints, and thus $C = 1$.

Modifying factors (D), are used to address other aspects of uncertainty:

1. Threatened, or listed, and endangered species—the bald eagle is a Federally Endangered Species, however, the toxicity data are for this species, so $d1 = 0$.
2. Relevance of endpoint to ecological health—the test endpoint in the field study was EROD induction in bald eagle chicks. The relevance of this endpoint to population sustainability is unknown, so we have conservatively set $d2 = 1$.
3. Extrapolation from laboratory to field—the study by Elliott et al. (1996) used bald eagle eggs collected from the wild and incubated them in the laboratory. Because the exposure doses were actual environmental (field) exposures, $d3 = 0$;
4. Study conducted with relevant co-contaminants—the eagle eggs contained a variety of environmental contaminants, including PCDDs and PCDFs, so $d4 = 0$.
5. Endpoint is mechanistically unclear (vs. clear)—the induction of EROD as an expression of P4501A induction, the mechanism of dioxin toxicity, is clear, so $d5 = 0$.
6. Because the study species is neither highly sensitive nor highly resistant and the bald eagle was both the study species and the wildlife species of interest, $d6 = 0$.

7. Ratios used to estimate whole body burden from tissue or egg—Elliott et al. (1996) measured TEQ concentrations in eagle eggs, which is the endpoint for this risk assessment, so $d7 = 0$.

8. Intraspecific variability—the field study investigated adverse effects on bald eagle hatchlings, the most sensitive life stage, so $d8 = 0$.

The TRV calculation for eagle eggs is therefore represented in equation 11-4:

$$\text{TRV} = \frac{134 \text{ pg TEQ}/\text{g egg, wet weight}}{[1*1*1*(0+1+0+0+0+0+0+0)]} \tag{11-4}$$

The calculated TRV is thus 134 pgTEQ/g egg (wet weight).

Fish

Effects of PCBs and Related Compounds on Fishes

It is against the complex and ever changing background of factors that the potential effects of synthetic organic chemicals need to be considered. There has been much study and debate about the relative effects of the concentrations of these compounds on fishes of the North American Great Lakes. It has been difficult work because of the factors already mentioned and because fishes are exposed to a complex mixture of compounds, which, like the environment and populations of the fishes, is ever changing.

There are literally thousands of synthetic halogenated compounds that have been identified in the flesh and tissues of fish from the North American Great Lakes (Schmitt et al., 1999). In addition, there are a number of natural products that have been released to the North American Great Lakes at abnormally high concentrations (Schmitt et al., 1999). These include hundreds of polycyclic aromatic hydrocarbons (PAHs) from the transport and combustion of petroleum hydrocarbons and a variety of classes of compounds of plant origin released from the forest products industry. It has been difficult from field studies, which are by nature correlational, to demonstrate cause–effect linkages between the occurrence of trace contaminants in fish and adverse effects. The lack of appropriate methods or authentic standards has made it difficult to identify and quantify the various pollutants. The study of the effects of these compounds under laboratory conditions has been complicated by the fact that many of the fish that are at the greatest potential risk are large and often long-lived. It is difficult to conduct controlled studies on fish like the lake trout that may not spawn, until they are more than 9 years old. It is difficult to conduct experiments of sufficient duration to demonstrate effects at realistic concentrations.

Diminished Plasma Steroid Hormone Levels in Great Lakes Salmonid Fishes

Concentrations of several plasma steroid hormones vary among stocks of salmonid fishes of the North American Great Lakes (Leatherland et al., 1982). Depending on the species, Pacific salmon live from 2 to several years, spawn once, and die. One stock, the Fairview stock of coho salmon (*Oncorhynchus kisutch*) in Lake Erie, exhibits significantly lower concentrations of plasma testosterone, cortisol, and gonadotropins in both male and female adult fish than coho salmon from Lakes Michigan and Ontario (Leatherland et al., 1982). The depressed plasma gonadotropin II concentrations in adult Fairview stock coho salmon in Lake Erie is probably the cause of depressed concentrations of several gonadal steroid hormones in both male and female adult fish (Leatherland et al., 1982), and it could explain the cause of the relatively poor reproductive performance of that stock. The cause for the lower concentrations of plasma gonadotropins in the Fairview coho stock is unknown. There are a number of possible causes of decreased hormonal concentrations in these fish, but as there seems to be no relationship with gonad size or concentrations of known contaminants, the cause could be genetic.

Thyroid Function in Great Lakes Salmonid Fishes

One of the effects in fish for which EDCs have been implicated is abnormal thyroid function. Thyroids of coho salmon from the North American Great Lakes are impaired (Leatherland and Sonstegard, 1980a, 1981, 1984; Leatherland, 1993). The epizootics of thyroid lesions are variously described as thyroid enlargement, thyroid hypertrophy and/or hyperplasia, thyroid tumors, or goiters. This condition was reported to occur in salmonid fishes from the central part of North America as early as the first part of the 20th century. Initially this condition was reported as simple goiter and thought to be due to the relatively small concentrations of iodine available in prey items in the Great Lakes. Thyroid hyperplasia was reported in steelhead trout from Lakes Michigan and Superior in the 1950s (Robertson and Chaney, 1952), for coho salmon from Lake Erie in the 1970s (Black and Simpson, 1974), and for coho salmon and steelhead trout from Lake Michigan (Drongowski et al., 1975). It has been reported that every salmonid fish examined from the Great Lakes since the mid-1970s has exhibited thyroid hyperplasia (Leatherland, 1993). It was concluded that the occurrence of goiter in a number of species from the Great Lakes and the fact that the iodine content of the fish and their diet were not different from marine fishes of the same species argues against an iodine deficiency etiology (Leatherland, 1993). It has been suggested that waterborne chemicals are a more likely etiology. Four primary reasons have been given for this conclusion (Leatherland, 1993). (1) The dietary iodine requirements of salmonid fishes are lower than that measured in the diets of fishes from the North American Great Lakes, and when coho salmon were fed an iodine-deficient diet, goiters could not be produced; (2) no correlation was observed between iodine concentrations in tissues and incidences of goiter, and iodine concentrations in Great Lakes salmon was similar to that of Pacific ocean-run salmon of the same species; (3) female salmon, which mobilize large amounts of iodine into the developing eggs, do not exhibit lower iodine concentrations in their tissues or a greater incidence of goiter; and (4) concentrations of both T_3 and T_4 are sufficient, which would not be expected for iodine-deficient fish. Leatherland (1993) indicates that although none of the proposed arguments is convincing alone, taken together they weigh against an iodine deficiency etiology.

It has been argued that a possible cause is xenobiotic goitrogens such as PCBs, PCDDs, and PCDFs that occur in the tissues of salmonid fishes and their prey in the Great Lakes and which are known to cause goiter in mammals (Leatherland, 1993). However, when studies were conducted in which coho salmon and lake trout were exposed to mixtures of the types of chemicals that were measured in Great Lakes fishes, such as Mirex and Aroclor 1254, there was no thyroid enlargement (Leatherland and Sonstegard, 1978). Thus, it is unlikely that the more commonly measured xenobiotics present in the Great Lakes salmonids are the active goitrogens (Leatherland, 1993). One piece of evidence that suggests that synthetic xenobiotics are responsible for goiter is a study in which fish from the Great Lakes were fed to rats, which subsequently developed hypothyroidism (Sonstegard and Leatherland, 1979). These effects could be caused either by nutritional deficiencies, exposure to xenobiotics, or a combination of both. Iodide is required for normal TH synthesis. Thyroid hormones, in turn, are required for growth and development, metabolism, and reproduction. Iodide occurs at lower concentrations in the Great Lakes than in the marine environment, which is the natural environment for some of the salmonids that have been introduced to the Great Lakes (Giesy et al., 1986). Feeding PCBs to coho salmon reduced plasma T_4 and T_3 (Leatherland and Sonstegard, 1978). Furthermore, when compounds such as PCBs, which are known to occur in the flesh of Great Lakes fishes, were fed to rats, thyroid activity was affected (Leatherland and Sonstegard, 1980b).

Altered concentrations of THs frequently co-occur with embryonic abnormalities in fish and wildlife populations exposed to PCDHs (Cecil et al., 1973; Cullum and Zile, 1985; Brouwer and van den Berg, 1986; Brouwer, 1987; Fox, 1993). Although much of the mechanistic toxicology has been studied in mammals, it is known that the system is similar among vertebrates and that the same types of effects of PCDHs observed in mammals are also observed in fishes (Brouwer and van den Berg, 1986; Brouwer et al., 1986, 1988). The lower fertility and fecundity of coho salmon in Lake Michigan has been attributed to their hypothyroid condition. Individual fish from the Great Lakes populations exhibit altered sexual development and sexual dysfunction as adults (Colborn et al., 1993). The observations on adult sexual dysfunction are especially significant because young that appear to be normal while raised by exposed parents may become reproductively dysfunctional when they mature (Brunström and Lund, 1988). Poor reproductive efficiencies and opportunistic diseases are characteristic of the wild animals in these exposed populations in the Great Lakes region (Beland et al., 1993). Adult female coho salmon from stock that exhibited the poorest embryo survival also had the lowest concentrations of TH in the plasma (Leatherland et al., 1984). The authors did not speculate on whether there was a causal relationship between plasma TH levels and embryo survival.

Extracts of water from the Great Lakes inhibit synthesis of iodine-containing compounds. Leatherland (1993) speculates that there is a waterborne goitrogen of microbial origin in the Great Lakes that may inhibit TH synthesis, resulting in secondary effects on steroid hormones and subsequently on reproduction (see chapter 6). As THs stimulate secretion of E_2 by ovarian follicles, there could be a relationship between TH deficiency, vitellogenesis, fertility, and fecundity. The argument that goitrogens are responsible for the thyroid hyperplasia is weakened by the fact that goiters were observed in fishes of the Great Lakes before many of these compounds were released into the environment. Furthermore, goiter incidence was not correlated with degree of contamination in the North American Great Lakes during the past 50 years. The effects seem to be the same in all of the lakes regardless of degree of contamination with PCDH. PCDHs and their metabolites can affect thyroid function and THs, but at this time xenobiotic effects on thyroid function and hormones have yet to be definitively demonstrated (Brouwer et al., 1990).

Precocious Sexual Maturation in Great Lakes Salmonid Fishes

Some salmonid fish stocks in the Great Lakes exhibit precocious sexual development, especially males (Leatherland, 1993). In these cases, salmonid fishes mature and make their spawning runs 1 or 2 years ahead of the normal 3-year cycle. This phenomenon has long been observed in natural and stocked populations of steelhead trout and Pacific salmon, but its occurrence in the Great Lakes is abnormally high. The rate of occurrence of precocious sexual development in some Pacific coho stocks is as great as 50% (Thorpe et al., 1989), and that of some populations of Great Lakes coho salmon is as great as 95% (Leatherland, 1993). Although this phenomenon has been studied intensely, the cause is still unknown. Based on what is known, the most likely cause of precocious sexual development is genetic (Naevdal, 1983; Gall et al., 1988), and environmental endocrine-modulating compounds may not be involved.

Loss of Secondary Sexual Characteristics in Great Lakes Salmonid Fishes

Sexually mature individuals of most species of salmon display a distinct sexual dimorphism that is probably under endocrine control. Specifically, sexually mature males develop the characteristic hooked protuberance on the lower jaw, known as a kype. In salmonid fishes, 11-ketotestosterone is the primary androgenic steroid hormone and seems to be responsible for development of secondary sexual characteristics in males (Idler et al., 1961). However, Fairview stock coho salmon in Lake Erie do not express a strong sexual dimorphism, possibly due to lower androgen levels (Leatherland et al., 1982). Under hatchery conditions, where selective pressures for secondary sexual characteristics are removed, expression of secondary sexual characteristics is diminished (Fleming and Gross, 1989). Suppression of secondary sexual characteristics in the Fairview stock of coho in Lake Erie may be greater than expected due to genetically related changes, and there might be an environmental etiology (Leatherland et al., 1982). However, there seems to be no evidence that the changes in secondary sexual characteristics observed in Lake Erie are caused by exogenous endocrine-modulating compounds.

Interactions between Vitamin A
and Xenobiotics in Fish

Concentrations of carotenoids that are important for survival, growth, and reproduction are different in freshwater than in marine ecosystems (Zile, 1992). Fish
cannot synthesize vitamin A (retinol) de novo but make it from carotenoid precursors obtained in the diet. These carotenoids are synthesized by plants and some
bacteria. When eaten by invertebrates, the carotenoids are retained and concentrated
in their bodies and subsequently become the dietary source of these compounds
for predators. Vitamin A is important in many functions in animals, including
embryonic development, vision, maintenance of the dermally derived tissues, immune competence, hematopoiesis, and reproductive functions (Zile and Cullum,
1983; Brouwer, 1989: Leith et al. 1989: Zile, 1992).Changes in the status of vitamin A in the plasma or liver may be responsible for the birth defects observed in
fish and wildlife of the Great Lakes region that have been exposed to PCDHs.

As described in chapter 4, TH functioning is closely related to vitamin A in fish
and other vertebrates. Exposure to xenobiotics can interfere with normal vitamin A
metabolism and TH function. Thyroid hormone can be influenced by exposure to
PCDHs (McKinney et al., 1985; Poellinger and Gustafsson, 1985; Brouwer, 1987;
Zile, 1992). There are several possible mechanisms for the observed effects on circulating T_3 and T_4. In fish T_4 is transported in blood bound to the TTR at a 1:1 molar
ratio with vitamin A. When the TTR–vitamin A–T_4 complex arrives at the target
tissue, T_4 is deiodinated to T_3, which typically binds to the TH receptor. This complex then forms a heterodimer with the RXR receptor to cause tissue-specific effects (see chapter 4). Disruption of T_4 transport and deiodination can result in
pleiotropic responses that are often difficult to link to the original cause (Leatherland
and Sonstegard, 1980b). First, hydroxy-substituted PCB congeners displace T_4 from
TTR, resulting in effects similar to TH deficiency (Cullum and Zile, 1985; Brouwer,
1989; Zile, 1992). PCDHs can induce hepatic enzymes such as acyl-CoA–retinol
acyltransferase and UDPGT activity in the liver, which then decrease the concentration of TTR in the blood (Brouwer et al., 1989). Concentrations of TTR are not
determined directly in the plasma, but instead T_4 binding capacity is measured.
Therefore, it is not possible to distinguish which of the two mechanisms is causing
the observed effects. An inverse correlation between the concentration of vitamin
A in serum and concentrations of PCDHs in tissues of animals, including fishes,
was observed when they were exposed to PCDHs (Zile, 1992). This correlation was
observed in birds of the Great Lakes region (Fox, 1993) and in salmonid fishes of
the Baltic Sea (Zile, 1992). Laboratory studies have determined that both vitamin
A and its storage form in the liver (retinol palmitate) are often depleted in seals
exposed to PCDHs (Brouwer et al., 1989) and levels of these compounds have been
suggested as biomarkers of exposure (Fox, 1993) to sublethal doses of the dioxinlike
planar PCB congener 77 in birds (Spear et al., 1985). PCDHs affect concentrations
of vitamin A in both the blood and liver of exposed organisms (Zile, 1992).

In conclusion, we must agree with Leatherland (1993) that "the evidence in support of a contaminant-related dysfunctional state in fish is not as convincing as the
evidence of the syndrome caused by PCDH in fish-eating mammals and birds"

(p. 747). While the recoveries of some species have paralleled the decline in concentrations of some of the more common halogenated hydrocarbons, these recoveries could be as much a result of better management practices as they are of a decline in concentrations of contaminants.

TRVs for Fishes

Adult fish are exposed to PCBs and related compounds via water, sediment, and food. Eggs and embryos may accumulate these highly lipophilic chemicals from the female during vitellogenesis. Bioaccumulation of PCBs by fishes depends on the physical and chemical characteristics of individual congeners and on the biotransformation and elimination rates of congeners. The log octanol/water partition coefficient (K_{ow}) of PCBs increases with molecular weight from approximately 4.5 to 8.2 (Mackay et al., 1992), which indicates that all but the most highly chlorinated congeners are efficiently bioaccumulated. As a result of these factors, fishes preferentially bioaccumulate highly chlorinated penta-, hexa-, and hepta-chlorinated biphenyls, but not deca-chlorinated biphenyls (Walker and Peterson, 1994a,b).

In fish, early life stages generally are the most sensitive developmental stage to chemical contaminants. Thus, accumulation of persistent chemicals during early life stages is critical to the characterization of risks posed by PCBs. Signs of toxicity and histopathologic lesions produced by PCBs and related compounds in juvenile fish are similar to those seen in tetrapods and include decreased food intake, wasting syndrome, delayed mortality, and lesions in epithelial and lymphomyeloid tissues. Toxicity and histopathologic lesions produced by PCBs during early development are characterized primarily by cardiovascular and circulatory changes, edema, hemorrhages, and mortality (Walker and Peterson, 1994a,b). As with tetrapods, there is great variability in species sensitivities of fishes to PCBs. Freshwater salmonid species, particularly lake trout and rainbow trout, are the most sensitive.

The adverse effects of PCBs on fish have been studied primarily by two experimental methods: (1) laboratory exposure of fish via water or the diet or via intraperitoneal or in ovo injections to single congeners or technical mixtures, or (2) correlation of concentrations of PCBs and related compounds in the environment with abnormalities in fish populations, such as mortality during early development or thyroid hyperplasia in adult fishes (Walker and Peterson, 1991, 1994b). Although field research can integrate the impact of multiple environmental contaminants on fishes, laboratory exposures can identify the specific responses associated with exposure to a single toxicant and determine the dose–response relationships for those responses. Thus, laboratory research can elucidate whether the body burden of a particular contaminant in fish in the environment is capable of producing the abnormalities observed in feral fish populations. Both field and laboratory research are vital to understanding the toxicity of PCBs to fish and to predicting the risk that these compounds pose to fishes in the environment. The available in vivo, in ovo, and in vitro toxic responses of fishes characterized in laboratory and field studies were evaluated. However, for the purposes of summarizing effect levels in fishes, only data from selected studies using the most relevant aquatic organisms are discussed here.

It is difficult to make accurate estimates of risk when the exposure pathways for field and laboratory studies are different. As an alternative approach, available studies reporting correlations between concentrations of PCBs in tissues and observed effects were evaluated. Jarvinen and Ankley (1999) compiled such data for a variety of chemicals including PCBs. Using this data set, effect levels were determined for marine fish species in which critical life stages were evaluated for effects, and tissue concentrations were measured (table 11-5). Studies were evaluated for comparability to the potential species of interest, strength of the cause–effect linkage, exposures to critical life stages (embryos, fry, and juveniles), and sensitive developmental toxicity endpoints, including decreased survival and/or decreased growth. These tissue-residue effect levels incorporate all of the possible exposure pathways. Because the data are so limited in number, a NOEL and LOEL could not be readily determined. Thus, a geometric mean of the available data was calculated to estimate a toxicity threshold in tissue (whole body or fillet). A similar table with greater tissue residue-based effect threshold values has been compiled (Niimi, 1996; table 11-6). To illustrate the relevance of a tissue-based effect threshold independent of exposure route, Walker et al. (1994) measured concentrations of TCDD in the eggs of rainbow trout after exposure by maternal transfer, water uptake, and injection. The results from this study show that the tissue-based effect level is consistent regardless of exposure route (table 11-7).

Concentrations of dissolved PCBs that are toxic to fishes also are very low, particularly after chronic exposure. The 96-h LC_{50} values (the concentration that is lethal to 50% of a test population) for fathead minnows are 8–15 µg/l for different PCB mixtures (Nebecker et al., 1974). The threshold levels for effects of TCDD in 96-h exposures to fathead minnows vary from 0.0001 µg/l (for retarded growth and development) to 0.01 µg/l for 100% mortality (Helder, 1980, 1981). A NOEC for TCDD has been estimated to be less than 38 pg/l in fingerling rainbow trout (one of the most sensitive fish species to TCDD) exposed for 28 days (Mehrle et al., 1988). However, such relatively short-term exposures do not generally address the chronic effects of dioxinlike PCBs, particularly because delayed mortality, usually after several weeks, occurs at lower concentrations than for acute effects.

Birds

Efficacy of PCBs on Thyroid Hormone

PCBs decrease circulating T_4 of birds in the laboratory and concentrations of PCBs in places like the North American Great Lakes and Baltic Sea are negatively correlated with plasma T_4 concentrations. Plasma T_4 of Japanese quail (*Coturnix coturnix japonica*) was decreased by exposure to PCBs under laboratory conditions (Grassle and Bressmann, 1982). The effects on T_4 were associated with histological changes in the thyroid such as hyperplasia. Similarly, PCBs caused a decrease in plasma T_4 in great blue herons (*Ardea herodias*; Janz and Bellward, 1997) and guillemots (*Cepphus grylle*; Jefferies and Parslow, 1976), eider ducks (*Somateria mollissima*), and common terns (*Sterna hurundo*; Murk et al., 1994). Reduced levels of T_4 in

Table 11-5. Tissue-based effect concentrations of total PCBs in early developmental stages of marine fishes.

Effect Level Species[a]	Life Stage	Exposure Route	Duration (Days)	Whole Body	Fillet	Effect	Reference
NOEL							
Sheepshead minnow, *Cyprinodon variegatus*	Embryo	Adult fish, 49 µg/g	5	27	9	No effect	Hansen et al., 1973
Cyprinodon variegatus	Embryo–larvae	Adult fish, 1.9–2.5 µg/g	28	0.88	0.3	No effect	Hansen et al., 1973
Pinfish, *Lagodon rhomboides*	Juvenile	Water, 100 µg/l	2	17	6	No effect	Duke et al., 1970
Spot, *Leiostomus xanthrus*	Juvenile	Water, 1 µg/l	33–56	27	9	No effect	Hansen et al., 1971
LOEL							
Sheepshead minnow, *Cyprinodon variegatus*	Embryo	Adult fish, 0.1–10 µg/l	5	170	59	Reduced survival	Hansen et al., 1973
	Embryo–larvae larvae	Adult fish, 9.3–9.7 µg/g	28	5.1	2	Reduced survival	Hansen et al., 1973
Pinfish, *Lagodon rhomboides*	Juvenile	Water, 5 µg/l	14	14	5	Reduced survival	Hansen et al., 1971
Spot, *Leiostomus xanthrus*	Juvenile	Water, 5 µg/l	20–26	46	16	Reduced survival	Hansen et al., 1971
Geometric mean of all NOEL and LOEL values for PCBs				9.2	3.2		

[a] NOEL, no-observed-effect level; LOEL, lowest-observed-effect level.
[b] A relationship between whole body and skin-off fillet was used to derive fillet concentrations.

Table 11-6. Summary of polychlorinated biphenyl (PCB) concentrations in fishes at which adverse and chronic effects, cytological change, and changes in biochemical activity levels may occur, based on short- and long-term laboratory studies.

Response	Concentration in Fish
Lethality	>100 mg/kg
Growth	>50mg/kg
Reproduction	
Female	>100 mg/kg
Progeny	>50 mg/kg
Behavior	>100 µg/l
Disease	mg/kg range
Cellular changes	High µg/kg to low mg/kg
Biochemical changes	High µg/kg to low mg/kg

A maximum concentration of >100 mg/kg of PCBs in tissues was used because higher concentrations may have limited environmental relevance. The threshold concentration for behavior is expressed on a waterborne exposure basis because the estimate of tissue concentrations for this response were poorly defined (from Niimi, 1996).

herring gulls (*Larus argentatus*) from the Great Lakes may be caused by PCB exposure (Moccia, 1986). There was also an inverse correlation between the concentration of vitamin A in serum and concentrations of planar PCDHs in tissues of birds from the upper Great Lakes (Fox, 1988). Similarly, when chicken (*Gallus gallus*; white leghorn strain) eggs were injected with Aroclor 1242 or Aroclor 1254, TH concentrations in the blood of developing chicks were lower than those in untreated chicks (Gould et al., 1999). Both Aroclors decreased type I monodeiodinase (5'DI) activity in the liver, but individual congeners including 2,2',6,6'-tetrachlorobiphenyl (TCB), 3,3',4,4'-TCB or 3,3',5,5'-TCB had no affect on the 5'DI activity at the concentrations tested.

Because of these conserved biochemical mechanisms, concentrations of TCDD-EQ correlate with egg lethality or birth defects in populations of colonial, fish-eating

Table 11-7. Effect of exposure route on the lethal potency of TCDD to rainbow trout eggs (data from Walker and Peterson 1994a).

Exposure Route	NOAEL µg/kg Egg	LOAEL µg/kg Egg	LD_{50} µg/kg Egg	LD_{100} µg/kg Egg
Maternal	0.023	0.05	0.058	0.145
Water uptake	0.034	0.04	0.069	0.119
Egg injection	0.044	0.055	0.080	0.154

NOAEL, no-observed-adverse-effect level; LOAEL, lowest-observed-adverse-effect level.

water birds, whereas concentrations of the total concentrations of PCBs, PCDF, and PCDD do not (Tillitt et al., 1991, 1992). TCDD is a potent T_4 agonist that may account for its capacity to cause wasting syndrome and developmental deformities.

Avian goiter resulting from iodine deficiency is characterized by epithelial cell hyperplasia (Blackmore, 1963). PCB treatment of birds results in hypothyroidism and a large-colloid goiter (Spear and Moon, 1985). DDT and DDE have produced hypothyroidism in Japanese quail (Richert and Prahlad, 1972), but produced follicle cell hyperplasia in the pigeon, as did dieldrin (Jefferies and French, 1972). Herring gulls in the Great Lakes had enlarged thyroids when compared to a control colony in the Bay of Fundy (Moccia, 1986). The goiter was characterized by follicle cell hyperplasia, few columnar epithelial cells, a diffuse microfollicular structure, and scant or absent luminal colloid. Because various etiologic agents produce different histological appearances in birds, histopathology can be useful in determining the cause of goiter. Based on these observations, it is unlikely that in the case of the Great Lakes herring gulls, PCBs were the cause of the observed goiter. While it is possible that DDE or dieldrin could have been a contributing factor, it is also possible that iodine deficiency or goitrogens in the forage could have been involved. Goiter is an endemic disease in some avian wildlife populations. It has been postulated that goiter may in some cases be caused by the modulation of TH homeostasis by xenobiotic chemicals in the environment. Goiter has been experimentally induced in birds by the administration of organochlorine contaminants (Feyk and Giesy, 1998). The histological presentation of goiter can vary substantially depending on the etiologic agent responsible. Moccia (1986) used histological evidence to infer that PCBs were not the cause of goiter observe in Great Lakes herring gulls. The observed goiter was characterized by follicle cell hyperplasia with scant or absent luminal colloid, in marked contrast to the large colloid goiter commonly observed in birds experimentally treated with PCBs.

TEFs

Endpoints used to estimate TEFs for birds include in vitro and in ovo EROD induction (Kennedy et al., 1996a,b; Yao et al., 1990; Bosveld et al., 1997) and embryo mortalities (Brunström, 1989). Most of the avian TEF data were derived from EROD induction potencies of PCB congeners. In chicken embryo hepatocytes, PCB 169 was less potent than PCB 77 (Kennedy et al, 1996a; Brunström, 1990; Brunström et al., 1995), which is different from the relative potency found in rodent bioassays (Safe, 1994). The lower potency of PCB 169 relative to PCB 77 in birds was further supported by embryo lethality data (Brunström, 1989). Similarly, TCD was more potent than TCDD in avian models based on EROD induction (Bosveld et al., 1997). Mono-*ortho* PCBs were less potent inducers of EROD than non-*ortho* congeners in bird models, but TEFs for mono-*ortho* congeners were relatively higher in birds compared with teleostean and rodent TEFs (table 11-2). For this reason, the PCB congener that contributes the greater proportion of TEQs based on avian TEFs in most environmental mixtures of PCBs is congener 77 (3,3',4,4'-tetra-CB). Therefore, more information on the TEF and environmental fate of this congener, particularly on its pharmacokinetics in birds, is necessary for an accurate risk assessment.

Avian TEFs are difficult to estimate because there is considerable interspecific variation in the toxicity of PCB congeners among birds (Sanderson et al., 1998b). Some TEFs for PCB congeners in birds are based on EROD induction with eggs and cell cultures from chicken. The preferred endpoint is embryo lethality based on in ovo exposure. Domestic chickens and their embryos are considerably more sensitive to AhR-mediated responses than other avian species (Kennedy et al., 1996a; Lorenzen et al., 1997). For example, based on in vitro EROD induction potency of coplanar PCB congeners in several bird species, the order of sensitivity was shown domestic chicken > ring-necked pheasant > turkey ≈ double-crested cormorant ≈ great blue heron ≈ ring-billed gull ≈ duck ≈ herring gull ≈ common tern > Foster's tern (Sanderson et al., 1998b). In general, fish-eating bird species examined so far are at least an order of magnitude less sensitive than the domestic chicken. Because the chicken is much more responsive than other birds, if chicken TEFs and RfDs are used as a surrogate for wild birds, no uncertainty factors should be applied because it is unlikely that other wild species would be more sensitive.

TRVs

Compared with other biota, there is a great deal of information available on the toxicity of PCBs to birds, including dietary and tissue residue-based effect levels of PCBs. Some toxic effects of PCBs and related compounds are listed in table 11-8. Although several earlier studies with juvenile and adult birds have shown lethal and biochemical effects of PCBs and related compounds, few studies were designed in such a way that TRVs could be determined reliably. Results of earlier studies,

Table 11-8. Toxic effects of PCBs and TCDD equivalents observed in birds (Blankenship and Giesy, 2001).

General toxicity
    Embryo lethality
    Decreased productivity
    Liver mixed-function oxidase
    Unabsorbed yolk sacs
    Vitamin A depletion
    Porphyria
Teratogenesis (birth defects):
    Gastroschisis
    Crossed bills
    Clubfoot
    Dwarfed appendages
    Edema/ascites
    Hemorrhaging
    Abnormal feathering
    Abnormal eyes
    Hydrocephaly
    Anencephaly

conducted before 1996, have been examined critically by Hoffman et al. (1996). Toxic effects of PCBs and related compounds in birds have been studied by in vivo exposure, in ovo exposure using egg injection, and in vitro exposure with cultured avian hepatocytes. Due to better control of exposure dose and timing, egg injection studies are often of more use in deriving TRVs than in vivo or in vitro exposure studies. In most cases, embryotoxic and teratogenic effects of PCBs seem to be the most sensitive and ecologically relevant endpoints in birds (Hoffman et al., 1998). Thus, results from in ovo studies are particularly relevant for developing tissue residue-based toxicity thresholds.

In ovo studies of dioxinlike compounds have been described (Powell et al., 1996b, 1997b; Hoffman et al., 1998). Several of these studies were well conducted and feature an adequate number of replications; several sensitive parameters were monitored. Values for LOAEL, NOAEL, and EC_{50} that are useful for derivation of TRVs have been estimated for certain PCB congeners from in ovo studies. Additionally, for many bird species, the most sensitive dose metric or effects predictor for PCBs and other dioxinlike chemicals is PCB concentrations in eggs rather than adult tissue concentrations (Giesy et al., 1994a). This is due in part to the sensitivity of developing embryos of many species (birds, fish, and mammals) and to the relative tolerance of adults to the effects of dioxinlike chemicals. For example, the LD_{50} for chicken eggs (Henshel et al., 1993) is 200-fold lower than the LD_{50} for an adult chicken on a wet weight basis (Greig et al., 1973). Furthermore, LOAEL values for developmental toxicity occur at doses that are approximately 10-fold lower than LD_{50} endpoints. Thus, when trying to characterize risk to avian species for dioxinlike compounds, the most sensitive endpoint is developmental toxicity.

Available toxicological studies that correlated effects with PCB concentrations in eggs were evaluated (tables 11-9–11-13). Similar tissue residue-based effect thresholds have been summarized elsewhere (Hoffman et al. 1996). Some of the dietary avian toxicological studies from the 1970s are still the most useful for deriving PCB reference doses. Alternatively, a tissue residue-based approach can be used in which observed effects are compared to a known dose to the egg or tissue residues of total PCBs (from a congener-specific analysis) or TEQs. The results of egg injection studies for predicting potential embryo toxicity of PCBs and TCDD compares favorably with that of feeding studies. In studies in which the same chemicals have been administered by both methods, the egg concentrations required to elicit effects are similar. The biological effects of Aroclor mixtures, individual PCB congeners, TCDD, and TCDD equivalents have been assessed with egg injection experiments. For risk assessment purposes, it is possible to model concentrations of PCBs in bird eggs using published biomagnification factors. The collection of site-specific foraging information and determination of PCB congener concentrations in prey and predator tissues can improve the accuracy of the predicted egg concentrations. The predicted or measured concentrations can then be compared to TRVs to estimate the magnitude of possible risks.

As with other organisms, birds demonstrate considerable differences in species sensitivities to PCBs and related dioxinlike chemicals. In particular, chickens, which are the most frequently used species for PCB exposures, are among the most sensitive of avian species to the effects of PCBs and dioxinlike chemicals. We recommended

Table 11-9. Avian PCB toxicity summary for dietary exposures to Aroclor in the laboratory.

Species	Adverse Effects Evaluated			Congener or Mixture	Critical Effect Level			Reference
	Hatching Success	Reproductive Success	Chick Growth		NOEL	LOEL	Units	
Chicken	✓			A 1242	980	9800	µg/kg/day	Lillie et al., 1974
Chicken			✓	A 1242	980		µg/kg/day	Lillie et al., 1974
Chicken	✓			A 1242	2440		µg/kg/day	Britton & Huston, 1973
Chicken	✓			A 1242	2440	4880	µg/kg/day	Lillie et al., 1975
Chicken	✓			A 1254	9760a		µg/kg/day	Cecil et al., 1974
Chicken	✓			A 1254	980	9800	µg/kg/day	Lillie et al., 1974
Chicken			✓	A 1254		980	µg/kg/day	Lillie et al., 1974
Chicken	✓			A 1254	2440		µg/kg/day	Lillie et al., 1975
Chicken		✓		A 1254	244	2440	µg/kg/day	Platonow & Reinhart, 1973
Pheasant	✓			A 1254	180	1800	µg/kg/day	Dahlgren et al., 1972
Mallard		✓		A 1254	1450		µg/kg/day	Custer & Heinz, 1980
Chicken	✓			A 1248	2440	4880	µg/kg/day	Lillie et al., 1975
Chicken	✓			A 1248	980	9800	µg/kg/day	Lillie et al., 1974
Chicken			✓	A 1248		980	µg/kg/day	Lillie et al., 1974
Chicken	✓			A 1248	490	4900	µg/kg/day	Scott, 1977
Screech owls	✓	✓		A 1248	410b		µg/kg/day	McLane & Hughes, 1980
Chicken	✓			A 1232	980	9800	µg/kg/day	Lillie et al., 1974
Chicken	✓			A 1232	980		µg/kg/day	Lillie et al., 1974
Chicken	✓			A 1232	2440	4880	µg/kg/day	Lillie et al., 1975
Chicken	✓			A 1016	2440		µg/kg/day	Lillie et al., 1975

NOEL, no-observed-effect level; LOEL, lowest-observed effect level.
a13200µg/kg in eggs.
b4000–18000 µg/kg in eggs

Table 11-10. Avian PCB toxicity summary for tissue residue effect levels for Aroclors.

Species	Labortory (L) or Field (F)	Adverse Effects Evaluated			Congener or Mixture	Tissue (T) Egg (E)	Critical Effect Levels		Reference
		Hatching Success	Chick Growth	Embryo Mortality			NOEL (µg/kg egg)	LOEL (µg/kg egg)	
Chicken[a]	L			✓	A 1242	E			Blazak & Marcum, 1975
Chicken	L		✓		A 1242	T	670	6700	Gould et al., 1997
Mallard	F				A 1242	T		105,000	Haseltime & Prouty, 1980
Chicken	L		✓		A 1254	T	670	6700	Gould et al., 1997
Ringed turtle dove	L	✓			A 1254	T		16,000	Peakall & Peakall, 1973

NOEL, no-observed-effect level; LOEL, lowest-observed effect level.
[a]The LD64 in this study was 10,000 µg/kg egg.

Table 11-11. Avian PCB toxicity summary for tissue residue effect levels for total PCBs determined by tissue injection experiments.

Species	Laboratory (L) Field (F) or Both (B)	Adverse Effects Evaluated				Critical Effect Levels		Reference
		Hatching Success	Reproductive Success	Deformities	Reproductive Behavior	NOEL (µg/kg egg)	LOEL (µg/kg egg)	
Chicken	L	✓				360	2500	Scott, 1977
Chicken	L	✓				950	1500	Britton & Huston, 1973
Chicken	L	✓	✓				4000	Tumasonis et al., 1973
Tree swallow	F				✓		5000–7000	McCarty & Secord, 1999
Bald eagle	F		✓				4000	Ludwig et al., 1993
Bald eagle	F		✓			1300	7200	Wiemeyer et al., 1984
Bald eagle	F		✓				13,000	Bosveld & van den Berg, 1994
Bald eagle	F	✓				400	4000	Bowerman et al., 1995
Double-crested cormorant	F	✓				350	3500	Tillit et al. 1992; Yamashita et al., 1993
Common tern	F		✓			7000	8000	Bosveld & van den Berg, 1994
Common tern	L	✓		✓		4800	10,000	Hoffman et al. 1993
Common tern	F	✓				5200–5600	7000	Becker et al., 1993
Forster's tern	B	✓				4500	22,200a	Kubiak et al., 1989
Forster's tern	F	✓				7000	19,000	Bosveld & van den Berg, 1994
Caspian terns	F	✓		✓		420	4200	Yamashita et al., 1993; Giesy et al., 1994b
Herring gulls	F	✓		✓		500		Weseloh et al., 1991; Giesy et al., 1994a

NOEL, no-observed-effect level; LOEL, lowest-observed-effect level.
aNOEL = 2.2 µg/kg TCDD-Eq in egg.

Table 11-12. Avian PCB toxicity summary for tissue residue effect levels for PCB congeners determined in laboratory studies.

Species	Hatching Success	Deformities	Embryo Mortality	Decreased Hatch Weight	Congener or Mixture	Tissue (T) or Egg (E) Injection	LD_{50} (µg/kg egg)	NOEL (µg/kg egg)	LOEL (µg/kg egg)	TCDD-EQ (LD_{50})	Reference
Chicken	✓				PCB 77	T	8.6			0.43	Brunstrom & Andersson, 1988
Chicken					PCB 77	E			30		Nikolaidis et al. 1988
Chicken					PCB 77	E	2.6	0.12	1.2		Hoffman et al., 1998
Chicken				✓	PCB 77	E	8.8	1.0	3.0		Powell et al., 1996a
Turkey	✓				PCB 77	T	~800			40	Brunström & Lund, 1988
Ring-necked pheasant	✓				PCB 77	T		100[a]			Brunström & Reutergardh, 1986
American kestrel			✓		PCB 77	E	316		100		Hoffman et al., 1998
Goldeneye	✓				PCB 77	T	>1000			>50	Brunström & Reutergardh, 1986
Domestic goose	✓				PCB 77	T	>1000			>50	Brunström, 1988
Mallard	✓				PCB 77	T	>5000			>250	Brunström, 1988
Black-headed gull	✓				PCB 77	T	<1000			<50	Brunström, 1988
Herring gull	✓				PCB 77	T	>1000			1–2	Brunström, 1988
Chicken	✓				PCB 126	T	3.2			0.32	Brunström, & Andersson, 1988
Chicken	✓				PCB 126	T	2.3			0.23	Powell et al., 1996
Chicken	✓				PCB 126	T	0.4			0.04	Hoffman et al., 1995
Chicken		✓	✓		PCB 126	E		0.5	1.0		Zhao et al., 1997
Chicken				✓	PCB 126	E	0.4		0.3		Hoffman et al., 1998

(continued)

Table 11-12. Continued

Species	Hatching Success	Deformities	Embryo Mortality	Decreased Hatch Weight	Congener or Mixture	Tissue (T) or Egg (E) Injection	LD$_{50}$ (µg/kg egg)	NOEL (µg/kg egg)	LOEL (µg/kg egg)	TCDD-EQ (LD$_{50}$)	Reference
Bobwhite	✓				PCB 126	T	24			2.4	Hoffman et al., 1995
American kestrel		✓			PCB 126	E	65	2.3	23	6.5	Hoffman et al., 1998
Double-crested cormorant	✓				PCB 126	T	158			16	Powell et al., 1997a
Double-crested cormorant			✓		PCB 126	E		200	400		Powell et al., 1997b
Common tern	✓		✓		PCB 126	E	104		44	10.4	Hoffman et al., 1998
Common tern			✓		PCB 126	E	45				Hoffman et al., 1995
Chicken			✓		PCB 157	E	1500				Brunström, 1990
Chicken	✓				PCB 105	T	2200			0.22	Brunström, 1990
Chicken				✓	PCB 105	E		100	300		Powell et al., 1996b
Chicken	✓				PCB 118	T	8000			0.08	Brunström, 1990
Chicken	✓				PCB 156	T	1500			0.15	Brunström, 1990
Chicken	✓				PCB 167	T	>4000			>0.04	Brunström, 1990
Chicken	✓				PCB 169	T	170			0.17	Brunström & Andersson, 1988

NOEL, no-observed-effect level; LOEL, lowest-observed-effect level.
a Noel = 5 µg/kg TCCD-EQ.

Table 11-13. Avian PCB toxicity summary for dietary exposures and tissue residue effect levels for TCDD and TCDD-equivalents (TCDD-Eq).

Species	Laboratory (L) or Field (F)	Hatching Success	Reproductive Success	Deformities	Embryo Mortality	Deceased Hatch Weight	Congener or Mixture	Dose (D), Tissue (T), or Egg (E) Injection	LD_{50} or (LD_x as Noted)	NOEL	LOEL	Units	Reference
Ring-neck pheasants	L				✓		TCDD	D		0.014	0.14	µg/kg/day	Nosek et al. 1992
Chicken	L				✓		TCDD	E	0.147			µg/kg egg	Verrett, 1976
Chicken	L				✓		TCDD	E	0.115			µg/kg egg	Henshel et al., 1993
Chicken	L				✓		TCDD	E	0.18			µg/kg egg	Henshel et al., 1993
Chicken	L				✓		TCDD	E	0.24			µg/kg egg	Allred & Strange, 1977
Chicken	L				✓		TCDD	E	LD_{100} = 1.0			µg/kg egg	Higginbotham et al., 1968
Chicken	L				✓		TCDD	E	0.15	0.08	0.16	µg/kg egg	Powell et al., 1996a
Chicken	L					✓	TCDD	E		0.06	0.01	µg/kg egg	Henshel et al., 1997a
Chicken	L					✓	TCDD	E		0.1	0.3	µg/kg egg	Henshel et al., 1997a
Chicken	L			✓			TCDD	E			0.32	µg/kg egg	Walker et al., 1997
Chicken	L			✓			TCDD	E		0.0006	0.01	µg/kg egg	Henshel et al., 1997b
Chicken	L						TCDD-EQ	T	LD_{100} = 1			µg/kg adult	Giesy et al., 1994a
Chicken	L						TCDD-EQ	T	0.14			µg/kg adult	Cheung et al 1981
Chicken	L			✓			TCDD-EQ	T	0.65			µg/kg egg	Giesy et al., 1994a
Chicken	L			✓			TCDD-EQ	T			0.006	µg/kg egg	Giesy et al., 1994a
Chicken	L				✓		TCDD-EQ	T	LD_{100} = 1.0		4	µg/kg egg	Higginbotham et al., 1986 Giesy et al., 1994a
Pheasant	L				✓		TCDD	E	1.4-2.2			µg/kg egg	Nosek et al., 1993
Wood duck	F		✓				TCDD-EQ	T			0.02	µg/kg egg	Walker et al., 1997; Giesy et al., 1994a
Wood duck	F						TCDD-EQ	T			>20-50	µg/kg egg	White & Hoffman, 1995
Double-crested cormorant	L	✓			✓		TCDD	E		1.0	4.0	µg/kg egg	Powell et al., 1997

(continued)

Table 11-13. Continued

Species	Laboratory (L) or Field (F)	Hatching Success	Reproductive Success	Deformities	Embryo Mortality	Deceased Hatch Weight	Congener or Mixture	Dose (D), Tissue (T), or Egg (E) Injection	LD$_{50}$ or (LD$_x$ as Noted)	NOEL	LOEL	Units	Reference
Double-crested cormorant	F	✓					TCDD-EQ	T	~0.55			µg/kg egg	Tillit et al., 1992
Double-crested cormorant	F	✓					TCDD-EQ	T	LD$_{100}$ = 0.1.03			µg/kg egg	Giesy et al., 1994a
Double-crested cormorant	F	✓					TCDD-EQ	T	LD$_{37}$ = 0.344			µg/kg egg	Giesy et al., 1994a
Double-crested cormorant	F	✓					TCDD-EQ	T	LD$_{27}$ = 0.217			µg/kg egg	Giesy et al., 1994a
Double-crested cormorant	F	✓					TCDD-EQ	T	LD$_8$ = 0.35			µg/kg egg	Giesy et al., 1994a
Double-crested cormorant	F				✓		TCDD-EQ	T	0.46			µg/kg egg	Tillit, 1989
Double-crested cormorant	F	✓					TCDD-EQ	T	0.46			µg/kg egg	Tillit et al., 1992
Caspian tern	F	✓					TCDD-EQ	T	0.75			µg/kg egg	Giesy et al., 1994a
Common tern	L	✓					TCDD-EQ	T		<1		µg/kg egg	Bosvold & van den Berg, 1994
Herring gull	F	✓					TCDD-EQ	T		1-2		µg/kg egg	Ludwig et al., 1993
Herring gull	F	✓					TCDD-EQ	T	LD$_{19}$ = 0.557			µg/kg egg	Giesy et al., 1994a
Osprey	F		✓				TCDD-EQ	T	0.14			µg/kg egg	Woodford et al., 1998
Bald Eagle	F		✓				TCDD-EQ	T	0.2			µg/kg egg	Elliott et al., 1996
Great Blue Heron	F			3	3		TCDD-EQ			0.02	0.245	µg/kg egg	Hart et al., 1991

NOEL, no-observed-effect level; LOEL, lowest-observed-level.

that, wherever possible, family-specific, if not species-specific, toxicity data (whether based on dietary or tissue residue-based TRVs) be chosen to most closely match the wildlife species of concern.

A number of studies with hepatocytes prepared from domestic and wild birds have examined the toxicity of PCBs and related compounds. EROD activity has been the most commonly measured endpoint in these studies. These studies suggest that, in general, EROD induction is not a toxic response per se but an adaptive biochemical response associated with exposure and some of the toxic effects of these chemicals. Excessive induction of mixed-function oxidase activity contributes to TCDD toxicity. Studies with cultured chicken embryo hepatocytes indicate that non-*ortho* PCB congeners (PCB-126, PCB-81, PCB-77, and PCB-169) are typically the most potent compounds, although the mono-*ortho* PCBs (PCB 66, 70, 105, 118, 122, 156, 157, and 167) and di-*ortho* PCBs (128, 138, 170, and 180) also can induce EROD activity (Kennedy et al., 1996a). Although in vitro studies can provide sound indications of the relative potency of different congeners, these experiments are of limited value for establishing defensible whole-organism TRVs for individual congeners in birds. In addition, in vitro studies do not account for pharmacokinetic and pharmacodynamic parameters, which could alter the toxic potential of a congener.

For birds, it is recommended that a combination of dietary exposure modeling and tissue residue-based effect levels be used. The potential for exposure is greatest for top-level predators; thus several lines of evidence should be evaluated (possibly in a phased approach).

There are considerable field and laboratory data available in which the concentrations of total PCBs, PCB congeners, TCDD, or TEQs in eggs have been measured and related to adverse effects. The main advantages of these studies, especially those conducted in the last 10 years, are that the total weathered PCBs (often by PCB congener analysis as either total PCBs or as TEQs) in the tissue of wildlife species was measured and that these concentrations were related to ecologically relevant endpoints. However, few studies have been conducted from which a dietary TRV for PCB exposure to birds can be derived.

Attempts to derive TRVs for birds at U.S. Navy sites in the San Francisco Bay have focused on a dietary study by Platonow and Reinhart (1973; see also Blankenship and Giesy, 2001). However, there appears to be confusion regarding the calculation of the daily intake rate from this study. In different parts of the same report (Blankenship and Giesy, 2001) and in a U.S. EPA document (U.S. EPA, 1995), four different values for the NOEL and LOEL values have been reported (table 11-14). Because the EPA document states clearly how their value was calculated, it appears to be the most valid. However, for a dietary exposure-based TRV, we recommend that the study by Dahlgren et al. (1971) be used because it was conducted on a wildlife species and it evaluated a sensitive life stage. This study was also selected by the EPA (1995) to provide a basis for water quality values protective of wildlife. The NOEL and LOEL values from this study are 0.18 and 1.8 mg/kg/day, respectively.

For a tissue residue-based TRV, several studies for wildlife have been considered. A specific recommendation cannot be made at this point because the TRVs are usually species specific and thus should be selected based on similarity to the

Table 11-14. Avian Tissue Residue Values (TRVs)

Origina Study	NOEL (mg/kg/day)	LOEL (mg/kg/day)	TRV Reference
Platonow & Reinhart, 1973	0.09	0.88	Blankenship & Giesy, 2001
Platonow & Reinhart, 1973	0.034	0.34	Blankenship & Giesy, 2001
Platonow & Reinhart, 1973	0.244	2.44	U.S. EPA, 1995
Dahlgren et al., 1972	0.18	1.8	U.S. EPA, 1995

wildlife species of concern. As an example, a tissue residue-based TRV to protect against embryo mortality for double-crested cormorants would be between 350 (NOAEL) and 3500 µg/kg (LOAEL) for total PCBs and between 1 (NOAEL) and 4 µg/kg (LOAEL) for TEQs (Giesy et al., 1994a,c).

Aroclor-based Data

Some of the most frequently cited laboratory bird studies for controlled, dietary exposures are Aroclor-based exposure studies (table 11-14). In particular, studies of the ring-necked pheasant (*Phasianus colchicus*; Dahlgren et al., 1971) and the chicken (Platonow and Reinhart, 1973) are often selected for development of TRVs. The Great Lakes Water Quality Initiative report (1995) used the ring-necked pheasant study of Dahlgren et al. (1972) to develop TRVs because the pheasant is a wildlife species that was evaluated at a critical life stage. However, to compare estimated exposures for wildlife species in a risk assessment to this type of benchmark, an estimate of total PCBs from a PCB congener-specific analysis, rather than an Aroclor-based method, is recommended because of the environmental weathering processes. Furthermore, although the predominant route of exposure for toxicity studies for Aroclors is dietary, few of the available dietary studies have evaluated the concentration of PCBs in eggs during and after the exposure (tables 11-9 and 11-10). Some of the limitations of these Aroclor-based data are that (1) few of these studies have been conducted on wildlife species, and (2) as discussed previously, it may not be appropriate to compare laboratory exposures with technical Aroclors (with potential contamination by other, more potent dioxinlike chemicals) to field exposures to weathered PCBs. These factors need to be considered when estimating TRVs and the uncertainty associated with them.

Total PCB Data

There have been many field and laboratory studies in which concentrations of total PCBs in eggs have been measured and related to adverse effects (table 11-12). The main advantages of these studies, especially those conducted in the last 10 years, are that the total weathered PCBs (often by PCB congener analysis as either total PCBs or as TEQs) were measured in the tissue of wildlife species and that these concentrations were related to ecologically relevant endpoints. Some of the limitations of total PCB data are that (1) in a few cases, the individual PCB congeners were not quantified (studies from the 1970s), and (2) there are potential cocontaminants that may confound the interpretation of effect levels from field studies; in many cases, however, data on some co-contaminants are available from these same studies.

PCB Congener and TCDD-Equivalent Data

There are considerable field and laboratory data available on the concentrations of PCB congeners, TCDD, or TEQs in eggs and their relationship to adverse effects (table 11-13). This includes several laboratory studies conducted under controlled conditions with wildlife species. Some of the limitations of these data are that (1) there are no dietary exposures, and (2) for the field studies, there are potential cocontaminants that may confound the interpretation of effect level; in many cases, however, cocontaminant data are available from these same studies.

Mammals

Toxic Effects

Immunotoxic effects of PCBs are well established in field and laboratory studies of mammals (Jepson et al., 1999; Lapierre et al., 1999). Due to their trophic status, geographic distribution, and other natural history characteristics, mammals accumulate high concentrations of PCBs in their tissues (Tanabe et al., 1994). Since 1968, 16 species of mammals have experienced population instability, major stranding episodes, reproductive impairment, endocrine and immune system disturbances, or have been afflicted with serious infectious diseases (Colborn and Smolen, 1996). Organochlorine contaminants, particularly PCBs and DDTs, are suspected to have caused reproductive and immunological disorders in mammals (Colborn and Smolen, 1996). The high prevalence of diseases and reduced reproductive capability of the Baltic gray seal (*Halichoerus grypus*) and the ringed seal (*Phoca hispida*), reproductive failure in Wadden Sea harbor seals (*Phoca vitulina*) and St. Lawrence estuary beluga whales (*Delphinapterus leucas*), and viral infection and mass mortalities of North American bottlenose dolphin (*Tursiops truncatus*), Baikal seal (*Phoca sibirica*), and Mediterranean striped dolphin (*Stenella coeruleoalba*) have all been associated with the presence of high concentrations of PCBs in tissues of affected individuals. However, due to the existence of several confounding factors that limit the ability to extrapolate results from field studies, unequivocal evidence of a cause–effect linkage between disease development and mass mortalities in mammals has not been established.

Apart from chemical contaminants, exposure to brevitoxin, a neurotoxin produced by a marine dinoflagellate, has been hypothesized as a possible cause for the mortality of bottlenose dolphins along the Atlantic Coast of North America (table 11-15; Anderson and White, 1989). However, later studies have indicated that this evidence is circumstantial (Lahvis et al., 1995). Similarly, other factors such as population density, migratory movement, habitat disturbance, and climatological

Table 11-15. Summary of NOAELs and LOAELs for total PCBs and TEQs in semifield investigations with seals, dolphins, and European otters.

Exposure	PCBs	TEQs
Seals/dolphins		
Daily dose NOAEL	5.2 µg/kg bw/d	0.58 ng/kg bw/day
Daily dose LOAEL	28.9 µg/kg bw/d	5.8 ng/kg bw/day
Dietary NOAEL	100 ng/g, wet wt	NA
Dietary LOAEL	200 ng/g, wet wt	NA
Seal blood NOAEL (lipid 0.05–0.32%)	5.2 µg/g, lipid wt	NA
Seal blood LOAEL	25 µg/g, lipid wt	NA
Seal blubber NOAEL	NA	90 pg/g, lipid wt
Seal blubber LOAEL	NA	286 pg/g, lipid wt
Dolphin blood LOAEL (in vitro)	26 ng/g, wet wt	NA
European otter		
Dietary NOAEL (lipid 6.2%)	12 ng/g, wet wt (or)	1 pg/g, wet wt or
	200 ng/g, lipid wt	16 pg/g, lipid wt
Dietary LOAEL	33 ng/g, wet weight (or)	2 pg/g, wet wt
	530 ng/g, lipid wt	33 pg/g, lipid wt
Otter liver NOAEL (lipid 4.2%)	170 ng/g, wet wt (or)	42 pg/g, wet wt
	4 µg/g, lipid wt	1 ng/g, lipid wt
Otter liver LOAEL	460 ng/g, wet wt (or)	84 pg/g, wet wt or
	11 µg/g, lipid wt	2 ng/g, lipid wt

NOAEL, no-observed-adverse-effect level; LOAEL, lowest-observed-adverse effect level. Data are from Kannan et al., 2000.

factors have been thought to play a role in mass mortalities of mammals (Lavigne and Schmitz, 1990). Another hypothesis is that synthetic chemicals such as PCBs initially cause immunosuppression, rendering mammals susceptible to opportunistic bacterial, viral, and parasitic infection (Lahvis et al., 1995). Debilitating viruses such as *Morbillivirus* may result in further immunosuppression, starvation, and death (Lahvis et al., 1995).

Due to ethical, logistical, and practical reasons, only a few controlled laboratory exposure studies with a few mammalian subjects have been conducted. A study was conducted in which immune function was compared in two groups of harbor seals that were fed herring originating from either the Baltic Sea ($n = 12$), which is an area of intense contamination with pollutants including PCBs, or fed fish from the Atlantic Ocean ($n = 12$), a less contaminated area (de Swart et al., 1994). Blood of seals fed Baltic Sea fish contained significantly lower concentrations of vitamin A, less natural killer cell activity, and exhibited less lymphocyte proliferation after exposure to mitogens compared to the seals fed Atlantic fish. The effect on immune function was observed within 4–6 months of the start of the experiment (Ross et al., 1996b).

The field studies conducted by de Swart et al. (1994) and Ross et al. (1995, 1996a) have the advantage of mimicking exposures to weathered mixtures of PCB congeners found under current field conditions. However, the presence of a variety of cocontaminants in the diet precludes the assumption that PCBs are the only cause for immune dysfunction. Nevertheless, based on the results of other laboratory studies involving exposure of rats to AhR-active compounds (Safe, 1994; Ross et al., 1997), reduction in the proliferative activity of lymphocytes exposed to mitogens in seals was consistent with an AhR-mediated mechanism of toxic action. PCBs accounted for 80–93% of the AhR-active compounds in the diet of seals (Ross et al., 1995, 1996a). Thus, the observed effects in seals were attributed primarily to PCBs (Ross et al., 1995, 1996a), even though other immunotoxic contaminants could have been present. For instance, di- and tributyltin (DBT and TBT) are potent immunotoxic compounds that suppress concanavalin-A–induced mitogenesis of peripheral blood mononuclear cells at several tens of ng/L concentrations in the blood of mammals (Nakata et al., 2000). Butyltin compounds could have been present in the fish diet of seals, but they were not measured. However, butyltins do not act through the AhR-mediated mechanism of immunotoxicity (Kannan et al., 2000).

The effects of in vitro exposure to different organochlorines, including PCB congeners (IUPAC nos. 138, 153, 180, and 169), have been evaluated on immune functions of beluga whale peripheral blood leukocytes and splenocytes (De Guise et al., 1998). When cells were exposed to a mixture consisting of a final concentration of 5 µg/g of each of three PCB congeners (138, 153, and 180), the PHA-stimulated proliferation of beluga whale splenocytes was significantly reduced relative to unexposed controls.

TRVs

Information on the effects of PCBs on marine mammals is limited, but a substantial amount of information is available, including chronic effects on reproduction and biochemical and immunological responses of otters and minks (tables 11-16–11-18). Because minks and otters are fish-eating mammals with physiologies and reproductive strategies that are in some ways similar to some marine mammals, we have relied on these species to develop TRV values for the most likely exposed types of mammals. Due to their sensitivities and potential for exposure because they eat fish and are at the top of the food chain, minks and otters are generally the critical species in risk assessments of PCB-contaminated sites. Therefore, extrapolation of threshold concentrations derived for PCBs in otters may be more relevant for wild or feral mammals than those based on studies of rodents.

A study that examined hepatic retinoids and corresponding total PCB concentrations in environmentally exposed feral and captive otters was used to derive threshold PCB concentrations for otters (Murk et al., 1998). Hepatic retinoids (retinol and retinyl palmitate) were significantly negatively correlated with hepatic TEQs, calculated based on the concentrations of non- and mono-*ortho* PCBs (Murk et al., 1998). The above results were further confirmed in a field study that examined the frequency and severity of diseases in feral European otters, which found that the disease symptoms were positively correlated with TEQ concentrations in the liver. At concentrations > 5 ng TEQ/g, lipid weight in the liver, there was a significant increase in the incidence of diseases. These results were supported by the fact that vitamin A is essential not only for normal growth and development but also to provide resistance against microbial infection. Moreover, vitamin A deficiency increases the risk of infection, which in turn decreases vitamin A

Table 11-16. LOAEC, NOAEC, or EC_{50} values for toxic effects of dietary exposure of commercial PCB mixtures or congeners in minks.

PCB Mixture/Congener	NOAEC, LOAEC, or EC_{50}	Reference
Commerial mixture in diet		
Aroclor 1016[a]	LOAEC = 2 µg/g	Bleavins et al., 1980
Aroclor 1254[b]	NOAEC = <1 µg/g	Aulerich and Ringer, 1977
Aroclor 1254	LOAEC = 0.1 mg/kg/day or 1 µg/g in diet	Wren et al., 1987
Aroclor 1254	LC_{50} 79 µg/g (28-d)	Aulerich et al., 1986
3,3',4,4',5,5'-(PCB 169)	LD_{50} = 0.05 µg/g; NOAEC = 0.01 µg/g	Aulerich et al., 1987
2,2',3,3',6,6'- (PCB 136)[c]	LOAEC = 5 µg/g	Aulerich et al., 1985
2,2',4,4',5,5'- (PCB 153)[c]	LOAEC = 5 µg/g	Aulerich et al., 1985
2,3,7,8-TCDD	LD_{50} = 4.2 ng/g body weight	Hochstein et al., 1988
Weathered PCBs/TEQs		
Aroclor 1254-fed-rabbit diet	LC_{50} = 47 µg/g (28 day)	Aulerich et al., 1986
Contaminated fish-diet total PCBs	NOAEC =72 ng/g	Giesy et al., 1994c
Contaminated fish-diet TEQs	NOAEC = 0.3 pg/g or 0.08 ng/kg body weight/day	Tillitt et al., 1996
Contaminated fish-diet TCDD-EQs (H4IIE- bioassay derived)	NOAEC = 2 pg/g or 0.54 ng/kg body weight/day	Giesy et al., 1994c
Body residues based PCBs/TEQs		
All technical PCB mixtures[d]	Relative litter size EC_{50} = 1.2 µg/g; kit survival EC_{50} = 2.4 µg/g	Leonards et al., 1995
TEQs[d]	Relative litter size EC_{50} = 0.16 ng/g; kit survival EC_{50} = 0.20 ng/g	Leonards et al., 1995

LOAEC, lowest-observed-adverse-effect concentration; NOAEC, no-observed-adverse-effect concentration.
[a]Aroclors 1242 and 1254 caused significant reproductive failure at 2 µg/g in the diet.
[b]Assessment based on several studies reported by Aulerich and co-workers.
[c]No effect on survival and reproduction, but slightly altered brain dopamine concentrations.
[d]Estimated based on several technical mixture-exposure studies, and the values were derived based on a bioaccumulation model.

concentrations. Thus, hepatic retinoids have been suggested as a sensitive indicator for PCB toxicity in otters (Murk et al., 1998).

Several studies have demonstrated that the mink is among the most sensitive species to the toxic effects of PCBs (Aulerich and Ringer, 1977; Tillitt et al., 1996; tables 11-17 and 11-18). For this reason, there have been several studies of the effects of PCBs on minks (Kihlström et al., 1992; Heaton et al., 1995a,b; Tillitt et al., 1996). These studies included technical PCB mixtures, weathered mixtures containing other toxicants, and individual congeners. Several authors have critically reviewed the toxic effects of PCBs to minks to derive NOAEL values (tables 11-17 and 11-18; Giesy et al., 1994c; Giesy and Kannan, 1998; Leonards et al., 1995).

Because minks are a sensitive species to the toxic effects of PCBs, toxic threshold concentrations could be used as a conservative estimate for the protection of

Table11-17. Toxic threshold concentrations of weathered PCBs (μg/g, wet weight) or TEQs (pg/g wet weight) in minks for reproductive effects.

Endpoint	Total PCBs	TEQs	Reference
Relative litter size (EC$_{50}$)	1.2 μg/g, wet wt	160 pg/g, wet wt	Leonards et al. (1995)
Kit survival (EC$_{50}$)	2.36 μg/g, wet wt	200 pg/g, wet wt	Leonards et al. (1995)
Mink liver NOAEL	2.03 μg/g, lipid wt	11 pg/g, wet wt	Heaton et al. (1995a); Tillitt et al. (1996)
Mink liver LOAEL	44.4 μg/g, lipid wt	324 pg/g, wet wt	Heaton et al. (1995a); Tillitt et al. (1996)
Dietary NOAEL	15 ng/g, wet wt	1.03 pg/g, wet wt	Heaton et al. (1995a)
Dietary LOAEL	720 ng/g, wet wt	19 pg/g, wet wt	Heaton et al. (1995a)
Daily dose NOAEL	4 μg /kg bw/d	0.25 ng/kg body weight/day	Heaton **et al.** (1995a)
Daily dose LOAEL	130 μg/kg bw/d	3.6 ng/kg body weight/day	Heaton et al. (1995a)
Threshold dietary concentration	250 ng/g, wet wt	1.9 pg/g, wet wt	Tillitt et al. (1996); Restum et al. (1998)
Threshold daily dose	NA	0.42 ng/kg body weight/day	Tillitt et al. (1996)
Threshold liver concentration	NA	60 pg/g, wet wt or 1.2 ng/g, lipid wt[a]	Tillitt et al. (1996)

NOAEL, no-observed-adverse-effect level; LOAEL, lowest-observed-adverse-effect level. EC$_{50}$ values refer to those for whole body concentrations. NA; data not available. Threshold values were estimated as geometric mean of NOAEL and LOAEL.

[a]Lipid content of mink liver was assumed to be 5% (Poole et al., 1995).

mammalian populations (Starodub et al., 1996; Giesy and Kannan, 1998). An analysis was conducted to determine whether AhR-mediated (dioxinlike) or non–AhR-mediated effects (neurotoxicity caused by di-*ortho*-substituted congeners) was the critical mechanism of toxicity (Giesy and Kannan, 1998). The purpose was to determine which of the TRVs would result in the lowest allowable concentration of the total weathered mixture of PCBs in the diet. To conduct this analysis, the weathered mixture of PCBs found in carp (*Cyprinus carpio*) from Saginaw Bay, Michigan, were added to the diet of minks held under laboratory conditions (table 11-18). In this way, the effects measures were for the same weathered mixture, and concentrations of AhR-active and inactive congeners in both the diet and bodies of the minks were known. World Health Organization TEF values (van den Berg et al., 1998) were used to calculate TEQs (table 11-18). In addition, TEF values were derived for the neurotoxicity of di-*ortho*-substituted congeners in PC12 cells (table 11-1). To determine the critical toxicant, hazard quotients (HQs) were calculated (table 11-19). Hazard quotients corrected for the relative potencies of the congeners to cause the two types of toxicity and accounted for changes in the absolute and relative proportions of the various congeners in the diet of the minks. Giesy and Kannan (1998) found that the critical toxic effect that resulted in the greatest HQ and that would allow the least exposure to the total concentration of PCBs in the weathered mixture was the AhR-mediated mechanism of action. This does not

Table 11-18. TEQs of non- and mono-*ortho* PCBs (pg/g, wet wt) in carp and mink.

IUPAC no.	Structure	TEFV	Carp	Mink
Non-*ortho*				
77	3,3',4,4'-	0.0001	0.23	0.32
126	3,3',4,4',5-	0.1	15	1050
169	3,3',4,4',5,5'-	0.01	0.3	104
Mono-*ortho*				
60	2,3,4,4'-	0.0001	4.1	0.08
66	2,3,4,4'-	0.0001	8.3	0.17
105	2,3,3',4,4'-	0.0001	6.0	72
118	2,3',4,4',5-	0.0001	13.5	203
156	2,3,3',4,4',5-	0.0005	10	300
Total			57	1730

Mink concentrations were calculated by multiplying fish concentration with BMFs from Leonards et al., 1997.

[a]WHO (1997) TEFs (van den Berg et al., 1998) for IUPAC nos. 60 and 66 were assigned as that of 105.

mean that the effects that *ortho*-substituted PCBs could cause are unimportant; it simply means that if minks were protected from AhR-mediated effects, they would likely be protected from the effects of the *ortho*-substituted PCBs. The critical concentration for AhR-mediated effects is less than that for the neurotoxic effects of the di-*ortho*-substituted congeners, but the concentration of di-*ortho*-substituted congeners is higher. Thus, there is a type of compensation. The HQ value for the AhR-mediated effects was approximately 32-fold lower than that for the non-AhR-mediated effects. Thus, basing risk assessments of the AhR-mediated effects on reproduction in minks should provide a 32-fold safety factor for neurobehavioral effects.

We recommend that a tissue-residue threshold effect concentration for PCBs in mammals be 11 µg/g, lipid weight (Kannan et al., 2000) to protect immune function. Based on immunotoxicological studies in seals, a threshold effect concentration for blubber TEQ of 520 pg/g, lipid weight, has been suggested (Ross et al., 1995).

Table 11-19. Hazard quotients (HQ) for total PCBs, TEQs and di- through tetra-*ortho* PCBs in mink based on concentrations in the diet.

Compound	NOAEC	HQ
Total PCBs (weathered)	72 ng/g	50
Total PCBs (technical mixtures)	200 ng/g	18
TEQ	0.3pg/g	190
Di- through tetra-*ortho* PCBs	500 ng/g	5.9

NOAEC, no-observed-adverse-effect concentration.

References

Abbott, B.D., Birnbaum, L.S., 1990. Effects of TCDD on embryonic ureteric epithelial EGF receptor expression and cell proliferation. Teratology 41, 71–84.

Abramowicz, D.A., 1990. Aerobic and anaerobic biodegradation of PCBs: A review. Crit. Rev. Biotechnol. 10, 241–251.

Abramowicz, D.A., 1994. Aerobic PCB biodegradation and anaerobic PCB dechlorination in the environment. Res. Microbiol. 145, 42–46.

Agrawal, A.K., Tilson, H.A., Bondy, S.C., 1981. 3,4,3',4'-tetrachlorobiphenyl given to mice prenatally produces long-term decreases in striatal dopamine and receptor binding sites in the caudate nucleus. Toxicol. Lett. 7, 417–424.

Ahlborg, U.G., 1989. Nordic risk assessment of PCDDs and PCDFs. Chemosphere 19, 603–608.

Ahlborg, U.G., Becking, G.C., Birnbaum, L.S., Brouwer, A., Derks, H.J.G.M., Feeley, M., Golor, G., Hanberg, A., Larsen, J.C., Liem, A.K.D., Safe, S.H., Schlatter, C., WFrn, F., Younes, M., Yrjanheikki, E., 1994. Toxic equivalency factors for dioxin-like PCBs, Chemosphere 28, 1049–1067.

Ahlborg, U.G., Brouwer, A., Fingerhut, M.A., Jacobson, J.L., Jacobson, S.W., Kennedy, S.W., Kettrup, A.A, Koeman, J.H., Poiger, H., Rappe, C., Safe, S.H., Seegal, R.F., Tuomisto, J., van den Berg, M., 1992. Impact of polychlorinated dibenzo-*p*-dioxins, dibenzofurans, and biphenyls on human and environmental health, with special emphasis on application of the toxic equivalency factor concept. Eur. J. Pharmacol. 228, 179–199.

Allred, P.M., Strange, J.R., 1977. The effects of 2,4,5-trichlorophenoxyacetic acid and 2,3,7,8-tetrachlorodibenzo-p-dioxin on developing chicken embryos. Arch. Environ. Contam. Toxicol. 6, 483–489.

Anderson, D.M., White, A.W., 1989. Toxic dinoflagellates and marine mammal mortalities. 89–3 (CRC-89-6). Woods Hole Oceanographic Institution, Woods Hole, MA.

Andersson, L., Nikolaidis, E., Brunström, B., Bergman, Å. Dencker, L., 1991. Effects of polychlorinated biphenyls with Ar receptor affinity on lymphoid development in the thymus and the bursa of Fabricius chick embryos *in ovo*. Toxicol. Appl. Pharmacol. 107, 183–188.

Angus, W.G., Contreras, M.L., 1996. Effects of polychlorinated biphenyls on dopamine release from PC12 cells. Toxicol. Lett. 89, 191–199.

Asplund, L., Grafström, A.-K., Haglund, P., Jansson, B., Järnberg, U., Mace, D., Strandell, M., De Wit, C., 1990. Analysis of non-*ortho* polychlorinated biphenyls and polychlorinated naphthalenes in Swedish dioxin survey samples. Chemosphere 20, 1481–1488.

Atlas, E., Giam, C.S., 1981. Global transport of organic pollutants: Ambient concentrations in remote marine atmosphere. Science 211, 163–165.

Aulerich, R.J., Bursian, S.J., Breslin, W.J., Olson, B.A., Ringer, R.K., 1985. Toxicological manifestations of 2,4,5,2',4',5'-, 2,3,6,2',3',6'- and 3,4,5,3',4',5'- hexachlorobiphenyl and Aroclor 1254 in mink. J. Toxicol. Environ. Health 15, 63–79.

Aulerich, R.J., Bursian, S.J., Evans, M.G., Hochstein, J.R., Koudele, K.A., Olson, B.A., Napolitano, A.C., 1987. Toxicity of 3,4,5,3',4',5'-hexachlorobiphenyl to mink. Arch. Environ. Contam. Toxicol. 16, 53–60.

Aulerich, R.J., Ringer, R.K., 1977. Current status of PCB toxicity to mink, and effect on their reproduction. Arch. Environ. Contam. Toxicol. 6, 279–292.

Aulerich, R.J., Ringer, R.K., Safronoff, J., 1986. Assessment of primary vs secondary toxicity of Aroclor 1254 to mink. Arch. Environ. Contam. Toxicol. 15, 393–399.

Ballschmiter, K., Zell, M., 1980. Analysis of polychlorinated biphenyls (PCB) by glass capillary gas chromatography. Fres. Z. Anal. Chem. 302, 20–31.

Bannister, R., Safe, S., 1987. Synergistic interactions of 2,3,7,8-TCDD and 2,2',4,4',5,5'-hexachlorobiphenyl in C57BL/6J and DBA/2J mice: Role of the Ah receptor. Toxicology 44, 159–169.

Bandiera, S., Safe, S., Okey, A.B., 1982. Binding of polychlorinated biphenyls classified as either phenobarbitone-, 3-methylcholanthrene- or mixed-type inducers to cytosolic Ah receptor. Chem. Biol. Interact. 39, 259–277.

Barnes, D.G., 1991. Toxicity equivalents and EPA's risk assessment of 2,3,7,8-TCDD. Sci. Total Environ. 104, 73–86.

Barrett, J.C., 1995. Mechanisms for species differences in receptor-mediated carcinogenesis. Mutat. Res. 333, 189–202.

Beattie, M.K., Gerstenberger, S., Hoffman, R., Dellinger, J.A., 1996. Rodent neurotoxicity bioassays for screening contaminated Great Lakes fish. Environ. Toxicol. Chem. 15, 313–318.

Becker, P.H., Schuhmann, S., Koepff, C., 1993. Hatching failure in common terns (Sterna hirundo) in relation to environmental chemicals. Environ Pollut. 79, 207–213.

Bedard, D.L., May, R.J., 1996. Characterization of the polychlorinated biphenyls in the sediments of woods pond: Evidence for microbial dechlorination of Aroclor 1260 in situ. Environ. Sci. Technol. 30, 237–245.

Bedard D. L., van Dort H., May R.J., Smullen, L.A., 1997. Enrichment of microorganisms that sequentially dechlorinate the residue of Aroclor 1260 in Housatonic River sediment. Environ. Sci. Technol. 31, 3308–3313.

Bedard, D.L., Bunnell, S.C., Smullen, L.A., 1996. Stimulation of microbial para-dechlorination of polychlorinated biphenyls that have persisted in Housatonic River sediment for decades. Environ. Sci. Technol. 30, 687–694.

Bedard, D.L., Haber, M.L., 1990. Influence of chlorine substitution pattern on the degradation of polychlorinated biphenyls by eight bacterial strains. Microb. Ecol. 20, 87–102.

Bedard, D.L., Quensen, J.F. III, 1995. Microbial reductive dechlorination of polychlorinated biphenyls. In: Microbial Transformation and Degradation of Toxic Organic Chemicals, (eds.) L.Y. Young and C.E. Cerniglia, Wiley-Liss Division, John Wiley & Sons, Inc., New York, pp. 127–216.

Beland, P., Deguise, S., Girard, C., Lagace, A., Martineau, D., Michaud, R., Muir, D.C.G., Norstrom, R.J., Pelletier, E., Ray, S., Shugart, L.R., 1993. Toxic compounds and health and reproductive effects in St. Lawrence Beluga whales. J. Great Lakes Res. 19, 766–775.

Bergman, Å., Athanasiadou, M., Bergek, S., Haraguchi, K., Jensen, S., Klasson-Wehler, E., 1992. PCB and PCB methyl sulphones in mink treated with PCB and various PCB fractions. Ambio 21, 570–576.

Bergman, Å., Klasson-Wehler, E., Kuroki, H., 1994. Selective retention of hydroxylated PCB metabolites in blood. Environ. Health Perspect. 102, 464–469.

Biegel, L., Harris, M., Davis, D., Rosengren, R., Safe, L., Safe, S., 1989. 2,2,4,4',5,5'-hexachlorobiphenyl as a 2,3,7,8-tetrachlorodibenzo-p-dioxin antagonist in C57BL/6J mice. Toxicol. Appl. Pharmacol. 97, 561–571.

Birnbaum, L.S., DeVito, M.J., 1995. Use of toxic equivalency factors for risk assessment for dioxins and related compounds. Toxicology 105, 391–401.

Birnbaum, L.S., Weber, H., Harris, M.W., Lamb, J.C., McKinney, J.D., 1985. Toxic interaction of specific polychlorinated biphenyls and 2,3,7,8-tetrachlorodibenzo-p-dioxin: Increased incidence of cleft palate in mice. Toxicol. Appl. Pharmacol. 77, 292–302.

Black, J.J., Simpson, C.L., 1974. Thyroid enlargement in Lake Erie coho salmon. J. Natl. Cancer Inst. 53, 725–729.

Blackmore, D.K., 1963. The incidence and aetiology of thyroid dysplasia in Budgerigars (*Melopsittacus undulates*). Vet. Rec. 75, 1068–1072.

Blankenship, A.L., Giesy, J.P., 2002. Use of biomarkers of exposure and vertebrate tissue residues in the hazard characterization of PCBs at contaminated sites—application to birds and mammals. In G.I. Sunahara, A.Y. Renoux, C. Thellen, and C.L. Gaudet (Eds.), *Environmental Analysis of Contaminated Sites: Tools to Measure Success or Failure.* John Wiley & Sons, Ltd., London, pp. 153–180.

Blankenship, A., Matsumura, F., 1994. Changes in Biochemical and Molecular Biological Parameters Induced By Exposure to Dioxin-Type Chemicals, in M.A. Saleh, J.N. Blancato, and C.H. Nauman, (Eds.), *Biomarkers of Human Exposure to Pesticides*, American Chemical Society Symposium Series No. 542, ACS Books: Washington, D.C.), pp. 37–50.

Blankenship, A., Matsumura, F., 1997a. 2,3,7,8-Tetrachlorodibenzo-p-dioxin-induced activation of a protein tyrosine kinase, pp60src, in murine hepatic cytosol using a cell-free system. Mol. Pharmacol. 52, 667–675.

Blankenship, A., Matsumura, F., 1997b. 2,3,7,8-Tetrachloro-p-dioxin (TCDD) causes an Ah receptor-dependent and ARNT-independent increase in membrane levels and activity of p60src. Environ. Toxicol. Pharmacol. 3, 211–220.

Blazak, W.F., Marcum, J.B., 1975. Attempts to induce chromosomal breakage in chicken embryos with Aroclor 1242. Poult. Sci. 54, 310–312.

Bleavins, M.R., Aulerich, R.J., Ringer, R.K., 1980. Polychlorinated biphenyls (Aroclors 1016 and 1242): Effects on survival and reproduction in mink and ferrets. Arch. Environ. Contam. Toxicol. 9, 627–635.

Bletchly, J.D., 1984. Polychlorinated biphenyls: Production, current use and possible rate of future disposal in OECD member countries. In: Barros, M.C., Koemann, H., Visser, R. (Eds.), Proceedings of PCB Seminar. Ministry of Housing, Physical Planning and Environment, Amsterdam, the Netherlands, pp. 343–372.

Boon, J.P., van der Meer, J., Allchin, C.R., Law, R.J., Klungsoyr, J., Leonards, P.E., Spliid, H., Storr-Hansen, E., Mckenzie, C., Wells, D.E., 1997. Concentration-dependent changes of PCB patterns in fish-eating mammals: Structure evidence for induction of cytochrome P450. Arch. Environ. Contam. Toxicol. 33, 298–311.

Bosveld, A.T., Kennedy, S.W., Seinen, W., van den Berg, M., 1997. Ethoxyresorufin-O-deethylase (EROD) inducing potencies of planar chlorinated aromatic hydrocarbons in primary cultures of hepatocytes from different developmental stages of the chicken. Arch. Toxicol. 71, 746–750.

Bosveld, A.T.C., van den Berg, M., 1994. Effects of polychlorinated biphenyls, dibenzo-p-dioxins, and dibenzofurans on fish-eating birds. Environ. Rev. 2, 147–166.

Bowerman, W.W., Giesy, J.P., Best, D.A., Kramer, V.J., 1995. A review of factors affecting productivity of bald eagles in the Great Lakes region: Implications for recovery. Environ Health Perspect. 103, 51–59.

Britton, W.M., Huston, T.M., 1973. Influence of polychlorinated biphenyls in the laying hen. Poult. Sci. 52, 1620–1624.

Brouwer, A., 1987. Interference of 3,4,3',4'-tetrachlorobiphenyl in vitamin A (retinoids) metabolism; possible implications for toxicity and carcinogenicity of polyhalogenated aromatic hydrocarbons. Radiobiological Institute, Division for Health Research, The Netherlands.

Brouwer, A., 1989. Inhibition of thyroid hormone transport in plasma of rats by polychlorinated biphenyls. Arch. Toxicol. Suppl. 13, 440–445.

Brouwer, A., Blaner, W.S., Kukler, A., Van Den Berg, K.J., 1988. Study on the mechanism of interference of 3,4,3',4'-tetrachlorobiphenyl with the plasma retinol-binding proteins in rodents. Chem. Biol. Interact. 68, 203–207.

Brouwer, A., Reijnders, P.J.H., Koeman, J.H., 1989. Polychlorinated biphenyl (PCB)-contaminated fish induces vitamin A and thyroid hormone deficiency in the common seal (*Phoca vitulina*). Aquat. Toxicol. 15, 99–106.

Brouwer, A., Klasson-Wehler, E., Bokdam, M., Morse, D.C., Traag, W.A., 1990. Competitive inhibition of thyroxin binding to transthyretin by monohydroxy metabolites of 3,4,',4'-tetrachlorobiphenyl. Chemosphere 20, 1257–1262.

Brouwer, A., Morse, D.C., Lans, M.C., Schuur, A.G., Murk, A.J., Klasson-Wehler, E., Bergman, Å., Visser, T.J., 1998. Interactions of persistent environmental organohalogens with the thyroid hormone system: Mechanisms and possible consequences for animal and human health. Toxicol. Ind. Health 14, 59–84.

Brouwer, A., van den Berg, K.J., 1986. Binding of a metabolite of 3,4,3',4'-tetrachlorobiphenyl to transthyrethin reduces serum vitamin A transport by inhibiting the formation of the protein complex carrying both retinol and thyroxin. Toxicol. Appl. Pharm. 85, 301–312.

Brouwer, A., van den Berg, K.J., Blaner, W.S., Goodman, D.S., 1986. Transthyretin (prealbumin) binding of PCBs, a model for the mechanism of interference with vitamin A and thyroid hormone metabolism. Chemosphere 15, 1699–1706.

Brown, A.P., Ganey, P.E., 1995. Neutrophil degranulation and superoxide production induced by polychlorinated biphenyls are calcium dependent. Toxicol. Appl. Pharmacol. 131, 198–205.

Brown, J.F., Wagner, R.E., Feng, H., Bedard, D.L., Brennan, M.J., Carnahan, J.C., May, R.J., 1987. Environmental dechlorination of PCBs. Environ. Toxicol. Chem. 6, 579–594.

Brunström, B., 1988. Sensitivity of embryos from duck, goose, herring gull, and various chicken breeds to 3,3',4, 4'-tetrachlorobiphenyl. Poult. Sci. 67, 52–67.

Brunström, B., 1989. Toxicity of coplanar polychlorinated biphenyls to avian embryos. Chemosphere 19, 765–768.

Brunström, B., 1990. Mono-ortho-chlorinated chlorobiphenyls: Toxicity and induction of 7-ethoxyresorufin-O-deethylase (EROD) activity in chick embryos. Arch. Toxicol. 64, 188–192.

Brunström, B., Andersson, L., 1988. Toxicity and 7-ethoxyresorufin O-deethylase-inducing potency of coplanar polychlorinated biphenyls (PCBs) in chick embryos. Arch. Toxicol. 62, 263–266.

Brunström, B., Engwall, M., Hjelm, K., Lindqvist, L., Zebhhr, Y., 1995. EROD induction in cultured chick embryo liver: A sensitive bioassay for dioxin-like environmental pollutants. Environ. Toxicol. Chem. 14, 837–842.

Brunström, B., Lund, J., 1988. Differences between chick and turkey embryos in sensitivity to 3,3',4,4'-tetrachlorobiphenyl and in concentration/affinity of the hepatic receptor for 2,3,7,8-tetrachlorodibenzo-p-dioxin. Comp. Biochem. Physiol. C 91, 507–512.

Brunström, B., Reutergardh, L. 1986. Differences in sensitivity of some avian species to the embryotoxicity of a PCB, 3, 3',4, 4'-tetrachlorobiphenyl, injected into eggs. Environ. Pollut. 42(ser. A), 37–45.

Bunn, H.F., Gu, J., Huang, L.E., Park, J.-W., Zhu, H., 1998. Erythropoietin: A model system for studying oxygen-dependent gene regulation. J. Exp. Biol. 201, 1197–1201.

Cahill, M.A., Ernst, W.H., Janknecht, R., Nordheim, A., 1994. Regulatory squelching. FEBS Lett. 344, 105–108.

Cantrell, S.M. Lutz, L.H., Tillitt, D.E., Hannink, M., 1996. Embryotoxicity of 2,3,7,8-tetrachlorodibenzo-p-dioxin (TCDD): The embryonic vasculature is a physiological target for TCDD-induced DNA damage and apoptotic cell death in Medaka (*Orizias latipes*). Toxicol. Appl. Pharmacol. 141, 23–34.

CCME (Canadian Council of Ministers of the Environment), 1998. Canadian tissue residue guidelines for polychlorinated biphenyls for the protection of wildlife consumers of biota. Prepared by the Guidelines and Standards Division, Science Policy and Environmental Quality Branch, Environment Canada, Hull, Quebec.

Cecil, H.C., Bitman, J., Lillie, R.J., Fries, G.F., Verrett, J., 1974. Embryotoxic and teratogenic effects in unhatched fertile eggs from hens fed polychlorinated biphenyls (PCBs). Bull. Environ. Contam. Toxicol. 11, 489–495.

Cecil, H.C., Harris, S.J, Bitman, J., Fries, G.F., 1973. Polychlorinated biphenyl-induced decrease in liver vitamin A in Japanese quail and rats. Bull. Environ. Contam. Toxicol. 9, 179–185.

Chan, W.K., Yao, G., Gu, Y.Z., Bradfield, C.A., 1999. Cross-talk between the aryl hydrocarbon receptor and hypoxia inducible factor signaling pathways—Demonstration of competition and compensation. J. Biol. Chem. 274, 12115–12123.

Chen, P.H., Hsu, S.T., 1986. PCB poisoning from toxic rice-bran oil in Tailan. In: Waid, J.S., (Ed.), PCBs and the Environment, vol. III, CRC Press, Boca Raton, FL, pp. 27–38.

Cheung, M.O., Gilbert, E.F., Peterson, R.E., 1981. Cardiovascular teratogenicity of 2, 3, 7, 8-tetrachlorodibenzo-p-dioxin in the chick embryo. Toxicol. Appl. Pharmacol. 61, 197-204.

Chishti, M.A., Fisher, J.P., Seegal, R.F., 1996. Aroclors 1254 and 1260 reduce dopamine concentrations in rat striatal slices. Neurotoxicology 17, 653–660.

Chishti, M.A., Seegal, R.F., 1992. Intrastriatal injection of PCBs decreases striatal dopamine concentrations in rats. Toxicologist 12, 320.

Colborn, T., Clement, C., 1992. Chemically-induced alterations in sexual development: The wildlife/human connection. In: Advances in Modern Environmental Toxicology. Princeton Scientific Publishing, Princeton, NJ, pp. 1–8.

Colborn T., Dumanoski, D., Myers, J.P., 1996. Our Stolen Future. Dutton Books, New York.

Colborn, T., Smolen, M.J., 1996. Epidemiological analysis of persistent organochlorine contaminants in cetaceans. Rev. Environ. Contam. Toxicol. 146, 91–172.

Colborn, T., Vom Saal, F.S., Soto, A.M., 1993. Developmental effects of endocrine-disrupting chemicals in wildlife and humans. Environ. Health Perspect. 101, 378–384.

Collins, W.T., Capen, C.C., 1980. Fine structural lesions and hormone alterations in thyroid glands of perinatal rats exposed in utero and by the milk to polychlorinated biphenyls. Am. J. Pathol. 99, 125–142.

Connor, K., Ramamoorthy, K., Moore, M., Mustain, M., Chen, I., Safe, S., Zacharewski, T., Gillesby, B., Joyeux, A., Balaguer, P., 1997. Hydroxylated polychlorinated biphenyls (PCBs) as estrogens and antiestrogens: Structure-activity relationships. Toxicol. Appl. Pharmacol. 145, 111–123.

Corrigan, F.M., French, M., Murry, L., 1996. Organochlorine compounds in human brain. Hum. Exp. Toxicol. 15, 262–264.

Cullum, M.E., Zile, M.H., 1985. Acute polybrominated biphenyl toxicosis alters vitamin A homeostasis and enhances degradation of vitamin A. Toxicol. Appl. Pharmacol. 81, 177–181.

Custer, T.W., Heinz, G.H., 1980. Reproductive success and nest attentiveness of mallard ducks fed Aroclor 1254. Environ. Pollut. (ser. A) 21, 313–318.

Dahlgren, R.B., Linder, R.L., 1971. Effects of polychlorinated biphenyls on pheasant reproduction, behavior and survival. J. Wildl. Manage. 35, 315.

Dahlgren, R.B., Linder, R.L., Carlson, C.W., 1972. Polychlorinated biphenyls: Their effects on penned pheasants. Environ. Heath Perspect. 1, 89–101.

Daly, H.B., 1993. Laboratory rat experiments show consumption of Lake Ontario salmon causes behavioral changes: Support for wildlife and human research results. J. Great Lakes Res. 19, 784–788.

Darnerud, P.O., Brandt, I., Klasson-Wehler, E., Bergman, A., d'Argy, R., Dencker, L., Sperber, G.O., 1986. 3,3',4,4'-tetrachloro-[^{14}C] -biphenyl in pregnant mice: Enrichment of phenol and methyl sulphone metabolites in late gestational fetuses. Xenobiotica 16, 295–306.

De Guise, S., Martineau, D., Beland, P., Fournier, M., 1998. Effects of in vitro exposure of beluga whale leukocytes to selected organochlorines. J. Toxicol. Environ. Health 55, 479–493.

De Vito, M.J., Birnbaum, L.S., Farland, W.H., Gasiewicz, T.A., 1995. Comparisons of estimated body burdens of dioxinlike chemicals and TCDD body burdens in experimentally exposed animals. Environ. Health Perspect. 103, 820–831.

De Vito, M.J., Ma, X., Babish, J.G., Menache, M., and Birnbaum, L.S. 1994. Dose-response relationships in mice following subchronic exposure to 2,3,7,8-tetrachlorodibenzo-p-dioxin: CYP1A1, CYP1A2, estrogen receptor and protein tyrosine phosphorylation, Toxicol. Appl. Pharmacol. 124, 82–90.

de Swart, R.L., Ross, P.S., Vedder, L.J., Timmerman, H.H., Heisterkamp, S., van Loveren, H., Vos, J.G., Reijnders, P.J.H., Osterhaus, A.D.M.E., 1994. Impaiment of immune function in harbor seals (*Phoca vitulina*) feeding on fish from polluted waters. Ambio 23, 155–159.

Delescluse, C., Lemaire, G., de Sousa, G., Rahmani, R., 2000. Is CYP1A1 induction always related to AHR signaling Pathway? Toxicology 153, 73–82.

Delzell, E., Doull, J., Giesy, J.P., Mackay, D., Monro, I.C., Williams, G.M., 1994. Interpretive review of the potential adverse effects of chlorinated organic chemicals on human health and the environment. Regul.Toxicol. Pharmacol. 20 (1, pt 2), S1–S1056.

Dewailly, E., Weber, J.P., Gingras, S., Lalibert, C., 1991. Coplanar PCBs in human milk in the province of Quebec, Canada: Are they more toxic than dioxin for breast fed infants? Bull. Environ. Contam. Toxicol. 47, 491–498.

Dohlen, K.D., Jarzab, B., 1992. The influence of hormones and hormone antagonists on sexual differentiation of the brain. In: Colborn, T. (Ed.), Advances in Modern Environmental Toxicology. Princeton Scientific Publishing, Princeton, NJ, pp. 231–260.

Drongowski, R.A., Wood, J.S., Bouch, G.R., 1975. Thyroid activity in coho salmon from Oregon and Lake Michigan. Trans. Am. Fish. Soc. 2, 349.

Duke, T.W., Lowe, J.L., Wilson, A.J., Jr., 1970. A polychlorinated biphenyl (Aroclor 1254) in the water, sediment, and biota of Escambia Bay, Florida. Bull. Environ. Contam. Toxicol. 5, 171–180.

Ebert, B.L., Bunn, H.F., 1998. Regulation of transcription by hypoxia requires a multiprotein complex that includes hypoxia-inducible factor 1, an adjacent transcription factor, and p300/CREB binding protein. Mol. Cell. Biol. 18, 4089–4096.

Elliott, J.E., Norstrom, R.J., Lorenzen, A., Hart, L.E., Philibert, H., Kennedy, S.W., Stegeman, J.J., Bellward, G., Cheng, K.M., 1996. Biological effects of polychlorinated dibenzo-p-dioxins, dibenzofurans, and biphenyls, in bald eagle chicks (*Haliaeetus leucocephalus*) chicks. Environ. Toxicol. Chem. 15, 782–793.

Enan, E., Matsumura, F., 1995. Evidence for a second pathway in the action mechanism of 2,3,7,8-tetrachloro-p-dioxin (TCDD). Significance of Ah-receptor mediated activation of protein kinase under cell-free conditions. Biochem. Pharmacol. 49, 249–261.

Erickson, M.D., 1997. Analytical Chemistry of PCBs. Lewis Publishers, Boca Raton, FL.

Eriksson, P., Fredriksson, A., 1996a. Developmental neurotoxicity of four ortho-substituted polychlorinated biphenyls in the neonatal mouse. Environ. Toxicol. Pharmacol. 1, 155–165.

Eriksson, P., Fredriksson, A., 1996b. Neonatal exposure to 2,2',5,5'-tetrachlorobiphenyl causes increased susceptibility in the cholinergic transmitter system at adult age. Environ. Toxicol. Pharmacol. 1, 217–220.

Eriksson, P., Fredriksson, A., 1998. Neurotoxic effects in adult mice neonatally exposed to 3,3',4,4',5-pentachlorobiphenyl or 2,3,3',4,4'-pentachlorobiphenyl. Changes in brain nicotinic receptors and behaviour. Environ. Toxicol. Pharmacol. 5, 17–27.

Eriksson, P., Lundkvist, U., Fredriksson, A., 1991. Neonatal exposure to 3,3',4,4'-tetrachlorobiphenyl: Changes in spontaneous behaviour and cholinergic muscarinic receptors in the adult mouse. Toxicology 69, 27–34.

Evangelista de Duffard, A.M., Duffard, R., 1996. Behavioral toxicology, risk assessment and chlorinated hydrocarbons. Environ. Health Perspect. 104 (suppl. 2) 353–360.

Fielden, M.R., Chen, I., Chittin, B., Safe, S.H., Zacharewski, T.R., 1997. Examination of the estrogenicity of 2,4,6,2',6'-pentachlorobiphenyl (PCB104), its hydroxylated metabolite 2,4,6,2',6'-pentachloro-4-biphenylol (HO-PCB 104) and a further chlorinated derivivative, 2,4,6,2',4',6'-hexachlorobiphenyl (PCB 155). Environ. Health Perspect. 105, 1238–1248.

Feyk, L.L., Giesy, J.P., 1998. Xenobiotic modulation of endocrine function in birds. In: Kendall, R.J., Dickerson, R.L., Giesy, J.P., Suk, W.A. (Eds.), Principles and Processes for Evaluating Endocrine Disruptors in Wildlife. SETAC Press, Pensacola, FL, pp. 121–140.

Fleming, I.A., Gross, M.R., 1989. Evolution of adult female life history and morphology in a Pacific salmon (coho: *Oncorhynchus kisutch*). Evolution 43, 141–157.

Foreman, M.M., Porter, J.C., 1980. Effects of catechol estrogens and catecholamines on hypothalamic and corpus striatal tyrosine hydroxylase activity. J. Neurochem. 34, 1175–1183.

Forsythe, J.A., Jiang, B-H., Iyer, N.V., Agani, F., Leung, S.W., Koos, R.D., Semenza, G.L., 1996. Activation of vascular endothelial growth factor gene transcription by hypoxia-inducible factor 1. Mol. Cell. Biol. 16, 4604–4613.

Fowles, J.R., Fairbrother, A., Trust, K.A., Kerkvliet, N.I., 1997. Effects of Aroclor 1254 on the thyroid gland, immune function, and hepatic cytochrome P450 activity in mallards. Environ. Res. 75, 119–129.

Fox, G.A., 1988. Porphyria in herring gulls : A biochemical response to chemical contaminants of Great Lakes food chains. Environ. Toxicol. Chem. 7, 831–839.

Fox, G.A., 1993. What have biomarkers told us about the effects of contaminants on the health of fish-eating birds in the Great Lakes? The theory and a literature review. J. Great Lakes Res. 19, 722–736.

Fox, L.L., Grasman, K.A., 1999. Effects of PCB 126 on primary immune organ development in chicken embryos. J. Toxicol. Environ. Health 58, 233–244.

Friend, M., Trainer, D.O., 1970. Polychlorinated biphenyls: Interaction with duck hepatitis virus. Science, 170, 1314–1316.

Froese, K.L., Verbrugge, D.A., Ankley, G.T., Niemi, G.J., Larsen, C.P., Giesy, J.P., 1998. Bioaccumulation of polychlorinated biphenyls from sediments to aquatic insects and tree swallow eggs and nestlings in Saginaw Bay, Michigan, USA. Environ. Toxicol. Chem. 17, 484–492.

Fry, M., Toone, K.C., 1981. DDT-induced feminization of gull embryos. Science 213, 922–924.

Gall, G.A.E., Baltodano, J., Huang, N., 1988. Heritability of age at spawning for rainbow trout. Aquaculture 68, 93–102.

Gasiewicz, T.A., 1997. Dioxins and the Ah receptor: Probes to uncover processes in neuroendocrine development. Neurotoxicology 18, 393–413.

Gassmann, M., Kvietikova, I., Rolfs, A., Wenger, R.H., 1997. Oxygen- and dioxin-regulated gene expression in mouse hepatoma cells. Kidney Int. 51, 567.

Georgakopoulou-Gregoriadou, E., Psyllidou-Giouranovits, R., Voutsinou-Taliadouri, F., Catsiki, U.A., 1995. Organochlorine residues in marine mammals from the Greek waters. Frese. Environ. Bull., 4, 375–380.

Gierthy, J.F., Crane, D., 1984. Reverse inhibition of *in vitro* epithelial cell proliferation by 2,3,7,8-tetrachlorodibenzo-*p*-dioxin. Toxicol. Appl. Pharmacol. 74, 91–98.

Giesy, J.P., Kannan, K., 1998. Dioxin-like and non-dioxin-like toxic effects of polychlorinated biphenyls (PCBs): Implications for risk assessment. Crit. Rev. Toxicol. 28, 511–569.

Giesy, J.P., Ludwig, J.P., Tillitt, D.E., 1994a. Dioxins, dibenzofurans, PCBs and colonial, fish-eating water birds. In: Schecter, A. (Ed.), Dioxin and Health. Plenum Press, New York, pp. 254–307

Giesy, J.P., Ludwig, J.P., Tillitt D.E., 1994b. Embryolethality and deformities in colonial, fish-eating, water birds of the Great Lakes region: Assessing causality. Environ. Sci. Technol. 28, 128A.

Giesy, J.P., Newsted, J., Garling, D.L., 1986. Relationships between chlorinated hydrocarbon concentrations and rearing mortality of chinook salmon *Oncorhynchus tschawytscha* eggs from Lake Michigan. J. Great Lakes Res. 12, 82–98.

Giesy, J.P., Verbrugge, D.A. Othout, R.A., Bowerman, W.W., Mora, M.A., Jones, P.D., Newsted, J.L., Vandervoort, C., Heaton, S.N., Aulerich, R.J., Bursian, S.J., Ludwig, J.P., Dawson, G.A., Kubiak, T.J., Best, D.A., Tillitt D.E., 1994c. Contaminants in fishes from Great Lakes-influenced sections and above dams on three Michigan rivers: II. Implications for the health of mink. Arch. Environ. Toxicol. Chem. 27, 213–223.

Gilbertson, M., Kubiak, T., Ludwig, J., Fox, G., 1991. Great lakes embryo mortality, edema, and deformities syndrome (GLEMEDS) in colonial fish-eating birds: Similarity to chick edema disease. J. Toxicol. Environ. Health 33, 455–520.

Gillman, A., Beland, P., Colborn, T., Fox, G., Giesy, J.P., Hesse, J., Kubiak, T., Piekavz, 1991. Environmental and wildlife toxicology of exposure to toxic chemicals. In: Flint, R.W., (Ed.), Human Health Risks from Chemical Exposure. The Great Lakes Ecosystem, Lewis Publishers, Chelsea, MI, pp. 61–69.

Goldberg, M.A., Schneider, T.J., 1994. Similarities between the oxygen-sensing mechanisms regulating the expression of vascular endothelial growth factor and erythropoietin. J. Biol. Chem. 269, 4355–4359.

Goldstein, J.A., Safe, S., 1989. Mechanism of action and structure-activity relationships for the chlorinated dibenzo-*p*-dioxins and related compounds. In: Halogenated biphenyls, Terphenyls, Naphthalenes, Dibenzodioxins and Related Products, 2nd ed., (Kimbrough, R.D., Jensen, A.A., Eds.), Elsevier, Amsterdam, The Netherlands, pp. 239–293.

Gould, J.C., Cooper, K.R., Scanes, C.G., 1997. Effects of polychlorinated biphenyl mixtures and three specific congeners on growth and circulating growth-related hormones. Gen. Comp Endocrinol. 106, 221–230.

Gould, J.C., Cooper, K.R., Scanes, C.G., 1999. Effects of polychlorinated biphenyls on thyroid hormones and liver type I monodeiodinase in the chick embryo. Ecotoxicol. Environ. Safety 43, 195–203.

Gradin, K., McGuire, J., Wenger, R. H., Kvietikova, I., Whitelaw, M. L., Toftgard, R., Tora, L., Gassmann, M., Poellinger, L., 1996. Functional interference between hypoxia and dioxin signal transduction pathways: Competition for recruitment of the Arnt transcription factor. Mol. Cell. Biol. 16, 5221–5231.

Grassle, B., Bressmann, A., 1982. Effects of DDT, polychlorinated biphenyls and thiouracil on circulating thyroid hormones, thyroid histology and eggshell quality in Japanese quail (*Coturnix coturnix Japonica*). Chemi. Biol. Interact. 42, 371–377.

Greene, L.A., Rein, G., 1977. Release, storage and uptake of catecholamines by a clonal cell line of nerve growth factor (NGF) responsive pheochromocytoma cells. Brain Res. 129, 247–263.

Greig, J.B., Jones, G., Butler, W.H., Barnes, J.M., 1973. Toxic effects of 2,3,7,8-tetrachlorodibenzo-*p*-dioxin. Food Cosmet. Toxicol. 11, 585–595.

Guillemin, K., Krasnow, M.A., 1997. The hypoxic response: Huffing and HIFing. Cell 89, 9–12.

Haake, J.M., Safe, S., Mayara, K., Phillips, T.D., 1987. Aroclor 1254 as an antagonist of the teratogenicity of 2,3,7,8-tetrachlorodibenzo-*p*-dioxin. Toxicol. Lett. 38, 299–306.

Hahn, M.E., 1998. The aryl hydrocarbon receptor: A comparative perspective. Comp. Biochem. Physiol. C Pharmacol. Toxicol. Endocrinol. 121C, 23–53.

Hahn, M.E., Poland, A., Glover, E., Stegeman, J.J., 1992. The Ah receptor in marine animals: Phylogenetic distribution and relationship to P4501A inducibility. Mar. Environ. Res. 34, 87–92.

Hansen, D.J., Parrish, P.R., Lowe, J.I., Wilson, Jr., A.J., Wilson, P.D., 1971. Chronic toxicity, uptake, and retention of Aroclor 1254 in two estuarine fishes. Bull. Environ. Contam. Toxicol. 6, 113–119.

Hansen, D.J., Schimmel, S.C., Forester, J., 1973. Aroclor 1254 in eggs of sheepshead minnows; effect on fertilization success and survival of embryos and fry. In: Proceedings of 27th Annual Conference, Southeastern Association of Game and Fish Commissioners, Hot Springs, AR, October 14–17, pp. 420–426.

Harper, N., Connor, K., Steinberg, M., Safe, S., 1995. Immunosuppressive activity of polychlorinated biphenyl mixtures and congeners: Nonadditive (antagonistic) interactions. Fundam. Appl. Toxicol. 27, 131–139.

Harris, G.E., Metcalfe, T.L., Metcalfe, C.D., Huestis, S.Y., 1994. Embryotoxicity of extracts from Lake Ontario rainbow trout (*Oncorhynchus mykiss*) to Japanese medaka (*Oryzias latipes*). Environ. Toxicol. Chem. 13, 1393–1403.

Harris, H.J., Erdman, T.C., Ankley, G.T., Lodge, K.B., 1993. Measures of reproductive success and polychlorinated biphenyl residues in eggs and chicks of Forster's terns on Green Bay, Lake Michigan, Wisconsin-1988. Arch. Environ. Contam. Toxicol., 25, 304–314.

Hart, L.E, Cheng, K.M., Whitehead, P.E., Shah, R.M., Ruschkowski, S.R., Blair, R.W., Bennet, D.C., Bandiera, S.M., Norstrom, R.J., 1991. Dioxin contamination and growth and development in great blue heron embryos. J. Toxicol. Environ. Health 32, 331–344.

Haseltine, S.D., Prouty, R.M., 1980. Aroclor 1242 and reproductive success of adult mallards (*Anas platyrhynchos*). Environ. Res. 23, 29–34.

Hauser, P., McMillin, J.M., Bhatara, V.S., 1998. Resistance to thyroid hormone: Implications for neurodevelopment research on the effects of thyroid hormone disruptors. Toxicol. Indust. Health 14, 85–101.

Heaton, S.N., Bursian, S.J., Giesy, J.P., Tillitt, D.E., Render, J.A., Jones, P.D., Verbrugge, D.A., Kubiak, T.J., Aulerich, R.J., 1995a. Dietary exposure mink to carp from Saginaw Bay, Michigan. 1. Effects on reproduction and survival, and the potential risks to wild mink populations. Arch. Environ. Contam. Toxicol. 28, 334–343.

Heaton, S.N., Bursian, S.J., Giesy, J.P., Tillitt, D.E., Render, J.A., Jones, P.D., Verbrugge, D.A., Kubiak, T.J., Aulerich, R.J., 1995b. Dietary exposure of mink to carp from

Saginaw Bay, Michigan: 2. Hematology and liver pathology. Arch. Environ. Contam. Toxicol. 29, 411–417.

Helder, T., 1980. Effects of 2,3,7,8-tetrachlorodibenzo-*p*-dioxin (TCDD) on early life stages of the pike (*Esox lucius* L.). Sci. Total Environ. 14, 255–264.

Helder, T., 1981. Effects of 2,3,7,8-tetrachlorodibenzo-p-dioxin (TCDD) on early life stages of rainbow trout (*Salmo gairdneri*, Richardson). Toxicology 19, 101–112.

Henshel, D.S., Hehn, B.M., Vo, M.T., Steeves, J.D., 1993. A short-term test for dioxin teratogenicity using chicken embryos. In: Gorsuch, J.W., Dwyer, F.J., Ingesoll, C.G., LaPoint, T.W. (Eds.), Environmental Toxicology and Risk Assessment. ASTM, Philadelphia, pp. 159–174.

Henshel, D.S., Hehn, B., Wagey, R., Vo, M., Steeves, J.D., 1997a. The relative sensitivity of chicken embryos to yolk- or air-cell-injected 2,3,7,8-tetrachlorodibenzo-*p*-dioxin. Environ. Toxicol. Chem. 16, 725–732.

Henshel, D.S., Martin, J.W., DeWitt, J.C., 1997b. Brain asymmetry as a potential biomarker for developmental TCDD intoxication: a dose-response study. Environ. Health Perspect. 105, 718–725.

Hertzler, D.R., 1990. Neurotoxic behavioral effects of Lake Ontario salmon diet in rats. Neurotoxicol. Teratol. 12, 139–143.

Higginbotham, G.R., Huang, A., Firestone, D., Verrett, J., Ress, J., Campbell, A.D., 1968. Chemical and toxicological evaluations of isolated and synthetic chloroderivatives of dibenzo-p-dioxin. Nature 220, 702–703.

Hilscherova, K., Machala, M., Kannan, K., Blankenship, A.L., Giesy, J.P., 2000. Cell bioassays for detection of aryl hydrocarbon (AhR) and estrogen receptor (ER) mediated activity in environmental samples. Environ. Sci. Pollut. Res. 7, 159–171.

Hochstein, J.R., Aulerich, R.J., Bursian, S.J., 1988. Acute toxicity of 2,3,7,8-tetrachlorodibenzo-*p*-dioxin to mink. Arch. Environ. Contam. Toxicol. 17, 33–37.

Hodson, P.V., McWhirter, M., Ralph, K., Gray, B., Thivierge, D., Carey, J.H., Van Der Kraak, G., Whittle, D.M., Levesque, M.-C., 1992. Effects of bleached kraft mill effluent on fish in the St. Maurice River, Quebec. Environ. Toxicol. Chem. 11, 1635–1651.

Hoffman, D.J., Melancon, M.J., Eisemann, J.D., Klein, P.N., 1995. Comparative developmental toxicity of planar PCB congeners by egg injection. In: Proceedings, 2nd SETAC World Congress, Society of Environmental Toxicology and Chemistry, Vancouver, BC, p. 207.

Hoffman, D.J., Melancon, M.J., Klein, P.N., Rice, C.P., Eisemann, J.D., Hines, R.K., Spann, J.W., Pendleton, G.W., 1996. Developmental toxicity of PCB 126 (3,3',4,4',5-pentachlorobiphenyl) in nesting American kestrels (*Falco sparverius*). Fundam. Appl. Toxicol. 34, 188–200.

Hoffman, D.J., Melancon, M.J., Klein, P.N., Eiseman, J.D., Spann, J.W., 1998. Comparative developmental toxicity of planar polychlorinated biphenyl congeners in chickens, American kestrels, and common terns. Environ. Toxicol. Chem. 17, 747–757.

Hoffman, D.J., Smith, G.J., Rattner, B.A., 1993. Biomarkers of contaminant exposure in common terns and black-crowned night herons in the Great Lakes. Environ. Toxicol. Chem. 17, 1095–1103.

Hogenesch, J.B., Chan, W.K., Jackiw, V.H., Brown, R.C., Gu, Y.-Z., Pray-Grant, M., Perdew, G.H., Bradfield, C.A., 1997. Characterization of a subset of the basic-helix-loop-helix-PAS superfamily that interacts with components of the dioxin signaling pathway. J. Biol. Chem. 272, 8581–8593.

Holene, E., Nafstad, I., Skaare, J.U., Bernhoft, A., Engen, P., Sagvolden, T., 1995. Behavioral effects of pre- and postnatal exposure to individual polychlorinated biphenyl congeners in rat. Environ. Toxicol. Chem., 14, 967–976.

Holmes, J.L., Pollenz, R.S., 1997. Determination of aryl hydrocarbon receptor nuclear translocator protein concentration and subcellular localization in hepatic and nonhepatic cell culture lines: Development of quantitative Western blotting protocols for calculation of aryl hydrocarbon receptor and aryl hydrocarbon receptor nuclear translocator protein in total cell lysates. Mol. Pharmacol. 52, 202–211.

Huang, Z.J., Curtin, K.D., Rosbash, M., 1995. PER protein interactions and temperature compensation of a circadian clock in Drosophila. Science 267, 1169–1172.

Huang, L.E., Gu, J., Schau, M., Bunn, H.F., 1998. Regulation of hypoxia-inducible factor 1α is mediated by an O2-dependent degradation domain via the ubiquitin-proteasome pathway. Proc. Natl. Acad. Sci. USA 95, 7987–7992.

Huisman, M., Koopman-Esseboom, C., Fidler, V., Hadders-Algra, M., van der Paauw, C.G., Tuinstra, L.G., Weisglas-Kuperus, N., Sauer, P.J., Touwen, B.C., Boersma, E.R., 1995. Perinatal exposure to polychlorinated biphenyls and dioxins and its effect on neonatal neurological development. Early Hum. Dev. 41, 111–127.

Hutchinson, T.H., Field, M.D.R., Manning, M.J., 1999. Evaluation of immune function in juvenile turbot *Scophthalmus maximus* (L.) exposed to sediments contaminated with polychlorinated biphenyls. Fish. Shellfish. Immunol. 9, 457–472.

Hutzinger, O., Safe, S., Zitko, V., 1974. The Chemistry of PCBs. CRC Press, Boca Raton, FL.

Idler, D.R., Bitners, I.I., Schmidt, P.J., 1961. 11-Ketotestosterone: an androgen for sockeye salmon. Can. J. Biochem. Physiol. 39, 1737.

Isaac, D.D., Andrew, D.J., 1996. Tubulogenesis in Drosophila: A requirement for the trachealess gene product. Gene Dev. 10, 103–117.

Ivanov, V., Sandell, E., 1992. Characterization of polychlorinated biphenyl isomers in Sovol and Trichlorodiphenyl formulations by high resolution gas chromatography with electron capture detection and high resolution gas chromatography-mass spectrometry techniques. Environ. Sci. Technol. 26, 2012–2017.

Iyer, N.V., Kotch, L.E., Agani, F., Leung, S.W., Laughner, E., Wenger, R.H., Gassmann, M., Gearhart, J.D., Lawler, A.M., Yu., A.Y., Semenza, G.L., 1998. Cellular and developmental control of O_2 homostasis by hypoxia-inducible factor 1α. Genes Dev. 12, 149–162.

Jacobson, J.L., Jacobson, S.W., 1997. Evidence for PCBs as neurodevelopmental toxicants in humans. Neurotoxicology 18, 415–424.

Janz, D.M., Metcalfe, C.D. 1991. Relative induction of aryl hydrocarbon hydroxylase by 2,3,7,8-TCDD and two coplanar PCBs in rainbow trout (*Oncorhynchus mykiss*), Environ. Toxicol. Chem. 10, 917–923.

Janz D.M., Bellward, G.D., 1997. Effects of acute 2,3,7,8-tetrachlorodibenzo-*p*-dioxin exposure on plasma thyroid and sex steroid hormone concentrations and estrogen receptor levels in adult Great Blue herons. Environ. Toxicol. Chem. 16, 985–989.

Jarvinen, A.W., Ankley, G.T., 1999. Linkage of Effects to Tissue Residues: Development of a Comprehensive Database for Aquatic Organisms Exposed to Inorganic and Organic Chemicals. SETAC Press, Pensacola, FL.

Jefferies, D.J., and French, M.C., 1972. Changes induced in the pigeon thyroid by p,p'-DDE and dieldrin. J. Wildl. Manage. 36, 24–30.

Jefferies, D.J., Parslow, J.L.F., 1976. Thyroid changes in PCB-dosed guillemots and their indication of one of the mechanisms of action of these materials. Environ. Pollut. 10, 293–311.

Jensen, S. 1966. Report of a new chemical hazard. New Scientist, 32, 612.

Jenssen, B.M., Skaare, J.U., Ekker, M. Vongraven, D., Lorentsen, S.H., 1996. Organochlorine compounds in blubber, liver and brain in neonatal gray seal pups. Chemosphere 32, 2115–2125.

Jepson, P.D., Bennett, P.M., Allchin, C.R., Law, R.J., Kuiken, T., Baker, J.R., Rogan, E., Kirkwood, J.K., 1999. Investigating potential associations between chronic exposure to polychlorinated biphenyls and infectious disease mortality in harbour porpoises from England and Wales. Sci. Total Environ. 243/244, 339–348.

Jiang, B.-H., Rue, E., Wang, G.L., Roe, R., Semenza, G.L., 1996. Dimerization, DNA binding, and transaction properties of hypoxia-inducible factor 1. J. Biol. Chem. 271, 17771–17778.

Kallio, P.J., Okamoto, K., O'Brien, S., Carrero, P., Makino, Y., Tanaka, H., Poellinger, L., 1998. Signal transduction in hypoxic cells: Inducible nuclear translocation and recruitment of the CBP/p300 coactivator by the hypoxia-inducible factor-1α. EMBO J. 17, 6573–6586.

Kallio, P.J., Pongratz, I., Gradin, K., McGuire, J., Poellinger, L., 1997. Activation of hypoxia-inducible factor 1α: Posttranscriptional regulation and conformational change by recruitment of the Arnt trancription factor. Proc. Natl. Acad. Sci. USA 94, 5667–5672.

Kannan, K., Blankenship, A.L., Jones, P.D., Giesy, J.P., 2000. Toxicity reference values for the toxic effects of polychlorinated biphenyls to aquatic mammals. Human and Ecological Risk Assessment 6, 181–201.

Kannan, K., Maruya, K., Tanabe, S., 1997. Distribution and characterization of polychlorinated biphenyl congeners in soil and sediments from a Superfund site contaminated with Aroclor 1268. Environ. Sci. Technol., 31, 1483–1488.

Kannan, K., Tanabe, S., Borrell, A., Aguilar, A., Focardi, S., Tatsukawa, R., 1993. Isomer-specific analysis and toxic evaluation of polychlorinated biphenyls in striped dolphins affected by an epizootic in the western Mediterranean Sea, Arch. Environ. Contam. Toxicol. 25, 227–233.

Kannan, N., Tanabe, S., Ono, M., Tatsukawa, R., 1989. Critical evaluation of polychlorinated biphenyl toxicity in terrestrial and marine mammals: Increasing impact of non-*ortho* and mono-*ortho* coplanar polychlorinated biphenyls from land to ocean. Arch. Environ. Contam. Toxicol. 18, 850–857.

Kannan, K., Tanabe, S., Tatsukawa, R., and Sinha, R.K., 1994. Biodegradation capacity and residue pattern of organochlorines in Ganges river dolphins from India. Toxicol. Environ. Chem., 42, 249–261.

Karchner, S.I., Kennedy, S.W., Trudeau, S., Hahn, M.E., 2000. Towards molecular understanding of species differences in dioxin sensitivity: Initial characterization of the Ah receptor cDNAs in birds and amphibians. Mar. Environ. Res. 50, 51–56.

Kendall, R.J., Dickerson, R.L., Giesy, J.P., Suk, W.A. (Eds.). 1998. Principles and Processes for Evaluating Endocrine Disruptors in Wildlife. SETAC Press, Pensacola, FL.

Kennedy, S.W., Lorenzen, A., Jones, S.P., Hahn, M.E., Stegeman, J.J., 1996a. Cytochrome P4501A induction in avian hepatocyte cultures: A promising approach for predicting the sensitivity of avian species to toxic effects of halogenated aromatic hydrocarbons. Toxicol. Appl. Pharmacol. 141, 214–230.

Kennedy, S.W., Lorenzen, A., Norstrom, R.J., 1996b. Chick embryo hepatocyte bioassay for measuring cytochrome P4501A-based 2,3,7,8-tetrachlorodibenzo-p-dioxin equivalent concentrations in environmental samples. Environ. Sci. Technol. 30, 706–715.

Kerkvliet, N.I., 1994. Immunotoxicology of Dioxins and Related Chemicals. In: Schecter, A. (Ed.), Dioxins and Health. Plenum Press, New York, pp. 199–225.

Kihlström, J.E., Olsson, M., Jensen, S., Johansson, C., Ahlbom, J., Bergman, C., 1992. Effects of PCB and different fractions of PCB on the reproduction of the mink (*Mustela vison*). Ambio 21, 563–569.

King, D.P., Zhao, Y., Sangoram, A.M., Wilsbacher, L.D., Tanaka, M., Antoch, M.P., Steeves,

T.D., Vitaterna, M.H., Kornhauser, J.M., Lowrey, P.L., Turek, F.W., Takahashi, J.S., 1997. Positional cloning of the mouse circadian clock gene. Cell 89, 641–653.

Kittner, B., Brautigam, M., Herken, H., 1987. PC12 cells: A model system for studying drug effects on dopamine synthesis and release. Arch. Intl. Pharamacodyn. Ther. 286, 181–194.

Klasson-Wehler, E., Bergman, Å., Athanasiadou, M., Ludwig, J.P., Auman, H.J., Kannan, K., Van den Berg, M., Murk, A.J., Feyk, L.A., Giesy, J.P., 1998. Hydroxylated and Methylsolfonyl PCBs in Albatrosses from Midway Atoll, North Pacific Ocean. Environ. Toxicol. Chem. 17, 1620–1625.

Knobil, E., Bern, H.A., Burger, J., Fry, D.M., Giesy, J.P., Gorski, J., Grossman, C.J., Guillette, L.J., Hulka, B.S., Lamb, J.C., McLachlan, J.A., Real, L.A., Safe, S.H., Soto, A.M., Stegeman, J.J., Swan, S.H., von Saal F.S., 1999. Hormonally Active Agents in the Environment. Committee on Hormonally Active Agents in the Environment, Board on Environmental Studies and Toxicology, Commission on Life Sciences, National Research Council, National Academy Press, Washington, DC.

Kobayashi, A., Numayama-Tsuruta, K., Sogawa, K., Fujii-Kuriyama, Y., 1997. CBP/p300 functions as a possible transcriptional coactivator of Ah receptor nuclear translocator (Arnt). J. Biochem. 122, 703–710.

Kodavanti, P.R., Mundy, W.R., Tilson, H.A., Harry, G.J., 1993a. Effects of selected neuro-active chemicals on calcium transporting systems in rat cerebellum and on survival of cerebellar granule cells. Fundam. Appl. Toxicol. 21, 308–316.

Kodavanti, P.R.S., Shafer, T.J., Ward, T.R., Mundy, W.R., Freudenrich, T., Harry, G.J., Tilson, H.A., 1994. Differential effects of polychlorinated biphenyl congeners on phosphoinositide hydrolysis and protein kinase C translocation in rat cerebellar granule cells. Brain Res. 662, 75–82.

Kodavanti, P.R., Shin, D.S., Tilson, H.A., Harry, G.J., 1993b. Comparative effects of two polychlorinated biphenyl congeners on calcium homeostasis in rat cerebellar granule cells. Toxicol. Appl. Pharmacol. 123, 97–106.

Kodavanti, P.R., Tilson, H.A., 1997. Structure-activity relationships of potentially neurotoxic PCB congeners in the rat. Neurotoxicology 18, 425–441.

Kodavanti, P.R.S., Ward, T.R., McKinney, J.D., Tilson, H.A., 1995. Increased [^{3}H]phorbol ester binding in rat cerebellar granule cells by polychlorinated biphenyl mixtures and congeners: Structure-activity relationships. Toxicol. Appl. Pharmacol. 130, 140–148.

Kodavanti, P.R.S., Ward, T.R., McKinney, J.D., Tilson, H.A., 1996. Inhibition of microsomal and mitochondrial Ca^{2+} sequestration in rat cerebellum by polychlorinated biphenyl mixtures and congeners: Structure-activity relationships. Arch. Toxicol. 70, 150–157.

Koga, N., Beppu, M., Yoshimura, H., 1990. Metabolism in vivo of 3,4,5,3',4'-pentachlorobiphenyl and toxicological assessment of the metabolite in rats. J. Pharmacobio-Dyn. 13, 497–506.

Kohle, C., Gschaidmeier, H., Lauth, D., Topell, S., Zitzer, H., Bock, K.W., 1999. 2,3,7,8-tetrachlorodibenzo-*p*-dioxin (TCDD)-mediated membrane translocation of c-Src protein kinase in liver WB-F344 cells. Arch. Toxicol. 73, 152–158.

Korach, K.S., Sarver, P., Chae, K., McLachlan, J.A., McKinney, J.D., 1988. Estrogen receptor-binding activity of polychlorinated hydroxybiphenyls: Conformationally restricted structural probes. Mol. Pharmacol. 33, 120–126.

Kozak, K.R., Abbott, B., Hankinson, O., 1997. ARNT-deficient mice and placental differentiation. Dev. Biol. 191, 297–305.

Kramer, V.J., Giesy, J.P., 1999. Specific binding of hydroxylated polychlorinated biphenyl metabolites and other substances to bovine calf uterine estrogen receptor: structure-binding relationships. Sci. Total. Environ. 233, 141–161.

Kramer, V.J., Klasson-Wehler, E., Bergman, Å., Helferich, W.G., Giesy, J.P., 1997. Hydroxylated polychlorinated biphenyl metabolites are anti-estrogenic in a stably transfected human breast adenocarcinoma (MCF7) cell line. Toxicol. Appl. Pharmacol. 144, 363–376.

Kubiak, T.J., Harris, H.J., Smith, L.M., Schwartz, T.R., Stalling, D.L., Trick, J.A., Sileo, L., Doherty, D.E., Erdman, T.C., 1989. Microcontaminants and reproductive impairment of the Forster's tern on Green Bay, Lake Michigan (U.S.A.) 1983. Arch. Environ. Contam. Toxicol. 18, 706–727.

Kuratsune, M., Yoshimura, T., Matsuzaka, J., Yamaguchi, A., 1972. Epidemiologic study on Yusho: A poisoning caused by ingestion of rice oil contaminated with a commerical brand of polychlorinated biphenyls. Environ. Health Perspect. 1, 119–128.

Kutz, F.W., Barnes, D.G., Bottimore, D.P., Greim, H., Bretthauer, E.W., 1990. The international toxicity equivalency factor (I-TEF) method of risk assessment for complex mixtures of dioxins and related compounds. Chemosphere 20, 751–757.

Lahvis, G.P., Wells, R.S., Kuehl, D.W., Stewart, J.L., Rhinehart, H.L., Via, C.S., 1995. Decreased lymphocyte responses in free-ranging bottlenose dolphins (*Tursiops truncatus*) are associated with increased concentrations of PCBs and DDT in peripheral blood. Environ. Health Perspect. 103, 67–72.

Lapierre, P., de Guise, S., Muir, D.C.G., Norstrom, R., BJland, P., Fournier, M., 1999. Immune functions in the Fisher rat fed beluga whale (*Delphinapterus leucas*) blubber from the contaminated St. Lawrence estuary. Environ. Res. 80, S104–S112.

Lavigne, D.M., Schmitz, O.J., 1990. Global warming and increasing population densities: a prescription for seal plagues. Mar. Pollut. Bull. 21, 280–284.

Lawrence, G.S., Gobas, F.A., 1997. A pharmacokinetic analysis of interspecies extrapolation in dioxin risk assessment. Chemosphere 35, 427–452.

Leatherland, J.F., 1993. Field observations on reproductive and developmental dysfunction in introduced and native salmonids from the Great Lakes. J. Great Lakes Res. 19, 737–751.

Leatherland, J.F., Copeland, P., Sumpter, J.P., Sonstegard R.A., 1982. Hormonal control of gonadal maturation and development of secondary sexual characteristics in coho salmon, *Oncorhynchus kisutch*, species from Lakes Ontario, Erie, and Michigan. Gen. Comp. Endocrinol 48, 196–204.

Leatherland, J.F., Sonstegard, R.A., 1978. Lowering of serum thyroxine and triiodothyronine levels in yearling coho salmon, *Oncorhynchus kisutch*, by dietary mirex and PCBs. J. Fish. Res. Board Can. 35, 1285–1289.

Leatherland, J.F., Sonstegard, R.A., 1980a. Effect of dietary mirex and PCB=s in combination with food deprivation and testosterone administration on thyroid activity and bioaccumulation of organochlorines in rainbow trout *Salmo gairdneri* Richardson. J. Fish Dis. 3, 115–124.

Leatherland, J.F., Sonstegard, R.A., 1980b. Structure of thyroid and adrenal glands in rats fed diets of Great Lakes coho salmon (*Oncorhynchus kisutch*). Environ. Res. 23, 77–86.

Leatherland, J.F., Sonstegard, R.A., 1981. Thyroid dysfunction in Great Lakes coho salmon, *Oncorhynchus kisutch* (Walbaum): seasonal and interlake differences in serum T3 uptake and serum total and free T4 and T3 levels. J. Fish Dis. 4, 413–423.

Leatherland, J.F., Sonstegard, R.A., 1984. Pathobiological responses of feral teleosts to environmental stressors: interlake studies of the physiology of Great Lakes salmon. In: Cairns, V.W., Hodson, P.V., Nriagu, J. (Eds.), Contaminant Effects on Fisheries. John Wiley and Sons, New York, pp. 115–149.

Leece, B., Denomme, M.A., Towner, R., Li, S.M., Safe, S., 1985. Polychlorinated biphe-nyls: Correlation between in vivo and in vitro quantitative structure-activity relation-ships (QSARs). J. Toxicol. Environ. Health 16, 379–388.

Leith, D., Holmes, J., Kaattari, S., 1989. Effects of vitamin nutrition on the immune re-sponse of hatchery-reared salmonids. Final report, July 1989, projects no. 84-45 and 84-945. U.S. Department of Energy. Bonneville Power Administration, Division of Fisheries and Wildlife, Portland, OR.

Leonards, P.E.G., de Vries, T.H., Minnaard, W., Stuijfzand, S., de Voogt, P., Cofino, W.P., van Straalen, N.M., van Hattum, B., 1995. Assessment of experimental data on PCB-induced reproduction inhibition in mink, based on an isomer- and congener-specific approach using 2,3,7,8-tetrachlorodibenzo-*p*-dioxin toxic equivalency. Environ. Toxicol. Chem. 14, 639–652.

Leonards, P.E.G., Zierikzee, Y., Brinkman, U.A.Th., Cofino, W.P., van Straalen, N.M., van Hattum, B., 1997. The selective dietary accumulation of planar polychlorinated biphe-nyls in the otter (*Lutra lutra*). Environ. Toxicol. Chem. 16, 1807–1815.

Lillie, R.J., Cecil, H.C., Bitman, J., Fries, G.F., 1974. Differences in response of caged White Leghorn layers to various polychlorinated biphenyls (PCBs) in the diet. Poult. Sci. 53, 726–732.

Lillie, R.J., Cecil, H.C., Bitman, J., Fries, G.F., 1975. Toxicity of certain polychlorinated and polybrominated biphenyls on reproductive efficiency of caged chickens. Poult. Sci. 54, 1550–1555.

Lloyd, T., Weisz, J., 1978. Direct inhibition of tyrosine hydroxylase activity by catechol estrogens. J. Biol. Chem. 243, 4841–4843.

Loganathan, B.G., Kannan, K., 1994. Global organochlorine contamination trends: An overview. Ambio 23, 187–191.

Lonky E., Riehman T.D., Mather S.J., Daly, H., 1996. Neonatal behavioral assessment scale performance in humans influenced by maternal consumption of environmentally con-taminated Lake Ontario fish. J. Great Lakes Res. 22, 198–212.

Lorenzen, A., Shutt, J.L., Kennedy, S.W., 1997. Sensitivity of common tern (*Sterna hirundo*) embryo hepatocyte cultures to CYP1A induction and porphyrin accumulation by halogenated aromatic hydrocarbons and common tern egg extracts. Arch. Environ. Contam. Toxicol. 32, 126–134.

Ludwig, J.P., Giesy, J.P., Summer, C.L., Bowerman, W., Heaton, S.N. Aulerich, R.J., Bursian, S.J., Auman, H.J., Jones, P.D., Williams, L.L., Tillitt, D.E., Gilbertson, M., 1993. A comparison of water quality criteria in the Great Lakes basin based on human and wildlife health. J. Great Lakes Res. 19, 789–807.

Mably, T.A., Moore, R.W., Peterson, R.E., 1992a. In utero and lactational exposure of male rats to 2,3,7,8-tetrachlorodibenzo-p-dioxin. 1. Effects on androgenic status. Toxicol. Appl. Pharmacol. 114, 97–107.

Mably, T.A., Moore, R.W., Goy, R.W., Peterson, R.E., 1992b. In utero and lactational ex-posure of male rats to 2,3,7,8-tetrachlorodibenzo-p-dioxin. 2. Effects on sexual behavior and the regulation of luteinizing hormone secretion in adulthood. Toxicol. Appl. Pharmacol. 114, 108–117.

Mably, T.A., Bjerke, D.L., Moore, R.W., Gendron-Fitzpatrick, A., Peterson, R.E., 1992c. In utero and lactational exposure of male rats to 2,3,7,8-tetrachlorodibenzo-p-dioxin. 3. Effects on spermatogenesis and reproductive capability. Toxicol. Appl. Pharmacol. 114, 118–126.

Mackay, D., Shiu, W.Y., Ma, K.C., 1992. Monoaromatic hydrocarbons, chlorobenzenes and PCBs. In: Illustrated Handbook of Physico-chemical Properties and Environmental Fate for Organic Chemicals, vol. 1. Lewis Publishers, Chelsea, MI.

Madhukar B.V., Brewster, D.W., Matsumura, F., 1984. Effects of *in-vivo* administered 2,3,7,8-tetrachlorodibenzo-*p*-dioxin on receptor binding of epidermal growth factor in the hepatic plasma membrane of rat, guinea pig, mouse, and hamster. Proc. Natl. Acad. Sci. USA 81, 7407–7411.

Maier, W.E., Kodavanti, P.R., Harry, G.J., Tilson, H.A., 1994. Sensitivity of adenosine triphosphatases in different brain regions to polychlorinated biphenyl congeners. J. Appl. Toxicol. 14, 225–229.

Maltepe, E., Schmidt, J.V., Baunoch, D., Bradfield, C.A., Simon, M.C., 1997. Anormal angiogenesis and response to glucose and oxygen deprivation in mice lacking the protein ARNT. Nature 386, 403–407.

Matsumura, F., 1995. Mechanism of action of dioxin-type chemicals, pesticides, and other xenobiotics affecting nutritional indexes. Am. J. Clin. Nutr. 61(suppl.), 695S-701S.

Maxwell, P.H., Dachs, G.U., Gleadle, J.M., Nicholls, L.G., Harris, A.L., Stratford, I.J., Hankinson, O., Pugh, C.W., Ratcliffe, P.J., 1997. Hypoxia-inducible factor-1 modulates gene expression in solid tumors and influences both angiogenesis and tumor growth. Proc. Natl. Acad. Sci. USA 94, 8104–8109.

McCarty, J.P., Secord, A.L., 1999. Nest-buildling behavior in PCB-contaminated tree swallows. Auk 116, 55–63.

McFarland, V.A., Clarke, J.U., 1989. Environmental occurrence, abundance, and potential toxicity of polychlorinated biphenyl congeners: Considerations for a congener-specific analysis. Environ. Health Perspect. 81, 225–239.

McKinney, J.D., Fawkes, J., Jordan, S., Chae, K., Outley, S., Coleman, R.E., Briner, D., 1985. 2,3,7,8-tetrachlorodibenzo-*p*-dioxin (TCDD) as a potent and persistent thyroxine agonist: a mechanistic model for toxicity based on molecular reactivity. Environ. Health Perspect. 61, 41–53.

McLane, M.A., Hughes, D.L., 1980. Reproductive success of screech owls fed Aroclor 1248. Arch. Environ. Contam. Toxicol. 9, 661–665.

Mehrle, P.M., Buckler, D.R., Little, E.E., Smith, L.M., Petty, J.D., Peterman, P.H., Stalling, D.L., De Graeve, G.M., Coyle, J.J., and Adams, W.J., 1988. Toxicity and bioconcentration of 2,3,7,8-tetrachlorodibenzodioxin and 2,3,7,8-tetrachlorodibenzofuran in rainbow trout. Environ. Toxicol. Chem. 7, 47–62.

Metcalfe, C.D., Haffner, G.D., 1995. The ecotoxicology of coplanar polychlorinated biphenyls. Environ. Rev. 3, 171–190.

Meyer, M.-E., Gronemeyer, H., Turcotte, B., Bocquel, M.-T., Tasset, D., Chambon, P., 1989. Steroid hormone receptors compete for factors that mediate their enhancer function. Cell 57, 433–442.

Moccia, R.D., Fox, G.A., Britton, A., 1986. A quantitative assessment of thyroid histopathology of herring gulls (*Larus argentatus*) from the Great Lakes and a hypothesis on the causal role of environmental contaminants. J. Wildl. Dis. 22, 60–70.

Moffett, P., Reece, M., Pelletier, J., 1997. The murine Sim-2 gene product inhibits transcription by active repression and functional interference. Mol. Cell. Biol. 17, 4933–4947.

Moore, M., Mustain, M., Daniel, K., Chen, I., Safe, S., Zacharewski, T., Gillesby, B., Joyeux, A., Balaguer, P., 1997. Antiestrogenic activity of hydroxylated polychlorinated biphenyl congeners identified in human serum. Toxicol. Appl. Pharmacol. 142, 160–168.

Morse, D.C., Klasson-Wehler, E., van der Pas, M., de Bie, A.T.H.J., van Bladeren, P.J., Brouwer A., 1995. Metabolism and biochemical effects of 3,3',4,4'-tetrachlorobiphenyl in pregnant and fetal rats. Chem.-Biol. Interact. 95, 41–56.

Morse, D.C., Seegal, R.F., Borsch, K.O., Brouwer, A., 1996a. Long-term alterations in regional brain serotonin metabolism following maternal polychlorinated biphenyl exposure in the rat. Neurotoxicology 17, 631–638.

Morse, D.C., Wehler, E.K., Wesseling, W., Koeman, J.H., Brouwer, A., 1996b. Alterations in rat brain thyroid hormone status following pre- and post-natal exposure to polychlorinated biphenyls (Aroclor 1254). Toxicol. Appl. Pharmacol. 136, 269–279.

Mousa, M.A., Ganey, P.E., Quensen, J.F., 3rd, Madhukar, B.V., Chou, K., Giesy, J.P., Fischer, L.J., Boyd, S.A., 1998. Altered biologic activities of commercial polychlorinated biphenyl mixtures after microbial reductive dechlorination. Environ. Health Perspect. 106 Suppl 6, 1409–1418.

Murk, A., Bosveld, A., van den Berg, M., Brouwer, A., 1994. Effects of PAHs on biochemical parameters in ducks and the common tern. Aquat. Toxicol. 30, 91–115.

Murk, A.J., Leonards, P.E.G., Van Hattum, B., Luit, R., Van der Weiden, M.E.J., Smit, M., 1998. Application of biomarkers for exposure and effect of polyhalogenated aromatic hydrocarbons in naturally exposed European otters (*Lutra lutra*). Environ. Toxicol. Pharmacol. 6, 91–102.

Naevdal, G., 1983. Genetic factors in conjunction with age at maturation. Aquaculture 33, 97–106.

Nagayama, J., Tsuji, H., Iida, T., Hirakawa, H., Matsueda, T., Okamura, K., Hasegawa, M, Sato, K., Ma, H.Y., Yanagawa, T., Igarashi, H., Fukushige, J., Watanabe, T., 1998. Postnatal exposure to chlorinated dioxins and related chemicals on lymphocyte subsets in Japanese breast-fed infants. Chemosphere 37, 1781–1787.

Nakata, H., Sakakibara, A., Kanoh, M., Kudo, S., Watanabe, H., Nagai, N., Miyazaki, N., Asano, Y., Tanabe, S., 2002. Evaluation of mitogen-induced responses in marine mammal and human lymphocytes by in-vitro exposure of butyltins and non-ortho coplanar PCBs. Environ. Pollut. 120, 245–253.

Nambu, J.R., Lewis, J.O., Wharton, K.A. Jr., Crews, S.T., 1991. The Drosophila single-minded gene encodes a helix-loop-helix protein that acts as a master regulator of CNS midline development. Cell 67, 1157–1167.

Nebecker, A.V., Puglisi, F.A., Defoe, D.L., 1974. Effect of polychlorinated biphenyl compounds on survival and reproduction of the fathead minnow and flagfish. Trans. Am. Fish Soc. 103, 562–568.

Nebert, D.W., 1990. The Ah locus: Genetic differences in toxicity, cancer, mutation, and birth defects. Crit. Rev. Toxicol. 20, 153–174.

Neubert, D., Golor, G., Neubert, R., 1992. TCDD-toxicity equivalencies for PCDD/PCDF congeners: Prerequisites and limitations. Chemosphere 25, 65–70.

Neumann, H.-G., 1996. Toxic equivalence factors, problems and limitations. Food Chem. Toxicol. 34, 1045–1051.

Newman, J.W., Becker, J.S., Blondina, G., Tjeerdema, R.S., 1998. Quantitation of Aroclors using congener-specific results. Environ. Toxicol. Chem. 17, 2159–2167.

Newsted, J.L., Giesy, J.P., 1991. Characterization of epidermal growth factor binding to hepatic plasma membranes of rainbow trout (*Oncorhynchus mykiss*). Gen. Comp. Endocrinol. 83, 345–352.

Newsted, J.L., Giesy, J.P., 2000. Epidermal growth factor receptor-mediated kinase interactions in hepatic membranes of rainbow trout (*Oncorhynchus mykiss*). Fish Physiol. Biochem. 22, 181–189.

Nikolaidis, E, Brunström, B., Dencker, L., 1988. Effects of the TCDD congeners 3,3',4, 4'-tetrachlorobiphenyl and 3, 3' tetrachloroazoxybenzene on lymphoid development in the bursa fabricus of the chick embryo. Toxicol. Appl. Pharmacol. 92, 315–323.

Niimi, A.J., 1996. PCBs in aquatic organisms. In: Beyer, W.N. (Ed.), Environmental Contaminants in Wildlife: Interpreting Tissue Concentrations, Lewis Publishers, Boca Raton, FL, pp. 117–152.

Nosek, J.A., Sullivan, J.R., Craven, S.R., Gendron-Fitzpatrick, A., Peterson, R.E., 1993.

Embryotoxicity of 2,3,7,8-tetrachlorodibenzo-*p*-dioxin in the ring-necked pheasant. Environ. Toxicol. Chem. 12, 1215–1222.

Okey, A.B., Riddick, D.S., Harper, P.A., 1994. Molecular biology of the aromatic hydrocarbon (dioxin) receptor. Trends Pharmacol. Sci. 15, 226–232.

Olie, K., Van den Berg, M., Hutzinger, O., 1983. Formation rate of PCDD and PCDF from combustion process. Chemosphere 12, 627–636.

Olson, E.N., 1990. MyoD family: A paradigm for development? Genes Dev. 4, 1454–1461.

Omara, F.O., Brochu, C., Flipo, D., Denizeau, F., Fournier, M., 1997. Immunotoxicity of environmentally relevant mixtures of polychlorinated aromatic hydrocarbons with methyl mercury on rat lymphocytes in vitro. Environ. Toxicol. Chem. 16, 576–581.

Opresko, D.M., Sample, B.E., Suter, G.W., 1994. Toxicological benchmarks for wildlife: 1994 revision. ES/ER/TM-86/R1. Oak Ridge National Laboratory, Health Sciences Research Division, Oak Ridge, TN.

Parkinson, A., Robertson, L., Safe, L., Safe, S., 1980. Polychlorinated biphenyls as inducers of hepatic microsomal enzymes: Structure-activity rules. Chem.-Biol. Interact. 30, 271–285.

Parkinson, A., Safe, S.H, Robertson, L.W, Thomas, P.E., Ryan, D.E., Reik, L.M., Levin, W., 1983. Immunochemical quantitation of cytochrome P-450 isozymes and epoxide hydrolase in liver microsomes from polychlorinated or polybrominated biphenyl-treated rats: A study of structure-activity relationships. J. Biol. Chem. 258, 2967–2976.

Peakall, D.B., Peakall, M.L., 1973. Effect of a polychlorinated biphenyl on the reproduction of artificially and naturally incubated dove eggs. J. Appl. Ecolog., 10, 863–868.

Peterson, R.E., Theobald, H.M., Kimmel, G.L., 1993. Developmental and reproductive toxicity of dioxins and related compounds: cross-species comparisons. Crit. Rev. Toxicol. 23, 283–335.

Platonow, N.S., Reinhart, B.S., 1973. The effects of polychlorinated biphenyls (Aroclor 1254) on chicken egg production, fertility and hatchability. Can. J. Comp. Med. 37, 341–346.

Pluess, N., Poiger, H., Hochbach, C., Schlatter, C., 1988. Subchronic toxicity of some chlorinated dibenzofurans (PCDFs) and a mixture of PCDFs and chlorinated dibenzodioxins (PCDDs) in rats. Chemosphere 17, 973–984.

Poellinger, L.L.J., Gustafsson, J.A., 1985. The rat liver receptor protein for 2,3,7,8-tetrachlorodibenzo-*p*-dioxin: a comparison with steroid hormone receptors. Chemosphere 14, 963–966.

Pohjanvirta, R., Unkila, M., Linden, J., Tuomisto, J.T., Tuomisto, J., 1995. Toxic equivalency factors do not predict the acute toxicities of dioxins in rats. Eur. J. Pharmacol. 293, 341–353.

Poland, A., Glover, E., 1977. Chlorinated biphenyl induction of aryl hydrocarbon hydroxylase activity: A study of the structure-activity relationship, Mol. Pharmacol. 13, 924–938.

Poland, A., Glover, E., Kende, A.S., 1976. Stereospecific, high affinity binding of 2,3,7,8-tetrachlorodibenzo-*p*-dioxin by hepatic cytosol: Evidence that the binding species is receptor for induction of aryl hydrocarbon hydroxylase. J. Biol. Chem. 251, 4936–4946.

Poland, A., Greenlee, W.F., Kende, A.S., 1979. Studies on the mechanism of action of chlorinated dibenzo-*p*-dioxins and related compounds. Ann. N.Y. Acad. Sci. 320, 214–230.

Poland, A., Knutson, J.C., 1982. 2,3,7,8-tetrachloro dibenzo-*p*-dioxin and related halogenated aromatic hydrocarbons: Examination of the mechanism of toxicity. Annu. Rev. Pharmacol. Toxicol. 22, 517–554.

Pongratz, I., Antonsson, C., Whitelaw, M.L., Poellinger, L., 1998. Role of the PAS domain in regulation of dimerization and DNA binding specificity of the dioxin receptor. Mol. Cell Biol. 18, 4079–4088.

Poole, K.G., Elkin, B.T., Bethke, R.W., 1995. Environmental contaminants in wild mink in the Northwest Territories, Canada. Sci. Total Environ. 160, 473–486.

Powell, D.C., Aulerich, R.J., Meadows, J.C., Tillitt, D.E., Giesy, J.P., Stromborg, K.L., Bursian, S.J., 1996a. Effects of 3,3',4,4',5-pentachlorobiphenyl (PCB 126) and 2,3,7,8-tetrachlorodibenzo-*p*-dioxin (TCDD) injected into the yolks of chicken (*Gallus domesticus*) eggs prior to incubation. Arch. Environ. Contam. Toxicol 31, 404–409.

Powell, D.C., Aulerich, R.J., Meadows, J.C., Tillitt, D.E., Stromborg, K.L., Giesy, J.P., Bursian, S.J., Kubiak, T.J., 1997a. Organochlorine contaminants in double-crested cormorants from Green Bay,Wisconsin: II. Effects of an extract derived from cormorant eggs on the chicken embryo. Arch. Environ. Contam. Toxicol 32, 316–322.

Powell, D.C., Aulerich, R.J., Meadows, J.C., Tillitt, D.E., Stromborg, K.L., Kubiak, T.J., Giesy, J.P., Bursian, S.J., 1997b. Effects of 3,3',4,4',5-Pentachlorobiphenyl (PCB 126), 2,3,7,8-tetrachlorodibenzo-p-dioxin (TCDD), or an extract derived from field-collected cormorant eggs injected into double-crested cormorant (*Phalacrocorax auritus*) eggs. Environ. Toxicol. Chem 16, 1450–1455.

Powell, D.C., Aulerich, R.J, Stromborg, K.L., Bursian, S.J., 1996b. Effects of 3,3',4,4'-tetrachlorobiphenyl, 2,3,3',4,4'- pentachlorobiphenyl, and 3,3',4,4',5-pentachlorobiphenyl on the developing chicken embryo when injected prior to incubation. J. Toxicol. Environ. Health 49, 319–338.

Puga, A., Maier, A., Medvedovic, M., 2000. The transcriptional signature of dioxin in human hepatoma HepG2 cells. Biochem. Pharmacol. 60, 1129–1142.

Putzrath, R.M., 1997. Estimating relative potency for receptor-mediated toxicity: Reevaluating the toxicity equivalence factor (TEF) model. Regul. Toxicol. Pharmacol. 25, 68–78.

Quensen, J.F., Mousa, M.A., Boyd, S.A., Sanderson, J.T., Froese, K.L., Giesy, J.P., 1998. Reduction of aryl hydrocarbon receptor-mediated activity of polychlorinated biphenyl mixtures due to anaerobic microbial dechlorination. Environ. Toxicol. Chem. 17, 806–813.

Restum, J.C., Bursian, S.J., Giesy, J.P., Render, J.A., Helferich, W.G., Shipp, E.B., Verbrugge, D.A, Aulerich, R.J., 1998. Multigenerational study of the effects of consumption of PCB-contaminated carp from Saginaw Bay, Lake Huron, on mink: 1. Effects on mink reproduction, kit growth and survival, and selected biological parameters. J. Toxicol. Environ. Health A, 54, 343–375.

Rhee, G.Y., Sokol, R.C., Bethoney, C.M., Bush, B., 1993a. A long-term study of anaerobic dechlorination of PCB congeners by sediment microorganisms: Pathways and mass balance. Environ. Toxicol. Chem. 12, 1829–1834.

Rhee, G.Y., Sokol, R.C., Bush, B., Bethoney, C.M., 1993b. Long-term study of the anaerobic cechlorination of aroclor 1254 with and without biphenyl enrichment. Environ. Sci. Technal. 27, 714–719.

Rice, C.D., Roszell, L.E., Banes, M.M., Arnold, R.E., 1998. Effects of dietary PCBs and nonylphenol on immune function and CYP1A activity in channel catfish, *Ictalurus punctatus*. Mar. Environ. Res. 46, 351–354.

Richert, E.P., Prahlad, K.V., 1972. Effects of DDT and its metabolites on thyroid of the Japanese Quail, *Coturnix coturnix japonica*. Poult. Sci. 51, 196–200.

Risebrough, R.W., Rieche, P., Peakall, D.B., Herman, S.G., Kirven, M.N., 1968. Polychlorinated biphenyls in the global ecosystem. Nature 220, 1098–1102.

Robertson, O.H., Chaney, A.L., 1952. Thyroid hyperplasia and tissue iodine content of spawning rainbow trout: A comparative study of Lake Michigan and California sea-run trout. Physiol. Zool. 25, 328–340.

Rogan, W.J., Gladen, B.C., 1992. Neurotoxicology of PCBs and related compounds. Neurotoxicology 13, 27–35.

Ross, P., de Swart, R., Addison, R., Van Loveren, H., Vos, J., Osterhaus, A., 1996a. Contaminant-induced immunotoxicity in harbour seals: Wildlife at risk? Toxicology 112, 157–169.

Ross, P.S., de Swart, R.L., Reijnders, P.J.H., van Loveren, H., Vos, J.G., Osterhaus, A.D., 1995. Contaminant-related suppression of delayed-type hypersensitivity and antibody responses in harbor seals fed herring from the Baltic Sea. Environ. Health Perspect. 103, 162–167.

Ross, P.S., de Swart, R.L., Timmerman, H.H., Reijnders, P.J.H., Vos, J.G., Van Loveren, H., Osterhaus, A.D., 1996b. Suppression of natural killer cell activity in harbour seals (*Phoca vitulina*) fed Baltic Sea herring. Aquat. Toxicol. 34, 71–84.

Ross, P.S., de Swart, R.L., Van der Vliet, H., Willemsen, L., de Klerk, A., van Amerongen, G., Groen, J., Brouwer, A., Schipholt, I., Morse, D.C., Van Loveren, H., Osterhaus, A.D., Vos, J.G., 1997. Impaired cellular immune response in rats exposed perinatally to Baltic Sea herring oil or 2,3,7,8-TCDD. Arch. Toxicol. 17, 563–574.

Safe, S., 1984. Polychlorinated biphenyls (PCBs) and polybrominated biphenyls (PBBs): Biochemistry, toxicology and mechanism of action. Crit. Rev. Toxicol. 13, 319–395.

Safe, S., 1990. Polychlorinated biphenyls (PCBs), dibenzo-*p*-dioxins (PCDDs), dibenzo-furans (PCDFs), and related compounds: Environmental and mechanistic considerations which support the development of toxic equivalency factors (TEFs). Crit. Rev. Toxicol. 21, 51–88.

Safe, S.H., 1994. Polychlorinated biphenyls (PCBs): Environmental impact, biochemical and toxic responses, and implications for risk assessment. Crit. Rev. Toxicol. 24, 87–149.

Salceda, S., Caro, J., 1997. Hypoxia-induced factor 1a (HIF-1a) protein is rapidly degraded by the ubiquitin-proteasome system under normoxic conditions. J. Biol. Chem. 272, 22642.

Sanderson, J.T., Giesy, J.P., 1998. Functional response assays in wildlife toxicology. In: Meyers, R.A., (Ed.), Encyclopedia of Environmental Analysis and Remediation. John Wiley and Sons, New York, pp. 5272–5297.

Sanderson, J.T., Janz, D.M., Bellward, G.D., Giesy J.P., 1997. Effects of embryonic and adult exposure to 2,3,7,8-tetrachlorodibenzo-*p*-dioxin on hepatic microsomal testosterone hydroxylase activities in Great Blue herons (*Ardea herodias*). Environ. Toxicol. Chem. 16, 1304–1310.

Sanderson, J.T., Kennedy, S.W., Giesy, J.P., 1998. *In vitro* induction of ethoxyresorufin-*o*-deethylase and porphyrins by halogenated aromatic hydrocarbons in avian primary hepatocytes. Environ. Toxicol. Chem. 17, 2006–2018.

Sawyer, T., Safe, S., 1982. PCB isomers and congeners: Induction of aryl hydrocarbon hydroxylase and ethoxyresorufin *O*-deethylase enzyme activities in rat hepatoma cells. Toxicol. Lett. 13, 87–93.

Schantz, S.L., Moshtaghian, J., Ness, D.K., 1992. Long-term perinatal exposure to PCB congeners and mixtures on locomotor activity of rats. Teratology 45, 524–525.

Schmitt, C.J., Zajicek, J.L., May, T.W., Cowman, D.F., 1999. Organochlorine residues and elemental contaminants in U.S. freshwater fish, 1976–1986: National Contaminant Biomonitoring Program. Rev. Environ. Contam. Toxicol. 162, 43–104.

Schulz, D.E., Petrick, G., Duinker, J.C., 1989. Complete characterization of polychlorinated biphenyl congeners in commercial Aroclor and Clophen mixtures by multidimensional gas chromatography-electron capture detection. Environ. Sci. Technol. 23, 852–859.

Scott, M.L., 1977. Effects of PCBs, DDT, and mercury compounds in chickens and Japanese quail. Fed. Proc. 36, 1888–1893.

Seed, J., Brown, R.P., Olin, S.S., Foran, J.A., 1995. Chemical mixtures: Current risk assessment methodologies and future directions. Regul. Toxicol. Pharmacol. 22, 76–94.

Seegal, R.F., 1996. Epidemiological and laboratory evidence of PCB-induced neurotoxicity. Crit. Rev. Toxicol. 26, 709–737.

Seegal, R.F., Brosch, K., Bush, B., Ritz, M., Shain, W., 1989. Effects of Aroclor 1254 on dopamine and norepinephrine concentrations in pheochromocytoma (PC-12) cells. Neurotoxicology 10, 757–764.

Seegal, R.F., Bush, B., Brosch, K.O., 1985. Polychlorinated biphenyls induce regional changes in brain norepinephrine concentrations in adult rats. Neurotoxicology 6, 13–23.

Seegal R.F., Bush, B., Brosch, K.O., 1994. Decreases in dopamine concentrations in adult, nonhuman primate brain persist following removal from polychlorinated biphenyls. Toxicology 86, 71–87.

Seegal, R.F., Bush, B., Shain, W., 1990. Lightly chlorinated *ortho*-substituted PCB congener decrease dopamine in nonhuman primate brain and in tissue culture. Toxicol. Appl. Pharmacol. 106, 136–144.

Seegal, R.F., Schantz, S.L., 1994. Neurochemical and behavioral sequelae of exposure to dioxins and PCBs. In: Schecter, A., (Ed.), Dioxins and Health, Plenum Press, New York, pp. 409–447.

Semenza, G.L., 1994. Regulation of erythropoietin production. Hematol. Oncol. Clin. N. Am. 8, 863–864.

Semenza, G.L., Nejfflt, M.K., Chi, S.M., Antonarakis, S.E., 1991. Hypoxia-inducible nuclear factors bind to an enhancer element located 3' to the human erythropoietin gene. Proc. Natl. Acad. Sci. USA 88, 5680–5684.

Shain, W., Bush, B., Seegal, R., 1991. Neurotoxicity of polychlorinated biphenyls: Structure-activity relationship of individual congeners. Toxicol. Appl. Pharmacol. 111, 33–42.

Shweiki, D., Itin, A., Soffer, D., Keshet, E., 1992. Vascular endothelial growth factor induced by hypoxia may mediate hypoxia-initiated angiogenesis. Nature 359, 843–845.

Silberhorn, E.M., Glauert, H.P., Robertson, L.W., 1990. Carcinogenicity of polyhalogenated biphenyls: PCBs and PBBs. Crit. Rev. Toxicol. 20, 440–496.

Sogawa, K., Nakano, R., Kobayashi, A., Kikuchi, Y., Ohe, N., Matsushita, N., Fujii-Kuriyama, Y., 1995. Possible function of Ah receptor nuclear translocator (Arnt) homodimer in transcriptional regulation. Proc. Natl. Acad. Sci. USA 92, 1936–1940.

Sonstegard, R.A., Leatherland, J.F., 1979. Hypothyroidism in rats fed Great Lakes coho salmon. Bull. Environ. Contam. Toxicol. 22, 779–784.

Spear, P. A., Moon, T.W., 1985. Low dietary iodine and thyroid anomalies in ring doves, *Streptopelia risoria*, exposed to 3,4,3',4'-tetrachlorobiphenyl. Arch. Environ. Contam. Toxicol. 14, 547–553.

Spear, P.A, Moon, T., Peakall, D., 1985. Liver retinoid concentrations in natural populations of herring gulls contaminated by 2,3,7,8-tetrachlorodibenzo-p-dioxin and in ring doves injected with a dioxin analogue. Can. J. Zool. 64, 204–208.

Starodub, M.E., Miller, P.A., Ferguson, G..M., Giesy, J.P., Willis, R.F., 1996. A risk-based protocol to develop acceptable concentrations of bioaccumulative organic chemicals in sediments for the protection of piscivorous wildlife. Toxicol. Environ. Chem. 54, 243–259.

Strang, C., Levine, S.P., Orlan, B.P., Gouda, T.A., Saner, W.A., 1984. High resolution GC analysis of cytochrome P-448 inducing polychlorinated biphenyl congeners in hazardous waste. J. Chromatogr. 314, 482–487.

Sundström, G., Hutzinger, O., Safe, S., 1976. The metabolism of chlorobiphenyls—A review. Chemosphere 5, 267–298.

Tanabe, S., 1988. PCB problems in the future: Foresight from current knowledge. Environ. Pollut. 50, 5–28.

Tanabe, S., Iwata, H., Tatsukawa, R., 1994. Global contamination by persistent organochlorines and their ecotoxicological impact on marine mammals. Sci. Total Environ. 154, 163–177.

Tanabe, S., Kannan, N., Subramanian, A., Watanabe, S., Ono, M., Tatsukawa, R., 1987a. Highly toxic coplanar PCBs: Occurrence, source, persistency and toxic implications to wildlife and humans. Environ. Pollut. 47, 147–163.

Tanabe, S., Kannan, N., Wakimoto, T., Tatsukawa, R., 1987b. Method for the determination of three toxic non-ortho chlorine substituted coplanar PCBs in environmental samples at parts-per-trillion levels. Intl. J. Environ. Anal. Chem. 29, 199–213.

Tanabe, S., Kannan, N., Wakimoto, T., Tatsukawa, R., Okamoto, T., Masuda, Y., 1989. Isomer-specific determination and toxic evaluation of potentially hazardous coplanar PCBs, dibenzofurans and dioxins in the tissues of "Yusho" PCB poisoning victim and in the causal oil. Toxicol. Environ. Chem. 24, 215–231.

Tanabe, S., Watanabe, S., Kan, H., Tatsukawa, R., 1988. Capacity and mode of PCB metabolism in small cetaceans. Mar. Mamm. Sci. 4, 103–124.

Tarhanen, J., Koistinen, J., Paasivirta, J., Vuorinen, P.J., Koivusaari, J., Nuuja, I., Kannan, N., Tatsukawa, R., 1989. Toxic significance of planar aromatic compounds in Baltic ecosystem—new studies on extremely toxic coplanar PCBs. Chemosphere, 18, 1067–1077.

Thorpe, J.E., Talbot, C., Miles, M., 1989. Research suggests a simple answer to grisle control. Fish Farmer 12, 10.

Tian, H., McKnight, S.L., Russell, D.W., 1997. Endothelial PAS domain protein 1 (EPAS1), a transcription factor selectively expressed in endothelial cells. Genes Dev. 11, 72–82.

Tilbury, K.L., Stein, J.E., Meador, J.P., Krone, C.A., Chan, S.-L., 1997. Chemical contaminants in harbor porpoise (*Phocoena phocoena*) from the North Atlantic coast: Tissue concentrations and intra- and interorgan distribution. Chemosphere, 34, 2159–2181.

Tillitt, D.E., 1989. Characterization studies of the H4IIE bioassay for assessment of planar halogenated hydrocarbons in fish and wildlife. Ph.D. dissertation, Michigan State University, East Lansing.

Tillitt, D.E., Gale, R.W., Meadows, J.C., Zajicek, J.L., Peterman, P.H., Heaton, S.N., Jones, P.D., Bursian, S.J., Kubiak, T.J., Giesy, J.P., Aulerich, R.J., 1996. Dietary exposure of mink to carp from Saginaw Bay. 3. Characterization of dietary exposure to planar halogenated hydrocarbons, dioxin equivalents, and biomagnification. Environ. Sci. Technol., 30, 283–291.

Tillitt, D.E., Giesy, J.P., Ankley, G.T., 1991. Characterization of the H4IIE rat hepatoma cell bioassay as a tool for assessing toxic potency of planar halogenated hydrocarbons in environmental samples. Environ. Sci. Technol., 25, 87–92.

Tillitt, D.E., Giesy, J.P., Ludwig, J.P., Kurita-Matsuba, H., Weseloh, D.V., Ross, P.S., Bishop, C.A., Sileo, L., Stromberg, K.L., Larson, J., Kubiak, T.J., Ankley, G.T., 1992. Polychlorinated biphenyl residues and egg mortality in double-crested cormorants from the Great Lakes. Environ. Toxicol. Chem. 11, 1281–1288.

Tillitt, D.E., Kubiak, T.J., Ankley, G.T., Giesy, J.P., 1993. Dioxin-like toxic potency in Forster's tern eggs from Green Bay, Lake Michigan, North America. Chemosphere 26, 2079–2084.

Tilson, H.A., Jacobson, J.L., Rogan, W.J., 1990. Polychlorinated biphenyls and the developing nervous system: Cross-species comparisons. Neurotoxicol. Teratol. 12, 239–248.

Tumasonis, C.F., Bush, B., Baker, F.D., 1973. PCB levels in egg yolks associated with embryonic mortality and deformity of hatched chicks. Arch. Environ. Contam. Toxicol. 1, 312–324.

U.S. Environmental Protection Agency, 1995. Great Lakes Water Quality Initiative Technical Support Document for Wildlife Criteria, EPA-820-B-95-009.

U.S. Environmental Protection Agency, 1996. Use of Uncertainty Factors in Toxicity Extrapolations Involving Terrestrial Wildlife. Technical Basis. Office of Research and Development, U.S. Environmental Protection Agency, Washington, DC.

U.S. EPA Region 9 Biological Technical Advisory Group, 1998. Use of PCB Congener and Homologue Analysis in Ecological Risk Assessments.

van Birgelen, A.P.J.M., van der Kolk, J., Poiger, H., van den Berg, M., Brouwer, A., 1992. Interactive effects of 2,2',4,4',5,5'-hexachlorobiphenyl and 2,3,7,8-tetrachlorodibenzo-p-dioxin, thyroid hormone, vitamin A, and vitamin K metabolism in the rat. Chemosphere 25, 1239–1244.

van den Berg, M., Birnbaum, L., Bosveld, A.T., Brumström, B., Cook, P., Feely, M., Giesy, J.P., Hanberg, A., Hasegawa, R., Kennedy, S.W., Kubiak, T., Larsen, J.C., van Leeuwen, F.X., Liem, A.K., Nolt, C., Peterson, R.E., Pollinger, L., Safe, S., Schrenk, D., Tillitt, D., Tusklind, M., Younes, M., Waern, F., Zacharewski, T., 1998. Toxic equivalency factors (TEFs) for PCBs, PCDDs, PCDFs for humans and wildlife. Environ. Health Perspect. 106, 775–792.

Verrett, M.J., 1976. Investigation of the toxic and teratogenic effects of halogenated dienzo-p-dioxins and dibenzofurans in the developing chicken embryo. Memorandum. U.S Food and Drug Administration, Washington, DC.

Ville, P., Roch, P., Cooper, E.L., Masson, P., Narbonne, J.F., 1995. PCBs increase molecular-related activities (lysozyme, antibacterial, hemolysis, proteases), but inhibit macrophage-related functions, phagocytosis—wound healing in earthworms. J. Invert. Pathol. 65, 217–224.

Villeneuve, D.L., Blankenship, A.L., Giesy J.P., 2000. Derivation and application of relative potency estimates based on in vitro bioassay results. Environ. Toxicol. Chem. 19, 2835–2843.

Voie, O.A., Fonnum, F., 1998. *Ortho* substituted polychlorinated biphenyls elevate intracellular [Ca^{2+}] in human granulocytes. Environ. Toxicol. Pharmacol. 5, 105–112.

Waid, J.S., 1987. PCBs and the Environment. Vols. I–III, CRC Press, Boca Raton, FL.

Walker, C.H., 1990. Persistent pollutants in fish-eating birds—bioaccumulation, metabolism and effects. Aquat. Toxicol. 17, 293–324.

Walker, M.A.K., Peterson, R.E., 1992. Toxicity of polychlorinated dibenzo-p-dioxins, dibenzofurans and biphenyls during early development in fish. In: Colborn, T. (Eds.), Advances in Modern Environmental Toxicology. Princeton Scientfic Publishing, Princeton, NJ, pp. 195–202.

Walker, M.K., Cook, P.M., Batterman, A.R., Butterworth, B.C., Berini, C., Libal, J.J., Hufnagle, L.C., Peterson, R.E., 1994. Translocation of 2,3,7,8-tetrachlorodibenzo-p-dioxin from adult female lake trout (*Salvelinus namaycush*) to oocytes—effects on early-life stage development and sac fry survival. Can. J.Fish. Aquat.Sci. 51, 1410–1419.

Walker, M.K., Cook, P.M., Butterworth, B.C., Zabel, E.W., Peterson, R.E., 1996. Potency of a complex mixture of polychlorinated dibenzo-p-dioxin, dibenzofuran, and biphenyl congeners compared to 2,3,7,8-tetrachlorodibenzo-p-dioxin in causing fish early life stage mortality. Fundam. Appl. Toxicol. 30, 178–186.

Walker, M.K., Peterson, R.E., 1991. Potencies of polychlorinated dibenzo-p-dioxin, dibenzofuran and biphenyl congeners, relative to 2,3,7,8-tetrachlorodibenzo-p-dioxin for producing early life stage mortality in rainbow trout (*Oncorhynchus mykiss*). Aquat. Toxicol. 21, 219–238.

Walker, M.K., Peterson, R.E., 1994a. Toxicity of 2,3,7,8-tetrachloriodibenzo-p-dioxin to brook trout during early development. Environ. Toxicol. Chem. 13, 817–820.

Walker, M.K., Peterson, R.E., 1994b. Aquatic toxicity of dioxins and related chemicals. In: Schecter, A.E. (Ed.), Dioxins in Health. Plenum Press, New York, pp. 347–387.

Walker, M.K., Pollenz, R.S., Smith, S.M., 1997. Expression of the aryl hycarbon receptor (AhR) and AhR nuclear translocator during chick cardiogenesis is consistent with 2,3,7,8-tetrachlorodibenzo-*p*-dioxin-induced heart defects. Toxicol. Appl. Pharmacol. 143, 407–419.

Wang, G.L., Semenza, G.L., 1993. Characterization of hypoxia-inducible factor 1 and regulation of DNA binding activity by hypoxia. J. Biol. Chem. 268, 21513–21518.

Wenger, R. H., Gassmann, M., 1997. Oxygen(es) and the hypoxia-inducible factor-1. Biol. Chem. 378, 609–616.

Weseloh, D.V., Bishop, C.A., Norstrom, R.J., Fox, G.A. 1991. Monitoring levels and effects of contaminants in herring gull eggs on the Great Lakes, 1974–1990. Abstracts of the Cause-Effects Linkages II Symposium, Traverse City, MI, September 27–28, 29–31. Michigan Audobon Society, Lansing, MI.

White, D.H., Hoffman, D.J., 1995. Effects of polychlorinated dibenzo-p-dioxins and dibenzofurans on nesting wood ducks (*Aix sponsa*) at Bayou Meto, Arkansas. Environ. Health Perspect. 103 (suppl. 4), 37–43.

Wiemeyer, S.N., Lamont, T.G., Bunck, C.M., Sindelar, C.R., Gramlicj, J., Fraser, J.D., Byrd, M.A., 1984. Organochlorine pesticide, polychlorobiphenyl, and mercury residues in bald eagle eggs—1969–79—and their relationships to shell thinning and reproduction. Arch. Environ. Contam. Toxicol 13, 529–549.

Williams, L.L., Giesy, J.P., 1992. Relationship among concentrations of individual polychlorinated biphenyl (PCB) congeners, 2,3,7,8-tetrachlorodibenzo-p-dioxin equivalents (TCDD-EQ) and rearing mortality of chinook salmon (*Oncorhynchus tshawytscha*) eggs from Lake Michigan. J. Great Lakes Res. 18, 108–124.

Wong, P.W., Joy, R.M., Albertson, T.E., Schantz, S.L., Pessah, I.N., 1997. Ortho-substituted 2,2',3,5',6-pentachlorobiphenyl (PCB 95) alters rat hippocampal ryanodine receptors and neuroplasticity in vitro: Evidence for altered hippocampal function. Neurotoxicology 18, 443–456.

Woodford, J.E., Karasov, W.H., Meyer, M.W., Chambers, L., 1998. Impact of 2,3,7,8,-TCDD exposure on survival, growth, and behavior of ospreys breeding in Wisconsin, USA. Environ. Toxicol. Chem 17, 1323–1331.

Wren, C.D., Hunter, D.B., Leatherland, J.F., Stokes, P.M., 1987. The effects of polychlorinated biphenyls and methyl mercury singly and in combination on mink. II. Reproduction and kit development. Arch. Environ. Contam. Toxicol. 16, 449–454.

Wright, P.J., Tillitt, D.E., 1999. Embryotoxicity of Great lakes lake trout extracts to developing trout. Aquat. Toxicol. 47, 77–92.

Yamashita, N., Tanabe, S., Ludwig, J.P., Kurita, H., Ludwig, M.E., Tatsukawa, R., 1993. Embryonic abnormalities and organochlorine contamination in double-crested cormorants (*Phalacrocorax auritus*) and Caspian terns (*Hydropogne caspia*) from the upper Great Lakes in 1988. Environ. Pollut. 79, 163–173.

Yao, C., Panigrahy, B., Safe, S., 1990. Utilization of cultured chick embryo hepatocytes as in vitro bioassays for polychlorinated biphenyls (PCBs): Quantitative-structure-induction relationships. Chemosphere 21, 1007–1016.

Yao, T.-P., Ku, G., Zhou, N., Scully, R., Livingston, D.M., 1996. The nuclear hormone receptor co-activator SRC-1 is a specific target of p300. Proc. Natl. Acad. Sci. USA. 93, 10626–10631.

Yoshimura, H., Yoshihara, S., Koga, N., Nagata, K., Wada, I., Kuroki, J., Hokama, Y., 1985. Inductive effect on hepatic enzymes and toxicity of congeners of PCBs and PCDFs. Environ. Health Perspect. 59, 113–119.

Yu, M.L., Hsin, J.Y., Hsu, C.C., Chan, W.C., Guo, Y.L., 1998. The immunologic evaluation of the Yucheng children. Chemosphere, 37, 1855–1865.

Zell, M., Ballschmiter, K., 1980. Baseline studies of the global pollution III. Trace analysis of polychlorinated biphenyls (PCB) by ECD glass capillary gas chromatography in environmental samples of different trophic levels. Fres. Z. Anal. Chem. 304, 337–349.

Zhao, F, Mayura, K., Kocurek, N., Edwards, J.F., Kubena, L.F., Safe, S.H., Phillips, T.D., 1997. Inhibition of 3,3',4,4',5-pentachlorobiphenyl-induced chicken embryotoxicity by 2,2',4,4',5,5'-hexachlorobiphenyl. Fundam. Appl. Toxicol. 35, 1–8.

Zile, M.H., 1992. Vitamin A homeostasis endangered by environmental pollutants. Proc. Soc. Exp. Biol. Med. 201, 141–153.

Zile, M.H., Cullum, M.E., 1983. The function of vitamin A: Current concepts. Proc. Soc. Exp. Biol. Med. 172, 139–152.

12

DDT and Its Analogues: New Insights into Their Endocrine-Disrupting Effects on Wildlife

Louis J. Guillette, Jr.
Stefan A.E. Kools
Mark P. Gunderson
Dieldrich S. Bermudez

From the 1940s to the beginning of the 1960s, pesticide use was seen as a major weapon against agricultural and public health pests with relatively few important detrimental consequences. However, with the publication of *Silent Spring*, by Rachel Carson in 1962, the public and scientific communities began to look at pesticide use differently. Pesticides did have beneficial effects, but there was a cost, and in many cases, the cost was unknown. These compounds were poisons, and DDT and related compounds were persistent pollutants capable of bioaccumulation and biomagnification within food chains. Although the initial concern about widespread DDT use focused on its possible carcinogenic actions, the most widely known effect was its association with eggshell thinning in many avian species. Although its use was restricted in 1972 in the United States and it is today restricted in many countries worldwide, it is still widely used in tropical regions for the control of the mosquito vectors of malaria. In fact, its use is still supported and promoted for malaria control (Roberts, 1999; Roberts et al., 2000), although the collateral effects of its use may be marginalized; alternatives appear to be available and economically reasonable (see Liroff, 2001).

As early as 1950, studies demonstrated that DDT and its metabolites were more than just pesticides. Burlington and Lindeman (1950) demonstrated that juvenile cockerels treated with DDT failed to develop male secondary sex characteristics. Data collected since the 1990s show that DDT and its metabolites have the potential to act as hormonal mimics or receptor antagonists and disrupt or alter the endocrine system of vertebrates (see Rooney and Guillette, 2000). Below, we provide a brief introduction to this class of compounds and review the recent data on the endocrine action of DDT and its metabolites.

DDT: Occurrence and Biotransformation

DDT (1,1,1-trichloro-2,2-bis[*p*-chlorophenyl]ethane) and its analogs are chemicals that have been, and continue to be, used around the world as commercial pesticides (figure 12-1). Various tradenames have been used for DDT or related compounds, such as TDE, dicofol (Kelthane), Bulan, chlorfenethol (DMC), chlorobenzilate, chloropropylate, DFDT, ethylan (Perthane), methoxychlor, and Prolan (figure 12-1). Together with four other chemical groups, these compounds form the organochlorine pesticides. The other chemicals include hexachlorocyclohexane, cyclodienes and similar structures, toxaphene and related chemicals, and caged structures such as mirex and chlordecone (Smith, 1991). This chapter focuses on DDT and its metabolites, but it must be noted that other compounds, such as methoxychlor and dicofol, have exhibited endocrine-disrupting effects (Peterle, 1991).

DDT has two primary metabolites that are commonly detected in the environment: DDD (1,1-dichloro-2,2-*bis*[*p*-chlorphenyl]ethane) and DDE (1,1-dichloro-2,2-*bis*[*p*-chlorophenyl]-ethylene) (figure 12-1). They all consist of two phenolic rings, with chlorine atoms bound between and/or on the rings. The structure permits several different isomeric forms. Commercial, technical grade DDT contains mainly two isomers: *p,p'*-DDT (85%) and *o,p'*-DDT (15%) (ATSDR, 1994).

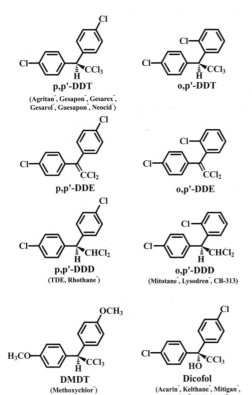

Figure 12-1. Forms of DDT and its common metabolites, DDE and DDD. Included are the structures of two related compounds, DMDT and dicofol.

DDT was first synthesized in 1874, but it was not until 1939 that Paul Müller discovered its insecticidal properties, for which he received a Nobel Prize. DDT has the ability to alter the movement of sodium and potassium ions across membranes of the nervous system, inhibiting neural transmission (ATSDR, 1994). Consequently, many exposed organisms suffer neurological symptoms, but lethality has not been reported for humans at normal environmental exposures. In addition to the acute effects on the nervous system, the liver is also affected. Due to its low relative toxicity in vertebrates, DDT was first widely used as a delousing agent and an insecticide for agricultural crops (ATSDR, 1994). Later, DDT was used to control insects that are vectors for disease, such as malaria and typhus. Reported effects on nontarget species (Cottam and Higgens, 1946) were of minimal concern until die-offs of many species of songbirds became well publicized (Carson, 1962). For example, the death of birds feeding on trees treated with DDT to control Dutch elm disease became a common urban event (Hickey and Hunt, 1960). Accumulating evidence of its persistence and its toxicity led to the ban of DDT for general use in the United States in 1972 (ATSDR, 1994).

Current use of DDT in the United States is restricted to emergency public health-related problems (ATSDR, 1994), but it is unknown to what extent it is used. The estimated global production of DDT in 1995 was 30,000 metric tons per year (Sharpe, 1995). DDT is commonly used in many tropical countries to control mosquitoes, the vectors of malaria. Given the stability of DDT and its metabolites, as well as its global use, average concentrations in many regions are 10 ppm (Simonich and Hites, 1995). Shipping records from the U.S. Customs Agency show export of DDT from the United States as late as 1992 (Smith and Root, 1999).

During application and production, DDT can be released to air, water, or soil. Although hundreds of tons were used for decades, there is little knowledge concerning the multiple pathways of degradation and accumulation of DDT in the environment. Free, unbound DDT in air or water is subject to photo-oxidation, and the half-life is as little as 2–6 days. In soil, however, DDT can bind strongly to particles and becomes more resistant to degradation. The half-life of DDT can therefore be greater than 50 years. Degradation of DDT results in more stable metabolites, of which DDE and DDD are most common (figure 12-2). DDE is the primary metabolite formed under aerobic conditions, whereas anaerobic conditions contribute to DDD. DDT and its metabolites leave soil or water by volatilization.

Due to water and atmospheric distribution, measurable levels of DDT and its metabolites are found all over the world, including polar regions (Iwata et al., 1993). DDT, DDD, and DDE are lipophilic and will bioaccumulate in fatty tissue, primarily as the very persistent DDE isomer. Biotransformation of DDT occurs rapidly through reductive dechlorination to DDD, with is further transformed and readily excreted (figure 12-2). Dehydrochlorination, another metabolic pathway, produces DDE. This pathway occurs less frequently. Metabolism of DDE is apparently very slow, leading to the bioaccumulation of DDE in fatty tissue. Transformation of both DDE and DDD results in 2,2-*bis*(chlorophenyl) acetic acid (DDA), which is readily excreted from the organism (Peterson and Robinson, 1964). However, more than 25 years after DDT use was greatly limited in Germany, DDA was still found as a major contaminant in surface water at $\mu g/l$ concentrations (Heberer and Dunnbier,

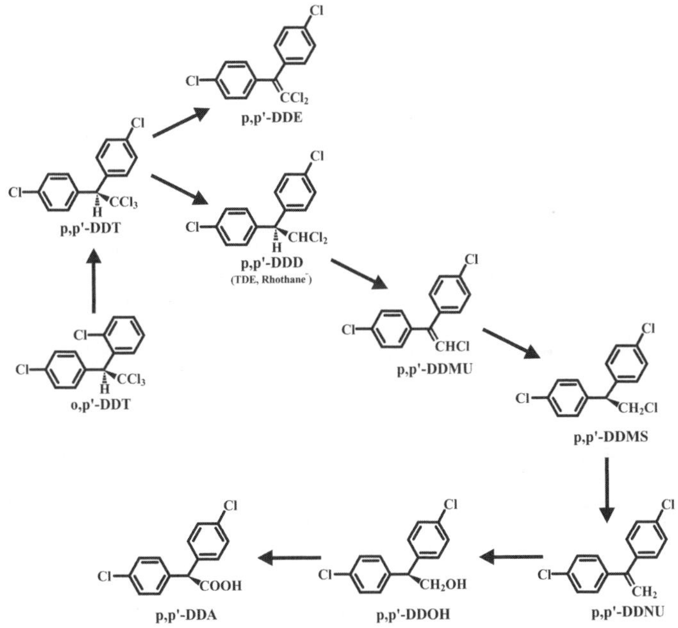

Figure 12-2. The degradation pathway of DDT as determined in mammalian and avian tissues. Abbreviations: DDD, 1,1-dichloro-2,2-bis(p-chlorophenyl)ethane; DDE, 1,1-dichloro-2,2-bis(p-chlorophenyl)ethylene; DDMU, 1-chloro-2,2-bis(p-chlorophenyl)ethylene; DDMS, 1-chloro-2,2-bis(p-chlorophenyl)ethane; DDNU, unsym-bis(p-chlorophenyl)ethylene; DDOH, 2,2-bis(p-chlorophenyl)ethanol; DDA, bis(p-chlorophenyl)acetic acid. See Peterson and Robinson (1964) for more details.

1999). DDA readily leaches into groundwater wells, but its biological effects, if any, are poorly understood due to an almost complete lack of research on this metabolite.

Levels of stored DDT, DDE, and DDD tend to be elevated in animals at the top of the food chain (see figure 12-3). For example, species living in polar regions and marine mammals, in which lipid storage is essential for energy balance and thermoregulation, accumulate considerable body burdens of organochlorine compounds, including DDT and its metabolites (for a review, see Colborn and Smolen, 1996). Storage of these compounds in fatty tissue can result in chronic exposure as well as significant acute exposure when contaminants are mobilized during periods when animals utilize fat stores, such as during lactation or migration. These features must be taken into account when reviewing endocrine effects on wildlife.

Although the body burden of DDT and its metabolites in many organisms, has decreased over the last two decades in North America, detectable levels are still observed in a high percentage of individuals of many species (see figure 12-4). Given the new data emerging on the actions of low doses of persistent pesticides, continued research is needed to determine the consequences of chronic, low level exposure.

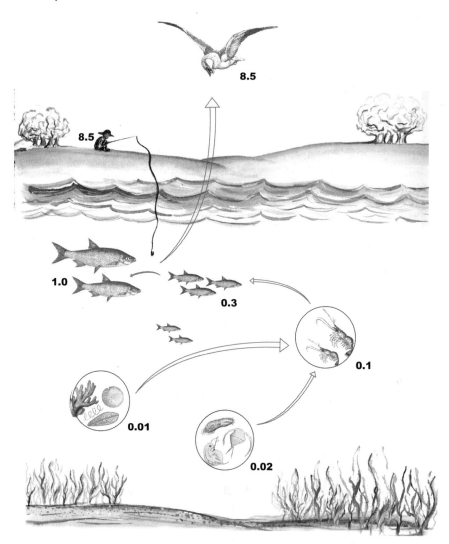

Figure 12-3. Bioaccumulation and biomagnification in the food chain of a lake. Values reported are in ppm concentrations. Data from Colborn et al. (1990). DDT and its metabolites are highly fat soluble and readily accumulate in food chains.

Endocrine Actions: Laboratory Studies

Sex Steroid Axis

DDT or its metabolites affect organisms via multiple mechanisms. Apart from acute neurological effects, these compounds influence various systems, such as the immune and endocrine systems (for a review, see ATSDR, 1994). Of concern here are effects related to the endocrine system. The effect of DDT and its metabolites

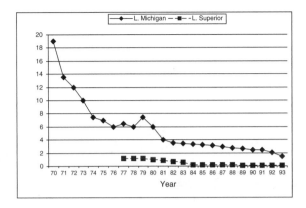

Figure 12-4. Mean total DDTs ($\Sigma DDTs$) concentrations (mg total DDT/kg wet weight) in whole lake trout collected from several of the Laurentian Great Lakes. Although $\Sigma DDTs$ declined rapidly after their use was stopped in the early 1970s, the absolute decrease currently is slow, approximately 3% per year. Thus, the current absolute change is very low, suggesting that we will have to manage and understand the biological consequences of these low dose exposures for decades, if not generations. Data from MDNR (1994).

on the reproductive system has been a primary focus of much research. In large part, this was in response to early studies showing an association between eggshell thinning in birds and DDT exposure (Ratcliffe, 1967, 1970; Cooke, 1973).

Laboratory studies in the 1950s and 1960s demonstrated the estrogenic action of DDT in roosters (Burlington and Lindeman, 1950) as well as in other species (Bitman et al., 1968; Welch et al., 1969). The isomer o,p'-DDT was the suspected estrogen and later was shown to interact with the mammalian estrogen receptor-α (ERα; Shelby et al., 1996; Klotz et al., 1997). More recent in vivo studies have documented the estrogenicity of DDT by injecting juvenile trout with different isoforms (o,p'-DDT, o,p'-DDE). Trout exhibited vitellogenesis, an effect normally initiated by estrogens (Donohoe and Curtis, 1996). Similar studies using an amphibian (*Xenopus laevis*) and freshwater turtle (*Trachemys scripta*) demonstrated vitellogenesis after experimental exposure to o,p'-DDT (Palmer and Palmer, 1995). Numerous other studies have documented the estrogenic action of DDT or its metabolites (see table 12-1). Much of the reported estrogenicity is hypothesized to be due to receptor-mediated mechanisms (see Rooney and Guillette, 2000). However, several studies have shown that pesticides can induce aromatase activity in gonadal, adrenal, or hepatic tissues (Crain et al., 1997; Sanderson et al., 2000; You et al., 2001). Aromatase converts testosterone to estradiol, thus providing a possible alternative estrogenic mechanism that does not require receptor binding.

In many species of fish, all crocodilians, and some turtles, lizards, and amphibians, sex determination is influenced by environmental factors such as incubation temperature (see Bull, 1983). In those turtles and crocodilians exhibiting environmental sex determination, treatment of eggs during a critical embryonic period with estrogenic compounds redirects sex determination so that females are produced (Matter et al., 1998; Willingham and Crews, 1999; table 12-2). Matter et al. (1998)

Table 12-1. Representative endocrine actions of DDT and its metabolites in vertebrates.

Compound	Label Effect	Effect	Study Type	Reference
Not specified DDT				
DDT [ns]	Antiandrogenic	Altered testosterone metabolism	Birds	Peakall, 1967
DDT [ns]		Altered thyroxine metabolism	Birds	Bastomsky, 1974
DDT [ns]		Altered progesterone metabolism	Birds	Peakall, 1967
Technical DDT				
DDT [tech]	Estrogenic	Estrogenic response cells	E-screen	Sonnenschein and Soto, 1998
DDT [tech]	Estrogenic	Increase uterine weight	Rat	Welch et al., 1969
DDT [tech]	Antiestrogenic	Antagonized estrogenic treatment	Tiger salamander	Clark et al., 1998
DDT [tech]		Increases steroid metabolism	Rat (liver)	Welch et al., 1969
DDT (2 isomers)				
o,p'-DDT	Estrogenic ?	ER interaction	Alligator ER	Vonier et al., 1996
o,p'-DDT	Estrogenic ?	Weakly binds ER	Human ER	Gaido et al., 1997
o,p'-DDT	Estrogenic	Induction of zona radiata proteins	Atlantic salmon	Arukwe et al., 1997
o,p'-DDT	Estrogenic	Estrogenic response cells	E-screen	Sonnenschein and Soto, 1998
o,p'-DDT	Estrogenic	ER binding + response	Mice ER	Shelby et al., 1996
o,p'-DDT	Estrogenic	Increase uterine weight	Rat	Welch et al., 1969
o,p'-DDT	Estrogenic	Vtg synthesis in males	Red eared turtle	Palmer and Palmer, 1995
o,p'-DDT	Estrogenic	Induction of Vtg/hepatic EBS	Trout	Donohoe, 1996
o,p'-DDT	Estrogenic	ER binding + response	YES	Arnold et al., 1996
o,p'-DDT	Estrogenic	ER binding + response	YES/CB/MCF-7	Klotz et al., 1996
o,p'-DDT	Antiestrogenic	Strong antagonist	Rabbit ER (CB)	Danzo, 1997
o,p'-DDT	Antiandrogenic	Antagonist	Human AR	Maness et al., 1998
o,p'-DDT	Antiandrogenic	Strong antagonist	Rat AR	Danzo, 1997
o,p'-DDT	Antiandrogenic	Antagonist	YES AR	Sohoni and Sumpter, 1998
o,p'-DDT	Hyperthyroidsm	Increased thyroid weight	Bengale finch	Jefferies, 1969
o,p'-DTT	Estrogenic/antiandrogenic	Skewed sex-ratio toward female	Seagull	Fry and Toone, 1981
o,p'-DTT	Estrogenic/antiandrogenic	Development of right oviduct	Seagull	Fry and Toone, 1981
o,p'-DTT	Estrogenic/antiandrogenic	Feminized testis	Seagull	Fry and Toone, 1981
p,p'-DDT	Estrogenic ?	Weak ER interaction	Alligator ER	Vonier et al., 1996
p,p'-DDT	Estrogenic	Estrogenic response cells	E-screen	Sonnenschein and Soto, 1998

338

Compound	Activity	Description	Test system	Reference
p,p'-DDT	Antiprogesterone	Antiprogesterone (antiprogestin)	HumanPR (CB)	Klotz et al., 1997
p,p'-DDT	Antiestrogenic	Weak antagonist	Rabbit ER (CB)	Danzo, 1997
p,p'-DDT	Antiandrogenic	Decrease testicular growth	Cockerel	Burlington and Lindeman, 1950
p,p'-DDT	Antiandrogenic	Antagonist	Human AR	Maness et al., 1998
p,p'-DDT	Antiandrogenic	Strong antagonist	Rat AR	Danzo, 1997
p,p'-DDT	"Corticosteronic"?	Induces stress related response	Senegal walking frog	Hayes et al., 1997
DDE (2 isomers)				
o,p'-DDE	Estrogenic ?	ER-interaction	Alligator ER	Vonier et al., 1996
o,p'-DDE	Estrogenic ?	Weakly binds ER	Human ER	Gaido, 1997
o,p'-DDE	Estrogenic	Estrogenic response cells	E-screen	Sonnenschein and Soto, 1998
o,p'-DDE	Estrogenic	Induction of Vg/hepatic EBS	Trout	Donohoe and Curtis, 1996
o,p'-DDE	Antiprogesterone	Antiprogesterone	Fowl PR (CB)	Lundholm, 1997
o,p'-DDE	Antiandrogenic	Antagonist	Fowl AR	Lundholm, 1997
o,p'-DDE	Antiandrogenic	Antagonist	Human AR	Maness et al., 1998
o,p'-DDE	Estrogenic ?	Weak ER interaction	Alligator ER	Vonier et al., 1996
p,p'-DDE	Estrogenic	Development of gonaducts	Tiger salamander	Clark et al., 1998
p,p'-DDE	Antiestrogenic?	Weak antagonist	Rabbit ER (CB)	Danzo, 1997
p,p'-DDE	Antiandrogenic	Inhibits transcription	Human AR/CB	Kelce et al., 1995
p,p'-DDE	Antiandrogenic	Reduction anogenital distance	Male rats (offspring)	Kelce et al., 1995
p,p'-DDE	Antiandrogenic	Retention of thoracic nipples	Male rats (offspring)	Kelce et al., 1995
p,p'-DDE	Antiandrogenic	Strong antagonist	Rat AR	Danzo, 1997
p,p'-DDE	Antiandrogenic	Antagonist	YES AR	Sohoni and Sumpter, 1998
p,p'-DDE	Androgenic?	Antagonist, agonist	Human AR	Maness et al., 1998
p,p'-DDE	Androgenic?.	AR interaction	Human AR	Gaido et al., 1997
DDD (2 isomers)				
o,p'-DDD	Estrogenic ?	ER interaction	Alligator ER	Vonier et al., 1996
o,p'-DDD	Estrogenic ?	Weakly binds ER	Human ER	Gaido et al., 1997
o,p'-DDD	Estrogenic	ER binding + response (weak)	YES/CB/MCF-7	Klotz et al., 1996
o,p'-DDD	Antiandrogenic	Antagonist	Human AR	Maness et al., 1998
p,p'-DDD	Estrogenic ?	Weak ER-interaction	Alligator ER	Vonier et al., 1996
p,p'-DDD	Estrogenic	Sex reversal	American alligator	Crain, 1997
p,p'-DDD	Estrogenic	ER binding + response	YES CB/MCF-7	Klotz et al., 1996

ER, estrogen receptor; Vtg, vitellogenin; YES, yeast estrogen screen; AR, androgen receptor; PR, progesterone receptor; EBS, estrogen binding sites.

Table 12-2. Representative examples of studies examining whether altered sex determination or sex reversal is induced in various species after exposure of embryos or breeding adults to DDT or its metabolites.

Species	Life Stage	Isomer	Exposure Dose	Effect	Reference
Medaka	Egg	o,p'-DDT	227 ng/egg	Skewed sex ratio toward females	Edmunds et al., 2000
Medaka	Adults	o,p'-DDT	0.3–1.94 µg/l	Skewed sex ratio toward females	Cheek et al., 2001
Alligator	Egg	o,p'-DDE	0.1–0.3 mg/kg[a]	Skewed sex ratio toward females	Matter et al., 1998
Alligator	Egg	p,p'-DDE	1–10 mg/kg[a]	Skewed sex ratio toward females	Matter et al., 1998
Alligator	Egg	p,p'-DDD	0.1–10 mg/ kg[a]	Skewed sex ratio toward females	Crain, 1997
Red-eared slider	Egg	p,p'-DDE	5.8 mg/kg[a]	Skewed sex ratio toward females	Willingham and Crews, 1999
Green sea turtle	Egg	p,p'-DDE	≤ 543 ng/g egg	No sex ratio effect	Podreka et al., 1998
Snapping turtle	Egg	p,p'-DDE	≤ 0.65 mg/ kg[a]	No sex ratio effect	Portelli et al., 1999

[a]Doses given as mg/kg initial egg weight.

demonstrated that treatment of alligator eggs with *p,p'*-DDE or *o,p'*-DDE induced sex reversal (male to female) of eggs incubated at male-producing temperatures. Both compounds show relatively weak affinity for the alligator ER (Vonier et al., 1996; figure 12-5). Concentrations used were ecologically relevant; that is, they were similar to concentrations previously reported in alligator eggs (Heinz et al., 1991). Several laboratory-based exposure studies have reported skewed sex ratios of offspring (female dominant) after exposure of medaka (*Oryzias latipes*) embryos or breeding adults to *o,p'*-DDT (Edmunds et al., 2000; Cheek et al., 2001). Other studies using *p,p'*-DDE in freshwater or marine turtles have observed no feminizing effect on sex determination (see table 12-2).

Antiestrogenic actions of *o,p'*-DDT (Danzo, 1997) or a technical mixture of DDT (Clark et al., 1998) also have been reported. For example, in larval tiger salamanders (*Ambystoma tigrinum*), technical DDT (mixture of 80% *p,p'*-DDT + 20% *o,p'*-DDT) blocked the estrogenic action of estradiol-17β on müllerian duct proliferation and hypertrophy (Clark et al., 1998). In contrast, *p,p'*-DDE exhibited estrogenic actions on the müllerian duct of females but not males. In another amphibian species, however, *o,p'*-DDT, *o,p'*-DDE, and *o,p'*-DDD are estrogenic. The reed frog, *Hyperolius argus*, exhibits estrogen-dependent sexually dimorphic coloration patterns (Hayes and Menendez, 1999). Female coloration can be induced in newly metamorphic froglets by exposing tadpoles to *o,p'*-DDT (0.1 μg/ml), *o,p'*-DDE (1.0 μg/ml) and *o,p'*-DDD (1.0 μg/ml), but *p,p'*-DDT, *p,p'*-DDE, or *p,p'*-DDD were not effective at inducing estrogen-dependent coloration (Noriega and Hayes, 2000). Thus, as described for sex determination above, variation in response will occur among species and needs to be further examined if we are to begin understanding the possible outcomes of exposure to DDT and its metabolites.

Antiandrogenic actions were suspected after the studies of Burlington and Lindeman (1950) showed a decrease of testicular growth in cockerels exposed experimentally to *p,p'*-DDT. Much of the antiandrogenic activity of DDT mixtures has been traced to *p,p'*-DDE. Ecologically relevant concentrations of *p,p'*-DDE inhibit androgen binding to the mammalian androgen receptor and depress androgen-induced transcriptional activity in vitro and androgen action during embryonic development in male rats (Kelce et al., 1995). Not only *p,p'*-DDE exhibits

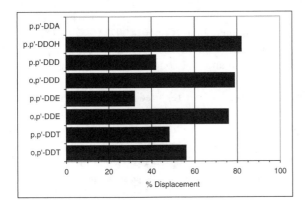

Figure 12-5. Displacement of estradiol-17β from an estrogen receptor purified from the alligator uterus. Displacement is based on a concentration of 33 μM for each compound tested. Data from Vonier et al. (1996).

antiandrogenic activity, as other metabolites have been reported to induce antiandrogenic effects in various systems and on different levels (for reviews, see Sohoni and Sumpter, 1998). Recent studies have shown additional antiandrogenic actions in vivo (see Gray et al., 2001; table 12-1).

DDT and DDE also appear to affect progesterone-related events. A number of studies failed to show significant affinity between DDT or its metabolites and the mammalian progesterone receptor (Gaido et al., 1997; Klotz et al., 1997). Pickford and Morris (1999) demonstrated that a related compound, methoxychlor, inhibited germinal vesicle breakdown in the oocytes of the frog *Xenopus laevis*. This process is progesterone dependent, but Pickford and Morris could not demonstrate an affinity between the *Xenopus* progesterone receptor and methoxychlor or its metabolite. These authors have suggested that other mechanisms could be important, such as alterations in metabolism, receptor cofactors, or receptor corepressors. Studies using the alligator oviducal progesterone receptor suggest that an affinity between DDT and the progesterone receptor can exist for at least this species (Vonier et al., 1996), although no direct in vivo actions has been demonstrated in this species to date. Further, Das and Thomas (1999) have demonstrated that kepone and *o,p'*-DDD were competitive inhibitors of a membrane progesterone receptor. In many fish, 17,20β, 21-trihydroxy-4-pregnen-3-one (20β-S), a progestogen, is an important factor in meiotic maturation of the oocyte. At a range of 10^{-4} to 10^{-7}, *o,p'*-DDD displaced 20β-S from its receptor. Further, *o,p'*-DDD was able to block final oocyte maturation at similar concentrations (Das and Thomas, 1999). These data suggest that DDT or one of its metabolites could influence other progestogen-dependent activities.

Thyroid Axis

Given their wide spectrum of steroid action, it is not surprising that DDT and its metabolites could affect other axes in the endocrine system. For example, DDT causes hyperthyroidism and enlargement of the thyroid (Jefferies, 1969). Xenobiotic compounds like DDT and its metabolites exert an effect on the thyroid by disrupting one of several possible steps in the biosynthesis and/or secretion of thyroid hormones. These steps include (1) inhibition of the iodine-trapping mechanism (thiocyanate or perchlorate), (2) blockage of organic binding of iodine and coupling of iodothyronines to form thyroxine (T_4) and triiodothyronine (T_3) (sulfonamides, thiourea, methimazole, aminotiazole), or (3) inhibition of T_3/T_4 secretion by an effect on proteolysis of active hormone from the colloid (Capen, 1992, 1994). DDT has been shown to disrupt thyroid hormone economy by increasing the peripheral metabolism of thyroid hormones through an induction of hepatic microsomal enzymes (Capen, 1992, 1994).

The effects of DDT and its metabolites have been reported in several species. Male juvenile alligators from Lake Apopka, Florida (dicofol and DDT spill in 1980s), have elevated plasma T_4 levels compared to male juvenile alligators in Lake Woodruff, Florida (reference site, National Wildlife Refuge) (Crain et al., 1998a). DDT-treated rats had increased thyroid mass as well as increased T_3 and T_4 concentrations in blood and displayed decreased thyroid iodine, serum iodine, and protein-bound iodine levels (Seidler et al., 1976; Goldman, 1981). The DDT metabolite *p,p'*-DDE has similar effects on thyroid hormones. There is a positive correlation between DDE and T_4/free T_4 in polar bears (Skaare et al., 2001). DDE-exposed tree swallows displayed slightly elevated T_3 plasma levels, although no significant difference was found among exposed and control groups (Bishop et al., 1998). Another DDT metabolite, o,p'-DDD, increases T_3, T_4, and free T_4 levels in dogs. This compound can be used to treat hyperadrenocorticism in canines (Ruppert and Kraft, 1999).

Although consistent effects are observed in the thyroid axis after exposure to DDT and its metabolites, variation within and among species has been demonstrated. Rats of the Crl:CDIGSBR strain show an increase in T_4 after treatment with DDE. The Long-Evans rat strain, when given the same treatment, shows an increase in plasma thyroid-stimulating hormone but a decrease in T_4 (O'Connor et al., 1999). Japanese quail exposed to DDT displayed a slight decrease in T_4 but a moderate increase in T_3 (Rattner et al., 1984). Humans seem to be the least affected by these xenobiotics. Schoolchildren from the Aral Sea region in Kazakhstan with high concentrations of DDE in their blood lipids showed no impairment in thyroid function (Mazhitova et al., 1989). DDT displayed a low affinity to human thyroid receptor-β in a competitive binding assay (Cheek et al., 1999).

Adrenal Axis

Hayes et al. (1997) suggested that DDT could have corticosteronelike actions in anurans. They described facial abnormalities in metamorphic froglets after exposure of *Kassina senegalensis* tadpoles to corticosterone. These effects are identical to those reported in other anuran larvae after DDT exposure (Cooke, 1970). Hayes (2000) has hypothesized that the effect of corticosterone is due to its stimulation of DDT release from fat stores.

Antiadrenal actions of o,p'-DDD and o,p'-DDE are well documented (Benecke et al., 1991). For example, o,p'-DDD inhibits adrenal steroidogenesis in a variety of species, including humans, dogs, various fish species, and birds (Benecke et al., 1991; Leblond and Hontela, 1999; Breuner et al., 2000) but does not affect adrenal steroidogenesis in sheep, tiger salamanders, or tree lizards (*Urosaurus ornatus*) (Thun et al., 1982; Breuner et al., 2000). Studies using fish adrenal cells in vitro suggest that o,p'-DDD impairs cortisol release by inhibiting a step between adenylyl cyclase activation and corticotropin binding. Unique but persistent metabolites such as 3-methylsulfonyl-DDE ($MeSO_2$-DDE) were first identified in Baltic gray seals exhibiting adrenocortical hyperplasia (Lindhe et al., 2001). A recent study demonstrated that $MeSO_2$-DDE exhibits irreversible binding and adrenocorticolytic activity in murine adrenal tissue in vitro. Within 24 h of treatment, adrenal corticosterone secretion had declined by 90% (Lindhe et al., 2001). These data clearly demonstrate that studies examining the biological effects of additional metabolites of DDT are required.

Hepatic Actions

Given the fact that DDT and its isomers interact with different receptors, other receptor-mediated events are likely to be affected. Besides direct effects on the endocrine system via receptor agonism or antagonism, different forms of DDT alter circulating hormone concentrations by affecting liver function. Liver enzymes play a central role in balancing circulating hormone levels, so changes in liver can result in a change in circulating hormone levels (for a review, see Guillette and Gunderson, 2001). Exposure to DDT or its metabolites results in induction of liver enzymes of the P450 family, comparable to pentobarbital exposure. An induction can accelerate detoxification of hormones in general or alter the metabolism of specific hormones such as androgens (see Gunderson et al., 2001). For example, as early as 1972, it was observed that pretreatment of chickens with DDT or PCBs enhanced hepatic metabolism of testosterone, 4-androstene-3,17-dione, and estradiol-17β (Nowicki and Norman, 1972). Recent studies have demonstrated that various contaminants can alter hepatic biotransformation of steroid hormones (Wilson and LeBlanc, 1998; Wilson et al., 1998, 1999; Gunderson et al., 2001).

Effects in Wildlife: Field Studies

Linking DDT to effects seen in the field is difficult and often impossible because organisms receive multiple doses of contaminants at the same time. Furthermore, DDT can have multiple effects, as discussed above. Here we present a few studies in which effects seen in the field appear to be linked to DDT exposure.

As already mentioned, one of the first and most extensively investigated effects of DDT is its effect on eggshell formation in birds. Eggshell formation is controlled by the endocrine system, and any effect on eggshells can be considered as endocrine disruption. In Scotland, low reproductive success of the golden eagle (*Aquila chrysaetos*) was reported due to egg breakage (Lockie and Rathcliffe, 1964). Subsequently, other reports showed that eagles, peregrine falcons (*Falco peregrinus*), and Eurasian sparrowhawks (*Accipiter nisus*) were suffering from decreased eggshell weights (Ratcliffe, 1967). Eggshell thinning was later shown for numerous predatory birds and several aquatic birds (mostly predatory), but rarely for seedeaters (Cooke, 1973). DDT was suspected, and a relationship between eggshell thickness and DDT residue levels in eggs of 14 bird species was found (Ratcliffe, 1970). However, it was for some time unclear whether DDT or its metabolites were responsible. Controlled studies revealed that DDE affected reproduction and eggshell thinning in a number of species (Feyk and Giesy, 1998). Linear inverse relationships between eggshell thinning and DDE levels in avian tissue and eggs indicated that p,p'-DDE had the strongest correlation of all environmental pollutants (Faber and Hickey, 1973).

An essential process in eggshell formation is the deposition of calcium on the egg while in the eggshell gland. The calcium is obtained from the bloodstream and replenished by absorption from the duodenum, jejunum, and medullary bone. Transport of calcium across the eggshell gland mucosa is apparently stimulated by prostaglandin-E_2. It is hypothesized that the probable mechanism for eggshell thinning is the inhibition of prostaglandin synthesis in the eggshell gland mucosa by p,p'-DDE, although the exact mechanism is still under investigation (Lundholm, 1997). The administration of p,p'-DDE was followed by decreased prostaglandin synthase activity, decreased prostaglandin E_2, and reduced uptake of radiolabeled calcium

(Lundholm, 1997). Supporting this association, administration of compounds inhibiting prostaglandin synthesis (e.g.,. indomethacin) exhibited similar effects (eggshell thinning) to those observed after p,p'-DDE treatment. Administration of o,p'-DDE, p,p'-DDT, o,p'-DDT, and p,p'-DDD to ducks does not result in eggshell thinning or in inhibition of prostaglandin synthesis (Lundholm, 1997). Other proposed mechanisms involve inhibition of carbonic anhydrase or hyperthyroidism (see Feyk and Giesy, 1998), although there are inconsistencies in these studies, and at least one study demonstrated that DDT or its metabolites do not influence carbonic anhydrase activity (see Pocker et al., 1971).

Another example of suspected interference with the endocrine system was seen on the coast of California. Between 1950 and 1970, the Los Angeles sewer system discharged large amounts of commercial DDT into Santa Monica Bay. High levels of mainly o,p'-DDT were found in water, sediment, and biota, even years after the discharge ceased. This resulted in high body burdens of DDT and metabolites in animals at the top of the food chain. Researchers later found severe problems associated with the reproductive system in these animals. Sea lions exhibited premature birth (Delong et al., 1973), and brown pelicans and double-crested cormorants exhibited decreased egg viability and eggshell thinning (see Rattner et al., 1984). Western gulls are insensitive to eggshell thinning, but reports indicated occurrence of homosexual pairings and a highly skewed sex ratio toward females (3.85 females for each male) (Hunt et al., 1980). Dosage experiments on eggs from California gulls (*Larus californicus*) and Western gulls (*Larus occidentalis*) revealed the capacity of o,p'-DDT to induce feminization at levels as low as 2 ppm, well below levels found in eggs from the field (Fry and Toone, 1981). Male gull embryos exposed to o,p'-DDT developed female properties that included a thickening of the cortex and the presence of primordial germ cells. It is not certain that DDT alone caused all the effects observed because other compounds also have the ability to induce feminization (see above) and skewed sex ratios. However, the occurrence of these effects close to a hot spot of DDT contamination makes it a suspect.

DDT and its metabolites are also suspected to play a role in the reproductive problems observed in Lake Apopka alligators in Florida. After the American alligator was put under federal protection, populations started to grow in the southeastern United States. Lake Apopka's population, however, showed an 80% decrease in juvenile recruitment throughout the 1980s and remained at this level through the early 1990s (Woodward, 1996). Further, egg viability was significantly lower than in other lakes frequently sampled in Florida (Woodward et al., 1993; Masson, 1995). The lake has a long history of receiving organochlorine contaminants from municipal waste discharged from the city of Winter Garden located on its south shore, from agricultural runoff from extensive muck farming operations on its northern shores, and from a pesticide spill from the Tower Chemical Company in 1980, which contained dicofol (plus 15% DDT, DDE, and DDD) (U.S. EPA, 1994), all of which contribute to the high levels of organochlorine compounds found in alligators and alligator eggs sampled from the lake (Heinz et al., 1991; Guillette et al., 1999a).

Gonadal abnormalities, reduced phallus size, and alterations in plasma concentrations of testosterone and estradiol have been documented in juvenile alligators from Lake Apopka when compared to alligators from Lake Woodruff, a relatively pristine lake with no known point source of organochlorine contamination, located within a national wildlife refuge in central Florida (Guillette et al., 1994, 1996, 1999b). Similar effects have been observed in alligators from Lake Okeechobee (Crain et al., 1998a; Guillette et al., 1999b), regions of which have documented histories of elevated sediment concentrations of organochlorine compounds (Pfeuffer, 1985, 1991, 1999; Pfeuffer and Matson, 2000). Histological examination of ovaries collected from juvenile females from Lake Apopka showed an increased number of polyovular follicles (more than one oocyte) and polynuclear follicles (more than one nucleus). It is interesting that similar abnormalities have been observed in mice injected with diethylstilbestrol, which is a potent synthetic estrogen (Iguchi et al., 1990).

Reduced phallus size has been observed in juvenile male alligators from Lake Apopka, Lake Griffin, and Lake Okeechobee compared to alligators from Lake Woodruff (Guillette et al., 1999b). Phallus size (an androgen-dependent structure in crocodilians; Forbes, 1939; Ramaswani and Jacob, 1965) was related to body size in Lake Woodruff and Lake Okeechobee males and to plasma dihydroxy-testosterone concentrations in Lake Woodruff males (Guillette et al., 1999b; Pickford et al., 2000). These relationships were not observed for males from Lake Apopka or Lake Griffin. p,p'-DDE is a potent androgen receptor antagonist (Kelce et al., 1995). This observation is important, as elevated p,p'-DDE occurs in the serum of juvenile alligators collected from Lake Apopka. One could hypothesize that DDE acts to block the growth of phallic tissue (as well as other androgen-dependent structures) in developing alligators (Guillette et al., 1996, 1999b).

Juvenile male alligators from Lakes Okeechobee and Apopka exhibited lower plasma testosterone concentrations than males from Lake Woodruff, whereas Okeechobee and Apopka females showed elevated estradiol compared to Lake Woodruff females (Guillette et al., 1994, 1999b). More recent studies have shown that other lakes in Florida with varying degrees of organochlorine contaminants exhibit similar patterns to those observed in Lake Apopka alligators (Guillette et al., 1999b). The differences in steroid concentrations and phallus sizes are suspected to be due to the endocrine activities (estrogenic, antiestrogenic, androgenic, anti-androgenic) of the compounds present in the lakes examined (Guillette et al., 2000). Proposed mechanisms for disruption in hormone homeostasis include alterations in (1) hormone biotransformation, which could affect the circulating half-life of the hormone (Wilson et al., 1998, 1999; Dickerson et al., 1999; Wilson and LeBlanc, 2000; Gunderson et al., 2001); (2) gonadal hormone synthesis (Guillette et al., 1995; Crain et al., 1997; Wilson and LeBlanc, 2000); (3) hormone excretion; (4) cytosolic and serum binding factors affecting the hormones' cellular bio-availability (Arnold et al., 1996; Nagel et al., 1997; Crain et al., 1998b); or (5) the regulation of the hypothalamus–pituitary–gonadal or liver axes.

Patterns of in vitro gonadal steroidogenesis in alligators collected from Lake Apopka and Lake Woodruff did not match the patterns observed for circulating plasma concentrations of estradiol 17β and testosterone, suggesting that altered hepatic biotransformation could explain the observed differences (Guillette et al., 1995).

An initial study has demonstrated that differences among alligators from different lakes exist in the enzymes involved in hepatic biotransformation of testosterone (Gunderson et al., 2001). Sexually dimorphic patterns of total testosterone hydroxylase and oxido-reductase activities were observed in Lake Woodruff alligators, with higher activity observed in females than males. This pattern did not exist in animals collected from Lake Apopka and Lake Okeechobee. Organochlorine compounds such as endosulfan and DDT differentially modulate hepatic enzyme activity with varying effects on plasma testosterone concentrations (Wilson et al., 1998, 1999; Sierra-Santoyo et al., 2000). It is interesting that the sexually dimorphic patterns of hepatic testosterone biotransformation enzyme expression observed in mice and rats is determined by the pattern of pulsatile secretions of growth hormone from the pituitary under feedback regulation from androgens and estrogens. These patterns are imprinted during development and are associated with sex steroid exposure during critical points in development (Gustafsson and Ingelman-Sundberg, 1974; Gustafsson et al., 1983; Gustafsson, 1994). Whether hepatic enzyme expression patterns are imprinted during development in crocodilians has yet to be determined, but the implications are intriguing given the fact that hormonelike activity has been attributed to several of the organochlorine compounds present in the sites where different patterns of hepatic enzyme activity are observed. These compounds could act as agonists or antagonist to the endogenous steroids involved in the feedback regulation of the pulsatile release of growth hormone or in the imprinting processes during development.

Implications for Wildlife and Human Health

Although the use of DDT was limited in 1972, largely because of its presumed carcinogenic potential and its effects on wildlife (ATSDR, 1994), only recently have we begun to understand the broad array of effects DDT has on the vertebrate reproductive and endocrine systems. Where do we go from here?

First, given data from a number of species demonstrating the affinity of DDT and its metabolites for sex steroid receptors from various species, more data are needed on the possible role of these compounds in disrupting or altering signals to systems other than the reproductive system. For example, what possible role could these compounds have in altering bone formation and function? It is well established that estrogens play an important role in calcium regulation in bone. Would DDT or its metabolites, capable of binding the estrogen receptor, alter osteocyte–calcium interactions? Studies demonstrating effects of other organochlorines on bone formation and function suggest these questions need further examination (see Lind et al., 1999). A number of studies have demonstrated that targets such as the prostate also can be affected by contaminant exposure, especially during embryonic development (vom Saal and Timms, 1999). These data, like those on bone, indicate that many additional systems and organs need to be examined.

Sex steroid receptors belong to a superfamily of nuclear receptors, many of which are transcription factors, that directly influence gene expression (Beato et al., 1995; Blumberg and Evans, 1998; Baker, 2001; see chapter 2). In addition to the receptors for androgens, estrogens, and progestogens, these receptors include those for glucocorticoids, aldosterone, vitamin D_3, thyroid hormones, all-*trans*-retinoic acid, and 9-*cis*-retinoic acid. Although a few studies have demonstrated sex steroid-receptor interactions with various DDT metabolites, our understanding of the influences resulting from these interactions is still limited. Further, studies examining the influence of DDT on actions induced by receptors for aldosterone, vitamin D_3, thyroid hormones, all-*trans*- retinoic acid, and 9-*cis*-retinoic acid, or most of the other nuclear receptors are unknown. At least one compound, methoprene (an insect growth regulator not related to DDT), binds to the mammalian retinoid X receptor (Harmon et al., 1995). Given the major role retinoic acid plays during development of many vertebrate species (see Blumberg et al., 1992), additional studies are needed given that the retinoic acid receptor is related to steroid hormone receptors with which DDT and its metabolites readily interact.

Endocrine alterations do not occur only through receptor binding. DDT and its metabolites alter uterine prostaglandin action, alter hepatic aromatase activity, and are associated with altered hepatic hormone biotransformation (see above). Many additional studies are required to examine the role of these compounds in altering hormone synthesis, storage, biotransformation, and clearance.

Given the persistence of these compounds in food chains worldwide, and its continued use, we cannot view DDT as an "old" pesticide, not worthy of future study. There are still many lessons this compound can teach us about the normal biology of the endocrine system and about human influence on the world and its organisms.

References

Arnold, S., Robinson, M.K., Notides, A.C., Guillette, L.J., Jr., McLachlan, J.A., 1996. A yeast estrogen screen for examining the relative exposure of cells to natural and xeno estrogens. Environ. Health Perspect. 104, 544–548.

Arukwe, A., Knudsen, F.R., Goksoyr, A., 1997. Fish zona radiata (eggshell) protein: A sensitive biomarker for environmental estrogens. Environ. Health Perspect. 105, 418–422.

ATSDR, 1994. Toxicological profile for 4,4'-DDT, 4,4'-DDE, 4,4'-DDD (update). U.S. Public Health and Human Services, Agency for Toxic Substances and Disease Registry, Division of Toxicology/Toxicology Information Branch, Atlanta, GA.

Baker, M.E., 2001. Adrenal and sex steroid receptor evolution: environmental implications. J. Mol. Endocrinol. 26, 119–125.

Bastomsky, C.H., 1974. Effects of a polychlorinated biphenyl mixture (Aroclor 1254) and DDT on biliary thyroxine excretion in rats. Endocrinology 95, 1150–1155.

Beato, M., Herrlich, P., Schutz, G., 1995. Steroid hormone receptors: Many actors in search of a plot. Cell 83, 851–857.

Benecke, R., Keller, E., Vetter, B., De Zeeuw, R.A., 1991. Plasma level monitoring of mitotane (o,p'DDD) and its metabolite (o,p'DDE) during long-term treatment of Cushing's disease with low doses. Eur. J. Clin. Pharmacol. 14, 421–426.

Bishop, C.A., Van Der Kraak, G.J., Smits, J.E.G., Hontela, A., 1998. Health of tree swallows (*Tachycineta bicolor*) nesting in pesticide-sprayed apple orchards in Ontario,

Canada. II. Sex and thyroid hormone concentrations and testes development. J. Toxicol. Environ. Health A 55, 561–581.

Bitman, J., Cecil, H.C., Harris, S.J., Fries, G.F., 1968. Estrogenic activity of o,p'-DDT in the mammalian uterus and avian oviduct. Science 162, 371–372.

Blumberg, B., Evans, R.M., 1998. Orphan nuclear receptors—new ligands and new possibilities. Genes Dev. 12, 3149–3155.

Blumberg, B., Mangelsdorf, D.J., Dyck, J.A., Bittner, D.A., Evans, R.M., Derobertis, E.M., 1992. Multiple retinoid-responsive receptors in a single cell-families of retinoid X receptors and retinoic acid receptors in the *Xenopus* egg. Proc. Natl. Acad. Sci. USA 89, 2321–2325.

Breuner, C.W., Jennings, D.H., Moore, M.C., Orchinik, M., 2000. Pharmacological adrenalectomy with mitotane. Gen. Comp. Endocrinol. 120, 27–34.

Bull, J.J., 1983. Evolution of sex determining mechanisms. Benjamin/Cummings, Menlo Park, CA.

Burlington, H., Lindeman, V., 1950. Effect of DDT on testes and secondary sex characters of White Leghorn cockerels. Proc. Soc. Exp. Biol. Med. 74, 48–51.

Capen, C.C., 1992. Pathophysiology of chemical injury of the thyroid-gland. Toxicol. Lett. 64-5, 381–388.

Capen, C.C., 1994. Mechanisms of chemical injury of the thyroid gland. Prog. Clin. Biol. Res. 387, 173–191.

Carson, R., 1962. Silent Spring. Houghton Mifflin, Boston.

Cheek, A.O., Brouwer, T.H., Carroll, S., Manning, S., McLachlan, J.A., Brouwer, M., 2001. Experimental evaluation of vitellogenin as a predictive biomarker for reproductive disruption. Environ. Health Perspect. 109, 681–690.

Cheek, A.O., Kow, K., Chen, J., McLachlan, J.A., 1999. Potential mechanisms of thyroid disruption in humans: Interaction of organochlorine compounds with thyroid receptor, transthyretin, and thyroid-binding globulin. Environ. Health Perspect. 107, 273–278.

Clark, E.J., Norris, D.O., Jones, R.E., 1998. Interactions of gonadal steroids and pesticides (DDT, DDE) on gonaduct growth in larval tiger salamanders, *Ambystoma tigrinum*. Gen. Comp. Endocrinol. 109, 94–105.

Colborn, T., Smolen, M.J., 1996. Epidemiological analysis of persistent organochlorine contaminants in cetaceans. Rev. Environ. Contam. Toxicol. 146, 91–172.

Colborn, T.E., Davidson, A., Green, S.N., Hodge, R.A., Jackson, C.I., Liroff, R.A., 1990. Great Lakes Great Legacy? The Conservation Foundation, Washington, DC.

Cooke, A.S., 1970. The effect of *p,p'*-DDT on tadpoles of the common frog (*Rana temporaria*). Environ. Pollut. 1, 57–71.

Cooke, A.S., 1973. Shell thinning in avian eggs by environmental pollutants. Environ. Pollut. 4, 85–1542.

Cottam, C., Higgens, E., 1946. DDT: Its Effects on Fish and Wildlife. U.S. Fish and Wildlife Service, Washington, DC.

Crain, D.A., 1997. Effects of endocrine-disrupting contaminants on reproduction in the American alligator, *Alligator mississippiensis*. Ph.D. dissertation, University of Florida, Gainesville.

Crain, D.A., Guillette, L.J., Jr., Pickford, D.B., Percival, H.F., Woodward, A.R., 1998a. Sex-steroid and thyroid hormone concentrations in juvenile alligators (*Alligator mississippiensis*) from contaminated and reference lakes in Florida. Environ. Toxicol. Chem. 17, 446–452.

Crain, D.A., Guillette, L.J., Jr., Rooney, A.A., Pickford, D.B., 1997. Alteration in steroidogenesis in alligators (*Alligator mississippiensis*) exposed naturally and experimentally to environmental contaminants. Environ. Health Perspect. 105, 528–533.

Crain, D.A., Noriega, N., Vonier, P.M., Arnold, S.F., McLachlan, J.A., Guillette, L.J., Jr., 1998b. Cellular bioavailability of natural hormones and environmental contaminants as a function of serum and cytosolic binding factors. Toxicol. Ind. Health 14, 261–273.

Danzo, B.J., 1997. Environmental xenobiotics may disrupt normal endocrine function by interfering with the binding of physiological ligands to steroid receptors and binding proteins. Environ. Health Perspect. 105, 294–301.

Das, S., Thomas, P., 1999. Pesticides interfere with the nongenomic action of a progestogen on meiotic maturation by binding to its plasma membrane receptor on fish oocytes. Endocrinology 140, 1953–1956.

Delong, R.L., Gilmartin, W.G., Simpson, J.G., 1973. Premature births in California sea lions: association with high organochlorine pollutant residue levels. Science 181, 1168–1170.

Dickerson, R.L., McMurry, C.S., Smith, E.E., Taylor, M.D., Nowell, S.A., Frame, L.T., 1999. Modulation of endocrine pathways by 4,4'-DDE in the deer mouse (*Peromyscus maniculatus*). Sci. Total Environ. 233, 97–108.

Donohoe, R.M., Curtis, L.R., 1996. Estrogenic activity of chlordecone, o,p'-DDT and o,p'-DDE in juvenile rainbow trout: Induction of vitellogenesis and interaction with hepatic estrogen binding sites. Aquat. Toxicol. 36, 31–52.

Edmunds, J.S.G., McCarthy, R.A., Ramsdell, J.S., 2000. Permanent and functional male-to-female sex reversal in d-rR strain medaka (*Oryzias latipes*) following egg microinjection of o,p'-DDT. Environ. Health Perspect. 108, 219–224.

Faber, R.A., Hickey, J.J., 1973. Eggshell thinning, chlorinated hydrocarbons and mercury in inland aquatic bird eggs. J. Pest. Monit. 7, 27–36.

Feyk, L.A., Giesy, J.P., 1998. Xenobiotic modulation of endocrine function in birds. In: Kendall, R., Dickerson, R., Giesy, J., Suk, W. (Eds.), Principles and Processes for Evaluating Endocrine Disruption in Wildlife. SETAC Press, Pensacola, FL, pp. 121–140.

Forbes, T.R., 1939. Studies on the reproductive system of the alligator. V. The effects of injections of testosterone proprionate in immature alligators. Anat. Rec. 75, 51–57.

Fry, D.M., Toone, C.K., 1981. DDT-induced feminization of gull embryos. Science 213, 922–924.

Gaido, K.W., Leonard, L.S., Lovell, S., Gould, J.C., Babai, D., Portier, C.J., McDonnell, D.P., 1997. Evaluation of chemicals with endocrine modulating activity in a yeast-based steroid hormone receptor gene transcription assay. Toxicol. Appl. Pharmacol. 143, 205.

Goldman, M., 1981. The effect of a single dose of DDT on thyroid function in male rats. Arch. Intl. Pharmacodyn. Ther. 252, 327–334.

Gray, L.E., Jr., Ostby, J., Furr, J., Wolf, C., Lambright, C., Parks, L., Veeramachanemi, D.N., Wilson, V.S., Price, M., Hotchkiss, A., Orlando, E., Guillette, L.J., Jr., 2001. Effects of environmental antiandrogens on reproductive development in experimental animals. Hum. Reprod. Update 7, 248–264.

Guillette, L.J., Jr., Brock, J.W., Rooney, A.A., Woodward, A.R., 1999a. Serum concentrations of various environmental contaminants and their relationship to sex steroid concentrations in juvenile American alligators. Arch. Environ. Contam. Toxicol. 36, 447–455.

Guillette, L.J., Jr., Crain, D.A., Galle, A., Gunderson, M., Kools, S., Milnes, M.R., Orlando, E.F., Rooney, A.A., Woodward, A.R., 2000. Alligators and endocrine disrupting contaminants: A current perspective. Am. Zool. 40, 438–452.

Guillette, L.J., Jr., Gross, T.S., Gross, D., Rooney, A.A., Percival, H.F., 1995. Gonadal steroidogenesis *in vitro* from juvenile alligators obtained from contaminated and control lakes. Environ. Health Perspect. 103(suppl. 4), 31–36.

Guillette, L.J., Jr., Gross, T.S., Masson, G.R., Matter, J.M., Percival, H.F., Woodward, A.R., 1994. Developmental abnormalities of the gonad and abnormal sex hormone concentrations in juvenile alligators from contaminated and control lakes in Florida. Environ. Health Perspect. 102, 680–688.

Guillette, L.J., Jr., Gunderson, M.P., 2001. Alterations in the development of the reproductive and endocrine systems of wildlife exposed to endocrine disrupting contaminants. Reproduction, 202, 857–864.

Guillette, L.J., Jr., Pickford, D.B., Crain, D.A., Rooney, A.A., Percival, H.F., 1996. Reduction in penis size and plasma testosterone concentrations in juvenile alligators living in a contaminated environment. Gen. Comp. Endocrinol. 101, 32–42.

Guillette, L.J., Jr., Woodward, A.R., Crain, D.A., Pickford, D.B., Rooney, A.A., Percival, H.F., 1999b. Plasma steroid concentrations and male phallus size in juvenile alligators from seven Florida lakes. Gen. Comp. Endocrinol. 116, 356–372.

Gunderson, M.P., LeBlanc, G.A., Guillette, L J., Jr., 2001. Alterations in sexually dimorphic biotransformation of testosterone in juvenile American alligators (*Alligator mississippiensis*) from contaminated lakes. Environ. Health Perspect. 109, 1257–1264.

Gustafsson, J.-A., 1994. Regulation of sexual dimorphism in the rat liver. In: Short, R.V., Balaban, E. (Ed.), The Differences Between the Sexes. Cambridge University Press, Cambridge, pp. 231–242.

Gustafsson, J.A., Ingelman-Sundberg, M., 1974. Regulation and properties of a sex-specific hydroxylase system in female rat liver microsomes active on steroid sulfates. J. Biol. Chem. 249, 1940–1945.

Gustafsson, J.A., Mode, A., Norstedt, G., Skett, P., 1983. Sex steroid induced changes in hepatic enzymes. Annu. Rev. Physiol. 45, 51–60.

Harmon, M.A., Boehm, M.F., Heyman, R.A., Mangelsdorf, D.J., 1995. Activation of mammalian retinoid X receptors by the insect growth regulator methoprene. Proc. Nat. Acad. Sci. USA 92, 6157–6160.

Hayes, T.B., 2000. The importance of comparative endocrinology in examining the endocrine disrupter problem. In: Guillette, L.J., Jr., Crain, D.A. (Eds.), Endocrine Disrupting Contaminants: An Evolutionary Perspective. Francis and Taylor, Philadelphia, pp. 22–51.

Hayes, T.B., Menendez, K.P., 1999. The effect of sex steroids on primary and secondary sex differentiation in the sexually dichromatic reedfrog (*Hyperolius argus*: Hyperolidae) from the Arabuko Sokoke Forest of Kenya. Gen. Comp. Endocrinol. 115, 188–199.

Hayes, T.B., Wu, T.H., Gill, T.N., 1997. DDT-like effects as a result of corticosterone treatment in an anuran amphibian: Is DDT a corticoid mimic or a stressor? Environ. Toxicol. Chem. 16, 1948–1953.

Heberer, T., Dunnbier, U., 1999. DDT metabolite bis(chlorophenyl)acetic acid: The neglected environmental contaminant. Environ. Sci. Technol. 33, 2346–2351.

Heinz, G.H., Percival, H.F., Jennings, M.L., 1991. Contaminants in American alligator eggs from lakes Apopka, Griffin and Okeechobee, Florida. Environ. Monit. Assess. 16, 277–285.

Hickey, J.J., Hunt, I.B., 1960. Initial song bird mortality following Dutch elm disease control program. J. Wildl. Manag. 24, 259–265.

Hunt, G.L., Jr., Wingfield, J.C., Newman, A., Farner, D.S., 1980. Sex ratio of Western Gulls on Santa Barbara Island. Auk 97, 473–479.

Iguchi, T., Fukazawa, Y., Uesugi, Y., Takasugi, N., 1990. Polyovular follicles in mouse ovaries exposed neonatally to diethylstilbestrol in vivo and in vitro. Biol. Reproduc. 43, 478–484.

Jefferies, D.J., 1969. Induction of apparent hyperthyroidism in birds fed DDT. Nature 222, 578–579.

Kelce, W.R., Stone, C.R., Laws, S.C., Gray, L.E., Kemppainen, J.A., Wilson, E.M., 1995. Persistent DDT metabolite p,p'-DDE is a potent androgen receptor antagonist. Nature 375, 581–585.

Klotz, D.M., Beckman, B.S., Hill, S.M., McLachlan, J.A., Walters, M.R., Arnold, S.A., 1996. Identification of environmental chemicals with estrogenic activity using a combination of in vitro assays. Environ. Health Perspect. 104, 1084–1089.

Klotz, D.M., Ladlie, B.L., Vonier, P.M., McLachlan, J.A., Arnold, S.F., 1997. o,p'-DDT and its metabolites inhibit progesterone-dependent responses in yeast and human cells. Mol. Cell. Endocrinol. 129, 63–71.

Leblond, V.S., Hontela, A., 1999. Effects of in vitro exposures to cadmium, mercury, zinc, and 1-(2-chlorophenyl)-1-(4-chlorophenyl)-2,2-dichloroethane on steroidogenesis by dispersed interrenal cells of rainbow trout (Oncorhynchus mykiss). Toxicol. Appl. Pharmacol. 157, 16–22.

Lind, P.M., Eriksen, E.F., Sahlin, L., Edlund, M., Orberg, J., 1999. Effects of the anti-estrogenic environmental pollutant 3,3',4,4',5-pentachlorobiphenyl (PCB #126) in rat bone and uterus: Diverging effects in ovariectomized and intact animals. Toxicol. Appl. Pharmacol. 154, 236–244.

Lindhe, O., Lund, B.-O., Bergman, A., Brandt, I., 2001. Irreversible binding and adrenocorticolytic activity of the DDT metabolite 3-methylsulfonyl-DDE examined in tissue-slice culture. Environ. Health Perspect. 109, 105–110.

Liroff, R.A., 2001. DDT risk assessments: response [letter to editor]. Environ. Health Perspect. 109, A302–A303.

Lockie, J., Rathcliffe, D., 1964. Insecticides and Scottish golden eagle eggs. Br. Birds 57, 89–102.

Lundholm, C.E., 1997. DDE-Induced eggshell thinning in birds: Effects of p,p'-DDE on calcium and prostaglandin metabolism of the eggshell gland. Comp. Biochem. Physiol. 118C, 113–128.

Maness, S.C., McDonnell, D.P., Gaido, K.W., 1998. Inhibition of androgen receptor-dependent transcriptional activity by DDT isomers and methoxychlor in HepG2 human hepatoma cells. Toxicol. Appl. Pharmacol. 151, 135–142.

Masson, G.R., 1995. Environmental influences on reproductive potential, clutch viability and embryonic mortality of the American alligator in Florida. Ph.D. dissertation, University of Florida, Gainesville.

Matter, J.M., Crain, D.A., Sills-McMurry, C., Pickford, D.B., Rainwater, T.R., Reynolds, K.D., Rooney, A.A., Dickerson, R.L., Guillette, L.J., Jr., 1998. Effects of endocrine-disrupting contaminants in reptiles: Alligators. In: Kendall, R., Dickerson, R., Giesy, J., Suk, W. (Eds.), Principles and Processes for Evaluating Endocrine Disruption in Wildlife. SETAC Press, Pensacola, FL, pp. 267–289.

Mazhitova, Z., Jensen, S., Ritzen, M., Zetterstrom, R., 1989. Chlorinated contaminants, growth and thyroid function in school children from Aral Sea region in Kazakhstan. Acta Paedia. 87, 991–995.

MDNR, 1994. Michigan fish contaminant monitoring program. 1994 Annual Report. Michigan Department of Natural Resources, Lansing.

Nagel, S.C., vom Saal, F.S., Thayer, K.A., Dhar, M.G., Boechler, M., Welshons, W.V., 1997. Relative binding affinity-serum modified access (RBA-SMA) assay predicts the relative in vivo bioactivity of the xenoestrogens Bisphenol A and octylphenol. Environ. Health Perspect. 105, 70–76.

Noriega, N., Hayes, T.B., 2000. DDT congener effects on secondary sex coloration in the reed frog Hyperolius argus: A partial evaluation of the Hyperolius argus endocrine screen. Comp. Biochem. Physiol. B 126, 231–237.

Nowicki, H.G., Norman, A.W., 1972. Enhanced hepatic metabolism of testosterone, 4-androstene-3,17-dione and estradiol-17β in chickens pretreated with DDT or PCB. Steroids 19, 85–90.

O'Connor, J.C., Frame, S.R., Davis, L.G., Cook, J.C., 1999. Detection of the environmental antiandrogen p,p-DDE in CD and long-evans rats using a tier I screening battery and a Hershberger assay. Toxicol. Sci. 51, 44–53.

Palmer, B., Palmer, S., 1995. Vitellogenin induction by xenobiotic estrogens in the red-eared turtle and African clawed frog. Environ. Health Perspect. 103 (suppl 4), 19–25.

Peakall, D.B., 1967. Pesticide-induced enzyme breakdown of steroids in birds. Nature 216, 505–506.

Peterle, T.J., 1991. Wildlife Toxicology. Van Nostrand Reinhold, New York.

Peterson, J.E., Robinson, W.H., 1964. Metabolic products of p,p'-DDT in the rat. Toxicol. Appl. Pharmacol. 6, 321–327.

Pfeuffer, R.J., 1985. Pesticide residue monitoring in sediment and surface water bodies within the South Florida Water Management District. SFWMD Technical Publication 85-2. South Florida Water Management District.

Pfeuffer, R.J., 1991. Pesticide residue monitoring in sediment and surface water within the South Florida Water Management District. SFWMD, Technical Publication 91-01. South Florida Water Management District.

Pfeuffer, R.J., 1999. Pesticide Surface Water & Sediment Quality Report: April 1999 Sampling Event. South Florida Water Management District.

Pfeuffer, R.J., Matson, F., 2000. Pesticide Surface Water & Sediment Quality Report: May 2000 Sampling Event. South Florida Water Management District.

Pickford, D.B., Guillette, L.J., Jr., Crain, D.A., Rooney, A.A., Woodward, A.R., 2000. Phallus size and plasma dihydrotestosterone concentrations in juvenile American alligators (*A. mississippiensis*) from contaminated and reference populations. J. Herpetol. 34, 233–239.

Pickford, D.B., Morris, I.D., 1999. Effects of endocrine-disrupting contaminants on amphibian oogenesis: methoxychlor inhibits progesterone-induced maturation of *Xenopus laevis* oocytes *in vitro*. Environ. Health Perspect. 107, 285–292.

Pocker, Y., Beug, W.M., Ainardi, V.R., 1971. Carbonic anhydrase interaction with DDT, DDE and Dieldrin. Science 174, 1336–1339.

Podreka, S., Georges, A., Maher, B., Limpus, C.J., 1998. The environmental contaminant DDE fails to influence the outcome of sexual differentiation in the marine turtle *Chelonia mydas*. Environ. Health Perspect. 106, 185–188.

Portelli, M.J., de Solla, S.R., Brooks, R.J., Bishop, C.A., 1999. Effect of dichlorodiphenyltrichloroethane on sex determination of the common snapping turtle (*Chelydra serpentina serpentina*). Ecotoxicol. Environ. Safety 43, 284–291.

Ramaswani, L.S., Jacob, D., 1965. Effect of testosterone proprionate on the urinogenital organs of immature crocodile *Crocodylus palustris* Lesson. Experientia 21, 206–207.

Ratcliffe, D., 1970. Changes attributable to pesticides in egg breakage frequency and eggshell thickness in some British birds. J. Appl. Ecol. 7, 67–115.

Ratcliffe, D.A., 1967. Decrease in eggshell weight in certain birds of prey. Nature 215, 208–210.

Rattner, B.A., Eroschenko, V.P., Fox, G.A., Fry, D.M., Gorsline, J., 1984. Avian endocrine responses to environmental pollutants. J. Exp. Zool. 232, 683–689.

Roberts, D.R., 1999. DDT and the global treat of re-emerging malaria. Pest. Safety News 2, 4–5.

Roberts, D.R., Manguin, S., Mouchet, J., 2000. DDT house spraying and re-emerging malaria. Lancet 356, 330–332.

Rooney, A.A., Guillette, L.J., Jr., 2000. Contaminant interactions with steroid receptors: Evidence for receptor binding. In: Guillette, Jr. L.J., Crain D.A. (Eds.), Endocrine Disrupting Contaminants: An Evolutionary Perspective. Francis and Taylor, Philadelphia, pp. 82–125.

Ruppert, C., Kraft, W., 1999. Examination in dogs with spontaneous hyperadrenocorticism. Part 3. Thyroid hormone concentrations before and after treatment with mitotane (Lysodren®). Tier. Prax. Aus. klein. Heimtiere 27, 167–171.

Sanderson, J.T., Seinen, W., Giesy, J.P., van den Berg, M., 2000. 2-Chloro-S-triazine herbicides induce aromatase (CYP19) activity in H295R human adrenocortical carcinoma cells: A novel mechanism for estrogenicity? Toxicol. Sci. 54, 121–127.

Seidler, H., Hartig, M., Engst, R., 1976. Effects of DDT and lindane on thyroid function in the rat. Nahrung 20, 399–406.

Sharpe, R.M., 1995. Reproductive biology—Another DDT connection. Nature 375, 538–539.

Shelby, M.D., Newbold, R.R., Tully, D.B., Chae, K., Davis, V.L., 1996. Assessing environmental chemicals for estrogenicity using a combination of in vitro and in vivo assays. Environ. Health Perspect. 104, 1296–1300.

Sierra-Santoyo, A., Hernandez, M., Albores, A., Cebrian, M.E., 2000. Sex-dependent regulation of hepatic cytochrome P-450 by DDT. Toxicol. Sci. 54, 81–87.

Simonich, S.L., Hites, R.A., 1995. Global distribution of persistent organochlorine compounds. Science 269, 1851–1854.

Skaare, J.U., Bernhoft, A., Wiig, O., Norum, K.R., Haug, E., Eide, D.M., Derocher, A.E., 2001. Relationships between plasma levels of organochlorines, retinol and thyroid hormones from polar bears (*Ursus maritimus*) at Svalbard. J. Toxicol. Environ. Health A. 62, 227–241.

Smith, A.G. 1991. Chlorinated hydrocarbon insecticides. In: Hayes, W., Jr., Laws, E., Jr. (Eds.), Handbook of Pesticide Toxicology. Academic Press, San Diego, California.

Smith, C., Root, D., 1999. The export of pesticides: Shipments from U.S. ports, 1995–1996. Intl. J. Occup. Environ. Health 5, 141–150.

Sohoni, P., Sumpter, J.P., 1998. Several environmental oestrogens are also anti-androgens. J. Endocrinol. 158, 327–339.

Sonnenschein, C., Soto, A.M., 1998. An updated review of environmental estrogen and androgen mimics and antagonists. J. Steroid Biochem. Mol. Biol. 65, 143–153.

Thun, R., Wild, P., Mettler, F., Djafarian, M., 1982. The effect of o,p'-DDD on the adrenal cortex of sheep. Experientia 38, 1494–1496.

U.S. EPA, 1994. Tower Chemical Company Superfund Site Biological Assessment March 1994. U.S. Environmental Protection Agency, Washington, DC.

vom Saal, F.S., Timms, B.G., 1999. The role of natural and manmade estrogens in prostate development. In: Naz, R.K. (Ed.), Endocrine Disruptors—Effects on Male and Female Reproductive Systems. CRC Press, Boca Raton, FL, pp. 307–327.

Vonier, P.M., Crain, D.A., McLachlan, J.A., Guillette, L.J., Jr., Arnold, S.F., 1996. Interaction of environmental chemicals with the estrogen and progesterone receptors from the oviduct of the American alligator. Environ. Health Perspect. 104, 1318–1322.

Welch, R.M., Levin, W., Conney, A.H., 1969. Estrogenic action of DDT and its analogs. Toxicol. Appl. Pharmacol. 14, 358–367.

Willingham, E., Crews, D., 1999. Sex reversal effects of environmentally relevant xenobiotic concentrations on the red-eared slider turtle, a species with temperature-dependent sex determination. Gen. Comp. Endocrinol. 113, 429–435.

Wilson, V.S., LeBlanc, G.A., 1998. Endosulfan elevates testosterone biotransformation and clearance in CD-1 mice. Toxicol. Appl. Pharmacol. 148, 158–168.

Wilson, V.S., LeBlanc, G.A., 2000. The contribution of hepatic inactivation of testosterone to the lowering of serum testosterone levels by ketoconazole. Toxicol. Sci. 54, 128–137.

Wilson, V.S., McLachlan, J.B., Falls, J.G., LeBlanc, G.A., 1998. The use of testosterone hydroxylase activities as a biomarker of endocrine disruption. In: *SETAC 19th Annual Meeting*. SETAC Press, Pensacola, FL.

Wilson, V.S., McLachlan, J.B., Falls, J.G., LeBlanc, G.A., 1999. Alteration in sexually dimorphic testosterone biotransformation profiles as a biomarker of chemically induced androgen disruption in mice. Environ. Health Perspect. 107, 377–384.

Woodward, A.R., 1996. Determination of appropriate harvest strategies for alligator management units. Final Report 7562. Florida Game and Fresh Water Fish Commission.

Woodward, A.R., Jennings, M.L., Percival, H.F., Moore, C.T., 1993. Low clutch viability of American alligators on Lake Apopka. Fl. Sci. 56, 52–63.

You, L., Sar, M., Bartolucci, E., Ploch, S., Whitt, M., 2001. Induction of hepatic aromatase by p,p'-DDE in adult male rats. Mol. Cell. Endocrinol. 178, 207–214.

13

Heavy Metals

Alice Hontela
Alexandra Lacroix

Structure and Physical Characteristics of Metals

Metals are a large family of elements with a complex chemistry, as they span several groups within the periodic table, each with its particular chemical and physical characteristics, from metalloids such as arsenic (As) and selenium (Se) to heavy metals such as cadmium (Cd), copper (Cu), zinc (Zn), and mercury (Hg). Some metals (Zn, Cu, Se, molybdenum [Mo], calcium [Ca], magnesium [Mg], manganese [Mn], iron [Fe], cobalt [Co]) are essential for normal physiological function. They are cofactors for a number of enzymes and proteins, and thus they must be supplied at low concentrations, as deprivation from these physiologically important elements will cause health problems. There is some evidence that body concentrations of certain metals (e.g., Cu, Zn, and Mn) are controlled by homeostatic mechanisms, at least in some vertebrates. Metals such as Cd, Hg, Pb, or As are referred to as nonessential because they do not have a known physiological function. Metals of concern to toxicologists are Cu, Zn, tin (Sn), Cd, and Hg, as well as chromium (Cr), Pb, nickel (Ni), As, and aluminium (Al). The biological half-life (i.e., the time it takes the body to excrete half of the accumulated amount) varies for different metals. In mammals, the half-life of Cd is 20–30 years and for As or Cr a few hours or days. All metals, essential and nonessential alike, when available at excess doses, have toxic effects. Differences in valence and ionization, as well as bonding capacity, influence the reactivity of metals toward macromolecules and therefore affect their toxicity.

The toxicity of metals depends not only on dose and length of exposure, as with other toxicants, but also on the ionic and chemical form (the species) of the metal and its bioavailability. Characteristics such as pH, concentrations of Ca^{2+}, and presence of organic ligands have an influence on metal speciation, a phenomenon of utmost importance in the evaluation of toxicity (Campbell and Tessier, 1996). These characteristics vary in the surrounding milieu, whether in the ecosystem and its surface waters, in extracellular fluid or tissues, or in vitro, and should be considered when evaluating the impact of metals on physiological systems and target cells within them.

The capacity to generate different species of metals through enzymatic transformations further influences the vulnerability of animals to metals, as these metabolic processes also vary between animal species. There are several well-known examples of metal speciation and different toxicities of individual metal species. Trivalent Cr^{3+} is less toxic than Cr^{6+}, a carcinogenic chromium species, and As^{3+} is significantly more toxic than As^{5+}. The two forms of arsenic are generated by different enzymes. Elemental mercury (Hg^o), inorganic mercury ($HgCl_2$), and methylmercury (CH_3HgCl) have different toxicokinetics, rates of membrane penetration, and half lives in the body and also different toxicities. Specific examples pertaining to the effects of metals on the physiological status of vertebrates will be discussed in this chapter. However, a general review of metal toxicity and speciation, as well as data on concentrations in the biotic and abiotic components in the environment, will not be presented here, as excellent reviews on this topic have been published (Sorensen, 1991; Heath, 1995). We focus here on the endocrine effects of metals, with particular emphasis on mechanisms of action through which bioavailable metals impair the integrity of endocrine processes.

Bioavailability of Metals

Metals are natural elements discharged into the environment by alterations of their geochemical cycles, either through human activities or natural processes such as volcanic eruptions or soil erosion. Mining and smelting activities, coal and petroleum combustion, and use of sludge from water treatment plants in agriculture are important sources of contamination. Specialized industries also may contribute to the contamination of the environment by metals. The glass industry generates Se, and electroplating and chloroalkali processes release Cd, Hg, and Zn. Uses of metal-based pesticides further contribute to environmental contamination. Arsenical pesticides used in cotton agriculture or Hg-containing slimicides used in papermills are examples of practices mostly discontinued now. Acidification of watersheds by acid rain influences metal distribution in the ecosystem by promoting lixiviation of metals from soils into the aquatic compartment. Flooding of large areas for reservoirs and dams releases Hg into the aquatic ecosystem through methylation of Hg from atmospheric depositions trapped in the vegetation.

Metals can travel long distances, either in watersheds or through the atmosphere. Their ecological half-life is long, and although their structure can be modified in the environment or in the animals by speciation and processes such as ionization, methylation, and binding to organic ligands, they are classified as persistent contaminants. Contamination of Minamata Bay by Hg in the 1950s and the severe health problems in the people consuming fish from the bay illustrates the insidious nature of metal contamination. Despite preventive measures and a better understanding of metal toxicity, the long ecological half-life of metals and their importance and widespread use in manufacture of many essential products make contamination by metals ubiquitous.

General Mode of Action of metals

Toxic effects of metals usually involve an interaction between the free metal ion and the toxicological target; specific biochemical processes, enzymes, membranes of cells and organelles. Metals may react with oxygen-, sulfur-, and nitrogen-containing ligands; the ligand acts as an electron donor. One of the mechanisms of metal toxicity is oxidative stress, a process that damages biological macromolecules via reactive oxygen species such as hydrogen peroxide (H_2O_2), the superoxide anion ($O_2^{\cdot}$), or hydroxyl radical ($OH^{\cdot}$). Glutathione, superoxide dismutase, catalase, and vitamin E are scavengers of the reactive oxygen species; they act as antioxidants and protect cells against oxidative stress (Congui et al., 2000; Pourahmad and O'Brien, 2000). The reactivity of the metal ion may be specific; as in the case of As^{5+} substituting for phosphorus ion P^{5+}, an example of molecular mimetism leading to faulty synthesis of ATP (de Master and Mitchell, 1973). Some metals react with many macromolecules, without any or limited specificity. A more general reactivity has been documented for several divalent metals, some interfering with Ca^{2+} homeostasis. Lead behaves like Ca^{2+} and is deposited into bones and other Ca^{2+}-rich structures such as fin rays or scales. Lead also interferes with Ca^{2+}-dependent release of neurotransmitters (Goldstein, 1993). Cadmium, in the form of Cd^{2+}, is another highly reactive ion. Complex interactions of Cd^{2+} and other divalent metals with Ca^{2+} have been described, including interference with the activity of calmodulin, a ubiquitous Ca^{2+}-binding protein. In fishes, Cd^{2+} enters branchial epithelial cells through Ca^{2+} channels in the apical membranes and may in fact block these channels (Verbost et al., 1989). In waters of high mineral content (high water hardness or salinity), Ca^{2+} in turn protects fish against the toxicity of Cd. Studies suggest that heavy metals may modulate the regulation of gene expression through structural alterations of the Zn finger proteins and their Zn motifs (Rasmiafshari and Zawia, 2000). Metals may cause structural alterations of proteins containing Zn motifs, such as protein kinase C, critical for endocrine processes.

Defense Mechanisms and Tissue Burdens

It is not surprising, given the ubiquity of metals in the natural abiotic environment and the role of essential metals in the organism, that defense mechanisms against toxicity of metals evolved. These mechanisms are efficient at low exposure, but toxicity ensues when their capacity is overwhelmed. Metallothionein is a protein with a capacity to sequester metals, making them less available to macromolecules. Metals can induce the expression of metallothioneins, and, in fact, metallothionein concentrations in key organs such as liver or kidney are used as indicators of exposure

Table 13-1. Concentrations (mean ± SE) of Zn, Cu, and Cd (μg/g dry weight) and metallothionein (MT; nmol. site g) in the liver of adult yellow perch, *Perca flavescens*, sampled in six lakes situated along a contamination gradient in northwestern Québec.

	N	[Zn]	[Cu]	[Cd]	[MT]
Reference lakes					
Opasatica	8	92.4±3.6[a]	10.4±1.8[a]	2.9±0.4[a]	250±20.5[a]
Dasserat	7	98.6±4.2[ab]	10.8±0.9[a]	5.3±0.6[b]	239±18.6[a]
Intermediate contamination lakes					
Bousquet	7	106.5±5.4[ab]	20.4±4.5[b]	20.3±2.9[c]	580±60.4[b]
Vaudray	8	108.9±1.6[b]	12.9±0.7[ab]	25.1±1.7[c]	568±40[b]
High contamination lakes					
Osisko	8	177.2±9.0[c]	246.5±29.8[c]	45.7±3.2[d]	2660±90[c]
Dufàult	8	151.1±3.7[c]	148.5±11.1[c]	61.3±5.3[d]	2800±75[c]

Means followed by the same letter are not significantly different; comparison between lakes only (*p* < .01, Tukey-Kramer HSD test). Correlation between concentrations of metallothionein and metals were > 0.91 at *P* < .001 (Pearson's test). Adapted from Laflamme et al. (2000).

to metals (Olsvik et al., 2000; table 13-1). The capacity to induce metallothioneins varies for different metals, with Cd > Zn > Cu, but corticosteroids as well as progesterone upregulate, while estrogens inhibit the induction of metallothionein. Acclimation to low levels of metals is thought to be mediated through induction of metallothioneins (Marr et al., 1995; Gerpe et al., 2000).

Ferritin and hemosiderin are also intracellular metal-binding proteins, which are particularly important for binding Fe. Heat shock proteins (hsp) also may be induced by metals such as Cd, thereby modulating metal toxicity (Wiegant et al., 1997). Clinical use of chelators such as DMPS (2,3-dimercapto-1-propanesulfonic acid), Ca^{2+} EDTA (ethylene diamine tetraacetic acid), or penicillamine (β,β-dimethylcysteine) to prevent or reverse metal toxicity is based on the principle that the chelators, electron donors with a high affinity for the metal ion, will bind the metals, thus preventing binding to biological macromolecules. Finally, some metals can be segregated into dense, metal–protein complexes that appear in the cytoplasm or nucleus as morphologically discernible inclusion bodies. This has been shown for As, Pb, and Se, mainly in the hepatocytes and renal tubular cells.

There are several key tissues where metals accumulate. The capacity to accumulate metals, particularly in aquatic animals, can be described in terms of bioconcentration factors (ratio of tissue metal concentration to water concentrations). Metals bound to organic ligands (e.g., CH_3HgCl) are more lipid soluble and traverse membranes more easily than inorganic forms of the same metal (e.g., $HgCl_2$). Animals can absorb metals through their gut, through lungs or gills, and through the skin. In fishes, gills accumulate large quantities of metals. Kidney and liver accumulate smaller quantities than gills but much more than other tissues. This is largely a consequence of easy accessibility related to blood-flow dynamics and the presence of transforming enzymes (Harrison and Klaverkamp, 1989). Hair, feathers, blood, and urine also are good indicators that can be used in routine and preferably nondestructive sampling for monitoring exposure to metals. For some metals, such as Cd, urine is particularly useful; urinary Cd is used as an indicator of kidney damage. However, endocrine organs that may accumulate relatively lower burdens of the contaminants compared to kidney or liver (table 13-1; figure 13-1) may be significantly impaired if their sensitivity to metals is high.

Effects of Metals on Endocrine Organs

Endocrine function can be perturbed by metals through either a direct effect on the hypothalamus, the pituitary axes, or on other endocrine organs such as the pancreas. Various disease states caused by metal contaminants in nonendocrine target organs also may produce secondary responses of the endocrine system.

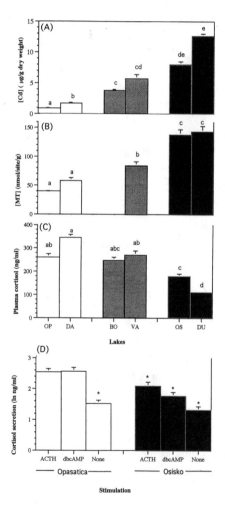

Figure 13-1. Relationship between (A) concentrations of Cd ($N = 8$) and (B) concentrations of metallothionein (MT) in the adrenocortical tissue ($N = 8$), (C) the capacity to increase plasma cortisol levels following a confinement for 1 h ($N = 32$), and (D) the capacity of the head kidney to respond in vitro to ACTH1-39 (2IU) or dbcAMP (4 mmol; $N = 17$) in yellow perch, *Perca flavescens*, sampled in lakes situated along a contamination gradient in northwestern Québec (reference lakes Opasatica, OP, and Dasserat, DA; intermediate lakes Bousquet, BO, and Vaudray, VA; highly contaminated lakes Osisko, OS, and Dufault, DU). The exposure of the fish and the gradient of contamination were characterized by the increasing Cd tissue burdens and metallothionein levels. Impairment of the capacity to elevate plasma cortisol levels in response to confinement stress and to respond to ACTH and dbcAMP in vitro was also evident. Means indicated by the same letters are not significantly different (Tukey-Kramer test, $P<.05$; *significantly different from the stimulated adrenocortical tissue of the reference lake, ANCOVA test, $P<.05$). All error bars refer to SEM. Adapted from Laflamme et al. (2000).

Effects of Metals on the Reproductive Axis

Metals have wide-ranging adverse effects on reproductive function. Young stages of fish and amphibians are highly sensitive to metals that decrease hatchability of eggs and interfere with development of embryos and larvae through embryotoxic and teratogenic effects. Fetal toxicity by high doses of Hg, As, or Pb in mammals, including humans, is well documented, although the effects depend not only on the dose but also on the time of exposure during embryonic or fetal development.

Antigonadal effects of Pb and Cd have been documented at concentrations not causing mortality (sublethal doses) in all vertebrates investigated thus far. The alterations are varied and range from a total loss of reproductive function to subtle alterations of the mechanisms involved in cell signaling and signal transduction in the synthesis of reproductive hormones. Lead toxicity in mammals has been associated with sterility and gametotoxic effects that are manifested as a decrease in sperm count and motility and abnormal sperm morphology (Assenato et al., 1986). Cd has similar adverse effects on reproductive function, including an increase in the permeability of the hemato-testicular barrier, abnormal vascularization of the testes, and an alteration of the biochemical composition of prostate fluid (Caflish and Dubose, 1991; Hew et al., 1993). However, it should be noted that the doses used in some of these studies were not representative of environmental exposures (table 13-2).

Exposures to sublethal doses of Pb or Cd result in abnormal episodic secretion of prolactin and leutinizing hormone (LH) in mammals (Sokol et al., 1998; Lafuente et al., 1999). However, the adverse effects of metals on gonadal function are not always mediated by metal-induced anomalies in the secretion of pituitary hormones. Direct effects of metals on spermatogenesis also have been reported, without any alterations of LH or follicle-stimulating hormone (FSH) secretion (Wadi and Ahmad, 1999).

Exposure to metals may impair gonadal steroidogenesis and gonadal development (table 13-2). A significant decrease in secretion of gonadal steroids has been detected in mammals as well as fish, often at doses lower than those causing effects on gonad size and sperm quality. Cd also increases pituitary sensitivity to gonadotropin-releasing hormone (GnRH) and can result in excess estrogen secretion (Thomas, 1989).

The effects of metals on reproductive function illustrate the rather nonspecific reactivity of metals to a wide array of macromolecules within the reproductive axis. Although these studies documented the adverse effects of metals on reproductive function and gonadal steroidogenesis, our understanding of the cellular mechanisms mediating these effects remains limited. However, several studies did investigate the cellular interactions of metals within steroidogenic cells and provided new insights into the mechanisms of endocrine toxicity exerted by metals. These studies are discusssed in the section "Effects of Metals at the Cellular Level."

Effects of Metals on the Adrenal Axis

Metals usually activate the neuroendocrine stress response after an acute exposure. The increase in plasma cortisol and glucose levels following a short-term (hours to

Table 13-2. The effects of metals on reproductive function.

Species	Metal	Dose	Exposure	Response	Reference
Rat	Pb	0.05–0.45% (w/v)	Lifetime, drinking water	↓ Birth wt, delayed sexual maturity, ↓ plasma T in early puberty	Ronis et al., 1994
Mouse	Pb	0.25 and 0.5%	6 Weeks, drinking water	↓ Sperm count and motility, ↑ sperm malformations, no effect on testis wt, no effect on plasma LH, FSH, T	Wadi & Ahmad, 1999
Rat	Cd	50 ppm	Day 30–60	↓ Plasma prolactin	Lafuente et al., 1999
			Day 60–90, drinking water	↓ Prolactin pulses	
Rat	Cd	2.7mg/kg	Single dose, subcutaneous	Changes in pH of luminal fluid in testis, ↓ testes wt, ↓ plasma T	Caflish & DuBose, 1991
Rat	Hg	0.8–80 μg/kg	19 Weeks	↑ Intratesticular T	Friedmann et al., 1998
			Gavage 2×/week	↓ Sperm count, ↓ fertility	
Rainbow trout	Cd	1.8–3.4μg/l	90 Weeks, water	Delayed oogenesis	Brown et al., 1994
Walleye	CH₃HgCl	0.1–1.0μg/kg	6 Months, food	↓ Gonadal and somatic growth, suppressed plasma cortisol	Friedmann et al., 1996
Catfish	Hg	16.7 ppb	90 Days, water	Delayed oogenesis	Dey & Bhattacharya, 1989

days) exposure to sublethal levels of heavy metals has been documented extensively in teleosts, mostly in salmonids exposed to Hg, Cd, or Cu (Hontela, 1997). Metals seem to be recognized as chemical stressors, and the organism attempts to maintain homeostasis through the increase in plasma cortisol and catecholamines in response. The elevated plasma cortisol may persist up to 30 days of exposure to Cd (Ricard et al., 1998), although an acclimation, characterized by a return to preexposure cortisol levels, has been reported in some experimental situations (Fu, 1990).

Chronic exposures, either to relatively high sublethal doses of metals in the laboratory or to metal-contaminated waters in the wild, seem to impair, at least in fishes, the capacity to secrete corticosteroids and to respond to additional stressors (see Hontela, 1998). A decreased capacity to secrete cortisol, compared to fish from a reference lake, was reported in Northern pike, *Esox lucius*, gill-netted from a Canadian Shield lake heavily polluted by Hg (Lockhart et al., 1975), in yellow perch (*Perca flavescens*) sampled in lakes in a mining region in northwestern Quebec (figure 13-1), and in stress-challenged brown trout (*Salmo trutta*) from the metal-contaminated portion of the Eagle River, Colorado (Norris et al., 1998).

To differentiate acclimated fish that have low plasma cortisol and functional adrenocortical tissue with a full capacity to respond to stressors from fish that have low plasma cortisol and dysfunctional adrenocortical tissue unable to respond, standardized functional corticotropin (ACTH) tests were used, in vivo and in vitro (figure 13-1; Laflamme et al., 2000). The ACTH challenge clearly demonstrated that yellow perch from the metal-contaminated lakes had impaired adrenocortical tissue, with a lower capacity to respond to ACTH. Our studies with yellow perch from lakes contaminated by heavy metals in northern Quebec also revealed a strong relationship between tissue burdens of heavy metals and degree of impairment of the cortisol stress response. The functional tests are highly relevant to the situation in the wild where fish from polluted waters must not only cope with the pollutant but also must react appropriately to predators, conspecifics, and various environmental stressors, either chronic or acute. The ability to activate the normal endocrine stress response and elevate plasma cortisol in response to a stressor is an integral part of the adaptive physiological response, and its disruption may reduce the survivorship of the fish.

To elucidate the mechanisms through which chronic exposures to metals impair the secretory capacity of the adrenocortical tissue, the ability of the adrenocortical steroidogenic cells to respond in vitro to a standardized stimulation by ACTH was assessed in rainbow trout (*Oncorhynchus mykiss*) exposed to Cd in water for 1, 7, 14, and 30 days. Adrenocortical tissue retained the ability to respond to ACTH, even in fish exposed to 1 μg Cd/l for 30 days (Brodeur et al., 1998). However, enzymatically dispersed head kidney cells exposed to metals for only 60 min in vitro lost their capacity to respond to ACTH in a concentration-related manner, at concentrations that did not cause cell mortality (Leblond and Hontela, 1999). The field studies where tissue burdens of metals were measured in the head kidneys of cortisol-impaired fish and the in vitro studies where adrenocortical cells were directly exposed to metals are consistent with the notion of a threshold of metal tissue burdens above which the normal secretory function is disrupted. This threshold may be surpassed by chronic environmental exposures in wild fishes or by in vitro exposures of the tissue. Further discussion of the mechanisms of action of adrenal disruption by metals is presented in the next section.

Effect of Heavy Metals at the Cellular Level

Until recently, our understanding of the mechanisms of action of heavy metals at the cellular level within the endocrine system was limited, as few studies attempted to identify the sites of actions of these toxic substances in endocrine cells. There is evidence, however, for a direct action of metals on steroidogenesis in adrenocortical tissue (Mgbonyebi et al., 1994; Mathias et al., 1998; Leblond and Hontela, 1999) and in the gonads (Laskey and Phelps, 1991). Although the characteristics of the toxic response depend on the animal species and the chemical nature of the toxicant (speciation, dose, length of exposure), two types of adverse effects have been reported by in vitro studies with steroidogenic tissues exposed to metals. First, some metals seem to interfere with the adrenal and gonadal functions and secretory processes by directly inhibiting specific steroidogenic enzymes. Second, many divalent heavy metals such as cadmium (Cd^{2+}) alter the intracellular pools of Ca^{2+} and target Ca^{2+}-dependent processes within the cell.

Intracellular Ca^{2+} and Heavy Metals

Ca^{2+} is essential for numerous biological processes. It plays an important role as a second messenger in intracellular signaling pathways in the endocrine systems of all vertebrate species examined thus far. Ca^{2+} is necessary for binding of ACTH to its receptor and is required to optimize steroid synthesis and secretion in the adrenal gland. Cadmium (Cd^{2+}), as a divalent heavy metal, may displace Ca^{2+} at many plasma membrane binding sites, block and permeate Ca^{2+} channels, or evoke, via stimulation of a cell-surface receptor, the release of Ca^{2+} stored in the endoplasmic reticulum or in mitochondria (figure 13-2; Mathias et al., 1998). Cd ions pass through the cell membrane of pheochromocytoma cells, a mammalian adrenal neurosecretory cell line, by Ca^{2+} channels (Hinkle and Osborne, 1994), and agonist-stimulated Ca^{2+} influx is affected by Cd and Ni (Benters et al., 1996). In a pituitary cell line, nimodipine, a Ca^{2+} channel antagonist, protected against Cd toxicity, suggesting that Cd may enter the cell by voltage-dependent Ca^{2+} channels, whereas BAY K 8644, a Ca^{2+} channel agonist, increased Cd toxicity (Hinkle et al., 1987).

In a study of adrenal steroidogenesis and the effects of Cd in Y-1 mouse adrenal tumor cells, inhibition of ACTH-stimulated steroidogenesis was reversed by using Ca^{2+}-supplemented media (Mathias et al., 1998). An increase of intracellular Ca^{2+} concentration by the ionophore A23187 provoked an increase in basal and stimulated steroid synthesis by Cd^{2+}-treated adrenal cells. These results suggest that extracellular Ca^{2+} may protect against Cd^{2+} effects on plasma membrane cAMP synthesis and on basal cholesterol metabolism by mitochondria. Measurement by fluorescence of Ca^{2+} entry into the cell suggests that Cd^{2+}, concurrently with Ca^{2+}, freely enters the cell under basal conditions and that its entry is accelerated by ACTH stimulation. These experiments provide new data on the cellular targets of Cd in endo-

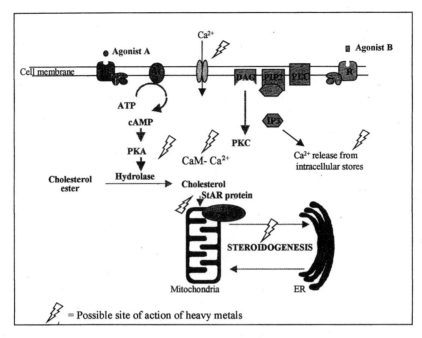

Figure 13-2. Schematic representation of the potential sites of action of heavy metals in steroidogenic cells. Agonist A, corticotropin for adrenal, growth hormone or luteinizing hormone for gonads; agonist B, angiotensin; cAMP, cyclic adenosine monophosphate; ATP, adenosine triphosphate; Ca^{2+}, ion calcium; CaM, calmodulin; DAG, diacyl glycerol; IP3, inositol triphosphate; PIP2, phosphatidyl inositol biphosphate; PKA, protein kinase A; PKC, protein kinase C; PLC, phospholipase C; R, receptor; StAR protein, steroidogenic acute regulatory protein; P450scc, side chain cleavage enzyme (rate-limiting enzymatic step in the conversion of cholesterol into pregnenolone).

crine cells. However, little is known about the effects of heavy metals on cellular mechanisms involving Ca^{2+} in other vertebrates. Calmodulin, a ubiquitous Ca^{2+}-binding protein, regulates many cellular processes and stimulates the activity of several target proteins. The stimulation of steroidogenesis by Ca^{2+}-calmodulin (Ca^{2+}/CaM) injected into Y-1 adrenal tumor cells suggested that Ca^{2+} has a regulatory effect, besides facilitating the binding of ACTH to its receptor. Formation of Ca^{2+}/CaM activates CaM, which then binds to enzymes such as phosphorylase kinase and cyclic AMP phosphodiesterase, causing changes in their activities. Ca^{2+} competes with metals for many intracellular sites, and CaM may be a potential cellular target of divalent heavy metals (figure 13-2). Cd can replace Ca^{2+} and thus interfere with the CaM function in human trophoblast cells (Powlin et al., 1997).

Effects of Heavy Metals on Steroidogenic Enzymes

As early as 1977, Freeman and Sangalang demonstrated that metals such as Cd, Hg, As as well as some organic pollutants inhibit the in vitro synthesis of steroid hormones by the adrenal tissue and testes in gray seal, *Halichoerus grypus*. Xenobiotics reduced the activity of the enzyme 3β-hydroxysteroid dehydrogenase, thereby modifying the final products of the synthesis. The mechanisms of action of heavy metals in the synthesis of corticosteroids were also investigated in vivo (Veltman and Maines, 1986). Activities of steroidogenic enzymes were measured in rats treated with Hg^{2+}, and 21α-hydroxylase was identified as a cellular target of the metal. Lower plasma corticosterone and an abnormal plasma steroid profile were linked to inhibition of the enzyme by Hg. Mgbonyebi and co-workers (1994) demonstrated an inhibitory effect of Cd on basal and ACTH-stimulated cortisol secretion from cultured rat adrenocortical cells, although as is often the case with in vitro studies, relevance of the Cd doses used was not established in terms of whole-animal toxicity. Nishiyama et al. (1985) used lower concentrations of heavy metals and did not detect an effect on ACTH-stimulated steroidogenesis in cultured rat adrenocortical cells. Reduced steroid production was observed only with Pb, and the inhibitory effect could be reversed by db-cAMP(dibutyryl cyclic AMP), suggesting that Pb interferes at the level of plasma membrane.

To investigate the toxicity and to determine the site of action of heavy metals on testosterone production by rat Leydig cells, Laskey and Phelps (1991) used biosynthetic substrates and analogs, such as hCG (human chorionic gonadotropin) and db-cAMP, respectively, to stimulate testosterone production. A concentration-dependent inhibition in both hCG- and db-cAMP-stimulated testosterone production by Leydig cells was detected with Cd^{2+}, Co^{2+}, Cu^{2+}, Hg^{2+}, Ni^{2+}, and Zn^{2+} treatments. However, in hydroxycholesterol- and pregnenolone-stimulated cells, an increase in testosterone secretion by Leydig cells treated with Cd^{2+}, Co^{2+}, Ni^{2+}, and Zn^{2+} was observed. These results indicate that these cations may act at multiple sites within the steroidogenic Leydig cells. The inhibitory site(s) appear to be situated beyond the membrane receptor and upstream from the mitochondria, where enzymatic steroid conversion occurs. Similarly, a study conducted by Leblond and Hontela (1999) investigated the effects of in vitro exposures to Cd^{2+}, Hg^{2+}, Zn^{2+}, and *o,p*'-DDD, an organochlorine pesticide, on steroidogenesis by dispersed adrenocortical cells of rainbow trout. The response to ACTH decreased in relation to the dose of the toxicant. However, stimulation with db-cAMP could not restore cortisol secretion in metal-exposed cells, but it did so in the *o,p*'-DDD-exposed cells. The results of this study suggest that in the signaling pathway leading to cortisol synthesis, metals disrupt the steps situated downstream from the step generating cAMP whereas *o,p*'-DDD seems to act upstream from the cAMP step (figure 13-3).

Despite these investigations in teleosts, most studies on the cellular mechanisms of action of heavy metals in endocrine cells were conducted with mammals. Although there is a vast literature on the physiological effects of heavy metals in aquatic species, there is a lack of knowledge of the mechanisms through which metals alter signaling pathways in endocrine cells of nonmammalian vertebrates. Experimental studies, using sound toxicological and endocrinological principles, will advance our knowledge in this relatively new area. Molecular techniques and approaches also will provide tools to understand and identify the site of action of heavy metals in regulatory processes of genes and protein expression in endocrine cells of nonmammalian vertebrates.

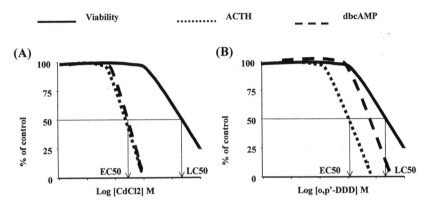

Figure 13-3. The effects of (A) CdCl₂ and (B) the organochlorine pesticide *o,p'*-DDD on ACTH- and dbcAMP-stimulated cortisol secretion and viability of enzymatically dispersed head kidney cells of rainbow trout. A concentration-dependent loss of the capacity to respond to ACTH and also to dbcAMP is observed in the metal exposed cells, with a significantly lower EC_{50} (concentration that inhibits secretion of cortisol by 50%) than LC_{50} (concentration that kills 50% of the cells). The response to *o,p'*-DDD is similar; however, stimulation with dbcAMP could restore the secretory response at doses of toxicant that inhibited the response to ACTH. The model suggests that metals disrupt multiple sites within the steroidogenic pathway, including sites downstream from the cAMP step, while *o,p'*-DDD acts upstream from the cAMP step.

Effects of Metals on the Thyroid Axis

The thyroid gland is another target of metals. Cd accumulates in the mammalian thyroid—specifically, in the mitochondria of thyroid follicle epithelial cells, where the metal seems to disrupt the synthesis of thyroid hormones (Yoshizuka et al., 1991). A decrease of the thyroid follicle epithelial height and a decrease in plasma thyroid hormone levels were documented in fish exposed to Cd in the laboratory and in the field (table 13-3). Oxidative stress may be a mechanism mediating the adverse effects of metals on the synthesis of thyroid hormones because, along with a decrease in plasma thyroid hormones levels and in the activity of the 5'-monodeiodinase, an increase in lipid peroxidation was demonstrated in the mouse (Gupta and Kar, 1997). Treatment with tocopherol, an antioxidant, restored thyroid function in Cd-treated animals.

Even though the adverse effects of metals on plasma thyroid hormones levels have been documented, few studies have investigated the impact of metal-induced disturbances of thyroid function on the physiological status of animals. Metal-induced alterations of thyroid status, and also of the secretion of corticosteroids, may be responsible, at least in part, for the effects of metals on intermediary metabolism and osmoionic homeostasis. This may be particularly relevant in fishes and amphibians where thyroid hormones and corticosteroids have roles in metabolism and in osmoregulation. The functional duality of these hormones and their permissive actions may increase the vulnerability of fishes and amphibians

Table 13-3. The effects of metals on thyroid function.

Species	Metal	Dose	Exposure	Response	Reference
Catfish	Cd	0.5.–1.5mg/l	30 days water	↓ Plasma T3, T4, ↑ Lipid peroxydation	Gupta et al., 1997
Lake trout	Cd	0.5 and 5 µg/l	8–9 months water	↓ Thyroid epithelial cell height	Scherer et al., 1997
Rainbow trout	HCN	0.01 mg/l	12 days	↓ Plasma T3 and T4	Ruby et al., 1993
Yellow perch	Cd, Zn, Cu	Field	Life long	↓ Plasma T3 and T4, ↓ Thyroid epithelial cell height	Levesque et al., 2003
Mouse	Cd	1.0–2.5mg/kg	15 days, daily sc injections	↓ Plasma T4, T3	Gupta and Kar, 1997

to toxicants such as metals that alter the secretory function of the thyroid and the adrenals.

Effects of Metals on Osmoregulation

Exposure to metals disturbs osmoionic homeostasis, and changes in the biochemical constituents of plasma and urine of animals exposed to metals are well documented, particularly in teleost fish and laboratory rodents. Alterations of ion concentrations in plasma or urine, anemia, hypocalcemia, and presence in urine of amino acids, proteins, glucose, and specific enzymes released from damaged cells are all symptoms of exposure to metals at relatively high concentrations used in laboratory exposures. Several of these changes are caused by damage to gills and kidneys, impairing the osmoregulatory capacity of these organs. Some of the adverse effects of metals on osmoregulation are secondary responses to metal-induced alterations in the secretion of key osmoregulatory hormones such as cortisol, thyroid hormones, or prolactin (Bleau et al., 1996). It has been demonstrated in tilapia, *Oreochromis mossambicus*, that while Cu induced necrosis in gill chloride cells, cortisol had a protective effect against Cu-induced necrosis and promoted apoptosis, possibly eliminating defective cells without inducing inflammation (Bury et al., 1998).

Metals also interfere with the fluxes of ions across epithelial surfaces of gills or kidney through physicochemical processes influencing electrochemical gradients and effects on ion pumps. A decrease in the activity of gill Na^+/K^+ ATPase in teleosts exposed to metals has been reported in recent studies. Metals such as Pb or Cd, which both interfere with Ca^{2+} actions, are well known for their osmoregulatory alterations in fish and mammals. Detailed studies on the effects of metals have been carried out with rainbow trout by Wood and colleagues (Webb and Wood, 1998). The relative importance of the intestinal tract and gills in metal assimilation was characterized through laboratory exposures. Interference of Cu, Ag, and also Cd with fluxes of Cl^- and Na^+ across the gills and subsequent failure of the fish to maintain constant concentrations of plasma electrolytes has been established as the main cause of acute metal toxicity in rainbow trout.

The effects of metals on osmoregulation illustrate the multiple actions of metals impacting diverse targets. Depending on the dose, the length of exposure, and the sensitivity of the animal species, metals can influence the morphology and function of osmoregulatory organs, as well as the secretion of osmoregulatory hormones and activities of regulatory enzymes and ion pumps.

Effects of Metals on Growth and Metabolism

Growth is a complex process regulated by physiological factors, such as hormones and enzymes controlling the cycling of energetic substrates for anabolic or catabolic reactions, as well as by external factors, mainly availability of food. Metals can influence all of these factors, directly or indirectly, making it difficult to establish mechanistic cause–effect relationships, particularly in the field.

Effects of metals on growth have been studied mostly in fishes under laboratory and field conditions. Most studies reported reduced growth or condition factor [(weight/length^3) × 100] in fish exposed to heavy metals (Marr et al., 1995; Laflamme et al., 2000). The effects of metals or other xenobiotics on growth in the field are difficult to interpret because the age–size relationships traditionally used for estimating growth are subject to the effects of density, the effects of metals on predators or prey, and the physiological status of the studied population. New methodologies using globally dispersed radioisotopes such as Cs^{137} can be used to assess, in combination with the age–size relationships, growth efficiency, an estimate of the efficiency of converting consumed food into body mass (Sherwood et al., 2000). These new approaches may provide new insight into the bioenergetics and growth capacity and elucidate the link between the effects of metals on hormonal and metabolic function and growth in fish from polluted waters.

The effects of metals on intermediary metabolism have been investigated, again mostly in fishes, with few studies in other vertebrate groups. Acute exposures to metals increase plasma cortisol levels, elevate plasma glucose, and decrease liver glycogen reserves. These responses may have an adaptive value because they augment the availability of energy substrates necessary for maintenance of homeostasis during the challenge that exposure to metals represent. These data also suggest that exposure to metals activates energetically costly homeostatic mechanisms, as has been suggested by Heath (1995). The effects of chronic exposures to metals on intermediary metabolism are less well understood. Seasonal changes in hepatic reserves of glycogen and triglycerides have been reported in yellow perch exposed to metals in lakes situated in a mining region (Levesque et al., unpublished data). Fish from the contaminated lakes had a decreased capacity to build up energy reserves compared to fish from reference sites, and they also exhibited important differences in the activities of key gluconeogenic and lipolytic enzymes. These metabolic anomalies may explain the lower growth efficiency and delayed gonadal recrudescence in fish from metal-contaminated lakes. Both somatic and gonadal growth are processes that require efficient use of energy substrates, particularly in contaminated environments where food availability may be limited as well.

Implications for Wildlife and Human Health

Metals remain ubiquitous contaminants in the environment despite preventive measures that have lowered the concentrations of metals in air, water, and food. The high reactivity of metals toward biological macromolecules, long biological half-life in organisms, and the tendency of some metals to bioaccumulate within specific organs make metals a family of environmental pollutants of concern. Mechanisms of action must be better understood to ensure that environmental norms for metals are compatible with physiological fitness and normal endocrine function.

There is already extensive evidence for sublethal metal toxicity within the endocrine system. Metals exert significant antigonadal actions through adverse effects on gametogenesis and the secretion of reproductive hormones, particularly steroids. They activate the hormonal neuroendocrine stress response after acute exposures. However, chronic exposures that result in adrenal tissue burdens of metals above the carrying capacity of detoxification mechanisms seem to impair the secretory capacity of adrenocortical cells. The thyroid axis is another endocrine target for metals.

Studies using mostly in vitro approaches have provided new insight into the cellular processes disrupted by metals. Disturbance of intracellular Ca^{2+} homeostasis is one mechanism of action associated with divalent cations such as Cd^{2+} and Pb^{2+}. Interactions of metals with the function of Ca^{2+} channels, with calmodulin, and with the mechanisms regulating intracellular stores of Ca^{2+} have been reported, although the number of animal models investigated thus far remains limited. Disruptive effects on steroidogenic enzymes also have been proposed as a mechanism of action of metals within gonadal and adrenal steroidogenic cells.

Important knowledge gaps concerning the effects of metals on health remain. Metals seem to disturb the osmoregulatory capacity of animals, particularly aquatic species, as well as alter metabolism and growth. Aquatic species are particularly vulnerable to metals because the aquatic environment remains a favored route of disposal of many industrial and agricultural waste products. Carefully designed studies, both in the field and in the laboratory, are required to increase our understanding of toxic effects of metals, particularly within the endocrine system, and to enable us to establish mechanistically based cause–effect relationships that can be used to diagnose metal-induced dysfunctions and to ensure that the environmental norms are compatible with maintenance of animal diversity and health.

References

Assenato, G., Paci, C., Molinini, R., 1986. Sperm count suppression without endocrine dysfunction in lead-exposed men. Arch. Environ. Health 41, 387–390.

Benters, J., Schäfer, T., Beyersmann, D., Hechtenberg, S., 1996. Agonist-stimulated calcium transients in PC12 cells are affected differentially by cadmium and nickel. Cell Calcium 20, 441–446.

Bleau, H., Daniel, C., Chevalier, G., van Tra H., Hontela, A., 1996. Effects of acute exposure to mercury chloride and methyl mercury on plasma cortisol, T3, T4, glucose and liver glycogen in rainbow trout, *O. mykiss*. Aquat. Toxicol. 34, 221–235.

Brodeur, J.C., Daniel, C., Ricard, A.C., Hontela, A., 1998. In vitro response to ACTH of the interrenal tissue of rainbow trout (*Oncorhynchus mykiss*) exposed to cadmium. Aquat. Toxicol. 42, 103–113.

Brown, V., Shurben, D., Miller, W., Crane, M., 1994. Cadmium toxicity to rainbow trout *Oncorhynchus mykiss* W. and brown trout *Salmo trutta* L. over extended exposure periods. Ecotoxicol. Environ. Safety 29, 38–46.

Bury, N.R., Jie, L., Flik, G., Lock, R.A.C., Wendelaar Bonga, S.E., 1998. Cortisol protects against copper induced necrosis and promotes apoptosis in fish chloride cells in vitro. Aquat. Toxicol. 40, 193–202.

Caflish, C.R., Dubose, T.D., Jr., 1991. Cadmium-induced changes in luminal fluid pH in testis and epididymis of the rat in vivo. J. Toxicol. Environ. Health 32, 49–57.

Campbell, P.G.C., Tessier, A., 1996. Ecotoxicology of metals in the aquatic environment: geochemical aspects. In: Newman, M.C., Jagoe, C. (Eds.), Ecotoxicology: A Hierarchical Treatment. CRC Press, Boca Raton, FL, pp. 11–58.

Congiu, L., Chicca, M., Pilastro, A., Turchetto, M., Tallandini, L., 2000. Effects of dietary cadmium on hepatic glutathione levels and glutathione peroxidase activity in starlings (*Sturnus vulfaris*). Arch. Env. Contam. Toxicol. 38, 357–361.

De Master, E.G., Mitchell, R.A.A., 1973. A comparison of arsenate and vanadate as inhibitors or uncouplers of mitochondrial and glycolytic energy metabolism. Biochemistry 12, 3616–3621.

Dey, S., Bhattacharya, S., 1989. Ovarian damage to Channa punctatus after chronic exposure to low concentrations of Elsan, mercury and ammonia. Ecotoxicol. Environ. Safety 17, 247–257.

Freeman, H.C., Sangalang, G.B., 1977. A study of the effects of methyl mercury, cadmium, arsenic, selenium, and a PCB (Aroclor 1254) on adrenal and testicular steroidogenesis *in vitro*, by the gray seal *Halichoerus grypus*. Arch. Environ. Contam. Toxicol. 5, 369–383.

Friedmann, A.S., Chen-Haolin, Rabuck, L.D., Zirkin, B.R., 1998. Accumulation of dietary methylmercury in the testes of the adult brown Norway rat: Impaired testicular and epididymal function. Environ. Toxicol. Chem. 17, 867–871.

Friedmann, A.S., Watzin, M.C., Brinck-Johnsen, T., Leiter, J.C., 1996. Low levels of dietary methylmercury inhibit growth and gonadal development in juvenile walleye (*Stizostedion vitreum*). Aquat. Toxicol. 35, 265–278.

Fu, H., 1990. Involvement of cortisol and MT-like proteins in the physiological responses of tilapia to sublethal Cd stress. Aquat. Toxicol. 16, 257–270.

Gerpe, M., Kling, P., Berg, A.H., Olsson, P.-E., 2000. Arctic char (*Salvelinus alpinus*) metallothionein: cDNA sequence, expression, and tisssue specific inhibition of cadmium-mediated metallothionein induction by 17β-estradiol, 4-OH-PCB 30, and PCB 104. Environ. Toxicol. Chem. 19, 638–645.

Goldstein, G.W., 1993. Evidence that lead acts as a calcium substitute in second messenger metabolism. Neurotoxicology 14, 97–102.

Gupta, P., Chaurasia, S., Kar, A., Maiti, P.K., 1997. Influence of cadmium on thyroid hormone concentrations and lipid peroxydation in a fresh water fish, *Clarias batrachus*. Fres. Environ. Bull. 6, 355–358.

Gupta, P., Kar, A., 1997. Role of testosterone in ameliorating the cadmium induced inhibition of thyroid function in adult male mouse. Bull. Environ. Contam. Toxicol. 58, 422–428.

Harrison, S.E., Klaverkamp, J.F., 1989. Uptake, elimination and tissue distribution of dietary and aqueous cadmium by rainbow trout (*Salmo gairdneri R.*) and lake whitefish (*Coregonus clupeaformis* M.). Environ. Toxicol. Chem. 8, 87–97.

Heath, A.G., 1995. Water Pollution and Fish Physiology, 2nd edition. CRC Press, Boca Raton, FL.

Hew, K.-W., Ericson, W.A., Welsh, M.J., 1993. A single low cadmium dose causes failure of spermiation in the rat. Toxicol. Appl. Pharmacol. 121, 15–21.

Hinkle, P.M., Kinsella, P.A., Osterhoudt, K.C., 1987. Cadmium uptake and toxicity via voltage-sensitive calcium channels. J. Biol. Chem. 262, 16333–16337.

Hinkle, P.M., Osborne, M.E., 1994. Cadmium toxicity in rat pheochromocytoma cells: Studies on the mechanism of uptake. Toxicol. Appl. Pharmacol. 124, 91–98.

Hontela, A., 1997. Endocrine and physiological responses of fish to xenobiotics: Role of glucocorticosteroid hormones. Reviews in Toxicology 1: 1–46. Environmental Toxicology, IOS Press.

Hontela, A., 1998. Interrenal dysfunction in fish from contaminated sites: *in vivo* and *in vitro* assessment. Environ. Toxicol. Chem. 17, 44–48.

Laflamme, J.-S., Couillard, Y., Campbell, P.G.C., Hontela, A., 2000. Interrenal metallothionein and cortisol secretion in relation to Cd, Cu, and Zn exposure in yellow perch, *Perca flavescens*, from Abitibi lakes. Can. J. Fish. Aquat. Sci. 57, 1692–1700.

Lafuente, A., Alvarez-Demanuel, E., Marquez, N., Esquifino, A.I., 1999. Pubertal dependent effects of cadmium on episodic prolactin secretion in male rats. Arch. Toxicol. 73, 60–63.

Laskey, J.W., Phelps, P.V., 1991. Effect of cadmium and other metal cations on *in vitro* Leydig cell testosterone production. Toxicol. Appl. Pharmacol. 108, 296–306.

Leblond, V., Hontela, A., 1999. Effects of in vitro exposures to cadmium, mercury, zinc, and 1-(2-chlorophenyl)-1-(4-chlorophenyl)-2,2-dichloroethane on steroidogenesis by dispersed interrenal cells of rainbow trout (*O. mykiss*). Toxicol. Appl. Pharmacol. 157, 16–22.

Levesque, H.M., Dorval, J., Van Der Kraak, G.J., Campbell, P.G.C., Hontela, A., 2003. Hormonal, morphological and physiological responses of yellow perch (*Perca flavescens*) to chronic enviornmental metal exposures. J. Toxicology and Environmental Health Part A, 657–676.

Lockhart, W.L., Uthe, J.F., Kenney, A.R., Mehrle, P.M., 1975. Methylmercury in northern pike (*Esox lucius*); distribution, elimination, and some biochemical characteristics of contaminated fish. J. Fish. Res. Board Can. 29, 1519–1523.

Marr, J.C.A., Bergman, H.L., Lipton, J., Hogstrand, C., 1995. Differences in relative sensitivity of naive and metals-acclimated brown and rainbow trout exposed to metals representative of the Clark Fork River, Montana. Can. J. Fish. Aquat. Sci. 52, 2016–2030.

Mathias, S.A., Mgbonyebi, O.P., Motley, E, Owens, J.R., Mrotek, J.J., 1998. Modulation of adrenal cell functions by cadmium salts. 4. Ca^{2+}-dependent sites affected by CdCl2 during basal and ACTH-stimulated steroid synthesis. Cell Biol. Toxicol. 14, 225–236.

Mgbonyebi, O.P., Smothers, C.T., Mrotek, J.J., 1994. Modulation of adrenal cell functions by cadmium salts: 3. Sites affected by $CdCl_2$ during stimulated steroid synthesis. Cell Biol. Toxicol. 10, 35–43.

Nishiyama, S., Nakamura, K., Ogawa, M., 1985. Effects of heavy metals on corticosteroid production in cultured adrenocortical cells. Ecotoxicol. Environ. Safety 81, 174–176.

Norris, D.O., Donahue, S., Dores, R., Lee, J.K., Maldonado, T., Ruth, T., Woodling, J.D., 1998. Impaired adrenocortical response to stress by brown trout, *Salmo trutta*, living in metal-contaminated waters of the Eagle River, Colorado. Gen. Comp. Endocrinol. 113, 1–8.

Olsvik, P.A., Gundersen, P., Andersen, R.A., Zachariassen, K.E., 2000. Metal accumulation and metallothionein in two populations of brown trout, *Salmo trutta*, exposed to different natural water environments during a run-off episode. Aquat. Toxicol. 50, 301–316.

Pourahmad, J., O'Brien, P.J., 2000. A comparison of hepatocyte cytotoxic mechanisms for Cu^{2+} and Cd^{2+}. Toxicology 143, 263–273.

Powlin, S.S., Keng P.C., Miller, R.K., 1997. Toxicity of cadmium in human trophoblast cells (Jar choriocarcinoma): Role of calmodulin and the calmodulin inhibitor, zaldaride maleate. Toxicol. Appl. Pharmacol. 144, 225–234.

Rasmiafshari, M., Zawia, N.H., 2000. Utilization of a synthetic peptide as a tool to study the interaction of heavy metals with the zinc finger domain of proteins critical for gene expression in the developing brain. Toxicol. Appl. Pharmacol. 166, 1–12.

Ricard, A.C., Daniel, C., Anderson, P., Hontela, A., 1998. Effects of subchronic exposure to cadmium chloride on endocrine and metabolic functions in rainbow trout, *O. mykiss*. Arch. Environ. Contam. Toxicol. 34, 377–381.

Ronis, M.J.J., Shahare, M., Mercado, C., Irby, D., Badger, T.M., 1994. Disrupted reproductive physiology and pubertal growth in rats exposed to lead during different developmental periods. Biol. Reprod. 50, 76.

Ruby, S., Idler, D.R., So, Y.P., 1993. Plasma vitellogenin, 17β-estradiol, T3 and T4 levels in sexually maturing rainbow trout *Oncorhynchus mykiss* following sublethal HCN exposure. Aquat. Toxicol. 26, 91–102.

Scherer, E., McNicol, R.E., Evans, R.E., 1997. Impairment of lake trout foraging by chronic exposure to cadmium: a black box experiment. Aquat. Toxicol. 37, 1–7.

Sherwood, G.D., Rasmussen, J.B., Rowan, D.J., Brodeur, J., Hontela, A., 2000. Bioenergetic costs of heavy metal exposure in yellow perch (*Perca flavescens*): *In situ* estimates with radiotracer (^{137}Cs) technique. Can. J. Fish. Aquat. Sci. 57, 441–450.

Sokol, R.Z., Berman, N., Okuda, H., Raum, W., 1998. Effects of lead exposure on GnRH and LH secretion in male rats: response to castration and alpha-methyl-p-tyrosine (AMPT) challenge. Reprod. Toxicol. 12, 347–355.

Sorensen, E.M., 1991. Metal Poisoning in Fish. CRC Press, Boca Raton, FL.

Thomas, P., 1989. Effects of Aroclor 1254 and cadmium on reproductive endocrine function and ovarian growth in Atlantic Croaker. Mar. Environ. Res. 28, 499–503.

Veltman, J.C., Maines, M.D., 1986. Alterations of heme, cytochrome P-450, and steroid metabolism by mercury in rat adrenal. Arch. Biochem. Biophys. 248, 467–478.

Verbost, P.M., Van Rooij, J., Flik, G., Lock, R.A.C., Wendelaar Bonga, S.E., 1989. The movement of cadmium through freshwater trout branchial epithelium and its interference with calcium transport. J. Exp. Biol. 145, 185–197.

Wadi, S.A., Ahmad, G., 1999. Effects of lead on the male reproductive system in mice. J. Toxicol. Environ. Health 56, 513–521.

Webb, N.A., Wood, C.M., 1998. Physiological analysis of the stress response associated with acute silver nitrate exposure in freshwater rainbow trout (*Oncorhynchus mykiss*). Environ. Chem. Toxicol. 4: 579–588.

Wiegant, F.A.C., van Rijn, J., van Wijk, R., 1997. Enhancement of the stress response by minute amounts of cadmium in sensitized Reuber H35 hepatoma cells. Toxicology 116, 27–37.

Yoshizuka, M., Mori, N., Hamasaki, K., Tanaka, I., Yokoyama, M., Hara, K., Doi, Y., Umezu, Y., Araki, H., Sakamoto, Y., 1991. Cadmium toxicity in the thyroid gland of pregnant rats. Exp. Mol. Pathol. 55, 97–104.

14

Effects of Alkylphenolic Compounds on Wildlife

Werner Kloas

During the past few decades emerging evidence has demonstrated that environmental pollutants can interfere with the endocrine systems of wildlife and humans (Colborn and Clement, 1992). These compounds are called endocrine disruptors, defined at the Weybridge workshop in 1996 (European Commission, 1996) as "an exogenous substance that causes adverse health effects in an intact organism, or its progeny, consequent to changes in endocrine function." A potential endocrine disruptor was defined as a "substance that possesses properties that might be expected to lead to endocrine disruption in an intact organism" (European Commission, 1996). Such a broad definition includes nearly all environmental pollutants because all classical toxicants have the potential to interfere with the endocrine system by acting as nonspecific stressors (see chapter 5). Thus, a more precise definition of endocrine disruptors is required. I propose that endocrine disruptors be defined as endocrine active compounds causing specific effects on endocrine systems at several levels without relevant toxic actions.

One substance class of major concern for potential endocrine-disrupting chemicals (EDCs) are alkylphenolic compounds, which consist of alkylphenols (APs), their polyethoxylates (APEOs), and their polyethoxycarboxylates (APECs). APEOs are used as nonionic surfactants, have been produced in large-volumes since the 1940s, and have been released into the environment in substantial amounts. APEOs cannot be completely degraded in sewage treatment plants and, together with their metabolites, APECs and APs, these APs are one of the major sources of chemicals in surface waters exhibiting endocrine-disrupting potential. AP compounds, especially nonylphenol (NP) and octylphenol (OP), have received great public attention because they exhibit estrogenic actions in various groups of animals (including mammals) and are abundant in the environment.

The global production of APEOs has been estimated at more than 500,000 tons in 1996, and about 100,000 tons are assumed to be released into the environment (Gies, 1996). APEOs are produced as mixtures that contain 80% nonylphenol polyethoxylates (NPEOs), 15% octylphenol polyethoxylates (OPEOs), and dodecylphenol and dinonylphenol ethoxylates (Naylor et al., 1992; White et al., 1994). The main APs used commercially are NP and OP, which are applied in plastic and rubber products as antioxidants and in manufacturing of APEOs. APEOs are used as nonionic surfactants for enhancement of products or processes wherever foaming, emulsification, solubilization, or dispersion is important. APEOs are applied mainly for production of plastics and elastomers, agricultural chemicals, paper, and textiles. Approximately 85% of APEOs are applied for several purposes in industry and agriculture, and the remaining 15% are used in household and personal care products (Talmage, 1994; Kiewiet and de Voogt, 1996). Release of APEOs into the environment is mainly due to applications for polymer stabilization of plastics, formulations of products for wool cleaning, leather manufacturing, lubrication oil, metal-working fluids, and pulping (Tanghe et al., 1998). In addition, APEOs and APs are directly distributed by spraying as integral parts of formulations of pesticides, herbicides, and fungicides, acting as adjuvants and emulsifiers. The use of APEOs for rig washes in offshore oil production is another pathway of direct contamination of the environment (Blackburn et al., 1999). In biotechnology, APEOs are used for several purposes including enhancement of stability of pharmaceuticals, emulsions, and dispersions. In the household APEOs are present in water-soluble paints, laundry detergents, cleaners, and personal care products, including contraceptive creams (Talmage, 1994; Toppari et al., 1996).

Structure and Physicochemical Characteristics

APEOs are amphipathic molecules consisting of an alkylated phenol derivative as a hydrophobic moiety and an ethylene oxide chain of variable length as a hydrophilic component. They are produced by reacting a branched-chain AP with ethylene oxide (Warhurst, 1995). The schematic nomenclature of APEOs is summarized in figure 14-1. In this nomenclature, R stands for the abbreviation of the latin number of carbons in the alkyl chain usually O (octyl = eight) or N (nonyl = nine), and the number of ethoxylate groups in the EO chain is denoted by n, as APnEO. For instance, AP1EO is alkylphenol monoethoxylate, AP2EO is alkyphenol diethoxylate, AP3EO is alkylphenol triethoxylate, and OP3EO is octylphenol triethoxylate. The nomenclature is similar for the APnEO metabolites, APnEC, in which the terminal ethoxylate group is carboxylated (figure 14-1): AP1EC is alkylphenoxy acetic acid, AP2EC is alkylphenoxyethoxy acetic acid, and AP3EC is alkylphenoxydiethoxy acetic acid.

APEOs commercially produced do not have well-defined chemical structures because they consist of complex mixtures of isomeric APs containing up to 10% chemical impurities. APnEO typically possess ethoxy chain lengths between 1 and

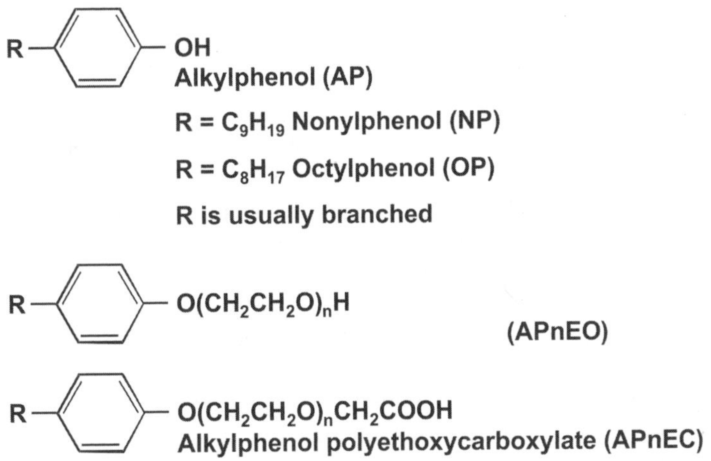

R⟨⟩— OH
Alkylphenol (AP)

R = C_9H_{19} Nonylphenol (NP)

R = C_8H_{17} Octylphenol (OP)

R is usually branched

R⟨⟩— $O(CH_2CH_2O)_nH$

(APnEO)

R⟨⟩— $O(CH_2CH_2O)_nCH_2COOH$
Alkylphenol polyethoxycarboxylate (APnEC)

Figure 14-1. Structures of and acronyms for alkylphenols, alkylphenol polyethoxylates, and alkylphenol polyethoxycarboxylates.

50, the majority being 9 or 10 (Swisher, 1987). The most abundantly used APnEO are NPnEO, in which a nonyl group opposes a chain containing n ethoxylate groups (n = 2 or more). More than 90% of NPnEO possess the two functional groups in a *para* position to one another, denoted as *para*-NP or 4-NP. The AP group occurs as a series of 25 isomers. Manufactured NP consists of a complex mixture of APs of approximately 3–6% 2-NP, 90–93% 4-NP consisting of about 18 isomers, and 2–5% decylphenol (Wheeler et al., 1997).

Due to the hydrophilic nature of the ethoxylate chain, the aqueous solubility, specific gravity, and viscosity of APEOs increase with the ethoxylate chain length. The octanol-water partition coefficients (K_{ow}) of the hydrophobic parts of AP are approximately 4.5 and 4.1 for NP and OP, respectively (Ahel and Giger, 1993). Assuming that log K_{ow} values > 4.0 characterize compounds being only weakly soluble in water, OPs and all OPnEO possess relatively good water solubility, whereas NPs and NPnEOs are only slightly soluble up to chain lengths of n = 6. The carboxylated metabolites APECs have log K_{ow} values of about 1–2, suggesting a much better solubility in water compared to APEOs under natural conditions because they are nearly completely ionized at the usual pH ranges.

Among the alkylphenolic compounds, OP and NP received the most attention for ecotoxicological research because they are semivolatile and more lipophilic compared with their corresponding APEOs and APECs. Furthermore, OP and NP are less biodegradable and display a greater bioaccumulation potential in aquatic organisms. The most abundant AP is NP because of the predominant use of its parent compound, NPEO, and, under natural conditions, NP is present mainly in its undissociated, unsolubilized form. Vapor pressure and Henry's law constant of NP, NPEOs, and NPECs are low, and therefore they should partition to air only to small extent. (Ahel et al., 1994a; Tyler et al., 1998).

Environmental Pathways

The main sources of AP compounds in the environment are products used for agricultural spraying and for cleaning purposes in several industries and most households. Spraying results in direct contamination of fruits and the environment, leading to enhanced concentrations of APEOs and metabolites in surface water as a result of run-off from rain (Fairchild et al., 1999). The source of APEOs in surface water, however, is sewage treatment plants. The sewage treatment process does not significantly biodegrade APEOs and their metabolites. As a result, substantial amounts of APEOs, APECs, and APs are released in sewage effluents or discharged in rural areas where sewage sludge is used for soil amendment (Giger et al., 1984).

The presence of AP compounds in the environment is a function of production, release into the environment, and elimination triggered by biodegradation and sorption processes. Microbial transformation, photo-oxidation, and physicochemical processes are the routes of elimination (Naylor et al., 1992). APEOs in surface waters are diluted and dissipated by vertical fluxes. Photolysis and volatilization contribute a small amount to loss of alkylphenolic compounds from surface water, but diffusion and sorption to organic and inorganic particulate matter are responsible for the main initial loss from the aqueous phase.

Microbial biodegradation of APEOs takes place in sewage sludge as well as in surface water. In summary, long-chain APEOs are transformed via sequential shortening of the ethoxy chain by ether cleavage under aerobic and anaerobic conditions yielding short-chain $APnEOs$ with $n = 1$ to 3 (John and White, 1998). Short-chain APEOs can be further transformed under aerobic conditions into APECs by oxidation of the terminal ethoxy units. Complete deethoxylation of APEC can only be obtained under anaerobic conditions leading to the corresponding APs, which are persistent and accumulate (Giger et al., 1984). However, aerobic conditions can lead to degradation and mineralization of AP, starting with a fission of the phenolic ring yielding alcohols, esters, and ethers as intermediary products (Tanghe et al., 1999; Fujii et al., 2000). The rate of AP degradation increases with increasing temperature. Biodegradability of NP has been investigated in sewage sludge (Banat et al., 2000), seawater (Ekelund et al., 1993), and soil (Topp and Starratt, 2000) resulting in half-life estimates for NP between 5 (soil) and 50 (seawater) days depending on the conditions used. Biodegradation of AP is accomplished by bacteria (Tanghe et al., 1999; Fujii et al., 2000), yeast (Corti et al., 1995), and fungi (Tsutsumi et al., 2001).

After reviewing several studies investigating the pathways of AP compounds in sewage treatment plants from various origins, some common conclusions can be drawn concerning the fate and distribution of alkylphenolic compounds. APEOs with long and medium ethoxy-chains are removed at rates of 90% or more during secondary biological treatment. The formation of more recalcitrant metabolites results in much lower elimination rates for the sum of alkylphenolic compounds than found for the parental APEOs. Short-chain APEOs, such as AP1EO and AP2EO, are the most abundant APEO oligomers in secondary effluents and are eliminated only modestly or may even demonstrate net formation during biological treatment. A net formation is observed for APECs during aerobic sewage treatment, and they become the major part of alkylphenolic compounds in the secondary effluents. AP measurements indicate their removal from the water to varying extent. Due to biodegradation processes, APs are produced from parent compounds in large amounts during sewage treatment but, due to their lipophilicity, partition and sorption to organic matter of sewage sludge is their major elimination route. However, formation of substantial amounts of AP occurs during anaerobic sewage sludge treatment, and that is the major route for release of AP into the environment: via sewage effluents or via distribution of sewage sludge used for soil amendment.

APEO-derived compounds are aquatic pollutants mainly introduced into surface waters by sewage treatment plants, pulp and paper mills, and textile industries as well as by run-offs from farms after application of sewage sludges to amend soil. Despite the ubiquitous presence of alkylphenolic compounds, the reported levels of APEO, APEC, and AP demonstrate a wide range of concentrations in surface waters ranging from a few nanograms per liter to several hundreds micrograms per liter due to close proximity to discharging points. The highest concentrations of AP have been measured in some rivers in Europe (Spain, Switzerland, UK) and in Asia (Taiwan), whereas in Northern America and Japan AP levels are substantially lower (Ahel et al., 1994b; Blackburn and Waldock, 1995; Espadaler et al., 1997; Renner, 1997; Ding et al., 1999; Bennett and Metcalfe, 2000; Hale et al., 2000; Tsuda et al., 2000; Isobe et al., 2001). Concentrations of NP, considered the most critical APEO-derived metabolite, are normally < 1 μg/l in surface waters but, due to some hot spots in European and Asian rivers, NP in waste water effluents often exceeds 10 or even several 100 μg/l. Concentrations of OP in surface waters are influenced by the lower production of OPEO and are usually one to two orders of magnitude lower compared to NP.

In estuarine and marine waters, alkylphenolic compounds are not usually detected and the concentrations of AP compared to river waters are considerably lower (Blackburn and Waldock, 1995; Blackburn et al., 1999; Dachs et al., 1999). These lower concentrations might be due to higher dilution and sorption processes (Blackburn et al., 1999; Marcomini et al., 2000).

Recent reports demonstrate the presence of NP in coastal and urban atmospheres in gas and aerosol phases at substantial concentrations (Dachs et al., 1999; van Ry et al., 2000). The main source of NP in the atmosphere seems to be water-to-air volatilization, leading to high concentrations that are suspected to be an important human and ecosystem health issue in areas receiving treated sewage effluents.

Mechanisms of Action

In general, AP compounds are much less toxic to terrestrial mammals than to aquatic organisms (Lewis, 1991; Servos, 1999), in which all surfactants increase epidermal mucus production, leading to impaired oxygen uptake and nonspecific narcotic effects (Cardellini and Ometto, 2000). Nonionic surfactants cause further adverse effects by interfering with membranes and bioactive macromolecules such as proteins, peptides, amino acids, and phospholipids (Cserhati, 1995; Kiewiet and de Voogt, 1996). Thus, in general, APs can modify membrane permeability of cells and cellular vesicles, leading to leakage between compartments and interfering with enzymes, resulting in metabolic alterations of cells.

The first concerns about the ecotoxicity of NP came after a study demonstrated NP toxicity to salmon, marine invertebrates, and a freshwater clam (McLeese et al., 1980, 1981). Across various test systems and different model organisms, a common pattern of toxicity appears to exist. In general, the shorter the ethoxylate chain of APEO, the greater the lipophilicity and potential acute toxicity for aquatic organisms from many different phyla, including plants (Naylor et al., 1992; Kahl et al., 1997; for review, see Servos, 1999). Thus, short-chain APEOs and APECs as well as APs (e.g., NP and OP), display toxic effects in wildlife.

In addition to the nonspecific narcotic effects of alkylphenolic compounds, much attention has been paid to the specific endocrine-disrupting effects of APs, APEOs, and APECs. NP and OP are some of the most important xenoestrogens, contributing to the major part of estrogenic environmental chemicals. Both NP and OP can affect reproductive biology, mainly by mimicking the effects of natural estrogens. Despite some reports of APs binding to progesterone receptors (PRs; Tran et al., 1996) and androgen receptors (ARs; Laws et al., 1995; Sohoni and Sumpter, 1998; Paris et al., 2002), the main pathway for endocrine signaling of APs is via binding to the estrogen receptor (ER; see chapters 2 and 6) and its transactivation, leading to altered expression of estrogen-responsive genes. Such changes are described in detail in chapter 6, and can include feminization phenomena such as behavioral changes, induction of estrogenic biomarkers, and alterations in sexual differentiation and reproductive parameters. The potential target sites for endocrine disruption by APs are schematically summarized in figure 14-2.

In vertebrates, APs enter the body through the diet or the environment due to their lipophilic nature and then reach the circulatory system. The blood contains sex hormone binding protein (SHBG), which reversibly binds natural 17β-estradiol (E_2) to about 90%, thus serving as both a transport and buffer system. Several studies demonstrate that SHBG has only weak specificity for APs in mammals, fishes, and amphibians (Milligan et al., 1998; Kloas et al., 2000; Kloas, 2002), indicating that APs are not buffered and are fully bioavailable after entering an individual. Thus, binding of AP to ER may occur more frequently compared to natural estrogens, which are buffered by SHBG, implicating a higher bioactivity under in vivo conditions than expected based on receptor binding studies. Many studies revealed specific binding of APs or APEOs to ER in all classes of vertebrates (White et al., 1994; Guillette and Crain, 1996; Lutz and Kloas, 1999; Routledge et al., 2000), in which AP display affinities usually at least three orders of magnitude lower compared to the natural E_2. As a consequence, ER–AP complexes bind as dimers to estrogen-responsive elements on DNA, leading to estrogen-specific gene expression encoding several estrogenic biomarkers such as ER itself, vitellogenin, and zona radiata protein. In species displaying sex reversal by exogenously induced alterations of sex steroids or their synthetic analogs, APs can exert estrogenic effects during embryonic or larval development, leading to feminization in fishes (Gray and Metcalfe, 1997) and amphibians (Kloas et al., 1999).

Another target site for estrogenic compounds is the endocrine feedback mechanism of E_2 on the hypothalamus and pituitary (see chapters 3 and 6), which in turn

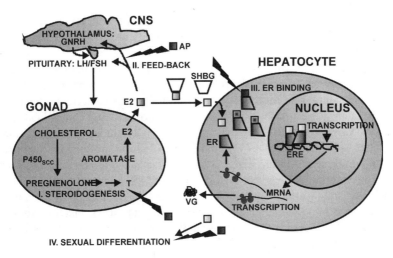

Figure 14-2. Hypothalamus–pituitary–gonad axis in vertebrates (amphibian) and potential target sites for endocrine disruption of alkylphenols (AP). AP interfere with I, steroidogenesis; II, feedback mechanisms on hypothalamus and pituitary; III, estrogen receptor (ER) binding; and IV, sexual differentiation. Besides the known nonspecific toxic effects of AP, which might be responsible for inhibitory action on I, steroidogenesis, AP cause mainly estrogenic actions on II, III, and IV by mimicking 17β-estradiol but do not bind to sexual hormone binding protein (see text for details). Abbreviations: E_2, 17β-estradiol; ERE, estrogen response element; FSH, follicle-stimulating hormone; GnRH, gonadotropin-releasing hormone; LH, luteinizing hormone; $P450_{SCC}$, side-chain cleavage enzyme; SHBG, sexual hormone binding protein; T, testosterone; Vg, vitellogenin.

affects circulating levels of the gonadotropins, LH (luteinizing hormone) and FSH (follicle-stimulating hormone). NP decreases LH and FSH in fish (Harris et al., 2002) and amphibians (Mosconi et al., 2002). Alterations in steroidogenic enzymes result in marked changes in sex steroids, affecting reproductive biology. Numerous attempts have been made to demonstrate effects of APs on sex steroid levels and steroidogenic enzymes in gonads. However, despite some inhibitory actions, especially on androgen-producing enzymes, the concentrations of AP needed to induce such effects are high (Laurenzana et al., 2002) and, at least for aquatic species, are in the range where acute toxicity occurs. Subtoxic concentrations of NP (5.4 µg/l) did not cause any effect on testosterone, E_2, and vitellogenin concentrations in mature male carp (Villeneuve et al., 2002).

Thus, it might be concluded that endocrine-disrupting effects of APs are mainly transmitted by their estrogenic effects via ER binding and estrogen-specific gene expression. Despite the low affinity of ER for AP, the fact that AP do not bind to SHBP may enhance the bioavailability of AP, causing greater biological significance in vivo. Estrogen receptor signaling by AP involves not only peripheral target organs, such as liver and gonads, but also the central nervous system via negative feedback mechanisms on hypothalamus and pituitary, leading to lower gonadotropin levels. Actions of AP via AR and PR seem to be negligible because the binding affinity is much lower in comparison to ER binding, and AP effects on steroidogenic enzymes seem to be associated to nonspecific toxicity rather than specific interference with the enzymes.

In general, it is impossible to clearly differentiate AP compounds as either EDCs or toxicants. In vivo, APs may act as EDCs at low concentrations and as a toxicant at higher concentrations. Moreover, endocrine and other toxic effects may overlap each other at the lowest-observed-effect concentration (NOEC), indicating that APs are specifically EDCs. Effects may vary substantially among groups of animals, and species-specific differences within the same class may account for the difficulties in determining if APs are EDCs or toxic surfactants. For ecotoxicological reasons, it does not matter whether AP actions start with endocrine-specific or with nonspecific narcotic effects because both are harmful to animals. However, for risk assessment it is important to get the most sensitive and earliest endpoint for detection of any adverse effect. The fluctuation of AP's effects on wildlife between endocrine disruption and other toxic effects is reviewed in the next section.

Effects on Wildlife

Most studies have focused on the toxicity of NP and OP, as these compounds are considered to be the most toxic of the AP compounds. Both NP and OP have similar acute toxicities, as shown for fishes (17–3000 µg/l), invertebrates (20–3000 µg/l), and algae (27–2500 µg/l). The NOEC for NP toxicity based on lethality in fishes and invertebrates has been reported as < 6 and 3.7 µg/l, respectively (Servos, 1999). The high toxicity of NP and OP might be due to the high bioconcentration factors (around 100–700 for NP in several fish species depending on exposure regime) and the fact that these compounds have biological half-lives of several hours to 2 days (Servos, 1999). It is noteworthy that acute toxicity levels in model organisms from different phyla such as fathead minnows (*Pimephales promelas*), waterfleas (*Daphnia magna*), and green algae (*Selenastrum capricornutum*) are nearly identical, with LC_{50} values for NP between 100 and 300 µg/l. Only green algae withstands a concentration of OP one order of magnitude higher than NP (Servos, 1999). Thus, toxic effects of APs play an important role in rivers that are highly polluted by APs (Spain, Switzerland, UK). In contrast to aquatic species, terrestrial plants are quite insensitive to NP at levels below 11 mg/l (Bokern and Harms, 1997). However, a relatively low concentration of NP (3.4 µg/g) in soil results in decreased reproductive rates in the earthworm *Apporectodea calignosa* (Krogh et al., 1996). In mammals, moderate toxic effects of APs on reproductive parameters in rats are reported only at high exposures of 2000 mg OP/kg (Tyl et al., 1999) or 50 mg NP/kg (Nagao et al., 2001). APs are much more harmful to aquatic species than they are to terrestrial mammals.

Most studies of the endocrine-disrupting effects of APs are restricted to vertebrates because our knowledge about endocrine systems in invertebrates is more limited. As mentioned above, endocrine disruption by AP is mainly triggered by ER-mediated estrogenic actions, which may be accompanied by toxic side effects involving actions on steroidogenic enzymes. The few studies that dealt with estrogenic effects in invertebrates show that invertebrates are much more sensitive to xenoestrogens. Freshwater and marine snails exposed to 1 µg/l OP became "super-females" and exhibited increased egg production (Oehlmann et al., 2000). The freshwater snail *Potamopyrgus antipodarum* produced significantly higher numbers of embryos after exposure to artificial sediments supplemented with OP and NP at 1 and 10 µg/kg, respectively, without any signs of toxic side effects (Duft et al., 2003). The cypris major protein (which is related to vitellin) is increased after exposure to NP concentrations as low as 1 µg/l in barnacle larvae (Billinghurst et al., 2000).

In vertebrates, the estrogenic effects of NP were found accidentally using estrogen sensitive MCF-7 cell lines (Soto et al., 1991). NP increased mitotic activity at a concentration of 1 µM (220 µg/l; Blom et al., 1998). Binding of AP and APEO to human ER revealed only low affinity compared to E_2 (Routledge and Sumpter, 1997). In mammals, OP and NP exhibit moderate estrogenic effects in various target tissues in vitro at relatively high concentrations that are close to levels that can cause toxic effects.

Aquatic vertebrates are more endangered by AP compounds. There have been many studies in fishes investigating potential estrogenic actions of AP using in vitro and in vivo approaches. Vitellogenin gene expression in trout hepatocytes was stimulated in vitro by OP and NP at 10^{-7} and 10^{-6} M (22 and 220 µg/l), respectively (White et al., 1994). AP had similar ER-binding affinities in trout (White et al., 1994) and in carp (Kloas et al., 2000), but was at least three orders of magnitude less efficient than E_2. The estrogenic biomarker vitellogenin was induced in vivo by 1–10 µg/l OP in trout and by 10–100 µg/l in roach fish (Routledge et al., 1998). Concentrations of NP must be somewhat higher, at about 10 µg/l, to increase vitellogenin induction (Jobling et al., 1995). NPEO administered at concentrations up to 7.9 µg/l affected neither vitellogenin nor sex steroid levels in both sexes of *Pimephales promelas*, demonstrating a lower estrogenic potency than NP (Nichols et al., 2001).

Exposure of rainbow trout to OP and NP reduced testicular growth in a concentration-dependent manner (Jobling et al., 1996), and similar effects were reported for NP2EO, which decreased gonadosomatic index and spermatogenesis (LeGac et al., 2001). NP hampered gonadal development and reproductive function in *Xiphophorus helleri* and *X. maculatus* at concentrations leading to increased mortalities (Magliulo et al., 2002). In the hermaphroditic teleost *Rivulus marmoratus*, NP treatment at 150 and 300 µg/l inhibited the development of testicular tissue but also inhibited oogenesis (Tanaka and Grizzle, 2002). It is still a matter of debate whether the effects of APs on testes are due to inhibition of androgen-synthesizing enzymes, negative feedback mechanisms acting on gonadotropins, nonspecific toxic effects, or complex interactions between these potential modes of action. In general, it seems reasonable that estrogenic and toxic effects of AP have some overlap, especially regarding interference with steroidogenic enzymes, but it must be kept in mind that chronic toxicity of NP for rainbow trout has a NOEC of only 6 µg/l, whereas acute toxicity tests resulted in an LC_{50} of around 200 µg/l (Servos, 1999).

Regarding sexual differentiation, 50% of male medaka exposed to 50 µg/l NP developed ovotestes. The incidence of ovotestes increased to 86% at the higher NP concentration of 100 µg/l and was accompanied by a shift in the sex ratio toward females (Gray and Metcalfe, 1997). Similar results were obtained for 50 µg/l OP, which induced feminization and to a lesser extent ovotestes in males (Knörr and Braunbeck, 2002). In male carps feminization of gonads has been reported for the pentylphenol at a concentration of 320 µg/l (Gimeno et al., 1996). In addition, the display of sexual behavior of male guppies could be suppressed by treatment with OP as well as with estradiol (Bayley et al., 1999).

In amphibians, APs had similar low ER-binding affinities as reported for other vertebrates (Lutz and Kloas, 1999). However, induction of vitellogenin-mRNA in vitro was stimulated by 10^{-8} M NP in *Xenopus laevis* hepatocytes (Kloas et al., 1999). In vivo exposure of male *Rana esculenta* to NP at a concentration of 10^{-10} M revealed significant elevations of vitellogenin in plasma (Mosconi et al., 2002). In addition, inhibitory effects of NP on the pituitary hormones LH, FSH, and prolactin have been shown at the same concentration. Sexual differentiation in *Xenopus laevis* was affected by exposure NP and OP during larval development. Both NP and AP caused significant feminization at 10^{-7} M (22 µg/l), whereas OP was effective at the lower concentration of 10^{-8} M (2.2 µg/l; Kloas et al., 1999). It is noteworthy that no toxic effects of NP and OP on development and performance of tadpoles were observed at the concentrations used. Amphibians seem to provide a more sensitive model for estrogenic effects of APs than do fishes, but toxicity data for endocrine disruption by APs in amphibian species are not yet available.

The impact of AP compounds on wildlife is characterized by two components—toxicity and endocrine disruption. It seems probable that some sentinel species might exist that would indicate endocrine disruption via estrogenic effects before any toxicity of APs were apparent. Snails and amphibians are probably the most sensitive models for assessing endocrine-disrupting effects of APs. In all other animal groups (e.g., in fishes) endocrine-disrupting and toxic activities of AP compounds seem to be tightly intertwined, defining APs as EDCs with remarkably toxic side effects.

Implications for Wildlife and Human Health

The fact that alkylphenolic compounds are preferentially toxic to aquatic organisms preceded the finding that they can act also as estrogenic EDCs. APEOs, APECs, and APs are abundant in the environment and hence are an important threat to wildlife and human health. The main sink of these compounds and their metabolites is surface waters, and thus aquatic species are endangered not only by their higher sensitivity to these substances but also by the higher exposure rates. However, AP pollution of surface waters varies greatly over the world. The most contaminated areas are rivers in Europe and Asia, which have concentrations of APs that could lead to acute toxicity in several sentinel species that might become locally extinct. In North America and Japan, only some hot spots might exist; the most abundant NP in rivers normally is found at concentrations below the threshold for adverse biological effects. However, it is obvious that AP compounds are toxic and can act as EDCs, which should lead to a phasing out of these chemicals and sustainable replacement by less toxic and endocrine-inert compounds wherever possible.

In Europe, the chemical industries voluntarily committed to phasing out APEOs in household cleaners in the late 1980s. APEOs have been replaced mainly by nonestrogenic alkyl ethoxylates, which also are more easily biodegraded. In Northern Europe, progress to phase out APEOs has occurred on a voluntary basis, resulting in a marked decrease of APs in the environment. However, it will take years to ban APEOs completely, as done by Norway in 2002. In the United States, APEOs in household detergents were voluntarily phased out but not in institutional and industrial applications. To date, no phase out comparable to that in Europe is planned for North America, perhaps due to the argument that AP contamination is much less than in Europe. Nevertheless, all North American rivers contain APs in measurable amounts, and alkylphenols contribute to the overall contamination and health risks through toxic and estrogenic actions. To eliminate any impact of AP compounds on wildlife and human health, a worldwide phase-out of APEOs is necessary as already started in Northern Europe.

References

Ahel, M., Giger, W., 1993. Partitioning of alkylphenols and alkylphenol polyethoxylates between water and organic solvents. Chemosphere 26, 1471–1478.

Ahel, M., Giger, W., and Koch, M., 1994a. Behaviour of alkylphenol polyethoxylate surfactants in the aquatic environment. I. Occurrence and transformation in sewage treatment. Water Res. 28, 1131–1142.

Ahel, M., Giger, W., Schaffner, C., 1994b. Behaviour of alkylphenol polyethoxylate surfactants in the aquatic environment: II. Occurrence and transformation in rivers. Water Res. 28, 1143–1152.

Banat, F.A., Prechtl, S., Bischof, F., 2000. Aerobic thermophilic treatment of sewage sludge contaminated with 4-nonylphenol. Chemosphere 41, 951–961.

Bayley, M., Nielsen, J.R., Baatrup, E., 1999. Guppy sexual behavior as an effect biomarker of estrogen mimics. Ecotoxicol. Environ. Safety 43, 68–73.

Bennett, E.R., Metcalfe, C.D., 2000. Distribution of degradation products of alkylphenol ethoxylates near sewage treatment plants in the Lower Great Lakes, North America. Environ. Toxicol. Chem. 19, 784–792.

Billinghurst, Z., Clare, C.S., Matsumura, K., Depledge, M.H., 2000. Induction of cypris major protein in barnacle larvae by exposure to 4-n-nonylphenol and 17β-oestradiol. Aquat. Toxicol. 47, 203–212.

Blackburn, M.A., Kirby, S.J., Waldock, M.J., 1999. Concentrations of alkylphenol polyethoxylates entering UK estuaries. Mar. Poll. Bull. 38, 109–118.

Blackburn, M.A., Waldock, M.J., 1995. Concentrations of alkylphenols in rivers and estuaries in England and Wales. Water Res. 29, 1623–1629.

Blom, A., Ekman, E., Johannisson, A., Norrgren, L., Pesonen, M., 1998. Effects of xenoestrogenic environmental pollutants on the proliferation of a human breast cancer cell line (MCF-7). Arch. Environ. Contam. Toxicol. 34, 306–310.

Bokern, M., Harms, H.H., 1997. Toxicity and metabolism of 4-n-nonylphenol in cell suspension cultures of different species. Environ. Sci. Technol. 31, 1849–1854.

Cardellini, P., Ometto, L., 2000. Teratogenic and toxic effects of alcohol ethoxylate and alcohol ethoxy sulfate surfactants on *Xenopus laevis* embryos and tadpoles. Ecotoxicol. Environ. Safety 25, 1–8.

Colborn, T., Clement, C., 1992. Chemically Induced Alterations in Sexual and Functional Development: The Wildlife/Human Connection. Princeton Scientific Publishing, Princeton, NJ.

Corti, A., Frassinetti, S., Vallini, G., D'Antone, S., Fichi, C., Solaro, R., 1995. Biodegradation of nonionic surfactants. I. Biotransformation of 4-(1-nonyl)phenol by a *Candida maltosa* isolate. Environ. Pollut. 90, 83–87.

Cserhati, T., 1995. Alkyl ethoxylated and alkylphenol ethoxylated nonionic surfactants: interaction with bioactive compounds and biological effects. Environ. Health Perspect. 103, 358–364.

Dachs, J., van Ry, D. A., Eisenreich, S.J., 1999. Occurence of estrogenic nonylphenols in the urban and coastal atmosphere of the lower Hudson river estuary. Environ. Sci. Technol. 33, 2676–2679.

Ding, W.H., Tzing, S.H., Lo, J.H., 1999. Occurrence and concentrations of aromatic surfactants and their degradation products in river waters of Taiwan. Chemosphere 38, 2597–2606.

Duft, M., Schulte-Oehlmann, U., Weltje, L., Tillmann, M., Oehlmann, J., 2003. Stimulated embryo production as a parameter of estrogenic exposure via sediments in the freshwater mudsnail *Potamopyrgus antipodarum.* Aquatic Toxicol. 64, 437–449.

Ekelund, R., Granmo, A., Magnusson, K., Berggren, M., Bergman, A., 1993. Biodegradation of 4-nonylphenol in seawater and sediment. Environ. Pollut. 79, 59–61.

Espadaler, I., Caixaxh, J., Om, J., Ventura, F., Cortina, M., Paune, F., Rivera, J., 1997. Identification of organic pollutants in Ter river and its system of reservoirs supplying water to Barcelona (Catalonia, Spain): A study by GC/MS and FAB/MS. Water Res. 31, 1996–2004.

European Commission, 1996. European workshop on the impact of endocrine disruptors on human health and wildlife. Weybridge, 2–4 December 1996. Report Eur 17549, Environment and Climate Research Programme, DG XII, European Commission, Brussels.

Fairchild, W.L., Swansburg, E.O., Arsenault, J.T., Brown, S.B., 1999. Does an association between pesticide use and subsequent declines in catch of Atlantic salmon (*Salmo salar*) represent a case of endocrine disruption? Environ. Health Perspect. 107, 349–358.

Fujii, K., Urano, N., Ushio, H., Satomi, M., Iida, H., Ushio-Sata, N., Kimura, S., 2000. Profile of a nonylphenol-degrading microflora and its potential for bioremedial applications. J. Biochem. (Tokyo) 128, 909–916.

Gies, A., 1996. Umweltbelastungen durch endokrin wirksame Stoffe. In: Bayerisches Landesamt für Wasserwirtschaft, Institut für Wasserforschung, München (Ed.), Stoffe mit endokriner Wirkung im Wasser, Oldenbourg, München, Wien, pp. 13–19.

Giger, W., Brunner, P.H., Schaffner, C., 1984. 4-Nonylphenol in sewage sludge: Accumulation of toxic metabolites from nonionic surfactants. Science 225, 623–625.

Gimeno, S., Gerritsen, A., Bowmer, T., 1996. Feminization of male carp. Nature 384, 221–222.

Gray, M.A., Metcalfe, C.D., 1997. Induction of testis-ova in Japanese medaka (*Oryzias latipes*) exposed to p-nonylphenol. Environ. Toxicol. Chem. 16, 1082–1088.

Guillette, L.J., Crain, D.A., 1996. Endocrine-disrupting contaminants and reproductive abnormalities in reptiles. Comm. Toxicol. 5, 381–399.

Hale, R.C., Smith, C.L., de Fur, P.O., Harvey, E., Bush, E.O., LaGuardia, M.J., Vadas, G.G., 2000. Nonylphenols in sediments and effluents associated with diverse wastewater outfalls. Environ. Toxicol. Chem. 19, 946–952.

Harris, C.A., Santos, E.M., Janbakhsh, A., Pottinger, T.G., Tyler, C.R., Sumpter, J.P., 2002. Nonylphenol affects gonadotropin levels in the pituitary gland and plasma of female rainbow trout. Environ. Sci. Technol. 35, 2909–2916.

Isobe, T., Nishiyama, H., Nakashima, A., Takada, H., 2001. Distribution and behavior of nonylphenol, octylphenol, and nonylphenol monoethoxylate in Tokyo metropolitan area: their association with aquatic particles and sedimentary distributions. Environ. Sci. Technol. 35, 1041–1049.

Jobling, S., Reynolds, T., White, R., Parker, M.G., Sumpter, J.P., 1995. A variety of environmentally persistent chemicals, including some phthalate plasticizers are weakly estrogenic. Environ. Health Perspect. 103, 582–587.

Jobling, S., Sheehan, D., Osborne, J.A., Matthiessen, P., Sumpter, J.P., 1996. Inhibition of testicular growth in rainbow trout (*Oncorhynchus mykiss*) exposed to estrogenic alkylphenolic chemicals. Environ. Toxicol. Chem. 15, 194–202.

John, D.M., White, G.F., 1998. Mechanisms for biotransformation of nonylphenol polyethoxylates to xenoestrogens in *Pseudomonas putida*. J. Bacteriol. 180, 4332–4338.

Kahl, M.D., Makynen, E.A., Kosian, P.A., Ankley, G.T., 1997. Toxicity of 4-nonylphenol in a life-cycle test with the midge *Chironomus tentans*. Ecotoxicol. Environ. Safety 38, 155–160.

Kiewiet, A.T., de Voogt, P., 1996. Chromatographic tools for analyzing and tracking nonionic surfactants in the aquatic environment. J. Chromatogr. 733A, 185–192.

Kloas, W., 2002. Amphibians as a model for the study of endocrine disruptors. Intl. Rev. Cytol. 216, 1–57.

Kloas, W., Einspanier, R., Lutz, I., 1999. Amphibians as model to study endocrine disruptors: II. Estrogenic activity of environmental chemicals *in vitro* and *in vivo*. Sci. Total Environ. 225, 59–68.

Kloas, W., Schrag, B., Ehnes, C., Segner, H., 2000. Binding of xenobiotics to hepatic estrogen receptor and plasma sex steroid binding protein in the teleost fish, the common carp (*Cyprinus carpio*). Gen. Comp. Endocrinol. 119, 287–299.

Knörr, S., Braunbeck, T., 2002. Decline in reproductive success, sex reversal, and developmental alterations in Japanese medaka (*Oryzias latipes*) after continuous exposue to octylphenol. Ecotoxicol. Environ. Safety 51, 187–196.

Krogh, P.H., Holmstrup, M., Jensen, J., Petersen, S.O., 1996. Okotoxikologisk vurdering af spildevandsslam i landbrugsjord [abstract in English]. Arbejdsrapport Nr. 43, 53 pp. Miljo-og Energiministeriet Miljostyrelsen, Danmark.

Laurenzana, E.M., Balasubramanian, G., Weis, C., Blaydes, B., Newbold, R.R., Delclos, K.B., 2002. Chem. Biol. Interact. 139, 23–41.

Laws, S.C., Carey, S.A., Kelce, W.R., 1995. Differential effects of environmental toxicants on steroid receptor binding. Toxicologist 15, 294.

LeGac, F., Thomas, J.L., Mourot, B., Loir, M., 2001. In vivo and in vitro effects of prochloraz and nonylphenol ehtoxylates on trout spermatogenesis. Aquat. Toxicol. 53, 187–200.

Lewis, M., 1991. Chronic and sublethal toxicities of surfactants to aquatic animals: A review and risk assessment. Water Res. 25, 101–113.

Lutz, I., Kloas, W., 1999. Amphibians as model to study endocrine disruptors: I. Environmental pollution and estrogen receptor binding. Sci. Total Environ. 225, 49–57.

Magliulo, L., Schreibman, M.P., Cepriano, J., Ling, J., 2002. Endocrine disruption caused by two common pollutants at "acceptable" concentrations. Neurotoxicol. Teratol. 24, 71–79.

Marcomini, A., Pojana, G., Sfriso, A., Quiroga Alonso, J.M., 2000. Behavior of anionic and nonionic surfactants and their persistent metabolites in the Venice lagoon, Italy. Environ. Toxicol. Chem. 19, 2000–2007.

McLeese, D.W., Zitko, V., Metcalfe, C.D., Seargant, D.B., 1980. Lethality of aminocarb and the components of the aminocarb formulation to juvenile Atlantic salmon, marine invertebrates and a freshwater clam. Chemosphere 9, 79–82.

McLeese, D.W., Zitko, V., Seargant, D.B., Burridge, L., Metcalfe, C.D., 1981. Lethality and accumulation of alkylphenols in aquatic fauna. Chemosphere 10, 723–730.

Milligan, S.R., Khan, O., Nash, M., 1998. Competitive binding of xenobiotic oestrogens to rat alpha-fetoprotein and to sex steroid binding proteins in human and rainbow trout (*Oncorhynchus mykiss*) plasma. Gen. Comp. Endocrinol. 112, 89–95.

Mosconi, G., Carnevali, O., Franzoni, M.F., Cottone, E., Lutz, I., Kloas, W., Yamamoto, K., Kikuyama, S., Polzonetti-Magni, A.M., 2002. Environmental estrogens and reproductive biology in amphibians. Gen. Comp. Endocrinol. 126, 125–129.

Nagao, T., Wada, K., Marumo, H., Yoshimura, S., Ono, H., 2001. Reproductive effects of nonylphenol in rats after gavage administration: a two-generation study. Reprod. Toxicol. 15, 293–315.

Naylor, C.G., Mieure, J.P., Adams, W.J., Weeks, J.A., Castaldi, F.J., Ogle, L.D., Romano, R.R., 1992. Alkylphenol ethoxylates in the environment. J. Am. Oil Chem. Soc. 69, 695–703.

Nichols, K., Snyder, E.M., Snyder, S.A., Pierens, S.L., Miles-Richardson, S.R., Giesy, J.P., 2001. Effects of nonylphenol ethoxylate exposure on reproductive output and bioindicators of environmental estrogen exposure in fathead minnows, *Pimephales promelas*. Environ. Toxicol. Chem. 20, 510–522.

Oehlmann, J., Schulte-Oehlamann, U., Tillmann, M., Markert, B., 2000. Effects of endocrine disruptors on prosobranch snails (Mollusca: Gastropoda) in the laboratory. Part I: Bisphenol A and octylphenol as xeno-estrogens. Ecotoxicology 9, 383–397.

Paris, F., Balaguer, P., Terouanne, B., Servant, N., Lacoste, C., Cravedi, J., Nicolas, J., Sultan, C., 2002. Phenylphenols, biphenols, bisphenol-A, and 4-tert-octylphenol exhibit a and β estrogen activities and antiandrogen activity in reporter cell lines. Mol. Cell. Endocrinol. 193, 43–49.

Renner, R., 1997. European bans on surfactant trigger transatlantic debate. Environ. Sci. Technol. 31, 316A–320A.

Routledge, E.J., Sheahan, D., Desbrow, C., Brighty, G.C., Waldock, M., Sumpter, J.P., 1998. Identification of estrogenic chemicals in STP effluents. 2. *In vivo* responses in trout and roach. Environ. Sci. Technol. 32, 1559–1565.

Routledge, E.J., Sumpter, J.P., 1997. Estrogenic activity of surfactants and some of their degradation products assessed using a recombinant yeast screen. Environ. Toxicol. Chem. 15, 241–248.

Routledge, E.J., White, R., Parker, M.G., Sumpter, J.P., 2000. Differential effects of xeno-estrogens on coactivator recruitment by estrogen receptor (ER) α and ERβ. J. Biol. Chem. 275, 35986–35993.

Servos, M.R., 1999. Review of the aquatic toxicity, estrogenic responses and bioaccumulation of alkylphenols and alkylphenol polythoxylates. Water Qual. Res. J. Canada 34, 123–177.

Sohoni, P., Sumpter, J.P., 1998. Several environmental oestrogens are also antiandrogens. J. Endocrinol. 158, 327–339.

Soto, A.M., Justicia, H., Wray, J.W., Sonnenschein, C., 1991. *p*-Nonylphenol: An estrogenic xenobiotic released from "modified" polystyrene. Environ. Health Perspect. 92, 167–173.

Swisher, R.D., 1987. Surfactant Biodegradation. Marcel Dekker, New York.

Talmage, S.S., 1994. Environmental and human safety of major surfactants: Alcohol-ethoxylates and alkylphenol ethoxylates. Lewis Publishers, Boca Raton, FL.

Tanaka, J.N., Grizzle, J.M., 2002. Effects of nonylphenol on the gonadal differentiation of the hermaphroditic fish, *Rivulus marmoratus*. Aquat. Toxicol. 57, 117–125.

Tanghe, T., Devriese, G., Verstraete, W., 1998. Nonylphenol degradation in lab scale activated sludge is temperature dependent. Water. Res. 32, 2889–2896.

Tanghe, T., Dhooge, W., Verstraete, W., 1999. Isolation of a bacterial strain able to degrade branched nonylphenol. Appl. Environ. Microbiol. 65, 746–751.

Topp, E., Starratt, A., 2000. Rapid mineralization of the endocrine-disrupting chemical 4-nonylphenol in soil. Environ. Toxicol. Chem. 19, 313–318.

Toppari, J., Larsen, J.C., Christiansen, P., Giwercman, A., Grandjean, P., Guillette, L.J., Jegou, B., Jensen, T.K., Jouannet, P., Kieding, N., Leffers, H., McLachlan, J.A., Meyer, O., Müller, J., Rajpert-De Meyts, E., Scheike, T., Sharpe, R., Sumpter, J.P., Skakkebaek, N.E., 1996. Male reproductive health and environmental xenoestrogens. Environ. Health Perspect. 104 (suppl. 4), 741–803.

Tran, D.Q., Klotz, D.M., Ladlie, B.L., Die, C.F., McLachlan, J.A., Arnold, S.F., 1996. Inhibition of progestrone receptor activity in yeast by synthetic chemicals. Biochem. Biophys. Res. Commun. 229, 518–523.

Tsuda, T., Takino, A., Kojima, M., Harada, H., Muraki, K., Tsuji, M., 2000. 4-Nonylphenols and 4-tert-octylphenol in water and fish from rivers flowing into Lake Biwa. Chemosphere 41, 757–762.

Tsutsumi, Y., Haneda, T., Nishida, T., 2001. Removal of estrogenic activities of bisphenol A and nonylphenol by oxidative enzymes from lignin-degrading basidiomycetes. Chemosphere 42, 271–276.

Tyl, R.W., Myers, C.B., Marr, M.C., Brine, D.R., Fail, P.A., Seely, J.C., Van Miller, J.P., 1999. Two-generation reproduction study with para-tert-octylphenol in rats. Regul. Toxicol. Pharmacol. 30, 81–95.

Tyler, C.R., Jobling, S., Sumpter, J.P., 1998. Endocrine disruption in wildlife: a critical review of the evidence. Crit. Rev. Toxicol. 28, 319–361.

Van Ry, D.A., Dachs, J., Gigliotti, C.L., Brunciak, P.A., Nelson, E.D., Eisenreich, S.J., 2000. Atmospheric seasonal trends and environmental fate of alkylphenols in the lower Hudson River estuary. Environ. Sci. Technol. 34, 2410–2417.

Villeneuve, D.L., Villalobos, S.A., Keith, T.L., Snyder, E.M., Fitzgerald, S.D., Giesy, J.P., 2002. Effects of waterborne exposure to 4-nonylphenol on plasma sex steroid and vitellogenin concentrations in sexually mature male carp (*Cyprinus carpio*). Chemosphere 47, 15–28.

Warhurst, A.M., 1995. An Environmental Assessment of Alkylphenol Ethoxylates and Alkylphenols. Friends of the Earth Scotland, Edinburgh, Scotland.

Wheeler, T.F., Heim, J.R., Latorre, M.R., Janes, A.B., 1997. Mass spectral characterization of p-nonylphenol isomers using high-resolution capillary GC-MS. J. Chromatogr. 35, 19–30.

White, R., Jobling, S., Hoare, S.A., Sumpter, J.P., Parker, M.G., 1994. Environmental persistent alkylphenolic compounds are estrogenic. Endocrinology 135, 175–182.

15

Endocrine-Active Phytochemicals: Environmental Signaling Context and Mechanisms

Alan M. Vajda
David O. Norris

Humans have long been familiar with the use of plants to modulate fertility, as is evident from the prevalence of contraceptives and abortificants in the *Materia Medica* of ancient Greek, Roman, Arabic, Ayurvedic, and Chinese physicians (see review by Riddle 1992). But it was not until the twentieth century, when livestock reproduction was impaired by certain pasture plants, that the active chemical constituents were identified and mechanistic studies demonstrated that phytochemicals could interact with vertebrate endocrine systems. In the early 1940s, sheep in Western Australia experienced an epidemic of infertility that progressed for 5 years until entire populations became sterile. The myriad reproductive problems exhibited by these sheep eventually were attributed to chemical compounds produced by an introduced European clover, *Trifolium subterraneum*, upon which the sheep grazed (Bennetts et al., 1946; Adams, 1998). These active plant compounds were given the name of "phytoestrogen" for their ability to mimic the biological effects of naturally occurring estrogenic vertebrate hormones (see review by Colborn et al., 1996; Gehm et al., 1997; Mitchell et al., 1998). Phytoestrogens are produced by at least 300 plant species representing at least 16 families (see Colborn et al., 1996). Many of these plants display seasonality in phytoestrogen production that is often coordinated with environmental stressors (e.g., drought, response to injury by grazing animals) and reproductive cycles of herbivorous species. For example, phytoestrogens in dry, summertime grasses reduce the number of offspring in wild populations of California quail and deer mice (Leopold et al., 1976).

Many species within the plant kingdom (seed plants, conifers, ferns, mosses, and algae) produce chemicals that can modulate the regulation of the vertebrate endocrine system, and hence the expression of vertebrate life histories, through actions on steroid hormone biosynthesis, transport, receptor action, and elimination. Modulation of hormonal functions may have effects on sexual maturation, gamete production and transport, sexual behavior, fertility, gestation, lactation, or modification of other traits that depend on endocrine system integrity. Although most research has focused on the phytoestrogens, modulation of androgenic, progestogenic, thyroidal, and corticosteroidal signaling pathways also have been reported (Rosenberg et al., 1998; Mesiano et al., 1999; Doerge and Sheehan, 2002). Consequently, we use a more inclusive term, endocrine-active phytochemicals (EAPs) to designate these plant compounds.

Diversity and Evolution of EAP Signaling

EAP Diversity

Bradbury and White (1954) listed 53 plants that have sufficient estrogenic activity to initiate estrus in animals. Within 20 years, this list was expanded to more than 300 plants (Farnsworth et al., 1975), and the number of phytochemicals identified as having endocrine activity is rapidly expanding. Although estradiol (E_2), estrone, pregnenolone, progesterone, and testosterone have been detected in some plant species (Grunwald, 1980; Agarwal, 1993; Milanesi et al., 2001), we focus here on five classes of nonsteroidal EAPs including flavonoids (e.g., genistein, coumestrol), lignans (e.g., enterolactone, enterodiol, matairesinol, secoisolariciresinol), phytosterols (e.g., β-sitosterol), terpenoids (e.g., tschimganidine), and the endocrine-active metabolites of fungi (mycotoxins such as zearalenone) and bacteria (e.g., androstenedione; see figure 15-1). At least some members of all of these structural classes of compounds elicit one or more endocrine responses in in vitro or in vivo assays (Lerner et al., 1963; Martin et al., 1978; Farmakalidis et al., 1985; Markaverich et al., 1988; Miksicek, 1993; Sathyamoorthy et al., 1994; Medlock et al., 1995; Scarlata and Miksicek, 1995; Ikeda et al., 2002).

The best characterized EAPs are the flavonoid isoflavones. Flavonoids are a class of approximately 4000 structurally related phenolic phytochemicals (Mann, 1978; Mazur and Adlercreutz, 1998) that include isoflavones (e.g., daidzein, genistein, biochanin A), isoflavanones (e.g., O-desmethylangolensin), coumestans (e.g., coumestrol), hydroxychalcones (e.g., 4,4'-dihydroxychalcone), flavones (e.g., flavone, apigenin, chrysin), flavonols (e.g., quercetin, kaemperfol), and flavonones (e.g., naringenin, 4',7-dihydroxyflavone; Whitten and Patisaul, 2001). Isoflavones differ structurally from other classes of flavonoids in having the phenyl ring attached at the 3 rather than the 2 position of the heterocyclic ring (see figure 15-1). They also differ from other flavonoids by their greater structural variation and the greater frequency of isoprenoid substitution (Mazur and Adlercreutz, 1998). Isoflavones are more restricted in the plant kingdom than other flavonoids. Although isoflavones have been reported from several classes of flowering plants (Compositae, Iridaceae, Myristicaceae, and Rosaceae), they are found regularly in only one subfamily of the Leguminosae, the Papilionoideae (Mazur and Adlercreutz, 1998).

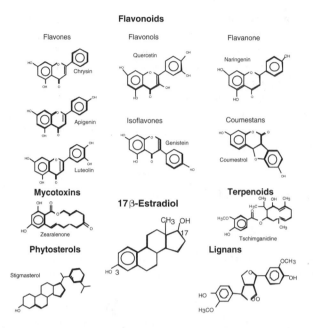

Figure 15-1. Structure of 17β-estradiol and representative endocrine-active phytochemicals.

EAP Evolution

Flavonoids are biosynthetically derived from the union of aromatic (hydroxycinnamyl coenzyme A ester) and aliphatic (malonyl coenzyme A) precursors. Flavonoids occur naturally in all plant families, although they have not yet been identified from the bryophyte hornworts (*Anthocerotae*) (Stafford, 2000). The primitive flavone apigenin is present in blue-green algae (cyanobacteria) and suggests a flavonoid origin at least 500 million years before present (figure 15-2; reviewed in Stafford, 2000). The introduction of lignans accompanied the course of lignification (the polymerization of cinnamyl alcohols) as plants stiffened internal and outer cells of stems for upright growth, some 420 million years ago. Dramatic diversification of flavonoids accompanied the rise of the angiosperms as flavonoids were exploited as compounds for color, pigmentation, and protection (Stafford, 2000).

The most important activities of flavonoids are dependent on the organisms in which they are present and also are related to flavonoid structure (Mazur and Adlercreutz, 1998). The actions of EAPs in vertebrate physiology are often not relevant to their function(s) in the producing organisms. As typical phenolic compounds, the flavonoids act as potent antioxidants (Gyorgy et al., 1964). As conjugated aromatic compounds, they can act as both potent screens against destructive UV light and as attenuators of visible light (Ryan et al., 2002). Individual members of this group are thought to serve as natural fungicides (phytoalexins: coumestrol, glyceollin), mediators of plant–microbe interactions, regulators of plant hormones, UV protectants, and precursors for pigment biosynthesis (Scandalios, 1990; Boue et al., 2000; Arts et al., 2002).

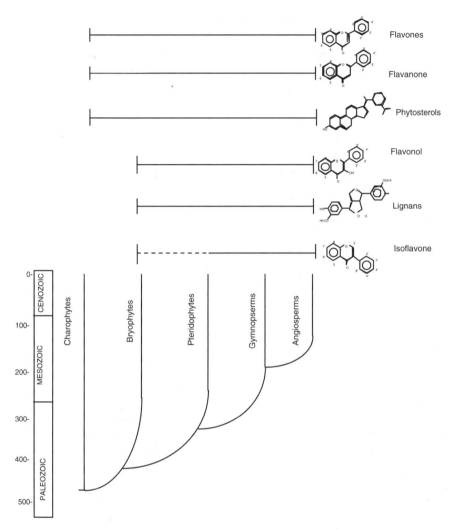

Figure 15-2. Phylogenetlc distribution of major flavonoids, phytosterols, and lignans. Flavonoids and other endocrine-active phytochemicals appeared early in the evolution of plants. Dramatic diversification of flavonoids accompanied the rise of the angiosperms as flavonoids were exploited as compounds for color, pigmentation, and protection. Data from Stafford (2000) and Mazur and Adlercreutz (1998).

EAP Signaling Context

Unlike the consequences of exposure to anthropogenic xenobiotics (Guillette et al., 1995; McLachlan, 2001), adverse outcomes of exposure to EAPs are not well known from field studies of interactions between free-living populations and their natural plant food sources (see, e.g., Leopold et al., 1976; Berger et al., 1981). The environmental and physiological persistence of naturally occurring EAPs is transient, and encounters with EAPs reflect the prevalence of environmental factors responsible for EAP induction in plants. In some adapted, coevolved systems, EAPs communicate meaningful information by providing important proximate environmental signals mediating adaptive responses in the timing and degree of reproductive expenditure in herbivores (Leopold et al., 1976; Berger et al., 1977, 1981). In contrast, commercial processing of plant material can result in the concentration of EAPs and cause persistent exposure in an evolutionarily novel signaling context to animal populations, including strictly carnivorous species (Setchell et al., 1987).

EAP Signal Content: Regulation of Plant EAP Expression

Many EAPs evolved as signal molecules between both leguminous and non-leguminous plants and nitrogen-fixing bacteria (Peters and Long, 1988; Koes et al., 1994) and fungi (Simon et al., 1993). Among these EAPs are compounds with potent estrogenic activity in vertebrates such as genistein, coumestrol, luteolin, and chrysin (Fox et al., 2001). Released from plant roots in response to low nitrogen conditions (Coronado et al., 1995), these EAPs act as chemoattractants and recruit nitrogen-fixing bacterial symbionts known collectively as rhizobium. An individual plant may exude a diverse suite of compounds that can synergistically induce bacterial nodulation (*nod*) gene expression (Srivastava et al., 1999), such as NodD proteins (Mulligan and Long, 1985). Nitrogen fixation is regulated by both agonistic and antagonistic phytochemical signaling to rhizobium (Peters et al., 1986).

Bacterial signal recognition is host specific; although some bacterial strains may establish symbioses with many different plant species (Perret et al., 2000), bacterial symbioses are established only through host-specific EAP expression (Srivastava et al., 1999). The particular structure of an EAP determines whether *nod* gene interactions are agonistic or antagonistic (Djordjevic and Weinman, 1991). While some EAPs, such as luteolin and apigenin (Peters et al., 1986; Djordjevic et al., 1987), stimulate NodD activation, others, such as chrysin, coumestrol, and genistein are potent NodD signaling inhibitors (Firmin, 1986; Peters and Long, 1988). Other flavonoids, such as quercetin, kaempferol, and biochanin A (Becard et al., 1992; Simon et al., 1993; Poulin et al., 1997), stimulate mutualistic arbuscular mycorrhizal fungi in an evolutionarily ancient relationship (Rerny et al., 1994). In anthropogenic contexts, as in the induction of coumestrol in alfalfa irrigated with sewage water (Shore et al., 1995), the environmental context of encounters with EAPs is similar to that of estrogenic pesticides (figure 15-3).

A possible homology has been identified between NodD and the estrogen receptor-α (ERα) (Gyorgypal and Kondorosi, 1991), and many flavonoid EAPs have been identified as NodD ligands. Two regions of NodD1 share 45% and 35% amino acid homology with two regions in the ligand-binding domain of the mammalian ERβ (Gyorgypal and Kondorosi, 1991).

Extracts from some dioecious male and female plants, including the osage-orange (*Maclura pomifera*) and mulberry (*Morus microphylla*), have differential estrogenic activity, with increased estrogenicity in extracts of female plants (Maier et al., 1995). Although the identity of the estrogenic EAP is unknown, yeast ex-

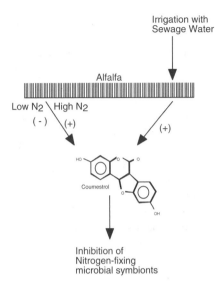

·Figure 15-3. Endocrine-active phytochemical expression in anthropogenic signaling context. In alfalfa, the flavonoid coumestrol communicates antagonistically with bacterial nodulation (NodD) proteins (Firmin et al., 1986; Peters and Long, 1988). Induced by high nitrogen levels, coumestrol inhibits nitrogen-fixation by bacterial symbionts. In alfalfa fields irrigated by sewage water, coumestrol induction is chronic and no longer reflects the proximate environmental availability of nitrogen (Shore et al., 1995).

pressing estrogen receptor assays demonstrated that EAP activity in both male and female plants varies with developmental stages and is highest before and during flowering and during the formation of new buds for the following year (Maier et al., 1997).

EAP Signaling in Adapted Vertebrate Herbivores

Many explanations for the expression of EAPs implicate these compounds as evolved defenses against vertebrate herbivores (Hughes, 1988; Wynne-Edwards, 2001). However, investigations of the interaction of free-living populations with EAPs in their natural plant food resources suggests that in adapted, coevolved systems, EAPs are interpreted as meaningful cues that yield adaptive modulation of vertebrate life histories (Leopold et al., 1976; Berger et al., 1977). Animals that are able to interpret these signals and modulate their reproductive efforts accordingly may ensure that the timing and expenditure of reproduction correspond optimally to available food resources.

EAP's may provide reliable cues that regulate the timing of both the initiation and termination of reproduction (Berger et al., 1977, 1981, 1987; Negus and Berger, 1977; Sanders et al., 1981). In free-living populations of the mountain vole, *Microtus montanus*, reproduction is stimulated by an EAP, 6-methoxybenzoxazolinone (6-MBOA), which is seasonally produced during the early growth cycle of their natural plant food resource (Negus and Berger, 1977; Sanders et al., 1981). 6-MBOA can stimulate breeding among nonbreeding populations (Negus and Berger 1977; Berger et al., 1981) and enhance reproduction in breeding animals (Berger et al., 1987). Reproductive effort is terminated by a different suite of compounds present in the latter stages of the plant growth cycle (Berger et al., 1977). Cinnamic acids and related vinylphenols expressed in drying grasses can inhibit follicular development, decrease uterine weight, and result in a cessation of breeding activity in *M. montanus* (Berger et al., 1977). Thus, interpretation of dietary EAPs as signals communicating either the commencement or termination of a high-quality food supply ensures adaptive modulation of the timing of reproduction to coincide with plant food availability for developing offspring (Berger et al., 1977).

Leopold et al. (1976) investigated the role that seasonal patterns in food plant EAP expression may play in mediating seasonal reproduction in California quail. An inverse correlation between EAP expression and quail reproductive success was observed. Plant expression of the flavonoid EAPs genistein, formonentin, and biochanin A was enhanced during a dry year when plants were stunted by inadequate rainfall. Quail breeding success was low, as was plant seed production. During a wet year when plant growth and seed production was abundant, EAP expression was low and quail breeding success was high. In California quail, dietary EAP content may function as a proximate environmental cue that communicates information about prospective food availability (Leopold et al., 1976).

EAPs as Defenses against Insect Herbivory

If EAPs represent an adaptation to limit herbivory, the likely targets of such chemical defenses are insect rather than vertebrate herbivores. Plants and insects have coevolved for much longer than have plants and mammals, and many EAPs that interact with vertebrate endocrine systems also interact with insect endocrine systems. Not surprisingly, there are many more classes of phytochemicals that interact with hormone systems of herbivorous insects, including the brassinosteroids and cucurbitacins (Dinan et al., 1997; Khripach et al., 1999; Oberdorster et al., 2001). Some endocrine-active flavonoids can modulate insect growth and metamorphosis through interactions with the ecdysteroid receptor (EcR), a member of the nuclear receptor superfamily (Oberdorster et al., 2001). Although none of the flavonoids is known to agonize the EcR independently, some flavones can significantly inhibit EcR-dependent gene transcription. Quercetin and coumestrol act as mixed agonists/antagonists, and genistein, tomatine, and apigenin act synergistically with ecdysteroid to reduce cell growth (Oberdorster et al., 2001).

EAPs as EDCs in Wildlife

Adverse outcomes of EAP exposure among wildlife are not well known from adapted systems. Most cases-studies are from agricultural (clover disease in livestock) and industrial (pulp-mill effluent and fish) systems where exposure to phytochemicals occurs within an evolutionarily novel signaling context.

The vulnerability of vertebrate endocrine systems to modulation by EAPs provides a context for understanding endocrine modulation by anthropogenic pollutants (McLachlan, 2001). Anthropogenic contaminants are messenger molecules of industrial civilization; their interpretation by evolved signaling networks in other organisms most often yields pathological outcomes that can contribute to population decline. Endocrine-modulation by anthropogenic xenobiotic endocrine-disrupting chemicals (EDCs) differs from that of EAPs in several fundamental ways, including signal persistence, context, pathway, and outcome (Berger et al., 1981; Genoni and Montague, 1995; Cunningham et al., 1997; Genoni, 1997; Diel et al., 2000). EDCs not only interfere with signal events within an organism but can corrupt the integrity of signaling networks evolved to communicate between organisms; i.e., pheromones and phytoallexins (Fox et al., 2001). Like xenoestrogenic EDCs (e.g. DDT), in many cases EAPs may be inadvertent estrogens, chemicals whose synthetic rationale is unrelated to their incidental (ultimately determined) hormonal activity (McLachlan, 2001).

The classic example of adverse outcomes of EAP exposure in animals is the economically important occurrence of clover disease among sheep and other livestock (Bennets et al., 1946; reviewed in Adams, 1998). Clover disease is characterized by the clinical manifestation of symptoms such as impaired fertility, uterine prolapse, and inappropriate mammary development and lactation in both unmated ewes and castrated males. Attributed to high expression of the isoflavones luteolin, genistein, and biochanin A in subterranean clover (*Trifolium subterraneum* L.), clover disease has its highest incidence in western Australia. Although sheep have been pastured on clover in Europe and Asia for centuries, sheep and clover were introduced to Australia only recently. The elevated expression of clover isoflavones in western Australia may reflect the low nitrogen content of the endemic nutrient-depleted soils (Adams, 1998). Clover isoflavonoid expression varies seasonally with plant growth cycles and can be induced in low-nitrogen conditions to recruit and mediate symbioses with nitrogen-fixing bacteria (Coronado et al., 1995). In this situation, signals developed for one communication system may be functionally misinterpreted by another system, resulting in pathological outcomes.

Fish living downstream from paper and pulp mills are exposed to endocrine-modulating ligands for the aryl hydrocarbon receptor (AhR), androgen receptor (AR), and ERs (Hewitt et al., 2000; Parks et al., 2001). There is evidence that pulp mill effluent can alter sexual differentiation (Howell et al., 1980; Larsson et al., 2000; Parks et al., 2001) and other physiological processes (Denton et al., 1985; Krotzer, 1990; Lehtinen et al., 1999), especially in sedentary fish (McMaster et al., 1992). The functional expressions of all fish life-stages are vulnerable to modulation by such wood-derived EAPs.

Trees processed for paper and pulp contain varying amounts of phytosterols, including β-sitosterol, which can affect fish egg size and mortality (Lehtinen et al., 1999; Mattson et al., 2001). β-Sitosterol also demonstrates estrogenlike activity for some estrogen-dependent endpoints in fish and breast cancer cell lines (Mellanen et al., 1996). Laboratory studies show that phytosterols found in pulp-mill effluent can affect embryological development of larvae before hatching, as well as the levels of circulating sex hormones of the parent fish (MacLatchy et al., 1997; Mattsson et al., 2001). Some of the endocrine-disrupting consequences of β-sitosterol in fish may be due to inhibition of steroidogenesis, either through inhibition of cholesterol uptake or impairment of the P450 cholesterol side-chain cleavage enzyme (MacLatchy et al., 1997). However, not all of the endocrine-disrupting consequences of exposure to pulp-mill effluent can be attributed solely to phytosterols (Dube and MacLatchy, 2001).

The androgenicity of pulp-mill effluent is due primarily to the bacterial metabolism of the natural plant sterol stigmasterol (Denton et al., 1985). Although stigmasterol has no intrinsic endocrine activity, it is efficiently metabolized to the androgenic steroid androstenedione by the bacterium *Mycobacterium smegmatis*, which forms extensive colonies at the effluent sites of pulp and paper mills. Female mosquitofish exposed to these androgens exhibit male secondary sex characteristics, such as a male gonapodium and anal fin morphology, as well as male sexual behavior (Howell et al., 1980; Denton et al., 1985; Davis and Bortone, 1992; Parks et al., 2001).

Upstream from pulp-processing factories, seasonal patterns of wood-derived EAP introduction into aquatic ecosystems may be miniscule (Parks et al., 2001) and presumably parallel terrestrial runoff from seasonal rainfall. Downstream, such runoff may dilute the chronic discharge of endocrine-active effluent (figure 15-4). Furthermore, any signal associated with terrestrial plant productivity that may be available for interpretation by fish endocrine systems may be concealed by the potent androgenic signal associated with bacterial metabolic preference. Chronic exposure of aquatic populations to phytosterols and their bacterial metabolites thus reflects a simplified, nonsense anthropogenic signaling context rather than the introduction of reliable and relevant environmental cues. Similar out-of-context adverse effects of EAPs have been reported in aquaculture (Markiewicz et al., 1993), laboratories (Thigpen et al., 1987, 1999), and zoos (Setchell et al., 1987), where animals were fed novel, simplified diets overreliant on industrially produced EAP-rich soy protein.

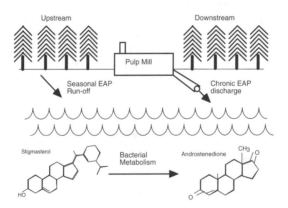

Figure 15-4. Proposed endocrine-active phytochemical (EAP) signaling context upstream and downstream from pulp-mill effluent. Upstream from pulp mills, fish in rivers are presumably exposed to terrestrial EAPs in a signaling context that reflects plant EAP production and seasonal patterns of rainfall and runoff. Downsteam from pulp mills, introduction of EAPs through effluent discharge is chronic and reflects rate of industrial production. Additionally, bacterial colonies at the site of effluent discharge metabolize non–endocrine-active phytochemicals such as stigmasterol into endocrine-active compounds such as androstenedione (via pregnenolone and DHEA) (Denton et.al., 1985).

Response of Vertebrate Endocrine Systems to EAP Signaling

Signaling through estrogen- and androgenlike pathways is an evolutionarily ancient means of communication both within and between organisms. EAPs can potentially interact with any protein-mediated interaction regulating the synthesis, availability, action, and disposition of steroid hormones. In vertebrates, individual EAPs exert endocrine-modulating action at multiple loci, including interactions with steroid-metabolizing enzymes, serum-binding proteins, and steroid receptors. Increasingly complex and possibly synergistic responses might be expected from natural dietary sources of EAPs; the diversity of EAPs encountered may vary seasonally and with environmental conditions and may increase with the diversity of the diet. EAP action at the cellular and molecular level is influenced by many factors, including concentration dependency, receptor status, presence or absence of endogenous steroids, and the type or target organ or cell.

The activity of EAPs exhibits specificity at many levels. Differential responsiveness has been observed between different species (Matthews et al., 2001) and extends to interindividual differences within a species and between different tissues within individuals (Whitten and Pautisaul, 2001). Some of the steroidlike effects of EAPs are due to their structural similarity with endogenous steroidal estrogens; others may have synergistic effects with hormone action independent of hormone-mimicking activity (Makela et al., 1995b; Shen and Weber, 1997).

Considerable evidence demonstrates that the molecular mechanisms involved in the action of steroids and steroidlike compounds are more complex than initially believed. The discovery of mechanisms such as cross-talk (Gottlicher et al., 1998), protein interactions between estrogen receptors and transcription factors (Routledge et al., 2000), novel ER isoforms, such as ERβ (Kuiper et al., 1996), and nongenomic pathways of steroid action (Wehling, 1997; Nadal et al., 2000) must be considered important modulators of steroid activity and candidate targets for dietary EAPs. EAPs are not known to bioaccumulate within an individual, but dosages that are subthreshold for acute responses may exert endocrine effects on target tissues during chronic exposure (Whitten et al., 1992; Whitten and Naftolin, 2001).

Estimates of endocrine potency vary across bioassays and between species (Whitten and Patisaul, 2001). Most EAPs display their hormonal activity over a concentration range of 0.01–10 μM (Miksicek, 1993, 1994, 1995; Nikov et al., 2000) and the doses reported to be biologically active in humans (0.4–10 mg/kg body weight/day) are lower than the doses generally reported to be active in rodents (10–100 mg/kg body weight/day; Whitten and Patisaul, 2001).

Nuclear Receptors and EAP Signaling

Many of the actions of EAPs in vertebrates are via interactions with nuclear receptor (NR) signaling pathways. NR signaling pathways are modulated by EAPs either through interactions with receptors as for the ER, androgen receptor (AR), progesterone receptor (PR), or the EcR, or by altering the biosynthesis of endogenous NR, as is known for the mineralocorticoids, thyroid hormones, estrogens, and androgens.

The best characterized actions of EAPs are their interaction with estrogen signaling pathways. Estrogens are associated with reproduction in many different organisms. Steroidal estrogens regulate reproduction in some plants (Geuns, 1978; Grunwald, 1980), many invertebrates (Atkinson and Atkinson, 1992; Baldwin and LeBlanc, 1994; Tarrant et al., 1999; LeBlanc and McLachlan, 2000), and all vertebrates. Cell cultures of the waxy-leaf nightshade (Solanaceae: *Solanum glaucophyllum*) were found to contain not only the steroidal estrogens E_2 and estrone but also abundant estrogen-binding sites (Milanesi et al., 2001). These estrogen-binding proteins were highly specific for E_2 and are structurally related to the mammalian ER. The concentration of estrogen-binding sites recovered in soluble fractions from the plant was comparable, or even higher, to that of ER-enriched mammalian target tissues (Milanesi et al., 2001; Monje and Boland, 2001). Nuclear receptors have yet to be identified in plants, and the actions of plant steroids (brassinosteroids) are mediated primarily by membrane-bound receptors (Mussig and Altmann, 2001; Marcinkowska and Wiedlocha, 2002).

Many of the effects of estrogens are mediated through binding to the ER, which then modulates estrogen-responsive gene expression. In vertebrates, the ER exists as at least two isoforms, ERα and ERβ, that differ in their tissue distribution and ligand preference (Mosselman et al., 1996; Kuiper et al., 1997). Kuiper et al. (1997) characterized the distribution of ERα and ERβ in tissue from 6- to 8-week-old male rats and 8-week-old female rats. The highest ERα expression was in the epididymis, testis, pituitary, uterus, kidney, and adrenal, which all showed moderate or no expression of ERβ. ERβ expression was greatest in the ovary and prostate, with the testis, uterus, bladder, and lung showing moderate expression. Less ERβ expression was detected in the pituitary, epididymis, thymus, brain sections, and spinal cord. In rat ovarian cells, ERβ is expressed in granulosa cells, and ERα is expressed in luteal tissue (Trembly et al., 1997; Telleria et al., 1998; Sharma et al., 1999). In the testes of humans and other primates, ERβ is found in Sertoli cells, Leydig cells, peritubular myoid cells, and in epithelial and stromal cell nuclei throughout the male reproductive system (Saunders et al., 2001). ERα is not found in the testes but does occur in the nonciliated epithelial cells of the efferent ductules (Saunders et al., 2001).

The ERβ has been sequenced, and its tissue distribution has been characterized in mammals (Kuiper et al., 1996; Vermeirsch et al., 1999; Martin de las Mulas et al., 2000; Muramatsu and Inoue, 2000; van Erdenburg et al., 2000; Zieba et al., 2000), birds (Bernard et al., 1999, Foidart et al., 1999), amphibians (Arenas et al., 2001), and fishes (Pakdel et al., 1989, Hawkins et al., 2000). Additionally, in rats, several splice variants of ERβ have been identified (Chu and Fuller 1997), adding further complexity to regulation of estrogen activity.

Diversification of ERα and ERβ occurred through gene duplication among the jawed vertebrate lineage, following the lamprey–gnathostome divergence (Thornton, 2001). These ERs share 95% homology in the DNA-binding region but share only 55% homology in the primary sequences of their ligand-binding domain/activation function-2 (AF-2) regions (Kuiper et al., 1997). This causes various environmental estrogens to exhibit different binding affinities and different agonist/antagonist characteristics for the two receptors. Although ERβ has not yet been identified in reptiles, its occurrence in bony fishes, amphibians, birds, and mammals suggests that future analyses will reveal this subtype in reptiles as well.

Nuclear Receptor Binding

Estrogen receptor binding is the best documented action of EAPs. The ER has been shown to bind several structurally diverse classes of chemicals, and this property appears to be unique among nuclear receptors (Kuiper et al., 1998). The promiscuity of the ER has been partially attributed to the size of the ligand-binding pocket, which is almost twice the volume of E_2 (Brzozowski et al., 1997; Elsby et al., 2000). Phylogenetic sequence analysis suggests that the ER is the most ancient of the steroid receptors (Thornton, 2001); its apparent ligand promiscuity may reflect an evolutionary constraint imposed by its contribution to fitness and high connectivity to other signaling systems (Fraser et al., 2002).

Estrogenic potency of individual ligands can vary between ER subtypes; E_2 has five times higher affinity for ERα than for ERβ (Kuiper et al., 1997, 1998). Structural requirements for ER-mediated flavonoid estrogenicity include the common diaryl ring and a minimum of one hydroxyl substituent on each aromatic ring (Miksicek, 1995). As with xenoestrogens (Mueller and Kim, 1978; Elsby et al., 2000), there appears to be a strong consensus in the optimal pattern of hydroxylation that gives rise to EAP estrogenic activity (Miksicek, 1995). This pattern of hydroxylation, however, is specific for each target protein (Kellis and Vickery, 1984; Makela et al., 1995b; Miksicek, 1995; Kao et al., 1998), and individual EAPs are not known to interact with all candidate loci.

Nuclear Receptor Binding Affinity

Environmental estrogenic pollutants such as polychlorinated biphenyls (PCBs), 1,1,1-trichloro-2,2-bis(*p*-chlorophenyl)ethane (DDT), alkylphenols, bisphenol A, methoxychlor, and chlordecone compete with E_2 for binding to both ER subtypes with a similar preference and degree, while other xenoestrogens such as 4-nonylphenol and 4-octylphenol, preferentially bind to ERα (Gutendorf and Westendorf, 2001). EAPs generally have a strong preference for ERβ (Kuiper et al., 1998; Nikov et al., 2000). For example, genistein preferentially binds with 30-fold greater affinity to human ERβ than ERα (Kuiper et al., 1998), and its ability to form estrogen response element complexes with human ERβ may equal that of E_2 (Nikov et al., 2000). The flavonoid coumestrol has higher relative binding to rat ERβ than does E_2 (Kuiper et al., 1997, 1998). Although coumestrol preferentially binds ERβ over ERα, studies of ERα knockout mice reveal that at least some of the estrogenic and antiestrogenic effects of coumestrol are mediated through ERα (Jacob et al., 2001).

It is interesting that diversification of the ERs contributed to the elaboration of a second ER (ERβ) that is more susceptible to modulation by exogenous EAPs. Differences in tissue distribution and ligand specificity of ERα and ERβ may mediate the relative regulatory contribution of endogenous and exogenous (e.g., dietary) estrogenic compounds.

Using bacterially expressed ER fusion proteins, Matthews et al. (2000) demonstrated species differences in ligand preferences and binding affinities for EAPs. In general, ER fusion proteins with greater sequence similarity exhibited similar relative binding affinities. The ER fusion protein from rainbow trout had the most divergent amino acid sequence, exhibited the greatest promiscuity in its ligand preference, and bound a significantly greater number of structurally diverse estrogenic compounds than human ERα, mouse ERα, chicken ER, or green anole ER. In general, EAPs had higher relative binding affinities across species than did the EDCs bisphenol A, o,p'-DDT, pthalates, and methoxychlor. The mycotoxin α-zearalenol bound rainbow trout ER fusion proteins with 2.6-fold greater affinity than E_2. Binding studies using purified receptor proteins show that some flavonoid EAPs are high-affinity ligands for ERs, especially ERβ. In whole-cell assays, however, their activity is significantly lower, probably as a result of interactions with binding proteins and other as-yet unidentified factors (reviewed in Whitten and Patisaul, 2001).

EAPs also can alter signaling through other hormonal pathways. Progestational activity of the flavonoid, apigenin, and the cinnamic acids (naringenin and syringic acid) was identified in cell cultures expressing androgen and progesterone receptors (breast cancer, T-47D and BT-474) (Rosenberg et al., 1998). Antiprogestogenic and antiandrogenic activities of 11 other EAPs, especially genistein (Zand et al., 2000), β-carotene, chlorophylline, α-tocopherol, and homocysteine, also have been reported (Rosenberg et al., 1998).

Whether an EAP acts as an agonist or an antagonist may depend on dosage. In cell proliferation assays, isoflavonoid concentrations of 1–100 nM stimulate, whereas doses of 5–100 μM suppress, proliferation (Miodini et al., 1999; Whitten and Patisaul, 2001). One proposed mechanism for the mixed agonist/antagonist phenomenon is through differential regulation of ER or PR expression (Katzenellenbogen, 2000). The isoflavonoid genistein downregulates the ER as it upregulates the PR; cross-talk between the two receptors may lead to the mixed estrogen agonist/antagonist activity of genistein (Miodini et al., 1999; Katzenellenbogen, 2000).

Receptor Coactivator Recruitment

Initiation of gene transcription by ligand-bound ER requires the recruitment of coactivators or transcription initiation factors (Danielian et al., 1992). Certain co-activators appear to be differentially expressed among tissues, potentially mediating tissue-selective gene expression. The ability of ERα and ERβ to recruit coactivator proteins that mediate transcription is ligand dependent (Routledge et al., 2000). The relative binding affinities of various xenoestrogens for ERα/β is not always consistent with their subsequent ability to recruit coactivator proteins. The actual conformational change in the tertiary structure of the ER induced by xenoestrogens may differ from that induced by E_2 due to differences in the steric and electrostatic properties of the various ligands (Paige et al., 1999a,b). This finding may explain the observation that there is not always a direct correlation between binding affinity and transcriptional potency with certain ER ligands. ERβ has a greater ability than ERα to recruit coactivators in the presence of xenoestrogens in vitro (Routledge et al., 2000). Using HeLa cells transfected with human ER and coactivators,

Routledge et al., (2000) showed that the 20-fold selective affinity of genistein for ERβ resulted in a 12,000- and 33-fold greater ability of ERβ to recruit the coactivators SRC-la and TIF2, respectively, compared with ERα. In contrast, the ability of ERα and ERβ to recruit SRC-la with octylphenol and bisphenol A were similar, despite the higher binding affinities of these two compounds for ERβ. These ligand-dependent differences in the ability of ERα and ERβ to recruit coactivator proteins may also contribute to the complex tissue-dependent responses observed with certain xenoestrogens and EAPs.

Nuclear Receptor Ligand Biosynthesis

EAP modulation of steroidogenic and other ligand biosynthesizing enzymes can alter substrate-product ratios and bioactive concentrations of endogenous signaling molecules. Flavonoids may compete with endogenous substrates for active sites of steroid biosynthesizing and metabolizing enzymes, especially cytochrome P450 aromatase (Kellis and Vickery, 1984; De Jong et al., 1990; Campbell and Kurzer, 1993; Wang et al., 1994; Jeong et al., 1999; Le Bail et al., 2000; Brueggemeier et al., 2001). P450 aromatase catalyzes the conversion of the androgens testosterone and androstenedione to E_2. Saarinen et al. (2001) compared the aromatase-inhibiting activity of various flavonoids in vitro and in vivo. Several flavonoids, including 7-hydroxyflavone, chrysin (4,7-dihydroxyflavone), luteolin (3',4',5,7-tetrahydroxyflavone), kaempferol (3,4',5,7-tetrahydroxyflavone), quercetin (3,3',4',5,7-pentahydroxyflavone), and apigenin (4',5,7-trihydroxyflavone) inhibit the formation of E_2 from androstenedione in vitro (Saarinen et al., 2001). Some flavonoids, such as chrysin, with relatively weak binding affinities for either ER subtype (Kuiper et al., 1998) show relatively high inhibitory capacity for steroid-metabolizing enzymes in vitro (Saarinen et al., 2001). However, none of the tested flavonoids showed aromatase-inhibiting activity in vivo, and they failed to reduce androgen- or estrogen-induced uterine growth (Saarinen et al., 2001). Differences between the in vitro and in vivo data may reflect the significant role metabolism and actions of binding proteins play in the activation or inactivation of EAPs.

Other steroidogenic enzymes susceptible to modulation by EAPs are 5α-reductase (Evans et al., 1995; Weber et al., 2001) and 3β- and 17β-hydroxysteroid dehydrogenase (HSD; Makela et al., 1995a; Le Bail et al., 2000; Krazeisan et al., 2001) in vitro but not always in vivo (Weber et al., 2001). EAP inhibitory potency can vary among steroidogenic enzymes and enzyme subtypes. In human placental microsomes, 3β-HSD is inhibited by genistein, daidzein, and biochanin A,whereas 17β-HSD activity is inhibited by coumestrol, genistein, biochanin A, 3'4'7-trihydroxyisoflavone, and daidzein, in order of decreasing inhibitory potency (Le Bail et al., 2000). For 17β-HSD, Krazeisan et al., (2001) found zearalenone, coumestrol, and quercetin to be the most potent inhibitors. Genistein may be a mixed agonist/antagonist of 17β-HSD, as it was identified as an inducer in MCF-7 breast cancer cells (Brueggemeier et al., 2001).

The susceptibility of steroidogenesis to modulation by EAPs may vary with age or life stage. Like E_2 (Voutilainen et al., 1979; Fujieda et al., 1982; Mesiano and Jaffe et al., 1993), the flavonoids genistein and daidzein can modulate adrenocortical steroidogenesis in vitro in an age specific manner. In cultured human fetal adrenocortical cells, genistein and daidzein decreased corticotropin stimulated cortisol production at micromolar concentrations (Mesiano et al., 1999). In cultured adult adrenocortical cells, this effect was accompanied by an increase in ACTH-stimulated androgen (dehydroepiandrosterone-sulfate) production. This modulation of adrenocortical steroidogenesis is likely mediated through the specific inhibition of the enzymes 21-hydroxylase (P450c21; Mesiano et al., 1999) and 3β-HSD (Byrne et al., 1986; Gell et al., 1998), shunting metabolites from glucocorticoid to androgen biosynthesis.

Genistein and daidzein have also been implicated in disruption of the synthesis of the nonsteroidal thyroid hormones (Divi et al., 1997), contributing to goiter and hypothyroidism among human infants drinking soy-containing formula (Van Wyk et al., 1959). In vitro, these isoflavones can inhibit thyroid peroxidase (TPO)-catalyzed reactions essential to thyroid hormone synthesis (Divi et al., 1997).

Serum Transport

Access of testosterone and E_2 to their target tissues is regulated in part by the blood plasma glycoprotein sex hormone binding globulin (SHBG; Hammond, 1995; Siiteri et al., 1982). Many EAPs, including the flavonoids, can bind to human SHBG steroid-binding sites in vivo with low affinities relative to E_2 (Martin et al., 1996; Jury et al., 2000). Binding to SHBG can increase ligand half-life and decrease tissue bio-availability. In dilute human serum, EAP relative binding affinities of 14–27% have been reported (Martin et al., 1978) but in partially purified SHBG, binding affinities range from 0.01 to 1% (Martin et al., 1996; Milhgan et al., 1998b) relative to E_2. The relative binding affinities of individual EAPs to SHBG varies between in vitro assays on diluted or undiluted serum, suggesting that additional plasma binding proteins may mediate EAP bioavailability (Jury et al., 2000). EAPs also can stimulate SHBG production in vitro (Adlercreutz et al., 1987; Mousavi and Adlercreutz, 1993; Loukovaara et al., 1995; Le Bail et al., 2000), reducing local levels of unbound, bioavailable sex steroids (Mendel, 1989). In vivo, a soy-based diet can increase (Brzezinski et al., 1997) or decrease (Nagata et al., 1997) serum SHBG levels or leave them unchanged (Schultz et al., 1991; Cassidy et al., 1994). Although EAPs have not been shown to significantly displace sex steroids from SHBG (Martin et al., 1996; Dechaud et al., 1999), low-affinity binding may be physiologically relevant because EAP blood levels can reach 100–1000 times that of E_2 (Kurzer and Xu, 1997; King and Bursill, 1998). The low affinity of some EAPs for serum proteins would be expected to enhance the numbers of molecules available for receptor occupancy (Mendel, 1989). The effective free fraction of coumestrol and genistein in human serum is 45–50%, whereas only 4% of E_2 is free (Nagel et al., 1998).

Although serum albumin does not appreciably modulate steroid hormone bio-availability, albumin may be an important regulator of EAP bioavailability (Baker, 1998; Nagel et al., 1998). Binding affinity of flavonoids for serum albumin generally increases with flavonoid hydrophobicity (Zhang et al., 1994). In humans, serum albumin is the primary protein responsible for binding 99% of quercetin in plasma (Boulton et al., 1998). Flavonoid binding to serum albumin can reduce interactions with the ER and other EAP targets (Arnold et al., 1996a,b; Boulton et al., 1998;

Dangles et al., 2001; Arts et al., 2002). Although binding to serum proteins significantly decreases the availability and estrogenic activity of E_2, coumestrol, and
genistein, the activity of other anthropogenic xenoestrogens (e.g., kepone, o,p'-DDT,
p,p'-DDD) may be only minimally reduced by extracellular proteins (Arnold et al.,
1996a,b).

Uptake, Assimilation, and Metabolism of EAP's in Vertebrates

EAP Metabolism and Bioavailability

In vivo studies demonstrate that EAPs differ significantly in absorption, pharmacokinetics, and metabolism (Kurzer and Xu, 1997). EAPs generally have much
shorter biological half-lives than do EDCs, and their biological activity is restricted
in time relative to EDCs (Sarver et al., 1997). The half-lives of the common EAPs
daidzein (9.34 h) and genistein (7.13 h) in humans (Setchell et al., 2001) is dramatically shorter than that of xenoestrogenic EDCs; 2,2'4,4'5,5'-PCB has a half-
life of 8,112 h in humans (Van der Berg et al., 1994) and tamoxifen has a half-life
of 264 h (Robinson et al., 1991). Furthermore, EAP pharmacokinetics may be gender specific. Following a single oral dose, mean and maximal concentrations of
plasma genistein are significantly higher in male than in female rats, with half-lives
of 12.4 h and 8.5 h, respectively (Coldham and Sauer, 2000).

Isoflavones and lignans show similar patterns of metabolism and disposition in
humans and other animals (figure 15-5). The metabolism of coumestans has not
been characterized. In most cases, metabolism of dietary EAPs by gastrointestinal
enzymes and bacteria yields bioavailable compounds with greater endocrine potency. Most of the flavonoids found in plants are present as glycosides, but only
the nonconjugated (aglycone) forms appear to exert endocrine activity in animals
(Liehr et al., 1998). The aglycones can readily be released from their sugar components by acid hydrolysis (Harborne, 1997).

Both isoflavones and lignans undergo significant metabolism by bacteria in the
gastrointestinal tract (Setchell and Adlercreutz, 1988). Glycosides of the isoflavones
genistein and daidzein are rapidly metabolized by various strains of *Escherichia
coli* bacteria into their respective aglycones (Hur et al., 2000). They also can be
derived from biochanin A and formonetin, which are converted to genistein and
daidzein, respectively, after breakdown by intestinal glycosidases (Hur and Rafii,
2000). Daidzein is further metabolized by various strains of intestinal bacteria to
equol, which has reduced steroidogenic inhibitory activity (Adlercreutz and Mazur,
1997; Whitten et al., 1997) but greater estrogenic potency than daidzein due to higher
ER binding affinity (Shutt and Cox, 1972; Shutt, 1976) and reduced serum SHBG
and albumin binding (Nagel et al., 1998). In vivo (Lampe et al., 1998), and in vitro
(Cassidy, 1991; Setchell, 1998; Setchell and Cassidy, 1999), intestinal conversion
of daidzein to the more biologically active equol is increased in animals fed a high-
carbohydrate diet. Thus, intestinal bacteria may play a key role in mediating the
endocrine activity of phytochemicals.

There is evidence of significant interindividual variation in the metabolism of
the isoflavones daidzein and genistein by intestinal microflora (Rowland et al.,

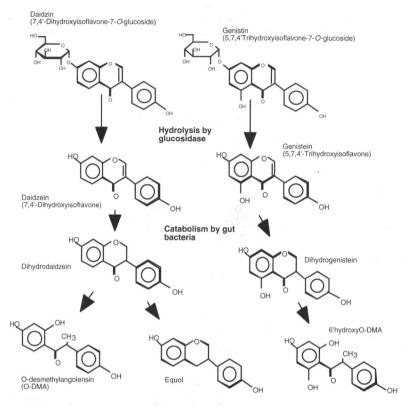

Figure 15-5. Intestinal metabolism of isoflavonoid glucosides. Hydrolysis of isoflavone glucosides by intestinal glucosidase yields bioactive isoflavonoid aglycone. Further metabolism by intestinal bacteria can alter endocrine activity of dietary flavonoids. Equol has less steroidogenic inhibitory activity than daidzein (Adlercreutz and Mazur, 1997; Whitten et al., 1997) but greater estrogenic potency due to higher estrogen receptor binding affinity (Shutt and Cox, 1972; Shutt, 1976) and reduced binding of sex hormone binding globulin (Nagel et al., 1998). Adapted from Hur et al. (2000) and Kurzur and Xu (1997).

2000). Furthermore, Setchell et al. (2001) noted that the proportion of individuals within sampled human populations that can metabolize daidzein to equol had decreased by 50% over a 20-year interval between studies. Borriello et al. (1985) and Setchell et al. (1984) first reported that two-thirds of adults who consumed soy-containing foods converted daidzein to equol; several later groups reported that this proportion was only one-third (Cassidy et al., 1994; Sathyamoorthy et al., 1994; Kelly et al., 1995; Setchell et al., 2001).

EAP Elimination

Most wild animals are well equipped to detoxify many of the phytochemicals present in the foliage within their habitat, and it is only in exceptional circumstances that a plant may gain complete protection from feeding through phytochemical defense. No species, family, or class of animals exhibits high sensitivity to all chemicals all the time (Mayer et al., 1986). After absorption in the small intestine, isoflavones and lignans may be conjugated with glucuronic acid and sulfate by hepatic phase II enzymes (UDP-glucuronosyltransferases and sulfotransferases). Lignan and isoflavone conjugate profiles in human urine suggest that glucuronic acid is the primary moiety (Adlercreutz et al., 1995). EAP conjugates also differ in tissue bioavailability. For example, in humans, daidzein conjugates are more bioavailable than genistein conjugates (King, 1998). Species differences in EAP metabolism and bioavailability have been reported (Gontier-Latonnelle, 2001). In Siberian sturgeon (*Acipenser baeri*), genistein is preferentially conjugated with sulfate, whereas in rainbow trout (*Oncorhynchus mykiss*), genistein is conjugated primarily with glucuronide (Gontier-Latonnelle, 2001). Differences in conjugation affect absorption and excretion and hence bioavailability and endocrine potency of EAPs.

Like endogenous estrogens, these conjugates are excreted through both urine and bile and also may undergo enterohepatic circulation. Many EAPs are substrates for cytochrome P450 phase-I metabolizing enzymes. The cytochrome P450 superfamily consists of more than 150 P450 genes, and its origins precede the evolution of multicellularity (Nelson et al., 1993). Cytochrome P450 proteins are present in bacteria, fungi, plants, and animals (Nelson et al., 1993). P450 diversity may have evolved as a mechanism by which organisms protected themselves from dietary and environmental toxicants (reviewed in Gonzalez and Nebert, 1993; Lewis et al., 1998). Using recombinant human cytochrome P450 enzymes, Roberts-Kirchhoff et al. (1999) showed that genistein can be metabolized in vitro by the P450s 1A1, 1A2, 1B1, and 2E1 to form predominantly 3',4',5,7-tetrahydroxyisoflavone. Metabolism by P450 3A4 yields two different, unidentified products. Cytochrome P450s in liver microsomes from Arochlor-treated male rats metabolize genistein to six metabolites, including four monohydroxylated and two dihydroxylated products (Kulling et al., 2000). The major products of genistein metabolism were identified as 5,6,7,4'-tetrahydroxyisoflavone and 5,7,8,4'-tetrahydroxyisoflavone.

EAPs may influence endocrine regulation through interactions with regulatory pathways not traditionally associated with the endocrine system, such as the AhR (Ciolino et al., 1999; Ashida et al., 2000). Through interactions with the AhR, quercetin, but not kaempferol or resveratrol, can induce transcription of CYP1A1 (Ciolino et al., 1999). Induction of cytochrome P450s is not well characterized for most EAPs. In rats, flavonoids can induce UDP-glucuronyl transferases at oral dosages of 20 ppm (Siess et al., 1992). Other drug-metabolizing enzymes are induced at higher dosages. In general, conjugating enzymes are induced at much lower doses of EAPs than are dealkylase and hydroxylase activities (Siess et al., 1992).

Regulation of EAP metabolism is complex, and EAPs may inhibit induction of the enzymes for which they are substrates. The flavonoids and furanocoumarins from grapefruit have strong inhibitory action on CYP3A4-mediated quinine 3-hydroxylase activity (Ho and Saville, 2001), and quercetin differentially inhibits sulfotransferase activities in vitro in human liver and duodenum (Marchetti et al., 2001). Numerous phytochemicals can inhibit the alkoxyresorufin-O-dealkylase activities of CYP2C11, CYP1A2 and CYP2C6 in vitro (Teel and Huynh, 1998) and may exert endocrine activity by inhibiting the metabolism of substrates for these cytochrome isoforms.

Adverse Effects of EAPs

The affects of EAPs on endocrine function reflect the interactions of complex, life-stage specific, target-dependent responses. Although the effect of individual EAPs at any given locus may appear miniscule relative to that of endogenous steroids, mild changes in the regulation of endogenous hormone action can result in significant physiological effects and biologically relevant outcomes (Brucker-Davis et al., 2001).

In vitro potencies of EAPs and xenobiotic estrogens may not be an accurate guide to in vivo potencies. Despite low ER binding affinities, the EDCs bisphenol A, nonylphenol, and *p-tert*-octylphenol produced effects in rat endometrial adenocarcinoma cells at only 100-fold higher concentrations than were needed for a similar level of induction by E_2 (Strunck et al., 2000). Furthermore, estimates of in vivo potency will depend on the assay used (Nagel et al., 1997; Steinmetz et al., 1997; Milligan et al., 1998a; Strunck et al., 2000; Ratna and Simonelli, 2002). Naturally occurring plant food resources contain variable concentrations of multiple EAPs, but laboratory investigations of the phenotypic effects of EAPs typically expose animals to relatively high concentrations of a single EAP (e.g., Levy et al., 1995). In some cases, the laboratory species used have evolved as carnivores (e.g., rainbow trout, mink) and have little evolutionary experience with EAPs. Pathological outcomes from such exposures are not surprising considering they occur within an evolutionarily novel signaling context.

Effects on Reproductive Tract

Although in vivo evidence suggests that both xenoestrogens and EAPs can significantly stimulate an increase in uterine weight, analyses of gene expression reveal compound-specific profiles of molecular action (Diel et al., 2000). For example, DDT increased uterine weight to a greater extent than coumestrol, yet the pattern of coumestrol-induced gene expression was more similar to that of E_2 than it was to that of DDT (Ashby et al., 1999; Diel et al., 2000). When coadministered orally with E_2, the effects of coumestrol are additive for uterine weight and cytosolic ER binding and antagonistic for nuclear ER binding, progestin receptor induction, and uterine protein (Whitten et al., 1994).

The effects of EAPs may differ depending on whether exposure occurs at a developmental or postdevelopmental stage. EAPs can cross the placenta in rats (Degen et al., 2002). However, EAP exposure during development had no effect on sex ratio, number of live pups, implantation sites, anogenital distance, vaginal opening, body weight, cell count in seminiferous tubules, or ovarian follicular development (Kang et al., 2002). Oral exposure to high doses of genistein (100 mg/kg) on postnatal days 1–5 affected postpubertal reproductive function in rats (Nagao et al., 2001). Genistein increased the frequency of females with estrous cycle irregularities, impaired fertility, and ovarian histopathology. Male reproductive function appeared to be unaffected by EAP treatment in this study (Nagao et al., 2001). However, decreased ventral prostate weights were reported from adult male rats maternally exposed to 10 mg/kg genistein, a dose similar to the human exposure level (Kang et al., 2002).

Male rainbow trout fed high-genistein diets for 1 year showed enhanced testicular development but decreased sperm motility and concentration (Bennetau-Pelissero et al., 2001). This was accompanied by significantly reduced plasma levels of $17\alpha,20\beta(OH)_2$-progesterone, slightly reduced plasma β-subunit concentrations of both follicle-stimulating hormone (βFSH) and luteinizing hormone (βLH), and a slight but constant induction of plasma vitellogenin. Reproductive consequences in female trout were less pronounced and occurred only at the highest genistein dosage.

Chronic dietary exposure to genistein can yield transient reductions in rat serum LH as well as testis and epididymal weights, without any significant effect on testicular sperm count (Roberts et al., 2000). In the prepubertal rat uterus, pharmacologic, but not physiologic, concentrations of genistein can alter sex steroid receptor expression (Cotroneo et al., 2001) and sustain endometriosis (Cotroneo and Lamartiniere, 2001). Female minks fed high doses of the mycotoxin zearalenone (10 mg/kg) showed impaired reproduction and reproductive tract pathology, including severely distended uteri, endometrial hyperplasia, uterine atrophy, endometritis, and pyometra (Yamini et al., 1997). Treatment with the selective estrogen receptor modulator tamoxifen did not alleviate the symptoms of this apparent hyperestrogenization (Yamini et al., 1997).

Neurobehavioral Effects

EAPs can reach the brain (reviewed in Setchell, 1998), and neurobehavioral effects of EAPs have been identified in rats and mice (reviewed in Lephart et al., 2002). Subtle yet significant differences in some sexually dimorphic behaviors have been reported for adult (Patisaul et al., 2001) and maternally exposed rats (Flynn et al., 2000). The expression of anxiety, learning, and memory are also vulnerable to modulation by flavonoid EAPs in rats and mice (Garey et al., 2001; Lephart et al., 2002). The flavonoid coumestrol can inhibit E_2-induced locomotion as well as modulate estrogen-independent behaviors (Garey et al., 2001). For example, in conditioned fear assays, E_2-treated animals do not differ from vehicle controls in fear responses, but coumestrol produced significantly less fearfulness (Garey et al., 2001). Coumestrol differs from E_2 in its in vivo effect on ERβ mRNA expression in the rat brain (Patisaul et al., 1999). Whereas E_2 yields a decrease in ERβ mRNA by 45% in the paraventricular nucleus of the hypothalamus, coumestrol increased ERβ expression by 48%. Neither E_2 nor coumestrol altered ERβ mRNA in other ERβ-rich brain regions such as the bed nucleus of the stria terminalis or the medial preoptic area (Patisaul et al., 1999).

EAPs can significantly alter brain morphology, especially that of the sexually dimorphic nucleus in the preoptic area of the hypothalamus (SDN-POA; Faber and Hughes 1991, 1993; Levy et al., 1995; Lund et al., 2001). The SDN-POA is usually two to five times larger in male than in female rats (Arnold and Gorskj, 1984;

Woodson and Gorski, 2000). Among male rats treated during adulthood, a high EAP diet yielded a significant increase in the volume of the SDN-POA (Lephart et al., 2002). The sexual dimorphism observed in rat SDN-POA volume was undetectable when both males and females were fed low EAP, soy-free diets (Lund et al., 2001; Lephart et al., 2002). The dietary actions on SDN-POA volume during adulthood were reversible, as SDN-POA volume declined upon removal of the high-EAP diet (Lund et al., 2001).

The effect of EAPs on SDN-POA volume in female rats depends on endogenous hormonal status. In gonadectomized females, genistein significantly increased SDN-POA volume (Faber and Hughes, 1991, 1993; Lephart et al., 2002), whereas in intact females genistein treatment contributed to a nonsignificant decrease in volume (Faber and Hughes, 1991; Lephart et al., 2002). These changes in male and female brain morphology were not due to significant changes in circulating serum testosterone or E_2, nor to changes in local conversion of testosterone to E_2, as rats fed high or low EAP diets do not differ in brain P450 aromatase levels (Lund et al., 2001; Lephart et al., 2002). Reductions in circulating testosterone were detected in male rats raised on the same high EAP diet (Weber et al., 2001). From these studies, it is unclear which experimental outcome (high EAP or EAP-free) represents an abnormal phenotype. EAPs occurring naturally in plant food sources may function as important modulators of the development of sexual dimorphism among herbivores and granivores.

Conclusions

Signaling through steroidlike pathways is an ancient and evolutionary conserved means of communication that exists in bacteria, fungi, plants, and animals; such signaling networks can regulate interactions both within and between organisms. The potential endocrine-modulating effects of EAPs depends on a number of factors, including evolutionary experience, level of expression, metabolism and pharmacokinetics, compound potency, serum concentrations, relative binding to serum proteins, levels of exposure during critical periods, and interactions or cross-talk with other endocrine-response pathways.

The vulnerability of the vertebrate endocrine system to exogenous chemicals may represent an adaptation to permit modulation of endocrine-mediated processes in response to diverse, meaningful environmental chemical signals, as the elaboration of an EAP-specific ER suggests. The exploitation of signaling pathways by anthropogenic EDCs may pose an especially daunting challenge to the integrity and complexity of biological signaling networks, which require correlation between information content and signaling pathway of messenger molecules. Through persistent, nonspecific signaling, anthropogenic pollutants reduce the complexity and information reliability of evolved chemical communication networks. The global magnitude of anthropogenic pollution and the ubiquity of diverse, persistent, high-signal-content molecules may obscure our understanding of the historical scale, diversity, complexity, and evolution of environmental chemical communication networks.

References

Adams, N.R., 1998. Clover phyto-oestrogens in sheep in Western Australia. Pure Appl. Chem. 70, 1855–1862.

Adlercretuz, H., Hockerstedt, K., Bannwart, C., 1987. Effect of dietary components, including lignans and phytoestrogens on enterohepatic circulation and liver metabolism of estrogens and on sex hormone binding globulin (SHBG). J. Steroid. Biochem. 27, 1135–1144.

Adlercreutz, H., Mazur, W., 1997. Phytoestrogens and western diseases. Ann. Med. 29, 95–120.

Adlercreutz, H.,Van der Wildtz, J., Kinzel, J., Attalla, H., Wahala, K., 1995. Lignan and isoflavonoid conjugates in human urine. J. Steroid. Biochem. Mol. Biol. 52, 97–103.

Agarwal, M.K. 1993. Receptors for mammalian steroid hormones in microbes and plants. FEBS. 322(3), 207–210.

Arenas, M.I., Royuela, M., Lobo, M.V.T., Alfaro, J.M., Fraile, B., Paniagua, R., 2001. Androgen receptor (AR), estrogen receptor-alpha (ER-α) and estrogen receptor-beta (ER-β) expression in the testis of the newt, Triturus marmoratus marmoratus during the annual cycle. J. Anat. 199, 465–472.

Arnold, A.P., Gorski, R.A., 1984. Gonadal steroid induction of structural sex differences in the central nervous system. Annu. Rev. Neurosci. 7, 413–422.

Arnold, S.F., Collins, B.M., Robinson, M.K., Guillette, L.J., McLachlan, J.A., 1996a. Differential interaction of natural and synthetic estrogens with extracellular binding proteins in a yeast estrogen screen. Steroids 61, 642–646.

Arnold, S.F., Robinson, M.K., Notides, A.C., Guillette, L.J., McLachlan, J.A., 1996b. A yeast estrogen screen for examining the relative exposure of cells to natural and xeno-estrogens. Environ. Health. Perspect. 104, 544–548.

Arts, M.J., Haenen, G.R., Wilms, L.C., Beetstra, S.A., Heijnen, C.G., Voss, H.P., Bast, A., 2002. Interactions between flavonoids and proteins: Effects on the total antioxidant capacity. J. Agric. Food Chem. 50, 1184–1187.

Ashby, J., Tinwell, H., Soames, A., Foster, J., 1999. Induction of hyperplasia and increased DNA content in the uterus of immature rats exposed to coumestrol. Environ. Health. Perspect. 107, 819–822.

Ashida, H., Fukuda, I., Yamashita, T., Kanazawa, K., 2000. Flavones and flavonols at dietary levels inhibit a transformation of aryl hydrocarbon receptor induced by dioxin. FEBS. Lett. 1476, 213–217.

Atkinson, S., and Atkinson, M.A., 1992. Detection of estradiol-17β during a mass coral spawn. Coral Reefs 11, 33–35.

Baker. M.E., 1998. Albumin's role in steroid hormone action and the origins of vertebrates: is albumin an essential protein? FEBS. Lett. 439, 9–12.

Baldwin, W.S., LeBlanc, G., 1994. In vivo biotransformation of testosterone by phase I and II detoxification enzymes and their modulation of 20-hydroxyecdysone in Daphnia magna. Aqua. Toxicol. 29, 103–117.

Becard, G., Douds, D.D., Pfeffer, P.E., 1992. Extensive in vitro hyphal growth of vesicular-arbuscular mycorrhizal fungi in the presence of CO_2 and flavonols. Appl. Environ. Microbiol. 58, 821–825.

Bennetau-Pelissero, C., Breton, B.. Bennetau, B., Corraze, G., LeMenn, F., Davail-Cuisset, B., Helou, C., Kaushik, S.J., 2001. Effects of genistein-enriched diets on the endocrine process of gametogenesis and on reproduction efficiency of the rainbow trout, Oncorhynchus mykiss. Gen. Comp. Endocrinol. 121, 173–187.

Bennetts, H.W., Underwood, E.J., Shier, F.L., 1946. A specific breeding problem of sheep on subterranean clover pastures in Western Australia. Aust. Vet. J. 22, 2–12.

Berger, P.J., Negus, N.C., Rowsemitt, C.N., 1987. Effects of 6-methoxybenzoxazolinone on sex ratio and breeding performance in Microtus montanus. Biol. Reprod. 36, 255–260.

Berger, P.J., Negus, N.C., Sanders, E.H., Gardner, P.D., 1981. Chemical triggering of reproduction in Microtus montanus. Science 2, 69–70.

Berger, P., Sanders, E.H., Gardner, P.D., Negus, N.C., 1977. Phenolic plant compounds functioning as reproductive inhibitors in Microtus montanus. Science 195, 575–577.

Bernard, D.J., Bentley, G.E., Balthazart, J., Turek, F.W., Ball, G.F., 1999. Androgen receptor, estrogen receptor α, and estrogen receptor β show distinct patterns of expression in forebrain song control nuclei of European starlings. Endocrinology 140, 4633–4643.

Borriello, S.P., Setchell, K.D.R., Axelson, M., and Lawson, A.M., 1985. Production and metabolism of lignans by the human fecal flora. J. Appl. Bacteriol. 58, 37~43.

Boue, S.M., Carter, C.H., Ehrlich, K.C., and Cleveland, T.E., 2000. Induction of the soybean phytoalexins coumestrol and glyceollin by Aspergillus. J. Agric. Food Chem. 48, 2167–2172.

Boulton, D.W., Walle, U.K., Walle, T., 1998. Extensive binding of the bioflavonoid quercetin to human plasma proteins. J. Pharm. Pharmacol. 50, 243–249.

Bradbury, R.B., White, D.E., 1954. Oestrogens and related substances in plants. Vit. Horm. 12, 207–233.

Brucker-Davis, F., Thayer, K., Colborn, T., 2001. Significant effects of mild endogenous hormonal changes in humans: Considerations for low-dose testing. Environ. Health. Perspect. 109 (suppl. 1), 21–26.

Brueggemeir, R.W., Gu, X., Mobley, J.A., Joomprabutra, S., Bhat, A.S., Whetstone, J.L., 2001. Effects of phytoestrogens and synthetic combinatorial libraries on aromatase, estrogen biosynthesis, and metabolism. Ann. N.Y. Acad. Sci. 948, 51–66.

Brzezinski, A., Adlercreutz, H., Shaoul, R., Rosler, A., Shmueli, A., Tanos, V., Schenker, J.G., 1997. Short-term effects of phytoestrogen-rich diet on postmenopausal women. Menopause 4, 89–94.

Brzozowski, A.M., Pike, A.C.W., Dauter, Z., Hubbard, R.E., Bonn, T., Engstrom, O., Ohman, L., Greene, G.L., Gustafsson, J.A., Carlquist, M., 1997. Molecular basis of agonism and antagonism in the estrogen receptor. Nature 389, 753–758.

Byrne, G.C., Perry, Y.S., Winter, J.S., 1986. Steroid inhibitory effects upon human adrenal 3β-hydroxysteroid dehydrogenase activity. J. Clin. Endocrinol. Metab. 62, 413–418.

Campbell, D.R., Kurzer, M.S., 1993. Flavonoid inhibition of aromatase enzyme activity in human preadipocytes. J. Steroid Biochem. Mol. Biol. 46, 381–388.

Cassidy, A., 1991. Plant oestrogens and their relation to hormonal status in women. Ph.D. dissertation, Darwin College, University of Cambridge.

Cassidy, A., Bingham, S., Setchell, K.D.R., 1994. Biological effects of a diet of soy protein rich in isoflavones on the menstrual cycle of premenopausal women. Am. J. Clin. Nutr. 60, 333–340.

Chu, S., Fuller, P.J., 1997. Identification of a splice variant of the rat estrogen receptor beta gene. Mol. Cell. Endocrinol. 132, 195–199.

Ciolino, H.P., Daschner, P.J., Yeh, G.C., 1999. Dietary flavonols quercetin and kaempferol are ligands of the aryl hydrocarbon receptor that affect CYP1A1 transcription differentially. Biochem. J. 340, 715–722.

Colborn,T., Dumanoski, D., Myers, J.P., 1996. Our stolen Future. Dutton, New York.

Coldham, N.G., Sauer, M.J., 2000. Pharmacokinetics of [C-14] Genistein in the rat: Gender related differences, potential mechanisms of biological action, and implications for human health. Toxicol. Appl. Pharmacol. 164, 206–215.

Coronado, C., Zuanazzi, J., Sallaud, D., Quirion, J.C., Esnault, R., Husson, H.P., Kondorosi, A., Ratet, P., 1995. Alfalfa root flavonoid production is nitrogen regulated. Plant. Phys. 108, 533–542.

Cotroneo, M.S., Lamartiniere, C.A., 2001. Pharmacologic, but not dietary, genistein supports endometriosis in a rat model. Toxicol. Sci. 61, 68–75.

Cotroneo, M.S., Wang, J., Elgoum, I.E.A., Lamartiniere, C.A., 2001. Sex steroid receptor regulation by genistein in the prepubertal rat uterus. Mol. Cell. Endocrinol. 173, 135–145.

Cunningham, A.R., Klopman, G., Rosenkranz, H.S., 1997. A dichotomy in the lipophilicity of natural estrogens, xenoestrogens, and phytoestrogens. Environ. Health. Perspect. 105 (suppl. 3), 665–668.

Dangles, O., Dufour, C., Manach, C., Morand, C., Remesy, C., 2001. Binding of flavonoids to plasma proteins. Meth. Enzymol. 335, 319–333.

Danielian, P.S., White, R., Lees, J.A., Parker, M.G., 1992. Identification of a conserved region required for hormone dependent transcriptional activity by steroid hormone receptors. Embo. J. 11, 1025–1033.

Davis, W.P., Bortone, S.A., 1992. Effects of kraft mill effluent on the sexuality of fishes: An environmental early warning? In: Colborn, T., Clements, C. (Eds.), Chemically-induced Alterations in Sexual and Functional Development: The Wildlife/Human Connection. Princeton Scientific Publishing, Princeton, NJ, pp. 113–127.

Dechaud, H., Ravard, C., Claustrat, F., de la Penier, A.B., Pugeat, M., 1999. Xenoestrogen interaction with human sex hormone-binding globulin (hSHBG). Steroids 64, 328–334.

Degen, H., Janning, P., Diel, P., Michna, H., Bolt, M., 2002. Transplacental transfer of the phytoestrogen daidzein in DA/Han rats. Arch. Toxicol. 76, 23–29.

De Jong, F.H., Oishi, K., Hayes, R.B., Bogdanowicz, J.F.A.T., Ibrahim, A.R., Abul Hajj, Y.J., 1990. Aromatase inhibition by flavonoids. J. Steroid. Biochem. Mol. Biol. 3, 257–260.

Denton, T.E., Howell, W.M., Allison, J.J., McCollum, J., Marks, B.J., 1985. Masculinization of female mosquitofish by exposure to plant sterols and Mycobacterium smegmatis. Bull. Environ. Contam. Toxicol. 35, 627–632.

Diel, P., Schulz, T., Smolnikar, K., Strunck, E., Vollmer, G., Michna, H., 2000. Ability of xeno- and phytoestrogens to modulate expression of estrogen-sensitive genes in rat uterus: Estrogenicity profiles and uterotrophic activity. J. Steroid Biochem. Mol. Biol. 73, 1–10.

Dinan, L., Whiting, P., Girault, J.P., Lafont, R., Dhadialla, T.S., Cress, D.E., Mugat, B., Antoniewski, C., Lepesant, J.A., 1997. Cucurbitacins are insect steroid hormone agonists acting at the ecdysteroid receptor. Biochem. J. 327, 643–650.

Divi, R.L., Chang, H.C., Doerge, D.R., 1997. Anti-thyroid isoflavones from soybean. Biochem. Pharm. 54, 1087–1096.

Djordjevic, M.A., Redmond, J.W., Batley, M., Rolfe, B.G., 1987. Clovers secrete specific phenolic compounds which either stimulate or repress nod gene expression in Rhizobium trifolium EMBO. J. 6, 1173–1179.

Djordjevic, M.A., Weinman, J.J., 1991. Factors determining host recognition in the clover-rhizobium symbiosis. Aust. J. Plant. Physiol. 18, 543–557.

Doerge, D.R., Sheehan, D.M., 2002. Goitrogenic and estrogenic activity of soy isoflavones. Environ. Health. Perspect. 110 (suppl. 3), 349–353.

Dube, M.G., MacLatchy D.L.,.2001. Identification and treatment of a waste stream at a bleached-kraft pulp mill that depresses a sex steroid in the mummichog (Fundulus heteroclitus). Environ. Toxicol. Chem. 20, 985–995.

Elsby, R.J., Ashby, J., Sumpter, J.P., Brooks, A.N., Pennie, W.D., Maggs, J.L., Lefevre,

P.A., Odum, J., Paton, D., Parker, B.K., 2000. Obstacles to the prediction of estrogenicity from chemical structure: assay-mediated metabolic transformation and the apparent promiscuous nature of the estrogen receptor. Biochem. Pharm. 60, 1519–1530.

Evans, B.A., Griffiths, K., Morton, M.S., 1995. Inhibition of 5α-reductase in genital skin fibroblasts and prostate tissue by dietary lignans and isoflavonoids. J. Endocrinol. 147, 295–302.

Faber, K.A., Hughes, C.L. Jr., 1991. The effect of neonatal exposure to dietliylstilbestrol, genistein, and zearalenone on pituitary responsiveness and sexually dimorphic nucleus volume in the castrated adult rat. Biol. Reprod. 45, 649–653.

Faber, K.A., Hughes, C.L. Jr., 1993. Dose-response characteristics of neonatal exposure to genistein on pituitary responsiveness to gonadotropin releasing hormone and volume of the sexually dimorphic nucleus of the preoptic area (SDN-POA) in postpubertal castrated female rats. Reprod. Toxicol. 7, 35–39.

Farmakalidis, E., Hathcock, J.N., Murphy, P.A., 1985. Oestrogenic potency of genistein and daidzin in mice. Food Chem. Toxicol. 23, 741–745.

Farnsworth, N.R., Bingel, A.S., Cordell, G.A., Crane, F.A., Fond, H.H.S., 1975. Potential value of plants as sources of new antifertility agents. J. Pharmacol. Sci. 64, 717–754.

Firmin, J.L., 1986. Flavonoid activation of nodulation genes in rhizobium is reversed by other compounds present in plants. Nature 324, 90–93.

Flynn, K.M., Ferguson, S.A., Delclos, K.B., Newbold, R.R., 2000. Effects of genistein exposure on sexually dimorphic behaviors in rats. Toxicol. Sci. 55, 311–319.

Foidart, A., Lakaye, B., Grisar, I., Ball, G.F., Balthazart, J., 1999. Estrogen receptor-β in quail: Cloning, tissue expression and neuroanatomical distribution. J. Neurobiol. 40, 327–342.

Fox, J.E., Starcevic, M., Kow, K.Y., Burrow, M.E., McLachlan, J.A., 2001. Endocrine disrupters and flavonoid signaling. Nature 413, 128–129.

Francis, W.J., 1970. Influence of weather on population fluctuations in California quail. J. Wild. Manage. 34, 249.

Fraser, H.B., Hirsh, A.E., Steinmetz, L.M., Scharfe, C., Feldman, M.W., 2002. Evolutionary rate in the protein interaction network. Science 296, 750–752.

Fujieda, K., Faiman, C., Reyes, F.I., Winter, J.S.D., 1982. The control of steroidogenesis by human fetal adrenal cells in tissue culture. IV. The effects of exposure to placental steroids. J. Clin. Endocrinol. Metab. 45, 89–94.

Garey, J., Morgan, M.A., Frohlich, J., McEwen, B.S., Pfaff, D.W., 2001. Effects of the phytoestrogen coumestrol on locomotor and fear-related behaviors in female mice. Horm. Behav. 40, 65–76.

Gehm, B.D., McAndrews, J.M., Chien, P.Y., Jameson, J.L., 1997. Resveratrol, a polyphenolic compound found in grapes and wine, is an agonist for the estrogen receptor. Proc. Natl. Acad. Sci. USA, 94, 14138–14143.

Gell, J.S., Oh, J., Rainey, W.E., Carr, B.R., 1998. Effects of estradiol on DHEAS production in the human adrenocortical cell line, H295R. J. Soc. Gynecol. Invest. 5, 144–148.

Genoni, G.P., 1997. Influence of the energy relationships of organic compounds on toxicity to the cladoceran Daphnia magna and the fish *Pimephales promelas*. Ecotoxicol. Environ. Safety 36, 27–37.

Genoni, G.P., Montague, C.L., 1995. Influence of the energy relationships of trophic levels and of elements on bioaccumulation. Ecotoxicol. Environ. Safety 30, 203–218.

Geuns, J.M.C., 1978. Steroid hormones and plant growth and development. Phytochemistry 17, 1–4.

Gontier-Latonnelle, K., 2001. Genistein pharmacokinetic analysis in rainbow trout

(*Oncorhynchus mykiss*) and Siberian sturgeon (*Acipenser baeri*). Genistein metabolism and estrogenic potency analysis. Cybium 25, 226.

Gonzales, F.J., Nebert, D.W., 1993. Evolution of the P450-gene superfamily: Animal plant warfare, molecular diversity and human genetic differences in drug oxidation. Trends. Genet. 6, 182–186.

Gottlicher, M., Heck, S., Herrlich, P., 1998. Transcriptional cross-talk, the second mode of steroid hormone receptor action. J. Mol. Med. 76, 480–489.

Grunwald, D., 1980. Steroids. In: Bell, E.D., Charlwood, B.V. (Eds.), Encyclopedia of Plant Physiology, vol. 8. Secondary Plant Products. Springer-Verlag, Berlin, pp. 221–256.

Guillette, L.J. Jr., Cram, D.A., Rooney, A.A., Pickford, D.B., 1995. Organization versus activation: the role of endocrine-disrupting contaminants (EDCs) during embryonic development in wildlife. Environ. Health. Perspect. 103(suppl. 7), 157–164.

Gutendorf, B., Westendorf, J., 2001. Comparison of an array of in vitro assays for the assessment of the estrogenic potential of natural and synthetic estrogens, phytoestrogens and xenoestrogens. Toxicology 166, 79–89.

Gyorgy, P., Murata, K., Ikehata, H., 1964. Antioxidant isolated from fermented soybeans. Nature 203, 870–872.

Gyorgypal, Z., Kondorosi, A., 1991. Homology of the ligand-binding regions of Rhizobium symbiotic regulatory protein NodD and vertebrate nuclear receptors. Mol. Gen. Genet. 226, 337–340.

Hammond, G.L., 1995. Potential functions of plasma steroid binding proteins. Trends Endocr. Metabl. 6, 298–304.

Harborne, J.B., 1997. Recent advances in chemical ecology. Nat. Prod Rep. 14, 83–98.

Hawkins, M.B., Thornton, J.W., Crews, D., Skipper, J.K., Dotte, A., Thomas, P., 2000. Identification of a third distinct estrogen receptor and reclassification of estrogen receptors in teleosts. Proc. Natl. Acad Sci. USA 97, 10751–10756.

Hewitt, L.M., Parrott, J.L., Wells, K.L., Calp, M.K., Biddiscombe, S., McMaster, M.E., Munkittrick, K.R., Van der Kraak, G.J., 2000. Characteristics of ligands for the Ah receptor and sex steroid receptors in hepatic tissues of fish exposed to bleached kraft mill effluent. Environ. Sci. Tech. 34, 4327–4334.

Ho, P-C., Saville, D.J., 2001. Inhibition of human CYP3A4 activity by grapefruit flavonoids, furanocoumarins and related compounds. J. Pharm. Pharmaceut. Sci. 4, 217–227.

Howell, W.M., Black, D.A., Bortone, S.A., 1980. Abnormal expression of secondary sex characters in a population of mosquitofish, Gambusia affinis holbrooki: Evidence for environmentally-induced masculinization. Copeia 4, 676–681.

Hughes, C.L. Jr., 1988. Phytochemical mimicry of reproductive hormones and modulation of herbivore fertility by phytoestrogens. Environ. Health. Perspect. 78, 171–175.

Hur, H.G., Lay, J.O. Beger, R.D., Freeman, J.P., Rafii, F., 2000. Isolation of human intestinal bacteria metabolizing the natural isoflavone glycosides daidzein and genistein. Arch. Microbiol. 174, 422–428.

Hur, H.G., Rafii, F., 2000. Biotransformation of the isoflavonoids biochanin A, formononetin, and glycitein by *Eubacterium limosum*. FEMS Microbiol. Lett. 192, 21–25.

Ikeda, K., Arao, Y., Otsuka, H., Nomoto, S., Horiguchi, H., Kato, S., Kayama, F., 2002. Terpenoids found in the Umbelliferae family act as agonists/antagonists for ERα and ERβ. Biochem. Biophys. Res. Comm. 291, 354–360.

Jacob, D.A., Temple, J.L., Patisaul, H.B., Young, L.J., Rissman, E.F., 2001. Coumestrol antagonizes neuroendocrine actions of estrogen via the estrogen receptor alpha. Exp. Biol. Med. 226, 301–306.

Jeong, H.J., Shin, Y.G., Kim, J.H., Pozzuto, J.M., 1999. Inhibition of aromatase activity by flavonoids. Arch. Pharm. Res. 22, 309–312.

Jury, H.H., Zacharewski, T.R., Hammond, G.L., 2000. Interactions between plasma sex hormone-binding globulin and xenobiotic ligands. J. Steroid. Biochem. Mol. Bio. 75, 167–176.

Kang, K.S., Che, J.H., Lee, Y.S., 2002. Lack of adverse effects in the F1 offspring maternally exposed to genistein at human intake dose level. Food Chem. Toxicol. 40, 43–51.

Kao, Y.C., Zhou, C., Sherman, M., Laughton, C.A., Chen, S., 1998. Molecular basis of the inhibition of human aromatase (estrogen synthetase) by flavone and isoflavone phyto-estrogens: A site-directed mutagenesis study. Environ. Health. Perspect. 106, 85–92.

Katzenellenbogen, B.S., 2000. Mechanism of action and cross-talk between estrogen receptor and progesterone receptor pathways. J. Soc. Gynecol. Invest. 7, S33–S37.

Kellis, J.T. Jr., and Vickery, L.E., 1984. Inhibition of human estrogen synthetase (aromatase) by flavones. Science 225, 1032–1034.

Kelly, G.E., Joannou, G.E., Reeder, A.Y., Nelson, C., Waring, M.A., 1995. The variable metabolic response to dietary isoflavones in humans. Proc. Soc. Exp. Biol. Med. J. 208, 40–43.

Khripach, V.A., Zhabinskii, V.N., de Groot, A.E., 1999. Brassinosteroids. Academic Press, San Diego, CA.

King, R.A., 1998. Daidzein conjugates are more bioavailable than genistein conjugates in rats. Am. J. Clin. Nutr. 68, 1496S–1499S.

King, R.A., Bursill, D.B., 1998. Plasma and urinary kinetics of the isoflavones daidzein and genistein after a single soy meal in humans. Am. J. Clin. Nutr. 67, 867–872.

Koes, R.E., Quattrocchio, F., Mol, J.N., 1994. The flavonoid biosynthetic pathway in plants: Function and evolution. Bioessays. 16, 123–132.

Krazeisen A., Breitling, R., Moller, G., Adamski, J., 2001. Phytoestrogens inhibit human 17β-hydroxysteroid dehydrogenase type 5. Mol. Cell. Endo. 171, 151–162.

Krotzer, M.J., 1990. The effects of induced masculinization on reproductive and aggressive behaviors on the female mosquitofish, Gambusizia affinis. Environ. Biol. Fish. 29, 127–134.

Kuiper, G.G., Carlsson, B., Grandien, K., Enmark, E., Haggblad, J., Nilsson, S., Gustafsson, J-A., 1997. Comparison of the ligand binding specificity and transcript tissue distribution of estrogen receptors α and β. Endocrinology. 138, 863–870.

Kuiper G.G., Enmark, E., Pelto-Huikko, M., Nilsson, S., Gustafsson, J.A., 1996. Cloning of a novel receptor expressed in rat prostate and ovary. Proc. Natl. Acad. Sci. USA 93, 5925–5930.

Kuiper, G.G., Lemmen, J.G., Carisson, B., Corton, J.C., Safe, S.H., van der Saag, P.T., van der Burg, B., Gustafsson, J-A., 1998. Interaction of estrogenic chemicals and phyto-estrogens with estrogen receptor β. Endocrinology. 139, 4252–4263.

Kulling, S.E., Honig, D.M., Simat, T.J., Metzler, M., 2000. Oxidative in vitro metabolism of the soy phytoestrogens daidzein and genistein. J. Agric. Food Chem. 48, 4963–4972.

Kurzer. M.S., Xu, X., 1997. Dietary phytoestrogens. Annu. Rev. Nutr. 17, 353–381.

Lampe, J.W., Karr, S.C., Hutchins, A.M., Slavin, J.L., 1998. Urinary equol excretion with a soy challenge: Influence of habitual diet. Proc. Soc. Exp. Biol. Med. 217, 335–339.

Larsson, D.G.J., Hallman, H., Forlin, L., 2000. More male fish embryos near a pulp mill. Environ. Toxicol. Chem. 19, 2911–2917.

Le Bail, J.C., Champavier, Y., Chulia, A.J., Habrioux, G., 2000. Effects of phytoestrogens on aromatase, 3β- and 17β-hydroxysteroid dehydrogenase activities and human breast cancer cells. Life. Sci. 66, 1281–1291.

LeBlanc, G.A., and McLachlan, J.A., 2000. Changes in the metabolic elimination profile of testosterone following exposure of the crustacean *Daphnia magna* to tributytin. Ecotoxicol. Environ. Safety 45, 296–303.

Lehtinen, K-J., Mattson, K., Tana, J., Engstrom, C., Lerche, O., Hemming, J., 1999. Effects of wood-related sterols on the reproduction, egg survival and offspring of brown trout (*Salmo trutta lacustris* L.). Ecotoxicol. Environ. Safety 42, 40–49.

Leopold, A.S., Erwin, M., Oh, J., Browning, B., 1976. Phytoestrogens: Adverse effects on reproduction in California quail. Science 191, 98–100.

Lephart, E.D., West, T.W., Weber, K.S., Rhees, R.W., Setchell, K.D.R., Adlercreutz, H., Lund, T.D., 2002. Neurobehavioral effects of dietary soy phytoestrogens. Neurotox. Teratol. 24, 5–16.

Lerner, L.J., Turkheimer, A.R., Borman, A., 1963. Phloretin, a weak estrogen and estrogen antagonist. Proc. Soc. Exp. Biol. Med. 114, 115–117.

Lewis, D.F.V., Watson, E., Lake, B.G., 1998. Evolution of the cytochrome P450 superfamily: sequence alignments and pharmacogenetics. Mutat. Res. 410, 245–270.

Levy, J.R., Faber, K.A., Ayyash, L., Hughes, C.L., 1995. The effect of prenatal exposure to the phytoestrogen genistein on sexual differentiation in rats. Proc. Soc. Exp. Biol. Med 208, 60–66.

Liehr, J.G., Somasunderam, A., Roy, D., 1998. Metabolism and fate of xeno-oestrogens in man. Pure Appl. Chem. 70, 1747–1758.

Loukovaara, M., Carson, M., Palotie, A., Adlercreutz., H., 1995. Regulation of sex hormone-binding globulin production by isoflavonoids and pattern of isoflavonoid conjugation in HepG2 cell cultures. Steroids 60, 656–661.

Lund, T.D., Rhees, R.W., Setchell, K.D.R., Lephart, E.D., 2001. Altered sexually dimorphic nucleus of the preoptic area (SDN-POA) volume in adult Long-Evans rats by dietary soy phytoestrogens. Brain. Res. 914, 92–99.

MacLatchy, D., Peters, L., Nickle, J., Van der Kraak, G.,.1997. Exposure to β-sitosterol alters the endocrine status of goldfish differently than 17β-estradiol. Environ. Toxicol. Chem. 16, 1895–1904.

Maier, C.G-A., Chapman, K.D., Smith, D.W., 1995. Differential estrogenic activities of male and female plant extracts from two dioescious species. Plant Sci. 109, 31–43.

Maier, C.G-A., Chapman, K.D., Smith, D.W., 1997. Phytoestrogens and floral development in dioecious Maclura pomifera (Raf.) Schneid. and Morus rubru L. (Moraceae). Plant. Sci. 130, 27–40.

Makela, S.R, Poutanen, M., Lehtimaki, J., Kostian, M., Santti, R., Vthko, R., 1995a. Estrogen specific 17β-hydroxysteroid oxidoreductase type I as a possible target for the action of phytoestrogens. Proc. Soc. Exp. Biol. Mol. 208, 51–59.

Makela, S., Santti, R., Salo, L., McLachlan, J.A., 1995b. Phytoestrogens are partial estrogen agonists in adult male mouse. Environ. Health Perspect. 103 (suppl. 7), 123–127.

Mann, J., 1978. Secondary Metabolism. Clarendon Press, Oxford, pp. 252–262.

Marchetti, F., De Santi, C., Vietri, M., Pietrabissa, A., Spisni, R., Mosca, F., Pacifici, G.M., 2001. Differential inhibition of human liver and duodenum.sulphotransferase activities by quercetin, a flavonoid present in vegetables, fruit and wine. Xenobiotica 31, 841–847.

Marcinkowska, E., Wiedlocha, A., 2002. Steroid signal transduction activated at the cell membrane: From plants to animals. Acta Biochim. Polon. 49(3), 735–745.

Markaverich, B.M., Roberts, R.R., Alejandro, M.A., Johnson, G.A., Middleditch, B.S., Clark, J.H., 1988. Bioflavonoid interaction with rat uterine type II binding sites and cell growth inhibition. J. Steroid. Biochem. 30, 71–78.

Markiewicz, L., Garey, J., Adlercreutz, H., Gurpide, E., 1993. In vitro bioassays of non-steroidal phytoestrogens. J. Steroid Biochem. Mol. Biol. 45, 399–405.

Martin, M.E., Haourigui, M., Pelissero, C., Benassayag, C., Nunez, E.A., 1996. Interactions between phytoestrogens and human sex steroid binding protein. Life Sci. 58, 439–436.

Martin, P.M., Horwitz, K.B., Ryan, D.S., McGuire, W.L., 1978. Phytoestrogen interaction with estrogen receptors in human breast cancer cells. Endocrinology 103, 1860–1867.

Martin de las Mulas, J., Millan, Y., Bautista, M.J., Perez, J., Carrasco, L., 2000. Oestrogen and progesterone receptors in feline fibroadenomatous change: An immunohistochemical study. Res. Vet. Sci. 69, 15–21.

Matthews, J., Celius, T., Halgren, R., Zacharewski, T., 2000. Differential estrogen receptor binding of estrogenic substances: A species comparison. J. Steroid. Biochem. Mol. Biol. 74, 223–234.

Mattson, K., Tana, J., Engstrom, C., Hemmings, J., Lehtinen, K-J., 2001. Effects of wood-related sterols on the offspring of the viviparous blenny Zoarces viviparous L. Ecotoxicol. Environ. Safety 49, 122–130.

Mayer, F.L., Mayer, K.S., Ellersieck, M.R., 1986. Relation of survival to other endpoints in chronic toxicity tests with fish. Environ. Toxicol. Chem. 5, 737–748.

Mazur, W., Adlercreutz, H., 1998. Naturally occurring oestrogens in food. Pure Appl. Chem. 70, 1759–1776.

McLachlan, J.A., 2001. Environmental signaling: What embryos and evolution teach us about endocrine disrupting chemicals. Endocri. Rev. 22, 319–341.

McMaster, M.E., Portt, C.B., Munkittrick, K.R., Dixon, D.G., 1992. Milt characteristics, reproductive performance and larval survival and development of white sucker exposed to bleached krafl mill effluent. Ecotoxicol. Environ. Safety 23, 103–117.

Medlock, K.L., Branham, W.S., Sheehan, D.M., 1995. The effects of phytoestrogens on neonatal rat uterine growth and development. Proc. Soc. Exp. Biol. Med. 208, 307–313.

Mellanen, P., Petanen, T., Lehtimaki, J., Makela, S., Bylund, G., Holmnom, B., Mannila, E., Oikari, A., Santti, R., 1996. Wood-derived estrogens: Studies in vitro with breast cancer cell lines and in vivo in trout. Toxicol. Appl. Pharmacol. 136, 381–388.

Mendel, C., 1989. The free hormone hypothesis: a physiologically based mathematical model. Endocr. Rev. 10, 232–274.

Mesiano, S., Jaffe, R.B., 1993. Interaction of insulin-like growth factor-II and estradiol directs steroidogenesis in the human fetal adrenal toward dehydroepiandrosterone sulfate production. J. Clin. Endocrinol. Metab. 77, 754–758.

Mesiano, S., Katz, S.L., Lee, J.Y., Jaffe, R.B., 1999. Phytoestrogens alter adrenocortical function: Genistein and daidzein suppress glucocorticoid and stimulate androgen production by cultured adrenal cortical cells. J. Clin. Endocr. Metab. 84, 2443–2448.

Miksicek, R.J., 1993. Commonly occurring plant flavonoids have estrogenic activity. Mol. Endocrinol. 44, 37–43.

Miksicek, R.J., 1994. Interaction of naturally occurring non-steroidal estrogens with the recombinant human estrogen receptor. J Steroid Biochem. Mol. Biol. 49, 153–160.

Miksicek, R.J., 1995. Estrogenic flavonoids: Structural requirements for biological activity. Proc. Soc. Exp. Biol. Med. 208, 44–50.

Milanesi, L., Monje, P., Boland, R., 2001. Presence of estrogens and estrogen receptor like-proteins in Solanum glaucophyllum. Biochem. Biophys. Res. Cornrnun. 21, 1175–1179.

Milligan, S.R., Balasubramanian, A.V., Kalita, J.C., 1998a. Relative potency of xenobiotic estrogens in an acute in vivo mammalian assay. Environ. Health Perspect. 106, 23–26.

Milligan, S.R., Khan, O., Nash, M., 1998b. Competitive binding of xenobiotic oestrogens to rat alpha-fetoprotein and sex steroid binding proteins in human and rainbow trout (*Oncorhynchus mykiss*) plasma. Gen. Comp. Endocrinol. 112, 89–95.

Miodini, P., Fioravanti, L., Di Fronzo, G., Cappelletti, V., 1999. The two phyto-oestrogens genistein and quercetin exert different effects on oestrogen receptor function. Br. J. Cancer. 80, 1150–1155.

Mitchell, J.H., Gardner, P.T., McPhail, D.B., Morrice, P.C., Collins, A.R., Duthie, G.G., 1998. Antioxidant efficacy of phytoestrogens in chemical and biological model systems. Arch. Biochem. Biophys. 360, 142–148.

Monje, P., Boland, R., 2001. Subcellular distribution of native estrogen receptor α and β isoforms in rabbit uterus and ovary. J. Cell. Biochem. 82, 467–479.

Mosselman, S.J., Polman, R. Dijkema, R., 1996. ER beta: Identification and characterization of a novel human estrogen receptor. FEBS Lett. 392, 49–53.

Mousavi, Y., Adlercreutz, H., 1993. Genistein is an effective stimulator of sex hormone binding globulin production in hepatocarcinoma human liver cancer cells and suppresses proliferation of those cells in culture. Steroids 58, 301–304.

Mueller, G.C., Kim, U.H., 1978. Displacement of estradiol from estrogen receptors by simple alkyl phenols. Endocrinology 102, 1429–1435.

Mulligan, J.T., Long, S.R., 1985. Induction of *Rhizobium meliloti* nodC expression by plant exudate requires nodD. Proc. Natl. Acad. Sci. USA 82, 6609–6613.

Muramatsu, M., Inoue, S., 2000. Estrogen receptors: how do they control reproductive and nonreproductive functions? Biochem. Biophys. Res. Commun. 270, 1–10.

Mussig, C., Altmann, T., 2001. Brassinosteroid signaling in plants. Trends Endocr. Metab. 12, 398–402.

Nadal, A., Ropero, A.B., Laribi, O., Maillet, M., Fuentes, E., Soria, B., 2000. Nongenomic actions of estrogens and xenoestrogens by binding at a plasma membrane receptor unrelated to estrogen receptor α and estrogen receptor β. Proc. Natl. Acad. Sci. USA 21, 11603–11608.

Nagao, T., Yoshimura, S., Nakagoi, M., Usumi, K., Ono, H., 2001. Reproductive effects in male and female rats of neonatal exposure to genistein. Reprod. Toxicol. 15, 399–411.

Nagata, C., Kabuto, M., Kurisu, Y., Shimizu, H., 1997. Decreased serum estradiol concentration associated with high dietary intake of soy products in premenopausal Japanese women. Nutr. Cancer 29, 228–233.

Nagel, S.C., vom Saal, F.S., Thayer, K.A., Dhar, M.G., Boechler, M., Welshons, W.V., 1997. Relative binding affinity-serum modified access (RBA-SMA) assay predicts the relative in vivo bioactivity of the xenoestrogens bisphenol A and octyiphenol. Environ. Health Perspect. 105, 70–76.

Nagel, S.C., vom Saal, F.S., Weishons, W.V., 1998. The effective free fraction of estradiol and xenoestrogens in human serum measured by whole cell euptake assays: Physiology of delivery modifies estrogenic activity. Proc. Soc. Exp. Biol. Med. 217, 300–309.

Negus, N.C., Berger, P.J., 1977. Experimental triggering of reproduction in a natural population of Microtus montanus. Science 196, 1230–1231.

Nelson, D.R., Kamataki, T., Waxman, D.J., Guengerich, F.P., Estabrook, R., Feyereisen, R., Gonzalez, F.J., Coon, M.J., Gunsalus, I.C., Gotoh, O., Okuda, K., Nebert, D.W., 1993. The P450 superfamily: update on new sequences, gene mapping, accession numbers, early trivial names of enzymes, and nomenclature. DNA Cell. Biol. 12, 1–51.

Nikov, G.N., Hopkins, N.E., Boue, S., Alworth, W.L., 2000. Interactions of dietary estrogens with human estrogen receptors and the effects on estrogen receptor-estrogen response element complex formation. Environ. Health. Perspect. 108, 867–872.

Oberdorstor, E., Clay, M.A., Cottam, D.M., Wilmot, F.A., McLachlan, J.A., Milner, M.J., 2001. Common phytochemicals are ecdysteroid agonists and antagonists: A possible evolutionary link between vertebrate and invertebrate steroid hormones. J. Steroid. Biochem. Mol. Biol. 77, 229–238.

Paige, L.A., Christensen, D.J., Gron, H., Norris, J.D., Gottlin, E.B., Padilla, K.M., Chang, C.Y., Ballas, L.M., Hamilton, P.T., McDonnell, D.P., Fowlkes, D.P., 1999. Estrogen receptor modulators each induce distinct conformational changes in ERα and ER,β. Proc. Natl. Acad. Sci. USA 96, 3999–4004.

Pakdel, F., Le Guellec, C., Caillant, C., LeRoux, M.G., Valotaire, Y., 1989. Identification and estrogen induction of two estrogen receptors (ER) messenger ribonucleic acids in the rainbow trout liver: sequence homology with other ERs. Mol. Endocrinol. 3, 44–51.

Parks, L.G., Lambright, C.S., Orlando, E.F., Guillette, L.J. Jr., Ankley, G.T., Gray, L.E. Jr., 2001. Masculinization of female mosquitofish in krafl mill effluent-contaminated Fenholloway river water is associated with androgen receptor agonist activity. Toxicol. Sci. 62, 257–267.

Patisaul, H.B., Dindo, M., Whitten, P.L., Young, L.J., 2001. Soy isoflavone supplements antagonize reproductive behavior and estrogen receptor α and β-dependent gene expression in the brain. Endocrinology 142, 2946–2952.

Patisaul, H.B., Whitten, P.L., Young, L.J., 1999. Regulation of estrogen receptor beta mRNA in the brain: opposite effects of 17β-estradiol and the phytoestrogen coumestrol. Mol. Brain. Res. 67, 165–171.

Perret, X., Staehelin, C., Broughton, W.J., 2000. Molecular basis of symbiotic promiscuity. Microbiol. Mol. Biol. Rev. 64, 180–194.

Peters, N.K., Frost, J.W., Long, S.R., 1986. A plant flavone, luteolin, induces expression of *Rhizobium meliloti* nodulation genes. Science 223, 977–980.

Peters, N.K., Long, S.R., 1988. Alfalfa root exudates and compounds which promote or inhibit induction of Rhizobium meliloti. Plant. Physiol. 88, 396–400.

Poulin, M-J., Simard, J., Catford, J-C., Labrie, F., Piche, Y., 1997. Response of symbiotic endomycorrhizal fungi to estrogens and antiestrogens. Mol. Plant Microbe Interact. 10, 481–487.

Ratna, W.N., Simonelli, J.A., 2002. The action of dietary phytochemicals quercetin, catechin, resveratrol and naringenin on estrogen-mediated gene expression. Life Sci. 70, 1577–1589.

Remy, W., Taylor, T.N., Hass, H., 1994. Four hundred-million year old vesicular arbuscular. Proc. Natl. Acad Sci. USA 91, 11841–11843.

Riddle, J.M., 1992. Contraception and Abortion from the Ancient World to the Renaissance. Harvard University Press, Cambridge, MA.

Roberts, D., Veeramachaneni, D.N.R., Schlaff, W.D., Awoniyi, C.A., 2000. Effects of chronic dietary exposure to genistein, a phytoestrogen, during various stages of development on reproductive hormones and spermatogenesis in rats. Endocrine 13, 281–286.

Roberts-Kirchhoff, E.S., Crowley, J.R., Hollenbert, P.F., Kim, H., 1999. Metabolism of genistein by rat and human cytochrome P450s. Toxicology 12, 610–616.

Robinson, S.P., Langan-Fahey, S., Johnson, D.A., Jordan, V.C., 1991. Metabolites, pharmacodynamics, and pharmacokinetics of tamoxifen in rats and mice compared to the breast cancer patient. Drug. Metab. Dispos. 19, 36–43.

Rosenberg, R.S., Grass, L., Jenkins, D.J.A., Kendall, C.W.C., Diamandis, E.P., 1998. Modulation of androgen and progesterone receptors by phytochemicals in breast cancer cell lines. Biochem. Biophys. Res. Commun. 248, 935–939.

Routledge, E.J., White, R., Parker, M.G., Sumpter, J.P., 2000. Differential effects of xenoestrogens on coactivator recruitment by estrogen receptor (ER) α and ERβ. J. Biol. Chem. 275, 35986–35993.

Rowland, I.R., Wiseman, H., Sanders, T.A., Adlercreutz, H., Bowey, E.A., 2000. Interindividual variation in metabolism of soy isoflavones and lignans: Influence of habitual diet on equol production by the gut microflora. Nutr. Cancer 36, 27–32.

Ryan, K.G., Swinny, E.E., Markham, K.R., Winefield, C., 2002. Flavonoid gene expression and UV photoprotection in transgenic and mutant Petunia leaves. Phytochemistry 59, 23–32.

Saarinen, N., Joshi, S.C., Ahotupa, M., Li, X., Ammala, J., Makela, S., Santti, R., 2001. No evidence for the in vivo activity of aromatase-inhibiting flavonoids. J. Steroid Biochem. Mol. Biol. 78, 231–239.

Sanders, E.H., Gardner, P.D., Berger, P.J., Negus, N.C., 1981. 6-methoxybenzoxazolinone: A plant derivative that stimulates reproduction in Microtus montanus. Science 214, 67–69.

Sarver, J.G., White, D., Erhardt, P., Bachmann, K., 1997. Estimating xenobiotic half-lives in humans and rat data: influence of log P. Environ. Health Perspect. 105, 1204–1209.

Sathyamoorthy, N., Wang, T.Y., Phang, J.M., 1994. Stimulation of pS2 expression by diet-derived compounds. Cancer. Res. 54, 957–961.

Saunders, P.T.K., Sharpe, R.M., Williams, K., Macpherson, S., Urquart, H., Irvine, D.S., Millar, M.R., 2001. Differential expression of oestrogen receptor α and β proteins in the testes and male reproductive system of human and non-human primates. Mol. Hum. Reprod. 7, 227–236.

Scandalios, J.G., 1990. Response of plant antioxidant defense genes to environmental stress. Adv. Genet. 28, 1–41.

Scarlata, S., Miksicek, R., 1995. Binding properties of coumestrol to expressed human estrogen receptor. Mol. Cell. Endocrinol. 115, 65–72.

Schultz, T.D., Bonorden, W.R., Seaman, W.R., 1991. Effects of short-term flaxseed consumption on lignan and sex hormone metabolism in men. Nutr. Res. 11, 1090–1100.

Setchell, K.D.R., 1998. Phytoestrogens: The biochemistry, physiology and implications for human health of soy isoflavones. Am. J. Clin. Nutr. 168, 1333S–1446S.

Setchell, K.D.R., Adlercreutz, H., 1988. Mammalian lignans and phytoestrogens. Recent studies on their formation, metabolism and biological role in health and disease. In: Rowland, I.R (Ed.), Role of the Gut Flora in Toxicity and Cancer. Academic Press, London, pp. 315–345.

Setchell, K.D., Borriello, S.P., Hulme, P., Kirk, D.N., Axelson, M., 1984. Nonsteroidal estrogens of dietary origin: Possible roles in hormone-dependent disease. Am. J. Clin. Nutr. 40, 569–578.

Setchell, K.D.R., Brown, N.M., Desai, P., Zimmer-Nechemias, L., Wolfe, B.E., Brashear, W.T., Kirschner, A.S., Cassidy, A., Heubi, J.E., 2001. Bioavailability of pure isoflavones in healthy humans and analysis of commercial soy isoflavone supplements. J. Nutr. 131, 1362S–1375S.

Setchell, K.D.R., Cassidy, A., 1999. Dietary isoflavones: Biological effects and relevance to human health. J Nutr. 129, 758S–767S.

Setchell, K.D.R., Gosselin, S.J., Welsh, M.B., Johnston, J.O., Balistreri, W.F., Kramer, L.W., Dresser, B.L., Tarr, M.J., 1987. Dietary estrogens—a probable cause of infertility and liver disease in captive cheetahs. Gastroenterology 93, 225–233.

Sharma, S.C., Clemens, J.W., Pisarska, M.D., Richards, J.S., 1999. Expression and function of estrogen receptor subtypes in granulosa cells: Regulation by estradiol and forskolin. Endocrinology 140, 4320–4334.

Shen, F., Weber, G., 1997. Synergistic action of quercetin and genistein in human ovarian carcinoma cells. Oncol. Res. 9, 597–602.

Shore, L.S., Kapulnik, Y., Gurevich, M., Wininger, S., Badamy, H., Shemesh, M., 1995. Induction of phytoestrogen production in Medicago sativa leaves by irrigation with sewage water. Environ. Exp. Bot. 35, 363–369.

Shutt, D., 1976. The effects of plant estrogens on animal reproduction. Endeavour 35, 110–113.

Shutt, D., Cox, R.I., 1972. Steroid and phyto-oestrogen binding to sheep uterine receptors in vitro. J. Endocrinol. 52, 299–310.

Siess, M.H., Lebon, A.M., Suschetet, M., 1992. Dietary modification of drug-metabolizing enzyme-activities: Dose-response effect of flavonoids. J. Toxicol. Environ. Health 35, 141–152.

Siiteri, P.K., Murai, J.T., Hammond, G.L., Nisker, J.A., Raymoure, W.J., Kuhn, R.W., 1982. The serum transport of steroid hormones. Rec. Prog. Horm. Res. 38, 457–510.

Simon, L., Levesque, R.C., Lalond, M., 1993. Identification of endomycorrhizal fungi colonizing roots by fluorescent single-stranded conformation polymorphism-polymerase chain reaction. Appl. Environ. Microbiol. 59, 4211–4215.

Srivastava, P., Sharma, P.K., Dogra, R.C., 1999. Inducers of nod genes of Rhizobium ciceri. Microbiol. Res. 154, 49–55.

Stafford, H.A., 2000. The evolution of phenolics in plants. In: Romeo, J. T., Ibrahim, R., Varin, L., DeLuca, V. (Eds.), Evolution of Metabolic pathways. Elsevier Science, Oxford, pp. 25–54.

Steinmetz, R., Brown, N.G., Allen, D.L., Bigsby, R.M., Ben-Jonathan, N., 1997. The environmental estrogen bisphenol A stimulates prolactin release in vitro and in vivo. Endocrinology 138, 1780–1786.

Strunck, E., Stemmann, N., Hopert, A-C., Wunsche, W., Frank, K., Vollmer, G., 2000. Relative binding affinity does not predict biological response to xenoestrogens in rat endometrial adenocarcinoma cells. J. Steroid Biochem. Mol. Biol. 74, 73–81.

Tarrant, A.M., Atkiknson, S., Atkinson, M.J., 1999. Estrone and estradiol-17β concentration in tissue of the scleractinian coral, Montipora verrucosa. Comp. Biochem. Physiol. A Mol. Integr. Physiol. 122, 85–92.

Teel, R.W., Huynh, H., 1998. Effects of phytochemicals on cytochrome P450-linked alkoxyresorufin O-dealkylase activity. Phytother. Res. 12, 89–93.

Telleria, C.M., Zhong, L., Deb, S., Srivastava, R.K., Park, K.S., Sugino, N., Park-Sarge, O-K., Gibori, G., 1998. Differential expression of the estrogen receptors α and β in the rat corpus luteum of pregnancy: regulation by prolactin and placental lactogens. Endocrinology 139, 2432–2442.

Thigpen, J.E., Li, L.A., Richter, C.B., Lebetkin, E.H., Jameson, C.W., 1987. The mouse bioassay for the detection of estrogenic activity in rodent diets. II. Comparative estrogenic activity of purified, certified and standard open and closed formula rodent diets. Lab. Anim. Sci. 49, 530–536.

Thigpen, J.E., Setchell, K.D.R., Ahlmark, K.B., Locklear, J., Spahr, T., Caviness, G.F., Goelz, M.F., Haseman, J.K., Newbold, R.R., Forsythe, D.B., 1999. Phytoestrogen content of purified, open- and closed-formula laboratory animal diets. Lab. Anim. Sci. 49, 530–536.

Thornton, J.W., 2001. Evolution of a vertebrate steroid receptors from an ancestral estrogen receptor by ligand exploitation and serial genome expansions. Proc. Natl Acad. Sci. USA 98, 5671–5676.

Trembly, G.B., Trembly, A., Copeland, N.G., Gilbert, D.J., Jenkins, N.A., Labrie, F., Giguere, V., 1997. Cloning, chromosomal localization and functional analysis of the murine estrogen receptor β. Mol. Endocrinol. 11, 353–365.

Van der Berg, M., Jongh, J., Poiger, H., Olson, J., 1994. The toxicokinetics and metabolism of polychlorinated dibenzo-p-dioxins (PCDDs) and dibenzofurans (PCDFs) and their relevance for toxicity. Crit. Rev. Toxicol. 24, 1–74.

van Erdenburg, F.J., Daemen, I.A., van der Beek, E.M., Leeuwen, F.W., 2000. Changes in estrogen-α receptor immunoreactivity during the estrous cycle in lactating dairy cattle. Brain. Res. 880, 219–223.

Van Wyk, J.J., Arnold, M.B., Wynn, J., Pepper, F., 1959. The effects of a soybean product on thyroid function in humans. Pediatrics 24, 752–760.

Vermeirsch, H., Simoens, P., Lauwers, H., Coryn, M., 1999. Immunohistochemical localization of estrogen receptors in the canine uterus and their relation to sex steroid hormone levels. Theriogenology 51, 729–743.

Voutilainen, R., Kahn, A.I., Salmenpera, M., 1979. The effects of progesterone, pregnenolone, estrol ACTH and hCG on steroid secretion of cultured human granulosa cells. J. Steroid Biochem. 10, 695–700.

Wang, C., Makela, T., Hase, T., Adlercreutz, H., Kurzer, M.S., 1994. Lignans and flavonoids inhibit aromatase enzyme in human adipocytes. J. Steroid Biochem. Mol. Biol. 50, 205–212.

Weber, K.S., Setchell, K.D.R., Stocco, D.M., Lephart, E.D., 2001. Dietary soy-phytoestrogens decrease testosterone levels and prostate weitht without altering LH, prostate 5α-reductase or testicular steroidogenic acute regulatory peptide levels in adult male Sprague-Dawley rats. J. Endocrinol. 170, 591–599.

Wehling, M., 1997. Specific, non-genomic actions of steroid hormones. Annu. Rev. Physiol, 59, 365–393.

Whitten, P.L, Kudo, S., Okubo, K.K., 1997. Isoflavonoids. In: D'Mello, J.P.F. (Ed.), Handbook of Plant and Fungal Toxicants. CRC Press, Boca Raton, FL, pp. 117–137.

Whitten, P.L., Naftolin, F., 2001. Effects of a phytoestrogen diet on estrogen-dependent reproductive processes in immature female rats. Steroids 57, 55–61.

Whitten, P.L., Patisaul, H.B., 2001. Cross-species and interassay comparison of phyto-estrogen action. Environ. Health. Perspect. 109(suppl. 1), 5–20.

Whitten, P.L., Russell, E., Naftolin, F., 1992. Effects of a normal, human-concentration phytoestrogen diet on rat uterine growth. Steroids 57, 98–106.

Whitten, P.L., Russell, E., Naftolin, F., 1994. Influence of phytoestrogen diets on estradiol action in the rat uterus. Steroids 59, 443–449.

Woodson, J.C., Gorski, R.A., 2000. Structural sex differences in the mammalian brain: Reconsidering the male/female dichotomy. In: Matsumoto, A. (Ed.), Sexual Differentiation of the Brain. CRC Press, Boca Raton, FL, pp. 230–247.

Wynne-Edwards, K.E., 2001. Evolutionary biology of plant defenses against herbivory and their predictive implications for endocrine disruptor susceptibility in vertebrates. Environ. Health Perspect. 109, 443–448.

Yamini, B., Bursian, S.J., Aulerich, R.J., 1997. Pathological effects of dietary zearalenone and/or tamoxifen on female mink reproductive organs. Vet. Hum. Toxicol. 39, 74–78.

Zand, R.S.R., Jenkins, D.J.A., Diamandis, E.P., 2000. Genistein: A potent antiandrogen. Clin. Chem. 46, 887–888.

Zhang, B.L., Wang, W.Q., Yuan, R.Y., 1994. Binding of anthraquinones and flavonoids to bovine serum-albumin. Acta Chim. Sin. 52, 1208–1212.

Zieba, D., Bilinska, B., Schmalz-Fraczek, B., Murawski, M., 2000. Immunohistochemical localization of estrogen receptors in the ovine corpus luteum throughout the estrous cycle. Folia Histochem. Cytobiol. 38, 111–117.

16

Triazines

Timothy S. Gross
R. Heath Rauschenberger

Triazines are a group of herbicides used worldwide to control weeds during the production of food crops such as corn (*Zea mays*), sorghum (*Sorghum vulgare*), wheat (*Triticum aestivum*), sugarcane, and pineapple (*Anana comsus*). In addition, triazines are used for weed control in the production of turf grass, conifers, guavas, and in roadside maintenance (Giddings et al., 2000). Triazines are among the top-selling herbicides in the United States. For example, the triazine herbicide known as atrazine is one of the most heavily used herbicides in the United States, with an estimated 76–85 million pounds produced annually (Rabert et al., 2001) and approximately 32 million kg of active ingredient applied in the United States annually (National Library of Medicine, 2001). Besides atrazine, some other commonly used triazines include cyanazine, propazine, and simazine. Other members of the triazine class include ametryn, atratone, aziprotyne, methoprotryne, metribuzin, prometon, prometryn, simetryn, terbutryn, and trietazine (Stevens and Sumner, 1991).

Despite the widespread use of these chemicals, relatively little is known regarding their potential effects on the health of wildlife and humans. In light of such widespread use and limited information about health effects, there has been an increasing interest in potential detrimental impacts due to unintended exposure. This chapter covers the physicochemical characteristics of triazine herbicides, as well as the potential mechanisms of action, environmental pathways, effects on vertebrate wildlife, and the overall implications for wildlife and human health. Discussions of potential effects include general toxicities, mutagenesis/carcinogenesis, and a primary focus on reproductive and endocrine toxicities. We concentrate on atrazine, cyanazine, propazine, and simazine because they are the most likely triazines to be encountered by wildlife. In addition, few or no data exist concerning potential reproductive or endocrine toxicities for triazines other than atrazine. Therefore, our primary focus is on atrazine and its potential as a hormonally active chemical.

General Characteristics of Triazines

Triazines are characterized by a symmetric, aromatic structure consisting of alternating carbon and nitrogen atoms in a six-membered ring. Of the three ring carbons, one carbon is bonded to a chlorine atom and the remaining two carbons are bonded to amino groups (Baird, 1995). The structures of atrazine, simazine, cyanazine, and propazine all exhibit the characteristic ring consisting of alternating nitrogen-carbon atoms but differ with respect to the attached functional groups (figure 16-1). Atrazine, simazine, cyanazine, and propazine also exhibit similar physicochemical properties, which is expected considering their similar structures and chemical formulas (table 16-1). For example, all four triazines (in technical form) are odorless, colorless crystalline solids. They exhibit low vapor pressures, low octanol-water coefficients (K_{ow}), and similar densities. Water solubility ranges from 5 mg/l at 20°C for propazine to 170 mg/l at 25°C for cyanazine, with atrazine having an intermediate solubility of 33 mg/l at 22°C (National Library of Medicine, 2001). These values indicate that triazines, in general, are water soluble and exhibit low volatility, low hydrophobicity, and low lipophilicity.

Triazine uptake occurs through the foliage and roots of herbaceous plants and through the surfaces of cells of unicellular plants (Hull, 1967). Once triazine is absorbed into a plant, triazines compete with plastoquinone II. Such competition prevents electron transport via photosystem II (Hull, 1967; Forney and Davis, 1981). Photosystem II utilizes the breakdown of water to acquire replacement electrons (Woolhouse, 1981). If replacement electrons are not available, electron flow from photons is arrested after all of the photosystem II molecules have been oxidized. As soon as electron flow is arrested, ATP production ceases, along with cyclic phosphorylation, fixation of CO_2 during cellular respiration, reduction of

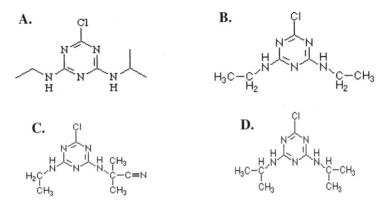

Figure 16-1. Structure of triazine compounds. A, atrazine. B, simazine. C, cyanazine. D, propazine.

Table 16-1. Physicochemical characteristics of selected triazine herbicides (Stevens and Sumner, 1991; National Library of Medicine, 2001).

	Atrazine	Simazine	Cyanazine	Propazine
CASRN no.	1912-24-9	122-34-9	21725-46-2	139-40-2
Formula	$C_8H_{14}ClN_5$	$C_7H_{12}ClN_5$	$C_9H_{13}ClN_6$	$C_9H_{16}ClN_5$
Molecular weight	216	202	241	230
Odor	Odorless	Odorless	Odorless	Odorless
Corrosivity	Noncorrosive	Noncorrosive	Noncorrosive	Noncorrosive
Dissociation constants: (pK_a)	1.68	1.62	0.87	1.7
Octanol-water coefficient (log K_{ow})	2.61	2.18	2.22	2.93
Physical state	White Crystals	White Crystals	White Crystals	White Crystals
Vapor pressure (mm Hg)	2.89×10^{-7} @ 25°C	2.2×10^{-8} @ 25°C	1.38×10^{-7} @ 25°C	2.9×10^{-8} @ 20°C
Melting point (°C)	173–175	225 to 227	167.5–169	230
Solubility in water (mg/l)	33 at 22°C	6.2 at 20°C	170 at 25°C	5.0 at 20°C
Density (g/cm³)	1.23 at 22°C	1.33 at 22°C	1.29 kg/l at 20°C	1.16 at 20°C

chlorophyll I molecules, and the production of NADPH. The interruption of photosynthesis ultimately causes carbohydrate production to terminate and leads to an increase in CO_2 within the cell and a reduction in the carbon reservoir (Shabana, 1987). In addition, damage to chlorophyll molecules will occur if plants are exposed to sunlight while under the influence of photosynthetic blockers (Giddings et al., 2000).

The plant toxicity of triazine is reversible if the triazine compound is removed from the active site (Jensen et al., 1977). For example, when plants are removed from triazine-treated media and replanted in uncontaminated media, photosynthesis will begin to increase (Klaine et al., 1996).

Atrazine

Atrazine was released to open market in 1958 by J.R. Geigy S.A. and is currently available as 50 and 80% wettable powders and in a flowable formulation. It is used on crops such as corn, sorghum, sugarcane, and pineapples as a selective pre- and postemergence herbicide. Atrazine also is used for selective control of aquatic plants, as well as general weed control.

Atrazine is used across a wide geographic range. For example, in 2000, atrazine was applied at a rate of 1 lb/acre to 68% of corn acreage in 18 states surveyed by the United States Department of Agriculture. The states surveyed were Colorado, Illinois, Indiana, Iowa, Kansas, Kentucky, Michigan, Minnesota, Missouri, Nebraska, New York, North Carolina, North Dakota, Ohio, Pennsylvania, South Dakota, Texas, and Wisconsin (National Agricultural Statistics Service, 2001). Besides its use on corn (83% of total yearly atrazine production), other uses include sorghum (11% of total yearly atrazine production) and sugarcane (4% of total yearly atrazine production). The remaining 2% of atrazine is split among turfgrass, roadside or right-of-way maintenance, and other uses (Stevens and Sumner, 1991; Rabert et al., 2001).

Physicochemical Characteristics

Atrazine ($C_8H_{14}CIN_5$) has a molecular weight of 216. The solubility of atrazine in methanol is 18,000 mg/l, in water it is 30 mg/l at 20°C, and in chloroform it is 52,000 mg/l. Atrazine forms white crystals with a vapor pressure of 3.0×10^{-7} mm Hg at 20°C and has a melting point of 175–177°C. It can be stored for a long period and still maintain its efficacy. Atrazine is hydrolyzed at high temperatures by alkali or mineral acids but is stable in slightly acidic, basic, and neutral material. Atrazine exhibits some sensitivity to sunlight and extreme temperatures (Stevens and Sumner, 1991; National Library of Medicine, 2001).

Chemical Name and Synonyms

Atrazine's technical names are 2-chloro-4-ethylamino-6-isoproplamine-*s*-triazine (CAS) and 6-chloro-*N*-ethyl-*N'*-isopropyl-1,3,5-triazinediyl-2-4-diamine (IUPAC). The CAS registry number for atrazine is 1912-24-9 and although the common name atrazine is generally used, several tradenames abound. Selected tradenames include Actinite, Argezin, Atrazin, Candex, Chromozin, Primase, Weedex, Wonuk, and Zeazine (Stevens and Sumner, 1991; National Library of Medicine, 2001).

Cyanazine

In 1971, Shell Chemical Company began marketing cyanazine. It is a herbicide used for pre- and postemergence control of annual grasses and broadleaf weeds. It is produced as a granular product, wettable powder, flowable concentrate, emulsifiable concentrate, and soluble concentrate. Cyanazine herbicide is registered for use on corn, cotton, grain sorghum, and wheat fallow. Cyanazine is offered in combination with atrazine, alachlor, metolachlor, paraquat, and butylate.

Physicochemical Characteristics

Cyanazine ($C_9H_{13}ClN_6$) has a molecular weight of 241. It is a white crystalline solid with a vapor pressure of 1.6×10^{-9} mmHg at 20°C and melts at 166.5–167°C. The solubility of cyanazine in water at 25°C is 171 mg/l, and it is very soluble in acetone, benzene, chloroform, and ethanol (Stevens and Sumner, 1991).

Chemical Name and Synonyms

Cyanazine is 2-[4-chloro-6- (ethylamino)-s-triazin-2-yl] amino-2-methylpropionitrile (IUPAC) or [2-chloro-4-(1-cyano-1-methylethylamino)-6-ethylamino-s-triazine] (CAS). The common name cyanazine is in general use, and tradenames include Bladex and Fortrol. Code designations include SD15418 and WL19805. The CAS registry number is 21725-46-2 (National Library of Medicine, 2001).

Propazine

Propazine was entered into the agrochemical market by J.R. Geigy S.A. in 1960. It is an herbicide for preemergence use on broadleaf and grassy weeds in millet and umbelliferous crops. Propazine is produced as 50 and 80% wettable powders as well as a 4 lb/gal emulsifiable concentrate and a 90% water-dispersible granule. Propazine is often distributed in combination with metolachlor (Stevens and Sumner, 1991).

Physicochemical Characteristics

The empirical formula for propazine is $C_9H_{16}ClN_5$ and the molecular weight is 230. Propazine forms colorless crystals that melt at 212–214°C, and it has a vapor pressure of 2.9×10^{-8} mm Hg at 20°C. It has more than 95% pure technical material. Propazine's solubility in water is 8.6 mg/l at 20°C. In organic solvents it is poorly soluble. Propazine is stable in neutral, slightly acid, or alkaline media but able to be hydrolyzed by stronger acids and alkalis (Stevens and Sumner, 1991).

Chemical Name and Synonyms

Propazine is 2-chloro-4,6-bis(isopropylamino)-*s*-triazine. Its common name is in general use in the United States. Commercial names include Gesamil, Milo-Pro, Milogard, Primatol, and Prozinex. Code designations include G-30028, and the CAS registry number is 139-40-2 (National Library of Medicine, 2001).

Simazine

J.R. Geigy S.A. introduced simazine in 1956. It is a herbicide used during preemergence to control broadleaf and grassy weeds in deep-rooted crops. It is also used on corn. Simazine shows potential for controlling submerged vegetation and algae. It is available as 50 and 80% wettable powders as well as a water-dispersible granule containing 90% of the active ingredient.

Physicochemical Characteristics

Simazine ($C_7H_{12}ClN_5$) has a molecular weight of 202. The white crystalline solid melts at 225–227°C, and its vapor pressure is 6.1 x 10^{-9}. Simazine's solubility in water is 5 mg/l at 20–22°C. In light petroleum, it is soluble at 2 mg/l and is soluble at 400 mg/l in methanol. It is stable in neutral and slightly basic or acidic media. Simazine is hydrolyzed by stronger acids and bases (Stevens and Sumner, 1991).

Chemical Name and Synonyms

The chemical name for simazine is 2-chloro-4,6-bis(ethylamino)-*s*-triazine. Its common name is in general use. Commercial names include Aquazine, Gesatop, Cekusan, Primatol/S, Princep, Simadex, and Simanex. The code designation is G-27692 and the CAS registry number is 122-34-9 (National Library of Medicine, 2001).

Mammalian Toxicity

Atrazine

There are few data regarding the effects of atrazine on mammalian wildlife. However, information regarding its effects on domestic and laboratory mammals is steadily increasing. Contemporary data suggest that mammals are comparatively resistant to atrazine and that atrazine itself is not carcinogenic, mutagenic, or teratogenic (Reed, 1982).

Atrazine exhibits low acute toxicity in mammals. The dermal LD_{50} for rabbits is 7500 mg/kg (Reinhardt and Britelli, 1982). In rats, oral LD_{50} values of approximately 1900–3000 mg/kg have been reported (Worthing, 1987). Hartley and Kidd (1983) reported oral LD_{50} values for mice and rabbits as 1750 and 750 mg/kg, respectively. Dermal LD_{50} and inhalation LC_{50} (1 h) values in rats are >3000 mg/kg and 700 mg/m^3, respectively (Worthing, 1987). Symptoms observed in rats that received a lethal oral dose include exicitation followed by depression with reduced respiratory rate, motor incoordination, clonic spasms, and hypothermia. The rats typically died within 12 to 24 h after being dosed. In rats fed atrazine for 6 months, dietary levels of 100 and 500 mg/l caused growth retardation, partly due to reduction in food intake. Histological examination revealed no lesions (Suschetet et al., 1974).

Rats that were orally dosed at a rate of 3000 mg/kg exhibited cardiac dilation, macroscopic hemorrhages in the liver and spleen, and lung edema with extensive hemorrhagic foci. These animals died within 6 h of dose administration. Rats observed on the second day following the same dosage showed dystrophic changes of the kidney tubules, hemorrhagic pneumonia, and hemorrhage in other organs (Stevens and Sumner, 1991).

In rats that received oral doses of atrazine for 6 months at a rate of 20 mg/kg/day, 4 out of every 10 animals that died showed respiratory dysfunction and paralysis of extremities. Respiratory dysfunction was the probable result of bronchitis and peribronchitis, along with edema of the brain (Nezefi, 1971). However, these data contradict other studies dealing with atrazine toxicity. For example, Suschetet et al. (1974) reported no pathology in rats that were fed 25 mg/kg/day for 6 months.

Atrazine has been reported to be an eye irritant (National Library of Medicine, 2001). However, technical atrazine has been reported as only slightly irritating to rabbit skin and nonirritating to the rabbit eye. Atrazine was found to be a sensitizer in the guinea pig, as demonstrated in a modified Buehler sensitization test (Ciba-Geigy, 1982a).

Reproductive Effects and Endocrine Modulation

Atrazine administered via subcutaneous injection at a rate 800 mg/kg/day during days 3, 6, and 9 of gestation resulted in death of a few to the majority of pups in each of the treated rat litters. Subcutaneous dosages of 200 mg/kg/day had no affect on the number of pups per litter or on their weight at weaning. Oral dosing of up to 1000 mg/l (about 50 mg/kg/day) also yielded no effect (Peters and Cook, 1973). In pigs, atrazine was mixed in feed at a rate of 1 mg/kg body weight and fed for 19 days before estrus onset. Pigs that were dosed exhibited lower estradiol-17β (E_2) in blood serum than control animals on day 2 before estrus began. However, on the day of expected estrus onset, E_2 blood serum levels were higher in treated animals than in control animals (Gojmerac et al., 1999).

In another study, pregnant and nonpregnant ewes died within 36–60 days after receiving a dose of 30 mg/kg. Some embryonic mortality was observed, but some embryos were alive when the ewe died. A dose of 15 mg/kg/day throughout pregnancy resulted in normal delivery at term. Offspring of the dosed females were nursed by their mothers for 30 days while the ewes continued to receive atrazine, and no toxicity was observed in either offspring or mother (Johnson et al., 1972). In mice, atrazine caused no significant increase in embryonic anomalies in three strains of mice after the pregnant female received an oral dose of 46.4 mg/kg/day during days 6–14 of gestation (Mrak, 1969). These studies suggest that mammalian wildlife are not at serious risk with respect to acute toxicity, as maximum concentrations in water bodies rarely exceed 1000 μg/l (Eisler, 1989). These elevated levels occur as a result of rainfall immediately after atrazine has been applied to a field.

Current evidence suggests that atrazine may induce endocrine modulation by acting as a steroid antagonist (antiandrogen and/or antiestrogen), probably through a non–receptor-mediated mechanism. Indeed, a number of in vivo and in vitro studies failed to detect estrogenic activity for triazines. In two independent studies, oral exposure to atrazine and simazine did not increase uterine weight in immature or ovariectomized female Sprague-Dawley rats (Connor et al., 1996; Tennant et al., 1994) and cell proliferation and binding studies found no evidence for either agonistic or antagonist activity for this herbicide (Connor et al., 1996). Furthermore, atrazine and related compounds failed to demonstrate estrogenic activity in human and yeast cells expressing the E_2 receptor (ER) and an estrogen-sensitive reporter gene (Tran et al., 1996; Balaguer et al., 1999), although the triazines have displaced radiolabeled E_2 from the ER in competition studies (Tran et al., 1996; Vonier et al., 1996). In addition, a study by Danzo (1997) showed that atrazine did not reduce the binding of radiolabeled E_2 to rabbit uterine ER, although it inhibited the binding of dihydrotestosterone (DHT) to androgen-binding protein by 40%. Triazines also may disrupt reproductive function by altering luteinizing hormone (LH) and prolactin concentrations (Cooper et al., 2000).

Metabolism

Analysis of atrazine metabolism in rat liver in vitro showed that conjugation was secondary to dealkylation. Conjugation and dealkylation were completed by the microsomal and soluble fractions, respectively. Isopropyl groups were more easily excised than the ethyl groups, and dechlorination of chloro-s-thiazines was not observed (Dauterman and Muecke, 1974). The major metabolites are suggested to be mono-or di-N-dealkylated products (in the rat model), and roughly 80% of a radiolabeled dose was eliminated within 72 h (Bakke et al., 1972a). Atrazine, and not its metabolites, were determined to be the main material excreted in a gas chromatographic–mass spectrophotometric analysis (Erickson et al., 1979).

Atrazine metabolites include hydroxyatrazine, dealkylated analogs, and cysteine and glutathione conjugates. The binding of atrazine and its metabolites in plants limits bioavailability to animals (Khan and Akhtar, 1983; Khan et al., 1985). Metabolic processing of atrazine in mammals is usually rapid and extensive; unchanged atrazine was recovered only from the feces (Geigy, 1963). Atrazine is metabolized in the liver of pigs, rats, and sheep by partial N-dealkylation and hydrolysis (Bakke et al., 1972b; Dauterman and Muecke, 1974; Foster et al., 1980). In male CD rats receiving 50 mg/kg of atrazine orally, 34% of the dose is reportedly converted in vivo to diaminoatrazine (Meli et al., 1992).

Mutagenesis and Carcinogenesis

Atrazine has been reported to be nonmutagenic as determined by studies that investigated gene mutation, chromosomal aberrations, and potential interaction with

DNA using animal, plant, and microbial models (Stevens and Sumner, 1991). In mice that received oral doses at a rate of 21.5 mg/kg/day from the age of 1–4 weeks and later a dose of 82 mg/kg for another 17 months, atrazine did not cause tumors (Innes et al., 1969).

Cyanazine

The oral LD_{50} of cyanazine for rats is 334 mg/kg and the dermal LD_{50} for rabbits is <2000 mg/kg. No mortality was noted in an inhalation toxicity evaluation in rats using an 80% wettable powder with 1-h exposure to 4.9 ml/l cyanazine (Beste, 1983). Another single-dose evaluation placed the oral LD_{50} for rats at 182–380 mg/kg and the oral LD_{50} in the mouse at 380 mg/kg (Worthing, 1987). Worthing (1987) also set the dermal LD_{50} for rats at >1200 mg/kg. No toxicological effects were noted in 2-year feeding studies in rats and dogs with dietary levels up to and including 25 mg/l cyanazine. Depression and inactivity were noted in laboratory animals given high doses of cyanazine (Beste, 1983).

Reproductive Effects

Head anomalies observed in fetuses of Fisher 344 rats exposed to cyanazine prompted the U.S. EPA to conduct a special review of the herbicide in 1985. The fetuses were taken after the mothers had received oral doses of cyanazine (25 mg/kg/day) during gestational days 6–15 (Office of Pesticides and Toxic Substances, 1985). This study established a clear no-effect level of 10 mg/kg/day. A similar study involving orally dosed Sprague-Dawley rats showed no teratogenic or developmental effects at 30 mg/kg/day. A no-observable-adverse-effect level (NOAEL) was not established for either developmental or maternal toxicity in a second teratology study conducted with the Fisher 344 rats receiving oral doses of 5, 25, and 75 mg/kg/day. In the New Zealand rabbit, a teratology study did not indicate any teratogenic potential, and an NOAEL was established at 1 mg/kg/day (oral administration). However, fetotoxicity was noted at 2 mg/kg/day (Office of Pesticide Programs, 1987). The teratogenic potential of cyanazine is still controversial (Stevens and Sumner, 1991).

Metabolism

In rats and dogs, cyanazine is rapidly metabolized and eliminated (within 4 days; Worthing, 1987). The oral administration of [^{14}C] cyanazine resulted in 40% of the administered dose being excreted in the urine and 47% in the feces (Hutson et al., 1970). Both studies suggested that the primary metabolic pathway for cyanazine in rats is N-deethylation to yield an amine. Dechlorination resulted in a 2-hydroxy triazine. The cyano group hydrolyzed to an amide and then further to a carboxyl analog. The 2-hydroxy compound was a major metabolite detected in the feces, and the bile contained glutathione conjugates. In addition, N-acetylcyteinyl derivatives were found in the urine.

Mutagenicity and Carcinogenicity

Studies covering the classes of gene mutation and chromosomal aberration, as well as other tests, indicate that cyanazine is not mutagenic (Office of Pesticide Programs, 1984). Cyanazine's oncogenic potential was evaluated in mice at oral dosing levels of 10, 25, and 1000 mg/l. There were no oncogenic responses noted at any level (Office of Pesticide Programs, 1986).

Propazine

In rats, the oral LD_{50} of propazine may range from 5000 mg/kg to >7700 mg/kg. The inhalation LC_{50} in the rat is >1 mg/l for a 4-h exposure, and the dermal LD_{50} is >3100 mg/kg (Ciba-Geigy, 1982b). Rats receiving lethal doses of propazine exhibit labored breathing, lethargy, diarrhea, emaciation, and rhinitis (Ciba-Geigy, 1982b). Propazine in its technical form is reported to be mildly irritating to rabbit skin and nonirritating to the rabbit eye. In standard guinea pig sensitization tests, propazine did not exhibit sensitization properties. Rabbits receiving oral doses of propazine at a rate of 500 mg/kg/day for 4 months exhibited effects in blood and liver, including hepatomegaly (liver) with focal necrosis and fatty degeneration, some atrophy of lymph nodes, and hypochromic, macrocytic anemia (Stevens and Sumner, 1991).

Metabolism

Ring-labeled propazine was orally administered to rats at doses of 41–56 mg/kg. Excretion of radiolabeled propazine was most rapid during the first 24 h and decreased to trace amounts by 72 h. By 72 h, 23% of the labeled propazine had been recovered in feces and 66% had been recovered in urine. Tissue concentrations at 96 h ranged from 19.8 to 39.3 mg/kg. At 8 days, the tissue concentrations ranged from 13 to 30 mg/kg. Analysis of urine using ion exchange chromatography identified 18 metabolites (Bakke et al., 1967).

Carcinogenesis and Mutagenesis

Propazine was not tumorigenic in studies on mice (Innes et al., 1969). However, increased occurrence of benign neoplasia was noted in rats receiving an extremely high dose (Ciba-Geigy, 1982b).

Simazine

In single doses the oral LD_{50} in the rat is >5000 mg/kg, and the acute dermal LD_{50} in the rabbit is >10,000 mg/kg (Worthing, 1987). The 4-h inhalation LC_{50} for rats is >2 mg/l (Ciba-Geigy, 1982c). Sheep are much more susceptible to poisoning by simazine with symptoms that include cyanosis and convulsions associated with acute toxicity (Stevens and Sumner, 1991). The reasons for the increased susceptibility are unknown. Palmer and Radeleff (1964) reported this in connection with repeated administrations of 1400 mg/kg or more. Other studies suggest that 500 mg/kg could be lethal to sheep. Drowsiness and irregular respirations are produced with oral administration of 5000 mg/kg in rats. Simazine is reported to be nonirritative to the skin and eyes of rabbits (Ciba-Geigy, 1982c).

Mutagenicity and Carcinogenicity

Simazine increased the frequency of X-linked lethality when injected into male fruit flies (*Drosophila melanogaster*). Other mutagenicity tests yielded negative results (Murnik and Nash, 1977). Mice given the highest tolerated dose of simazine did not develop tumors, but subcutaneous injection of simazine did produce sarcoma at the injection site in rats and mice (Innes et al., 1969). In Sprague-Dawley rats, females developed mammary tumors when fed doses at the 100 mg/l level (Office of Pesticide Programs, 1989).

Effects of Triazines in Vertebrate Wildlife

Fishes

Laboratory studies have revealed considerable variability with respect to atrazine toxicity among fish species. For example, rainbow trout (*Salmo gairdnerii*) exhibited an LC_{50} of 4.5–8.8 mg/l during a 96-h exposure period, whereas the carp (*Cyprinus carpio*) LC_{50} was 76–100 mg/l during the same exposure timeframe. Guppies (*Lebistes cyanellus*) exhibited sensitivity (LC_{50}: 4.3 mg/l) similar to that of rainbow trout, while bluegill (*Lepomis macrochirus*) showed slightly greater tolerance (16.0 mg/l at 96-h exposure) than rainbow trout or guppies (National Library of Medicine, 2001). Toxicokinetics of atrazine were examined in zebra fish at concentrations of 4, 10, and 20 mg/l. Normal development at the long pec stage (48 h after fertilization) were affected at the 4 mg/l concentration, and 10 and 20 mg/l caused retardations in organogenesis, a reduction in movements, and cardiac and circulatory system dysfunction (Wiegand et al., 2001). In addition, although normal development at the long pec stage was affected, no acute toxicity was evident at the 4 mg/l level (Wiegand et al., 2000). These data, when compared to ecologically relevant concentrations (ranging from 0.00003 to 4.9 mg/l; Eisler, 1989), suggest that toxic effects are unlikely, except for areas receiving runoff from fields immediately after application.

Few studies have examined endocrine or reproductive function in fish exposed to atrazine or other triazine pesticides. Channel catfish (*Ictalurus punctatus*) and gizzard shad (*Dorosoma cepedianum*) maintained for 4.5 months in ponds containing 20 µg/l atrazine failed to reproduce, and reproductive success of bluegills was reduced more than 95% (Kettle et al., 1987). Since the dietary habits of bluegill were largely affected by the herbicide treatment, the authors suggested that impaired reproduction might have been due to impoverishment rather than to a direct effect of atrazine exposure. In mature male Atlantic salmon (*Salmo salar*), short-term exposure of atrazine at 1.0 µg/l reduced response to the female pheromone pros-taglandin $F_{2\alpha}$ (Moore and Lower, 2001). Results from our laboratories have shown that atrazine affects sex steroids in male and female largemouth bass (*Micropterus salmoides*; Gross and Wieser, 2002). After 20 days of exposure, plasma 11-ketotestosterone (11-KT) concentrations were elevated in males exposed to 100 µg/l atrazine, and E_2 concentrations were increased in females exposed to 50 and 100 µg/l atrazine. Studies with largemouth bass also have shown that incubation of ovarian follicles 10 mg/l atrazine results in increased E_2 and decreased testosterone production. Furthermore, in situ testosterone synthesis is greatly reduced when go-nads are incubated with a combination of atrazine and floridone or atrazine and chlor-dane (Shrestha et al., 1997).

Amphibians and Reptiles

Relative to mammals, few studies have dealt with the toxicological effects of atra-zine on amphibians and reptiles. However, due to the apparent global decline in amphibian populations, the effects of environmental contaminants on amphibians have received increased attention. Morgan et al. (1996) reported that atrazine ex-hibited significant teratogenic effects at 8 mg/l using the FETAX (Frog Embryo Teratogenesis Assay–*Xenopus*) test. The LOAEC (lowest-observed-adverse-effect concentration) was extrapolated to be 1.1 mg/l. Allran and Karasov (2001) reported similar results in that the occurrence of abnormalities (i.e., wavy tail, lateral tail flexure, and facial edema) in the developing larvae of the northern leopard frog (*Rana pipiens*), the wood frog (*Rana sylvatica*), and the American toad (*Bufo americanus*) showed a concentration-dependent pattern. In addition, adults exposed to the highest concentrations would not eat during the 14-day exposure period. However, it was suggested that 0–20 mg/l concentrations did not affect hatchabil-ity or 96-h posthatch mortality in the larvae of these three species.

Several studies have reported potential effects of atrazine on sex differentiation in the South African clawed frog (*Xenopus laevis*) but with differing results. Tavera-Mendoza et al. (2002) reported reduced testicular volume, testicular resorption, and a reduction in spermatogonial cells and nurse cells after a single-pulse exposure for 48 h during sexual differentiation at 21 µg atrazine/l. Similar results were re-ported by Hayes et al. (2002) with exposures as low as 1 µg atrazine/l. This study reported demasculinization and hermaphroditic metamorphs following continuous exposure to low doses of atrazine throughout metamorphosis in *X. laevis*. How-ever, another study performed in the same species (Carr et al., 2003) did not dem-onstrate effects at these low concentrations. Indeed, this study did not indicate a significant increase in the incidence of intersex metamorphs until atrazine concen-trations exceeded 20 µg atrazine/l. Nonetheless, these studies suggest a potential for altered sex differentiation in frogs exposed to atrazine during metamorphosis, and additional studies will be needed to document and verify these potential effects.

Alligators and turtles have received some attention related to effects involving endocrine modulation. Studies with alligators have shown that atrazine might in-duce differential responses in developing embryos depending on timing of expo-sure. Crain et al. (1997) reported that topical application of an atrazine-ethanol

(14 mg atrazine /L-ethanol) mixture induced gonadal aromatase activity in male hatchling alligators exposed in ovo. In a later study, however, incubation of alligator eggs with atrazine before the critical period of gonadal differentiation did not influence sex determination and had no apparent effect on gonadal structure (measured as sex cord diameter in males, Müllerian duct epithelial cell height in females, and medullary regression of the ovaries in females), nor on hepatic aromatase activity (Crain et al., 1999). Since most endocrine changes associated with atrazine have been reported in normally organized reptilian reproductive systems, the authors hypothesized that the lack of noticeable effects in the latter study was the result of exposing embryos during very early developmental stages, before or during the development of the reproductive system. Recent studies evaluated the potential for atrazine to alter sex differentiation in alligators and freshwater turtles exposed to atrazine at ecologically relevant doses and exposures (Gross 2001a,b). These studies did not demonstrate any effects of atrazine on sex differentiation in either species at concentrations of 100 µg/l or less. In addition, these studies indicated that surface exposure of reptile eggs (trans-ovo exposure) was not a significant route of exposure in reptiles.

Birds

At ecologically relevant levels, atrazine is not acutely toxic to birds. Studies report oral LD_{50} values of >5000 mg/kg (30% active ingredient) for the mallard duck (*Anas platyrhynchos*) and >2000 mg/kg for 6-month-old female mallard ducks. For bobwhite quail (*Colinis virginianus*), oral LD_{50} values are >5000 mg/kg for an 8-day dietary dose, and LD_{50} values for pheasants are >2000 mg/kg (80% active ingredient) (National Library of Medicine, 2001). In addition, the possibility of atrazine exposure causing chronic effects on wetland aquatic organisms, as well as for biomagnification to result in toxic residues through waterfowl food chains, is unlikely (Huckins et al., 1986). Indirect effects of atrazine on insect- and seed-eating birds are not well understood, and such impacts may alter the survival of certain species during nesting and brood rearing.

Metabolism

Atrazine is rapidly metabolized in domestic chickens (*Gallus* sp.) through partial *N*-dealkylation along with hydrolysis. Dealkylation occurs through the ethylamino group, which results in degradation products (Foster and Khan, 1976). Goose liver (*Anser* sp.) homogenates contained enzyme systems that metabolized atrazine by hydrolysis, resulting in the formation of hydroxyatrazine. The atrazine metabolites deethylatrazine and deisopropylatrazine were further hydrolyzed to their corresponding hydroxy analogs (Foster et al., 1980). Chickens (in vivo) replace the chlorine with a hydroxyl group, and the ethyl rather than the isopropyl group is removed. Atrazine and its metabolites were excreted for 4 days after atrazine dosing ceased. A lesser amount of metabolites were identified in urine than in the tissues. After chickens had been fed atrazine (100 mg/l) for 7 days, the highest concentration of atrazine (38.8 mg/kg) was in abdominal fat, and liver contained the highest concentrations of hydroxyatrazine (16.2 mg/l) and of deethylhydroxyatrazine (15.5 mg/l) (Foster and Khan, 1976).

Environmental Pathways and Ecological Effects of Triazines

As noted above, atrazine was introduced in 1958 (Baird, 1995) and is the most heavily used triazine in the United States. Thus, atrazine is the focus of this section because it has received the most attention concerning environmental fate and effects. Atrazine may be introduced to the environment either through effluents at manufacturing sites, rainfall, accidental spills during transportation or handling, or through weed-control-related applications. The latter constitutes the pathway by which the majority of atrazine enters the environment. Atrazine is usually applied in a single application using ground equipment to disperse it in a water solution or in liquid fertilizer. Applications are usually made during preemergence, although preplant and postemergent applications are occasionally used (National Library of Medicine, 2001). Annually, 31,000 to 33,000 metric tons of active ingredient (a.i.) atrazine is used for all purposes within the United States (Aspelin, 1997).

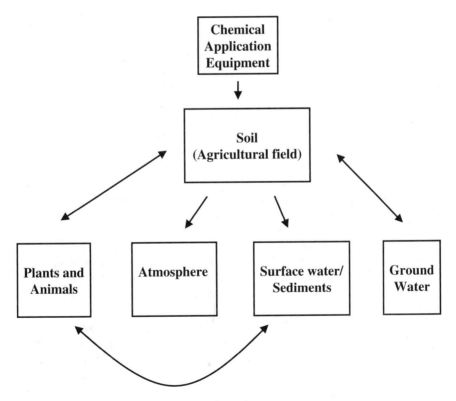

Figure 16-2. Environmental pathways of atrazine.

More than 80% of all atrazine applications occur during corn production. Sorghum receives approximately 10% of all atrazine applications, sugarcane in Florida and Louisiana account for 2.5%, and other crop and turfgrass production make up the remaining 2.5%. Because corn is the major crop that receives atrazine applications, it is no surprise that the major corn-producing areas (the Midwest) receive the majority of atrazine application.

The amount of atrazine used on corn has dropped slightly since 1990. Although the number of acres of corn may vary from year to year, there is little variation in the percentage of those acres receiving atrazine. Atrazine field use rates averaged 2.2 kg a.i./ha, and corn applications in the United States totaled approximately 44,000 metric tons in 1985. By 1987, use rates dropped to slightly >1.6 kg a.i./ha and then ranged between 1.2 and 1.6 kg a.i./ha through 1997. Total atrazine applied to corn during this period ranged from 23,000 to 33,000 tons. In the 2000 growing season, atrazine was applied at a rate of 1.0 lbs a.i./acre (National Agricultural Statistics Service, 2001). This application rate is lower than the 1990 registered application rates of 3.0 lbs a.i./acre of atrazine. A maximal combined pre- and postemergence application rate of 2.5 lbs a.i./acre for corn and sorghum is currently allowed by the label. The application rate for sugarcane is 11.2 kg a.i./ha (Ciba-Geigy, 1993); however, this rate is at least four times the per-acre rate of application for corn and might serve as a significant source of exposure for species inhabiting sugarcane fields. The drop in actual use rates is likely due to the introduction of grass and broadleaf products that combine atrazine with other active ingredients.

After entering the environment, atrazine is leached into the soil by irrigation or precipitation. Such leaching is affected by soil constituents that have the ability to bind atrazine and prevent its vertical movement and reduce the risk of groundwater contamination (Geigy, 1963). In general, atrazine is lost more rapidly from moist soils than from dry soils and is more rapidly lost during periods of high temperatures than low temperatures. Atrazine loss is slower in soils with sandy mineral content than in soils with high organic and/or clay content. Furthermore, the rate of atrazine loss is increased in soils with higher microbial densities and in acidic soils. Other factors that increase the loss of atrazine from soils include storm runoff events, increased ultraviolet irradiation, and the shallowness of the atrazine in the soil.

Since the log K_{ow} of atrazine is relatively low (2.68 at 25°C; Carpenter, 1986) and atrazine exhibits moderate water solubility (log K_{ow} = 40–394) (Giddings et al., 2000), the movement of the chemical occurs primarily by atrazine being dissolved from treated soil and transported via runoff to surface or subsurface waters during irrigation or rain. As a result of its chemical properties, atrazine is not expected to adsorb strongly to sediments and may partition only moderately from the water column.

Chemical degradation of atrazine may occur via hydrolysis or photolysis. Atrazine may break down through hydrolysis at carbon 2 by splitting the triazine ring, and by N-dealkylation at carbon 4 (Kneusli et al., 1969). Li and Feldbeck (1972) reported atrazine exhibited a half-life ($T_{1/2}$) of 244 days at a pH of 4 and a temperature of 25°C. They reported that the addition of 2% humic acid decreased the half-life to 1.73 days, suggesting a catalytic increase in hydrolysis rates (Howard, 1991).

When 5 mg/l of fulvic acid (a concentration naturally occurring in surface waters) was added, $T_{1/2}$ was 34.8, 174, 398, and 742 days at pH values of 2.9, 4.5, 6.0, and 7.0, respectively. Hydrolysis of atrazine follows first-order kinetics and produces hydroxyatrazine as the transformation product (Khan, 1978). Another study reported atrazine exhibited limited change in concentration after a 30-day period in which the temperature was kept at 25°C and pH values of 5–9 (Giddings et al., 2000).

A laboratory study of aqueous photolysis conducted using natural sunlight suggested the $T_{1/2}$ to be 335 days (Spare, 1988). Others have reported that photolysis of atrazine did not occur in water at wavelengths >300 nm (Pape and Zabik, 1970). Burkhard and Guth (1976) reported that the photolysis $T_{1/2}$ of atrazine was 25 h at a concentration of 10 mg/l at 15°C and a wavelength of 290 nm. Jones et al. (1982) reported that the $T_{1/2}$ values of ring-labeled atrazine exposed to sunlight, at concentrations of 0.1 mg/l, were 15 and 20 days for two estuarine sediments and 3–12 days for estuarine water. The major short-term degradate was hydroxyatrazine, which was identical to the results of studies by Khan (1978). Atrazine $T_{1/2}$ was estimated to be 45 days in a study using a xenon lamp as an artificial light source and a sandy-loam soil as a substrate (Das, 1989).

An estimated 0.1–3% of atrazine applied to crop fields find its way into nearby surface waters (Jones et al., 1982). In this study, concentrations of atrazine in runoff from treated fields have been reported to exceed 740 μg/l. The high concentrations were correlated with increased initial treatment rates, storm events after application, and conventional tillage. Other factors included high flow rates, high levels of suspended solids, and dissolved nitrates and nitrites. The elevated atrazine levels observed in runoff declined significantly within a couple of days (Eisler, 1989). Although uncommon, groundwater contamination involving atrazine has been reported in parts of Colorado, Iowa, and Nebraska where cornfields dominate the landscape (Wilson et al., 1987). For example, the amount of atrazine entering the Wye River estuary in Maryland was positively correlated with the volume of atrazine applied in the watershed and the amount and time period when rainfall occurred. In years of high precipitation and runoff, atrazine amounts equivalent to 2–3% of the amount applied to the watershed were transported to the estuary within 2 weeks of application. After 6 weeks, atrazine levels dropped significantly (Glotfelty et al., 1984). In the Chesapeake Bay, atrazine from agricultural soils created concentrations averaging 17 μg/l, such levels are potentially harmful to a variety of estuarine plants (Jones et al., 1982).

Atmospheric transport of atrazine is another way that this triazine is introduced into ecosystems. Aerosol particulates and soils transported via wind likely contribute to atrazine contamination of terrestrial and aquatic ecosystems. For example, Rhode River, Maryland, received an estimated annual atmospheric input of atrazine in rainfall of 1016 mg/surface ha in 1977 and 97 mg/ha in 1978 (Wu, 1981). A similar result occurs in fog, whereby plant surfaces become saturated with liquid for the duration of the fog (Glotfelty et al., 1987).

Indirect deleterious effects on freshwater aquatic fauna, such as fish (Kettle et al., 1987) and benthic organisms (Dewey, 1986) have been noted at environmental concentrations as low as 20 μg atrazine/l. For example, the emergence of aquatic insects such as chironomids (*Labrundinia pilosella*) declined at 20 μg atrazine/l

(Dewey, 1986). When atrazine was applied, the richness of benthic insect species and total emergence declined significantly. The mechanism appears to be indirect, in that the reduction of macrophytes and food sources of nonpredatory insects is the suggested cause of decreased species richness and emergence.

In three species of fish, reproductive efficiency and feeding habits were impacted adversely after being exposed for 136 days in ponds containing 20 µg atrazine/l (Kettle et al., 1987). The amount of atrazine present in the ponds at the end of the study was about 70% of the original concentration. The ability of channel catfish (*Ictalurus punctatus*) and gizzard shad (*Dorosoma cepedianum*) to reproduce was deleteriously affected, and the number of bluegill (*Lepomis macrochirus*) offspring per pond was decreased by more than 95%. A possible cause for the decrease in bluegill numbers may have been related to the decrease in the number of prey items in the stomachs of bluegill from treated ponds (3.8) as compared to bluegill from control ponds (25.6). The quantities of prey items were significantly higher in control ponds (25.6), where, in addition, the number of taxa represented was significantly increased. Aquatic macrophytes were reduced more than 60% in treated ponds within a 2-month period. Kettle et al. (1987) surmised that the dietary effects of atrazine on bluegills were likely indirect and that macrophyte reduction resulted in poor diets, which increased cannibalism in the bluegill population.

Bioconcentration of atrazine is unlikely from information related to uptake and bioconcentration factors (BCF) in aquatic organisms. Biomagnification through the food chain also is expected to be low. The low potential for bioconcentration and biomagnification is related to the relatively low K_{ow} for atrazine and its susceptibility for metabolism and depuration. Brook trout exhibited a low BCF of <0.27 when subjected to dietary atrazine exposure (National Library of Medicine, 2001). However, Giddings et al. (2000) reported a BCF of 12 for bluegill sunfish (*Lepomis macrochirus*), and other studies report bioconcentration values ranging from 0.9 for fathead minnows (*Pimephales promelas*) (Veith et al., 1979) to 4–5 for whitefish (*Coregonus fera*) fry (Ellgehausen et al., 1980) and 4–8 for largemouth bass (Gross and Wieser, 2002). In a farm pond treated once with 300 µg atrazine/l, residues at 120 days posttreatment were not detectable in bullfrog (*Rana catesbeiana*) tadpoles, and levels ranged between 204 and 286 µg/kg in mud and water. In addition, bluegills exhibited atrazine residues of 290 µg/kg. One year after treatment, no residues were detected in the biota, although atrazine residues were <21 µg/kg in water and mud (Klaassen and Kadoum, 1979). Maximal accumulation factors ranged from about 4-fold in annelids to 480-fold in mayfly nymphs in an artificial stream treated four times with 25 µg atrazine/l for 30 days, although after a 60-day washout period began, residues began to decline. Maximal atrazine residue concentrations (mg/kg whole organism fresh weight) were 3.4 in mayflies, (*Baetis* sp.), 2.4 in the mottled sculpin (*Cottus bairdi*), 0.2 in a clam (*Strophitis rugosus*), 0.4 in a snail (*Physa* sp.), 0.9 in a crayfish (*Orconectes* sp.), and 3.0 in an amphipod (*Gammarus pseudolimnaeus*) (Lynch et al., 1982). Atrazine concentrations in a freshwater snail (*Ancylus fluviatilis*) and in whitefish (*Coregonus fera*) were rapidly accumulated from the ambient environment, and saturation was reached within 12 to 24 h. Atrazine bioaccumulation was not reported in mollusks, leeches, cladocerans, or fish when contamination was by way of the diet (Gunkel and Streit, 1980; Gunkel, 1981).

In summary, atrazine is released into the environment primarily through agricultural activities. Initially, atrazine is deposited in the soil by ground equipment, where it exhibits a $T_{1/2}$ that ranges from 20 to 385 days, depending on soil composition (Jones et al., 1982). Once in the soil, atrazine may remain there long enough to be degraded by hydrolysis, photolysis, or N-dealkylation (Gunther and Gunther, 1970; Reed, 1982; Menzer and Nelson, 1980). If degradation does not occur in the soil, atrazine may be transported to surface waters and/or groundwater by either storm event runoff or by irrigation. Jones et al. (1982) reported that 0.1–3% of atrazine applied to fields is transported to nearby aquatic environments. Once in surface waters, atrazine's $T_{1/2}$ ranges from 34 to 335 days. In the aquatic compartment atrazine may infiltrate sediment or be taken up in the food chain, where it is expediently metabolized from atrazine to hydroxyatrazine (Spare, 1988; National Library of Medicine, 2001).

Potential Measures of Endocrine Modulation in Amphibians and Reptiles

Because acute toxicity is not likely at ecologically relevant levels, subtoxic effects such as endocrine modulation appear to be a more serious risk. Environmental scientists have searched intensively for biomarkers that can detect early physiological or biochemical changes in organisms exposed to environmental contaminants. Although a number of effective indicators have been developed for various conditions in humans (e.g., numerous tumor markers for early detection of cancer), few effective indicators are available for wildlife, especially for animals exposed to environmental contaminants. For those chemicals that are potential endocrine modulators, the effects of primary concern (and normally monitored) are reproductive. Damage to reproductive and endocrine functions from environmental contamination has been reported on the scientific and popular literature for many years. Extensive laboratory studies show strong correlations between specific impairments of reproductive activity and elevated tissue concentrations of xenobiotic agents (Hose and Guillette, 1995; Di Giulio and Tillit, 1999). The reproductive damages reported to date include reduced fertility, reduced hatchability, reduced viability of offspring, impaired hormone activity, and altered sexual behavior. There are also reports of poor growth, wasting, and lower rates of metabolic activity (Colborn and Clement, 1992). For this discussion, we will, likewise, concentrate on effects associated with reproduction. However, other endocrine effects such as metabolic, immune function, and stress are also of physiological significance and warrant the development and identification of effects and biomarkers.

Several aspects of amphibian development are controlled by steroids and thyroid hormones and can be monitored for indications of endocrine-disrupting activity. The vitellogenin gene is induced directly by E_2. This gene is not typically expressed in males. The presence of this protein in males or the expression of the mRNA for the gene in the livers of males can be used as a laboratory screen to measure estrogenicity of xenobiotic compounds after treatment or to monitor environmental exposure by examining wild animals. The expression of estrogen-dependent, female-typical development, such as female coloration in the sexually dimorphic *Hyperolius argus* (in which males are green and females are red with white spots) or the expression of androgen-dependent male-typical developmental patterns in females (such as thumb pad or gular pouch development) can be used as an indicator of estrogenic or androgenic activity (Hayes, 1997). To measure antagonism, animals can be treated in the laboratory with androgens or estrogen in combination with xenobiotics to determine if these compounds block the effects of the endogenous hormones.

Metamorphosis provides a unique measure to monitor interactions with the thyroid axis. Larvae can be treated with compounds and metamorphic rates measured either by monitoring time to foreleg emergence or tail resorption in frogs, or gill or tail fin resorption in salamanders. Limb and digit development may not be a good measure because this part of development is independent of thyroid hormones in salamanders, and its role in frogs varies between species (some frog species show thyroid dependency, whereas others show limb development even in the absence of thyroid hormones) (Hayes, 1997). In addition, metamorphosis offers the potential for treatments and subsequent evaluation of other endocrine-mediated effects such as sex differentiation (Hayes et al., 2002; Tavera-Mendoza et al., 2002; Carr et al., 2003).

Effective reproductive endpoints of contaminant effects in reptiles, include egg hatchability (Gross et al., 1994; Guillette et al., 1994; Masson, 1995; Rauschenberger, 2004), nest numbers (Masson, 1995; Gross et al., 1997), and secondary sex characteristics (Guillette et al., 1996). Monitoring population and ecosystem endpoints of reproductive success (pod size, age class analyses, population numbers, etc.) also has been used to indicate potential contaminant effects in alligators (Woodward et al., 1993; Percival et al., 1994). However, establishing a cause–effect association of these organismal, population, and ecosystem-level effects to endocrine modulation and contaminants is often difficult. Indeed, it is often advisable to incorporate cellular, biochemical, and molecular effects measures along with the measures of reproductive endpoints.

The hormonal factors that control reproduction in reptiles have been extensively studied, but an examination of the modulation of these endocrine factors/hormones by environmental contaminants in reptiles has been limited. Systemic concentrations of sex steroids, estrogens, and androgens have been used as effects measures of endocrine modulation in turtles (Gross et al., 1999) and alligators (Gross et al., 1994; Guillette et al., 1994). Additionally, in vitro secretion of gonadal sex steroids has been used to evaluate reproduction and potential contaminant effects in alligators (Guillette et al., 1995).

Chorioallantoic sex steroid concentrations have been used as effect measures in both turtles and alligators to indicate endocrine-modulation effects as well as sex differentiation (Gross et al., 1994, 1999; Guillette et al., 1994). These chorioallantoic analyses indicate the endocrine parameters during embryonic sex differentiation and have been conducted as part of sex-reversal studies. Sex differentiation in many reptiles, primarily turtles and crocodilians, is temperature dependent and is easily manipulated in vitro. Studies have indicated the ability of endocrine-modulating contaminants to alter sex differentiation in freshwater turtles (Bergeron et al., 1994; Gross et al., 1999) and alligators (Gross et al., 1994; Crain et al., 1997). Sex-reversal /sex-determination effects measures in reptiles may be useful for testing the endocrine-modulating effects of contaminants; however, the mechanisms controlling sex differentiation on reptiles is not well understood, and egg availability could be limited. Most important, eggs used for sex-reversal /sex-differentiation analyses must be at critical-sensitive early embryonic ages, which could greatly limit the use of this procedure as a screening method. Other reproductive effects measurements of interest in reptiles have not been used for evaluation of contaminant effects, but monitoring follicular development, spermatogenesis, and neuroendocrine function might be useful as valid physiological monitoring methods are developed for reptiles.

Another biochemical measure of interest in reptiles is vitellogenin. Vitellogenin is an egg-yolk precursor protein under the control of estrogen. Natural vitellogenin production is limited to oviparous female fish, amphibians, and reptiles (Carnevali and Belveldere, 1991). Oviparous species have natural vitellogenic cycles corresponding to the production of eggs, during which the synthesis of vitellogenin is at its highest level. Many of the potential endocrine-modulating contaminants might mimic or alter estrogen function and thereby induce or inhibit the natural expression of vitellogenin. These effects might be seen in both females and males of each species. Although valid techniques for the measurement of vitellogenin in reptiles are not yet available, vitellogenin might serve as an effective bioindicator of exposure (Chen and Sonstegard, 1984; Chakravorty et al., 1992).

Alterations in endocrine-reproductive processes at the molecular level might also be useful as biomarkers of endocrine modulation in reptiles. Voiner et al. (1996) examined estrogen and progestin receptors in alligator oviductal tissue. An ability of contaminants to bind with or alter the binding properties and interactions of hormonal receptors could be a useful indictor of potential endocrine-modulation effects. However, characterization of receptors and/or other molecular mechanisms of endocrine function in reptiles is currently limited.

Implications for Wildlife and Humans

The risks of triazines having direct, acute toxic effects on mammals, birds, reptiles, amphibians, and fish are low given the current application rates. Indirect effects on wildlife are of much more concern and have been illustrated in a few studies (Gunkel, 1981; Kettle et al., 1987). In addition, the effects of low levels of triazine exposure on the endocrine systems of wildlife are an emerging concern. Although the endocrine-modulating effects of triazine exposure have been documented in laboratory studies (Connor et al., 1996; Vonier et al., 1996), few studies have documented endocrine modulation of wild populations exposed to triazines at ecologically relevant levels.

With respect to public health, the rarity of human toxicity incidents involving farm workers or production workers underscores the hypothesis that direct toxicity is not the major concern with respect to triazine exposure. Future studies should focus on the indirect effects of triazines at the ecosystem level and how atrazine's endocrine modulating effects may potentially alter populations of wildlife and humans subjected to extended triazine exposure.

References

Allran, J.W., Karasov W.H., 2001. Effects of atrazine on embryos, larvae, and adults of anuran amphibians. Environ. Toxicol. Chem. 20, 769–775.

Aspelin, A.L., 1997. Pesticides industry sales and usage: 1994 and 1995 market estimates. EPA/733-r-97-002. U.S. Environmental Protection Agency, Washington, DC.

Baird, C., 1995. Environmental Chemistry. W.H. Freeman, New York.

Bakke, J.E., Larson, J.D., Price, C.E., 1972a. Metabolism of atrazine and 2-hydroxyatrazine by the rat. J. Agric. Food Chem. 20, 602–607.

Bakke, J.E., Robbins, J.D., Feil, V.J., 1967. Metabolism of 2-chloro-4. 6-bis(isopropylamino)-s-triazine (propazine) and 2-methoxy-4,4-bis(isopropylamino)-s-triazine (prometone) in the rat. Balance study and urinary metabolite separation. J. Agric. Food Chem. 15, 628–631.

Bakke, J.E., Shimabukuro, R.H., Davison, K.L., Lamoureux, G.L., 1972b. Sheep and rat metabolism of the insoluble 14C-residues present in 14C-atrazine-treated sorghum. Chemosphere 1, 21–24.

Balaguer, P., Francois, F., Comunale, F., Fenet, H., Boussioux, A.M., Pons, M., Nicolas, J.C., Casellas, C., 1999. Reporter cell lines to study the estrogenic effects of xeno-estrogens. Sci. Total Environ. 233, 47–56.

Bergeron, J. M., Crews, D., McLachlan, J.A., 1994. PCBs as environmental estrogens: Turtle sex determination as a biomarker of environmental contamination. Environ. Health Perspect. 102, 780–781.

Beste, C.E. (Ed.), 1983. Herbicide Handbook of the Weed Science Society of America. Weed Science Society of America, Champaign, IL.

Burkhard, N., Guth, J.A., 1976. Photodegradation of atrazine, atraton and ametryne in aqueous solution with acetone as photosensitizer. Pestic. Sci. 7, 65.

Carnevali, O., Belveldere, P., 1991. Comparative studies of fish, amphibian and reptilian vitellogenins. J. Exp. Zool. 259, 18–25.

Carpenter, M., 1986. Determination of octanol-water partition coefficient of atrazine. Report no. 34616. ABC Laboratories, Columbia, MO.

Carr, J.A., Gentles, A., Smith, E.E., Goleman, W.L., Urquidi, L.J., Thuett, K., Kendall, R.J., Giesy, J.P., Gross, T.S., Solomon, K.R., Van Der Kraak, G., 2003. Response of larval *Xenopus laevis* to atrazine: Assessment of growth, metamorphosis, and gonadal and laryngeal morphology. Environ. Toxicol. Chem. 22, 396–405.

Chakravorty, S., Lal, B., Singh, T.P., 1992. Effect of endosulfan (Thiodan) on vitellogenesis and its modulation by different hormones in the vitellogenic catfish *Clarias batrachus*. Toxicology 75, 191–198.

Chen, T.T., Sonstegard, R.A., 1984. Development of a rapid, sensitive and quantitative test for the assessment of the effects of xenobiotics on reproduction in fish. Mar. Environ. Res. 14, 429–430.

Ciba-Geigy, 1982a. Aatrex herbicides. Toxicology data. Agricultural Division, Department of Toxicology, Ciba-Geigy, Greensboro, NC.

Ciba-Geigy, 1982b. Milogard herbicides. Toxicology data. Agricultural Division, Department of Toxicology, Ciba-Geigy, Greensboro, NC.

Ciba-Geigy, 1982c. Pricep herbicides. Toxicology data. Agricultural Division, Department of Toxicology, Ciba-Geigy, Greensboro, NC.

Ciba-Geigy, 1993. Aatrex Nine-0 label. EPA registration no. 100-585. Ciba Crop Protection, Ciba-Geigy, Greensboro, NC.

Colborn, T., Clement, C.R., 1992. Chemically-induced Alterations in Sexual and Functional Development: The Wildlife-Human Connection. Princeton Publishing, Princeton, NJ.

Connor, K., Howell, J., Chen, I., Liu, H., Berhane, K., Sciaretta, C., Safe, S., Zacharewski, T., 1996. Failure of chloro-s-triazine-derived compounds to induce estrogen receptor-mediated responses in-vivo and in-vitro. Fundam. Appl. Toxicol. 30, 93–101.

Cooper, R., Stoker, T., Tyrey, L., Goldman, J., McElroy, W., 2000. Atrazine disrupts the hypothalamic control of pituitary-ovarian function. Toxicol. Sci. 53, 297–307.

Crain, D.A., Guillette, L.J., Jr., Rooney, A.A., Pickford, D.B., 1997. Alterations in the steroidogenesis in alligators (*Alligator mississipiens*) exposed naturally and experimentally to environmental contaminants. Environ. Health Perspect. 105, 528–533.

Crain, D.A., Spiteri, I.D., Guillette L.J., 1999. The functional and structural observations of the neonatal reproductive system of alligators exposed in-ovo to atrazine, 2,4-D, or estradiol. Toxicol. Ind. Health 15, 180–185.

Danzo, B.,1997. Environmental xenobiotics may disrupt normal endocrine function by interfering with the binding of physiological ligands to steroid receptors and binding proteins. Environ. Health Perspect. 105, 294–301.

Das, Y.T., 1989. Photodegradation of [triazine (U)-14C]atrazine on soil under artificial sunlight. Project no. 98070. Innovative Science Services, Piscataway, NJ.

Dauterman, W.C., Muecke, W., 1974. In-vitro metabolism of atrazine by rat liver. Pestic. Biochem. Physiol. 4, 212–219.

Dewey, S.L.,1986. Effects of the herbicide atrazine on aquatic insect community structure and emergence. Ecology 67, 148–162.

Di Giulio, R.T., Tillit, D.E., 1999. Reproductive and Developmental Effects of Contaminants in Oviparous Vertebrates. SETAC Press, Pensacola, FL.

Eisler, R., 1989. Atrazine hazards to fish, wildlife, and invertebrates: A synoptic review. Contaminant Hazard Reviews report no. 18. U.S. Fish and Wildlife Service, Washington, DC.

Ellgehausen, H., Gut, J.A., Esser, D.O., 1980. Factors determining the bioaccumulation potential of pesticides in the individual compartments of aquatic food chains. Ecotoxicol. Environ. Safety 4, 134–157.

Erickson, M.D., Frank, C.W., Morgan, D.P., 1979. Determination of s-triazine herbicide residues in urine. Studies of excretion and metabolism in swine as a model to human metabolism. J. Agric. Food Chem. 27, 743–746.

Forney, D.R., Davis, D.E., 1981. Effects of low concentrations of herbicides on submersed aquatic plants. Weed Sci. 29, 677–685.

Foster, T.S., Khan, S.U., 1976. Metabolism of atrazine by the chicken. J. Agric. Food Chem. 24, 566–570.

Foster, T.S., Khan, S.U., Akhtar, M.H., 1980. Metabolism of deethylatrazine, deisopropyl-atrazine, and hydroxyatrazine by the soluble fraction (105000 g) from goose liver homogenates. J. Agric. Food Chem. 28, 1083–1085.

Geigy, 1963. Atrazine. Geigy Agricultural Chemical technical bulletin 63-1. Geigy Chemical Corporation, Yonkers, NY.

Giddings, J.M., Anderson, T.A., Hall, L.W. Jr., Kendall, R.J., Richards, R.P., Solomon, K.S., Williams, W.M., 2000. Aquatic ecological risk assessment of atrazine: A tiered probabilistic approach. Novartis Crop Protection and Ecorisk, Greensboro, NC.

Gojmerac, T., Uremovic, M., Uremovic, Z., Curic, S., Bilandzic, N., 1999. Reproductive disturbance caused by s-triazine in pigs. Acta Vet. Hung. 47, 129–135.

Glotfelty, D.E., Seizer, J.W., Liljedahl, L.A., 1987. Pesticides in fog. Nature 325, 602–605.

Glotfelty, D.E., Taylor, A.W., Isensee, A.R., Jersey, J., Glenn, S., 1984. Atrazine and simazine movement to Wye River estuary. J. Environ. Qual. 12, 115–121.

Gross, T.S., 2001a. Determination of potential effects of 10 day neonatal exposure of atrazine on histological and hormonal sex determination in incubated American alligator (*Alligator mississipiensis*). Syngenta Crop Protection report, #NOVA9802a, Greensboro, NC.

Gross, T.S., 2001b. Determination of potential effects of 10 day neonatal exposure of atrazine on histological and hormonal sex determination in incubated red-eared slider (*Psuedemys elegans*) eggs. Syngenta Crop Protection report, #NOVA9802b, Greensboro, NC.

Gross, T.S., Guillette, L.J., Percival, H.F., Masson, G.R., Matter J.M., Woodward, A.R., 1994. Contaminant-induced reproductive anomalies in Florida. Comp. Pathol. Bull. 4, 2–8.

Gross, T.S., Wiebe, J., Centonze, V., Centonze, L., Schoeb, T., Hosmer, A.J., 1999. Effects of atrazine treatments on freshwater turtle eggs (*Trachemys scripta elegans*): An evaluation of endocrine disruption, sex reversal, and developmental toxicity effects. In: 20th Annual Meeting of SETAC. SETAC Press, Pensacola, FL.

Gross, T.S., Wiebe, J., Wieser, C., Shrestha, S., Reiskind, M., Denslow, N., Johnson, W.E., Stout, R., 1997. Endocrine-disrupting effects of pesticides in largemouth bass: An examination of potential synergistic effects. In: 18th Annual Meeting of SETAC. SETAC Press, Pensacola, FL, pp. 138–139.

Gross, T.S., Wieser, C.M., 2002. Determination of potential effects of 20 day exposure of atrazine on endocrine function in adult largemouth bass. Syngenta Crop Protection report, #1168-99. Greensboro, NC.

Guillette, L.J., Jr., Crain, D.A., Rooney A.A., Pickford, D.B., 1995. Organization versus activation: The role of endocrine-disrupting environmental contaminants (EDCs) during embryonic development in wildlife. Environ. Health Perspect. 103 (suppl. 7), 157–164.

Guillette, L.J., Jr., Gross, T.S., Masson, G.R., Matter, J.M., Percival, H.F., Woodward, A.R., 1994. Developmental abnormalities of the gonad and abnormal sex hormone concentrations in juvenile alligators from contaminated and control lakes in Florida. Environ. Health Perspect. 102, 680–688.

Guillette, L.J., Jr., Pickford, D.B., Crain, D.A., Rooney, A.A., Percival, H.F., 1996. Reduction in penis size and plasma testosterone concentrations in juvenile alligators living in a contaminated environment. Gen. Comp. Endocrinol. 101, 32–42.

Gunkel, G., 1981. Bioaccumulation of herbicide (atrazine, s-triazine) in whitefish (*Coregonus fera* J.): Uptake and distribution of the residue in fish. Arch. Hydrobiol. (suppl.) 59, 252–287.

Gunkel, G., Streit, B., 1980. Mechanisms of a herbicide (atrazine, s-triazine) in a freshwater mollusk (*Ancylus fluviatilis* Mull.) and a fish (*Coregonus fera* Jurine). Water Res. 14, 1573–1584.

Gunther, F.A., Gunther, J.D., 1970. The triazine herbicides. Residue Rev. 32, 1–413.

Hartley, D., Kidd, H., 1983. The Agrochemicals Handbook. Royal Society of Chemistry, Nottingham, England.

Hayes, T.B., 1997. Steroids as potential modifiers of thyroid hormone action in anuran metamorphosis. Am. Zool. 37, 185.

Hayes, T.B., Collins, A., Lee, M., Mendoza, M., Noriega, N., Stuart, A.A., Vonk, A., 2002.

Hermaphroditic, demasculinized frogs after exposure to the herbicide atrazine at low ecologically relevant doses. Proc. Natl. Acad. Sci. USA 99, 5476–5480.

Hose, J.E., Guillette, L.J., Jr., 1995. Defining the role of pollutants in the disruption of reproduction in wildlife. Environ. Health Perspect. 103 (suppl. 4), 87–92.

Howard, P.H., 1991. Handbook of Environmental Fate and Exposure Data for Organic Chemicals, vol. III. Lewis Publishers, Chelsea, MI.

Huckins, J.N., Petty, J.D., England, D.C., 1986. Distribution and impact of trifluralin, atrazine, and fonofos residues in microcosms simulating a northern prairie wetland. Chemosphere 15, 563–588.

Hull, H.M., 1967. Herbicide Handbook of the Weed Society of America. W.F. Humphrey Press, Geneva, NY.

Hutson, D.H., Hoadley, E.C., Griffiths, M.H., Donninger, C., 1970. Mercapturic-acid formation in the metabolism of 2-chloro-4-ethylamino-6-1-methyl-1-cyanoethylamino-s-triazine in the rat. J. Agric. Food Chem. 18, 507–512.

Innes, J.R.M., Ulland, B.M., Valerio, M.G., Petrucelli, L., Fishbein, L., Hart, E.R., Pallotta, A.J., Bates, R.R., Falk, H.L., Gart, J.J., Klein, M., Mitchell, I., Peters, J., 1969. Bioassay of pesticides and industrial chemicals for tumorgenicity in mice: A preliminary note. J. Natl. Cancer Inst. 42, 1101–1114.

Jensen, K.I.N., Stephenson, G.R., Hunt, L.A., 1977. Persistence and movement of atrazine in a salt marsh sediment microecosystem. Bull. Environ. Contam. Toxicol. 39, 516–523.

Johnson, A.E., Van Kampen, K.R., Binnis, W., 1972. Effects on cattle and sheep of eating hay treated with triazine herbicides, atrazine and prometone. Am. J. Vet. Res. 7:1433–8.

Jones, T.W., Kemp, W.M., Stevenson, J.C., Means, J.C., 1982. Degradation of atrazine in estuarine water/sediment systems and soils. J. Environ. Qual. 11, 632–638.

Kettle, W.D., DeNoyelles, F., Jr., Heacock, B.D., Kadoum, A.M., 1987. Diet and reproductive success of bluegill recovered from experimental ponds treated with atrazine. Bull. Environ. Contam. Toxicol. 38, 47–52.

Khan, S.U., 1978. Kinetics of hydrolysis of atrazine in aqueous fulvic acid solution. Pest. Sci. 9, 39–43.

Khan, S.U., Akhtar, M.H., 1983. *In-vitro* release of bound (nonextractable) atrazine residues from corn plants by chicken liver homogenate and bovine rumen liquor. J. Agric. Food Chem. 24, 768–771.

Khan, S.U., Kacew, S., Molnar, S.J., 1985. Bioavailability in rats of bound ^{14}C residues from corn plants treated with [^{14}C]atrazine. J. Agric. Food Chem. 33, 712–717.

Klaassen, H.E., Kadoum, A.M., 1979. Distribution and retention of atrazine and carbofuran in farm pond ecosystems. Arch. Environ. Contam. Toxicol. 8, 345–353.

Klaine, S., Dixon, K., Florian, J.D., 1996. Characterization of *Selenastrum capricornatum* response to episodic atrazine exposure. TIWET study 09524. Novartis Crop Protection, Greensboro, NC.

Kneusli, E., Berrer, D., Depuis, G., Esser H., 1969. s-Triazines. In: Kearney, P.C., Kaufman, D.D. (Eds.), Degradation of Herbicides. Marcel Dekker, NY, pp. 51–78.

Li, G.C., Feldbeck, G.T., 1972. Atrazine hydrolysis as catalyzed by humic acids. Soil Sci. 114, 201–209.

Lynch, T.R., Johnson, H.E., Adams, W.J., 1982. The fate of atrazine and a hexachlorobiphenyl isomer in naturally-derived model stream ecosystems. Environ. Toxicol. Chem. 4, 399–413.

Masson, G.R., 1995. Environmental influences on reproductive potential, clutch viability and embryonic mortality of the American alligator in Florida. Ph.D. Dissertation, University of Florida, Gainesville.

Meli, G., Bagnati, R., Fanelli, R., Benfenati, E., Airoldi, L., 1992. Metabolic profile of atrazine and N-nitrosoatrazine in rat urine. Bull. Environ. Contam. Toxicol. 48, 701–708.

Menzer, R.E., Nelson, J.O., 1980. Water and soil pollutants. In: Klaassen, C.D., Amdur, M.O., Doull, J. (Eds.), Casarett and Doul's toxicology, 2nd edition. Macmillan Publishing, New York, pp. 632–658.

Moore, A., Lower, N., 2001. The impact of two pesticides on olfactory-mediated endocrine function in mature male Atlantic salmon (*Salmo salar* L.) parr. Comp. Biochem. Physiol. B 129, 269–276.

Morgan, M.K., Scheuerman, P.R., Bishop, C.S., Pyles, R.A., 1996. Teratogenic potential of atrazine and 2,4-D using FETAX. J. Toxicol. Environ. Health 48, 151–168.

Mrak, E.M., 1969. Report of the Secretary's Commission on pesticides and their relationship to environmental health. U.S. Department of Health, Education, and Welfare, Washington, DC.

Murnik, M.R., Nash, C.L., 1977. Mutagenicity in the triazine herbicides atrazine, cyanazine, and simazine in *Drosophila melanogaster*. J. Toxicol. Environ. Health 3, 691–697.

National Agricultural Statistics Service. 2001. Agricultural Chemical Usage 1998 Field Crops Summary. U.S. Department of Agriculture, Beltsville, MD.

National Library of Medicine. 2001. National Institutes of Health. Available: toxnet.nlm. nih.gov/cgi-bin/sis/search/f?./temp/~HD1DWD:1, accessed Nov 8, 2004.

Nezefi, T.A., 1971. Morphological alterations in the organs of white rats during chronic treatment with atrazine. Zdravookhr. Turkm. 15, 9–12.

Office of Pesticide Programs. 1984. Guidance for the reregistration of products containing alachlor as an active ingredient. U.S. Environmental Protection Agency, Washington, DC.

Office of Pesticide Programs. 1986. Alachlor special review technical support document. U.S. Environmental Protection Agency, Washington, DC.

Office of Pesticide Programs. 1987. Cyanazine special review technical support document. U.S. Environmental Protection Agency, Washington, DC.

Office of Pesticide Programs. 1989. Peer review of simagene. U.S. Environmental Protection Agency, Washington, DC.

Offices of Pesticides and Toxic Substances. 1985. Cyanazine special review position document 1. U.S. Environmental Protection Agency, Washington, DC.

Palmer, J.S., Radeleff, R.D., 1964. The toxicological effects of certain fungicides and herbicides on sheep and cattle. Ann. N.Y. Acad Sci. 111, 729–736.

Pape, B.D., Zabik, M.J., 1970. Photochemistry of bioactive compounds—photochemistry of selected 2-chloro-and 2-methyl-4,6-di(alkylamino)-s-triazine herbicides. J. Agric. Food Chem. 18, 202–207.

Percival, H.F., Rice, K., Woodward, A., Jennings, M., Masson, G., Abercrombie, C., 1994. Depressed alligator clutch viability on Lake Apopka, Florida. In: Fifth Annual Lake Management Society Symposium, World Meeting no. 942. Florida Wildlife Conservation Commission, p. 15.

Peters, J.W., Cook, R.M., 1973. Effects of atrazine on reproduction in rats. Bull. Environ. Contam. Toxicol. 9, 301–304.

Rabert, W., Lin, J., Frankenberry, M., Nelson, H., Urban, D., 2001. Reregistration eligibility science chapter for atrazine: Environmental fate and effects. U.S. Environmental Protection Agency, Office of Pesticide Programs, Washington, DC.

Rauschenberger, R.H., 2004. Developmental mortality in American alligators (*Alligator mississippiensis*) exposed to organochlorine pesticides. Ph.D. dissertation, University of Florida.

Reed, D., 1982. Atrazine (unpubl. Mimeo.). Available from U.S. Food and Drug Administration, Bureau of Foods, Washington, DC.

Reinhardt, C.F., Britelli, M.R., 1982. Heterocyclic and miscellaneous nitrogen compounds. In: Clayton, G.D., Clayton, G.E. (Eds.), Patty's Industrial Hygiene and Toxicology. Wiley Interscience, New York, pp. 2778–2822.

Shabana, E.F., 1987. Use of batch assays to assess the toxicity of atrazine to some selected cyanobacteria. 1. Influence of atrazine on the growth, pigmentation and carbohydrate contents of *Aulosira fertilissima, Anabaena oryzae, Nostoc muscorum* and *Tolypothrix tenuis*. J. Basic Microbiol. 2, 113–119.

Shrestha, S., Gross, T.S., Wieser, C., Wiebe, J., Johnson, W.E., Douglas, D., 1997. Effects of some commonly used pesticides on *in vitro* production of reproductive hormones by largemouth bass gonads. In: 18th Annual Meeting SETAC. SETAC Press, Pensacola, FL.

Spare, W.C., 1988. Aqueous photolysis of ^{14}C-atrazine under natural sunlight. Report no. 12112-A. Agrisearch, Frederick, MD.

Stevens, J.T., Sumner, D.D., 1991. Herbicides. In: Hayes, W.J., Jr., Laws, E.R., Jr. (Eds.), Handbook of Pesticide Toxicology, vol. 3. Classes of Pesticides. Academic Press, San Diego, CA, pp. 1317–1408.

Suschetet, M., Leclerc, J., Lhuissier, M., 1974. The toxicity and nutritional effects for the rat of two herbicides; Picloram (4-amino-3,5,6-trichloropicolinic acid) and atrazine (2-chloro-4-ethylamino-6-isopropylamino-s-triazine). Ann. Nutr. Ailment. 28, 29–47.

Tavera-Mendoza, L., Ruby, S., Brousseau, P., Fournier, M., Cyr, D., Marcogliese, D., 2002. Response of the amphibian tadpole (*Xenopus laevis*) to atrazine during sexual differentiation of the testis. Environ. Toxicol. Chem. 21, 527–531.

Tennant, M., Hill, D., Eldridge, J., Wetzel, L., Breckenridge, C., Stevens, J., 1994. Possible antiestrogenic properties of chloro-s-triazines in rat uterus. J. Toxicol. Environ. Health 43, 183–196.

Tran, D., Kow, K., McLachlan, J., Arnold S., 1996. The inhibition of estrogen receptor-mediated responses by chloro-s-triazine-derived compounds is dependent on estradiol concentration in yeast. Biochem. Biophys. Res. Commun. 227, 140–146.

Veith, G.D., Defoe, D.L., Bergstedt, B.V., 1979. Measuring and estimating the bioconcentration factor of chemicals in fish. J. Fish. Res. Biodiv. Can. 36, 1040–1048.

Vonier, P., Crain, D., McLachlan, J., Guillette, L., Arnold, S., 1996. Interaction of environmental chemicals with the estrogen and progesterone receptors from the oviduct of the American alligator. Environ. Health Perspect. 104, 1318–1322.

Wiegand, C., Krause, E., Steinberg, C., Pflugmacher, S., 2001. Toxicokinetics of atrazine in embryos of the zebrafish (*Danio rerio*). Ecotoxicol. Environ. Safety 49, 199–205.

Wiegand, C., Plumacher, S., Giese, M., Frank, H., Steinberg, C., 2000. Uptake, toxicity, and effects on detoxication enzymes of atrazine and trifluoroacetate in embryos of zebrafish. Ecotoxicol. Environ. Safety 45, 122–131.

Wilson, M.P., Savage, E.P., Adrian, D.D., Aaronson, M..J., Keefe, T.J., Hamar, D.H., Tessari, J.T., 1987. Groundwater transport of the herbicide, atrazine, Weld County, Colorado. Bull. Environ. Contam. Toxicol. 39, 807–814.

Woodward, A.R., Jennings, M..L., Percival, H.F., Moore, J.F., 1993. Low clutch viability of American alligators on Lake Apopka. Fl. Sci. 56, 52.

Woolhouse, H.W., 1981. Aspects of the carbon and energy requirements of photosynthesis considered in relation to environmental constraints. In: Townsend, C.R., Calow, P. (Eds.), Physiological Ecology. Sinauer Associates, Sunderland, MA, pp. 51–85.

Worthing, C.R. (Ed.), 1987. The Pesticide Manual: A World Compendium, 8th ed. British Crop Protection Council, Thorton Heath, UK.

Wu, T.L., 1981. Atrazine in estuarine water and the aerial deposition of atrazine into Rhode River, Maryland. Water, Air Soil Pollut. 15, 173–184.

Appendices

Appendix A: Books about or Related to Endocrine Disruption

This is a selected listing of available publications and is not an endorsement of any of these sources.

Popularized Accounts

Berkson, D.L., Berkson, L., McLachlan, J.A., 2001. Hormone Deception. McGraw Hill/ Contemporary Books, New York.

Cadbury, D., 2000. The Estrogen Effect. Griffen Trade Paperbacks.

Carson, R., 1962. Silent Spring. Houghton Mifflin, Boston.

Colburn, T., Dumanoski, D., Myers, J.P., 1996. Our Stolen Future. Penguin Books, New York.

Krimsky, S., 2000. Hormonal Chaos: The Scientific and Social Origins of the Environmental Endocrine Hypothesis. The Johns Hopkins University Press, Baltimore, MD.

Schettler, T., Solomon, G., Valenti, M., Huddle, A., 1999. Generations at Risk: Reproductive Health and the Environment. MIT Press, Cambridge, MA.

Thornton, J., 2000. Pandora's Poison: Chlorine, Health, and a New Environmental Strategy. MIT Press, Cambridge, MA.

General Sources on Endocrine Disruptors

Damstra, T., Barlow, S., Bergman, A., Kavlock R., van der Kraak, G., 2002. Global Assessment of the State-of-the-Science of Endocrine Disruptors. World Health Organization, Geneva.

Guillette, L., Jr., Crain, D.A., 2000. Environmental Endocrine Disrupters: An Evolutionary Perspective. Taylor and Francis, New York.

Keith, L.H., 1997. Environmental Endocrine Disruptors: A Handbook of Property Data. John Wiley & Sons, New York.

Kime, D.E., 2001. Endocrine Disruption in Fish. Kluwer Academic Publishers, Dordrecht, the Netherlands.

Naz, R.K., 1999. Endocrine Disruptors: Effects on Male and Female Reproductive Systems. CRC Press, Boca Raton, FL.

Biomarkers/Sentinel Species

Wilson S.H., Suk, W.A., 2002. Biomarkers of Environmentally Associated Disease. Lewis Publishers, CRC Press, Boca Raton, FL.

Toxicology and Ecotoxicology

Albers, P.H., Heinz, G.H., Ohlendorf, H.M., 2000. Environmental Contaminants and Terrestrial Vertebrates. SETAC Press, Pensacola, FL.

Braunbeck, T., Hinton, D.E., Streit, B., 1998. Fish Ecotoxicology. Birkhäuser Verlag, Basel.

Di Guilio, R.T., Tillitt, D.E., 1999. Reproductive and Developmental Effects of Contaminants in Oviparous Vertebrates. SETAC Press, Pensacola, FL.

Heath, A.G., 1995. Water Pollution and Fish Physiology, 2nd ed. Lewis Publishers, CRC Press, Baton Rouge, LA.

Hoffman, D.J., Rattner, B.A., Burton Jr., G.A., Cairns, J., 2002. Handbook of Ecotoxicology, 2nd ed., CRC Press, Boca Raton, FL.

Jezierska, B., Witeska, M., 2001. Metal Toxicity to Fish. University of Podlasie, Siedlce, Poland.

Manahan, S.E., 2002. Toxicological Chemistry and Biochemistry. CRC Press, Boca Raton, FL.

Newman, M.C., Unger, M.A., 2002. Fundamentals of Ecotoxicology, 2nd ed. Lewis Publishers, CRC Press, Boca Raton, FL.

Posthuma, L, Sutter, G.W. II, Traas, T.P., 2001. Species Sensitivity Distributions in Ecotoxicology. Lewis Publishers, CRC Press, Boca Raton, FL.

Rolland, R.M., Gilbertson, M., Peterson, R.E., 1997. Chemically Induced Alterations in Functional Development and Reproduction of Fishes. SETAC Press, Pensacola, FL.

Sheppard, S., Bembridge, J., Homstrup, M., Posthuma, L., 1998. Advances in Earthworm Ecotoxicology. SETAC Press, Pensacola, FL.

Sparling, D.W., Linder, G., Bishop, C.A., 2000. Ecotoxicology of Amphibians and Reptiles. SETAC Press, Pensacola, FL.

Yu, M.-H., 2000. Environmental Toxicology. Lewis Publishers, CRC Press, Boca Raton, FL.

Risk Assessment/Monitoring

deFur, P.L., Crane, M., Ingersoll, C., Tattersfield, L., 1999. Endocrine Disruption in Invertebrates: Endocrinology, Testing, and Assessment. SETAC Press, Pensacola, FL.

Ferenc, S.A., and Foran, J.A., 2000. Multiple Stressors in Ecological Risk and Impact Assessment: Approaches to Risk Estimation. SETAC Press, Pensacola, FL.

Hart, A., Balluff, D., Barlknecht, R., et al., 2001. Avian Effects Assessment: A Framework for Contaminants Studies. SETAC Press, Pensacola FL.

Kendall, R., Dickerson, R., Giesey, J., Suk, W., 1998. Principles and Processes for Evaluating Endocrine Disruption in Wildlife. SETAC Press, Pensacola, FL.

Lewis, M.A., Mayer, F.L., Powell, R.L. et al., 1999. Ecotoxicology and Risk Assessment for Wetlands. SETAC Press, Pensacola, FL.

Munkittrick, K.R., McMaster, M.E., van der Kraak, G., Portt, C., Gibbons, W.N., Farewell, A., Gray, M., 2000. Developing of Methods for Effects-Driven Cumulative Effects Assessment Using Fish Populations: Moose River Project. SETAC Press, Pensacola, FL.

Pastorok, R.A., Bartell, S.M., Ferson, S., Ginzburg, L.R., 2002. Ecological Modeling in Risk Assessment. Lewis Publishers, CRC Press, Boca Raton, FL.

Pacquin, P.R., Di Toro, D.M., Farley, K., Mooney, K., Winfield, R., Wu, B., 2002. A Review: Exposure, Bioaccumulation, and Toxicity Models for Metals in Aquatic Systems. SETAC Press, Pensacola, FL.

Stahl, R.G. Jr., Bachman, R.A., Barton, A.L. et al., 2001. Risk Management: Ecological Risk-Based Decision-Making. SETAC Press, Pensacola, FL.

Sutter, G.W., II., Efroymson, R., Jones, D.S., Sample, B.E., 2000. Ecological Risk Assessment for Contaminated Sites. Lewis Publishers, CRC Press, Boca Raton, FL.

Tattersfield, L., Matthiessen, Campbell, P., Grandy, N., Länge, R., 1997. Endocrine Modulators and Wildlife: Assessment and Testing. SETAC Press, Pensacola, FL.

Human Health Issues

Di Guilio, R.T., Bensen, W.H., 2002. Interconnections Between Human Health and Ecological Integrity. SETAC Press, Pensacola FL.

Philp, R.B., 2001. Ecosystems and Human Health: Toxicology and Environmental Hazards, 2nd ed. Lewis Publishers, CRC Press, Boca Raton, FL.

Appendix B

Abbreviation	Definition
11-KT	11-Ketotestosterone
1α-HC	1α-Hydroxycorticosterone
3β-HSD	3β-Hydroxysteroid dehydrogenase
5D	5-Deiodinase
5'D	5'-Deiodinase
5-HT	5-Hydroxytryptamine (= serotonin)
6-MBOA	6-Methoxybenzoxazolinone
ACE	Angiotensin converting enzyme
ACh	Acetylcholine
ACTH	Corticotropin (adrenocorticotropic hormone)
ADI	Allowable daily intake
ADP	Adenosine diphosphate
AHH	Aryl hydrocarbon hydroxylase
AhR	Aryl or aromatic hydrocarbon receptor
α-MSH	Melanotropin
AMH	Anti-Müllerian hormone
ANGI	Angiotensin 1
ANGII	Angiotensin 2
ANP	Atrial natriuretic peptide
AP	Activation protein
AP-1	Activation protein-1
apaf-1	Apoptosis protease activating factor 1
APEC	Alkyphenol polyethoxycarboxylate
APEO	Alkyphenol polyethoxylate
AR	Androgen receptor
Arc	Arcuate nucleus
Arnt	Aryl hydrocarbon receptor nuclear translocator
ATP	Adenosine triphosphate
ATPase	Adenosine triphosphatase

AVP	Arginine vasopressin
AVT	Arginine vasotocin
BAT	Brown adipose tissue
BBP	Butyl benzyl phthalate
BCF	Bioconcentration factor (= concentration in the organism/concentration in the matrix at equlibrium)
β-LPH	β-lipotropin
B_{max}	Total receptor number
BMD	Benchmark dose
BMR	Basal metabolic rate
BSD	Behavioral sex determination
CaM	Calmodulin
cAMP	Cyclic adenosine monophosphate
Casp 3	caspase 3
Casp9	caspase 9
CAT	Computerized axial tomography
CAT	Chloramphenicol acetyl transferase
CBG	Corticosteroid-binding globulin
CBP	cAMP response element binding protein
CG	Chorionic gonadotropin
cGnRH-I	Chicken-I GnRH
cGnRH-ll	Chicken-II GnRH
CMA	Chemical Manufacturers Association
CoA	Coenzyme A
CRE	Corticosteroid response element
CRF	Corticotropin-releasing factor
CRH	Corticotropin-releasing hormone
CYP	Cytochrome P450 gene family
CYP1A1	Gene encoding the enzyme aryl hydrocarbon hydroxylase; cytochrome P450, aromatic compound-inducible
CYP2B	Gene encoding cytochrome P450, phenobarbital-inducible
CYP2D7	CYP2D (debrisoquine 4-hydroxylase) pseudogene
DA	Dopamine
DAX- 1	Dosage-sensitive sex-reversal adrenal hypoplasia congenital, critical region on the X chromosome gene-1
DBD	DNA binding domain
DBH	Dopamine β-hydroxylase
DCVC	Isocitrate dehydrogenase
DCT	Dopachrome tautomerase
DDA	2,2-*bis* (chlorophenyl) acetic acid
DDC	DOPA decarboxylase
DDD	1,1-Dichloro-2,2-*bis* (*p*-chlorophenyl)ethane
DDE	1,1-Dichloro-2,2-*bis* (*p*-chlorophenyl)ethylene
DDT	1,1,1-Trichloro-2,2-*bis* (*p*-chlorophenyl)ethane
DES	Diethylstilbestrol
Dfx	desferrioxamine
DHEA	Dehydroepiandrosterone
DHI	5,6-Dihydroxyindole
DHICA	5,6-Dihydroxyindole-2-carboxylic acid
DHP	17,20β-Dihydroxyprogesterone
DHT	Dehydrotestosterone
DI	Type 1 deiodinase
DII	Type 2 deiodinase
DIT	Diiodotyrosine
DMN	Dorsomedial nucleus

DNA	Deoxyribose nucleic acid
DOPA	Dihydroxyphenylalanine
DRE	Dioxin response element (=XRE)
E	Epinephrine
E_2	Estradiol
EAP	Endocrine-active phytochemical
EC_{50}	Effective concentration to kill 50% of subjects
ECOFRAM	Ecological Committee on FIFRA Risk Assessment Methods
EcR	Ecdysone receptor
ED_{50}	Effective dose to kill 50% of subjects
EDC	Endocrine-disrupting chemical
EDSTAC	Endocrine disruptor screening and testing advisory committee
EGF	Epidermal growth factor
EGFR	Epidermal growth factor receptor
EPA	U.S. Environmental Protection Agency
ER	Estrogen receptor
ERα	Estrogen receptor α
ERβ	Estrogen receptor β
ERE	Estrogen response element
EROD	Ethoxyresorufin-o-deethylase
FETAX	Frog Embryo Teratogenesis Assay–*Xenopus*
FSH	Follicle-stimulating hormone
$GABA_A$	Gamma amino butyric acid-A
GAS	General adaptation syndrome
GH	Growth hormone
GHRH	GH receptor
GHRIH	Growth hormone release-inhibiting hormone (= SS)
Glu	Glutamate
GnRH	Gonadotropin-releasing hormone
GR	Glucocorticoid receptor
GSD	Genetic sex determination
GSU	Glycoprotein subunit
GTH	Gonadotropin
GTH-I	Gonadotropin-I
GTH-II	Gonadotropin-II
HAH	Halogenated aromatic hydrocarbons
HAT	Histone acetyltransferase
hCG	Human chorionic gonadotropin
HDAC	Histone deacetylase
HDL	High-density lipoprotein
HHA	Hypothalamo-hypophysial axis
HIOMT	Hydroxy-O-methyl transferase
Hist	Histamine
HMG-CoA	Hydroxymethylglutaryl-CoA
HPA	Hypothalamus–pituitary–adrenal (axis)
HPG	Hypothalamus–pituitary–gonad (axis)
HPT	Hypothalamus–pituitary–thyroid (axis)
HPTE	2,2-*bis* (p-hydroxyphenyl)-1,1,1-trichloroethane
HRHs	Hypothalamic releasing hormones
HSD	Hydroxysteroid dehydrogenase
HSP	Heat shock protein
IC_{50}	Initial concentration
IGF-I	Insulinlike growth factor-I
IL	Interleukin
IST	Isotocin

JAK/STAT	Janus kinase
K_d	Equilibrium dissociation constant
K_i	Equilibrium dissociation constant for inhibitor
K_{ow}	Octanol-water partition coefficient
LBD	Ligand binding domain
LD_{50}	Lethal dose for 50%
LDL	Low-density lipoprotein
LH	Luteinizing hormone
LOEL	Lowest-observed-effect level
LPH	Lipotropin
LR	Ligand-receptor complex
MC	Melanocortin
MDR	Multidrug resistance protein
MALDI-TOF	Matrix-assisted laser desorption ionization-time of flight
MIH	Müllerian-inhibiting hormone (= MIS, Müllerian-inhibiting substance; = AMH)
MIT	Monoiodotyrosine
MOE	Margin of exposure (= MOS)
MPO	Medial preoptic nucleus
MOS	Margin of safety (= MOE)
MR	Mineralocorticoid receptor
MRI	Magnetic resonance imaging
mRNA	Messenger ribonucleic acid
MSH	Melanotropin
MST	Mesotocin
MTD	Maximum tolerated dose
NADH	Reduced nictotinamide adenine dinucleotide
NCoR	Nuclear receptor corepressor
NE	Norepinephrine
NEFA	Nonesterified fatty acid
NIS	Sodium/iodide symporter
NOAEL	No-observable-adverse-effect level
NOEC	No-observed-effect concentration
NOEL	No-observable-effect level
NP	Nonylphenol
NPEC	Nonylphenol polyethoxycarboxylate
NPEO	Nonylphenol polyethoxylate
NSB	Nonspecific binding
NTCP	Na^+/taurochlorate cotransporting polypeptide
OATP	Organic anion transporter protein
OP	Octylphenol
OPEC	Octylphenol polyethoxycarboxylate
OXY	Oxytocin
p53	Tumor-suppressor gene P53
P_{450}	Cytochrome enzyme family
PACAP	Pituitary adenylate cyclase-activating peptide
PAH	Polynuclear aromatic hydrocarbons
PAPS	Phosphoadenosylphosphorylsulfate
PAS	per-ARNT-Sim protein family
PBPK	Physioiogically-based pharmacokinetic model
p/CAF	p300/CBP-associated factor
PCB	Polychlorinated biphenyl
PCDD	Polychlorinated dibenzo-p-dioxin
PCDF	Polychlorinated dibenzofuran
PD	Pars distalis
PDS	Pendred syndrome gene

PEPCK	Phosphoenol pyruvate carboxykinase
PET	Positron emission tomography
PI	Pars intermedia
PN	Pars nervosa
PNMT	Phenylethanolamine-N-methyltransferase
POA	Preoptic area
POMC	Proopiomelanocortin
PPD	Proximal pars distalis
PR	Progesterone receptor
PRIH	Prolactin release-inhibiting hormone
PRL	Prolactin
PT	Pars tuberalis
PTU	Propylthiouracil
PVN	Paraventricular nucleus
RAS	Renin-angiotensin system
RBA	Relative binding affinity
redox	Reduction-oxidation reactions
RPD	Rostral pars distalis
RXR	Retinoic acid X receptor
SCC	Cholesterol side-chain cleavage
SDN	Sexually dimorphic nucleus
SER	Smooth endoplasmic reticulum
SERM	Selective estrogen receptor modulator
SF-1	Steroid factor-1
SGOT	Serum glutamic oxaloacetic transaminase
SGPT	Serum glutamic pyruvate transaminase
SHBG	Sex hormone binding globulin
SL	Somatolactin
SMRT	Silencing mediator for retinoid and thyroid-hormone receptors
SNS	Sympathetic nervous system
SON	Supraoptic nucleus
SS	Somatostatin (= GHRIH)
StAR	Steroidogenic acute regulatory protein
T/EBP	Thyroid-specific enhancer-binding protein
T_3	Triiodothyronine
T_4	Tetraiodothyronine (= thyroxine)
TBG	Thyroid hormone binding globulin
TCDD	2,3,7,8-tetrachlorodibenzo-p-dioxin
TEF	Toxic equivalency factor
TG	Thyroglobulin
TGF	Transforming growth factor
TH	Thyroid hormone
TI	Therapeutic index
TPO	Thyroid peroxidase
TR	Thyroid hormone receptor
TRAP	Thyroid hormone-associated protein
TRE	Thyroid response element
TRH	Thyrotropin-releasing hormone
TSD	Temperature-dependent sex determination
TSH	Thyrotropin = thyroid-stimulating hormone
TTF	Thyroid transcription factor
TTR	Transthyretin
TyR	Tyrosinase
UI	Urotensin I
UCP	Uncoupling protein (= thermogenin)

V_1	AVT receptor type 1
V_2	AVT receptor type 2
Vtg	Vitellogenin
XRE	Xenobiotic response element
ZF	Zona fasciculata
ZG	Zona glomerulosa
ZR	Zona reticularis

Index